Lecture Notes on Data Engineering and Communications Technologies

201

Series Editor

Fatos Xhafa, *Technical University of Catalonia, Barcelona, Spain*

The aim of the book series is to present cutting edge engineering approaches to data technologies and communications. It will publish latest advances on the engineering task of building and deploying distributed, scalable and reliable data infrastructures and communication systems.

The series will have a prominent applied focus on data technologies and communications with aim to promote the bridging from fundamental research on data science and networking to data engineering and communications that lead to industry products, business knowledge and standardisation.

Indexed by SCOPUS, INSPEC, EI Compendex.

All books published in the series are submitted for consideration in Web of Science.

Leonard Barolli

Editor

Advanced Information Networking and Applications

Proceedings of the 38th International Conference on Advanced Information Networking and Applications (AINA-2024), Volume 3

 Springer

Editor
Leonard Barolli
Department of Information and Communication
Engineering
Fukuoka Institute of Technology
Fukuoka, Japan

ISSN 2367-4512 ISSN 2367-4520 (electronic)
Lecture Notes on Data Engineering and Communications Technologies
ISBN 978-3-031-57869-4 ISBN 978-3-031-57870-0 (eBook)
https://doi.org/10.1007/978-3-031-57870-0

This Springer imprint is published by the registered company Springer Nature Switzerland AG
The registered company address is: Gewerbestrasse 11, 6330 Cham, Switzerland

Paper in this product is recyclable.

Welcome Message from AINA-2024 Organizers

Welcome to the 38th International Conference on Advanced Information Networking and Applications (AINA-2024). On behalf of AINA-2024 Organizing Committee, we would like to express to all participants our cordial welcome and high respect.

AINA is an International Forum, where scientists and researchers from academia and industry working in various scientific and technical areas of networking and distributed computing systems can demonstrate new ideas and solutions in distributed computing systems. AINA is a very open society and is always welcoming international volunteers from any country and any area in the world.

AINA International Conference is a forum for sharing ideas and research work in the emerging areas of information networking and their applications. The area of advanced networking has grown very rapidly and the applications have experienced an explosive growth, especially in the area of pervasive and mobile applications, wireless sensor and ad-hoc networks, vehicular networks, multimedia computing, social networking, semantic collaborative systems, as well as IoT, big data, cloud computing, artificial intelligence, and machine learning. This advanced networking revolution is transforming the way people live, work, and interact with each other and is impacting the way business, education, entertainment, and health care are operating. The papers included in the proceedings cover theory, design and application of computer networks, distributed computing, and information systems.

Each year AINA receives a lot of paper submissions from all around the world. It has maintained high-quality accepted papers and is aspiring to be one of the main international conferences on the information networking in the world.

We are very proud and honored to have two distinguished keynote talks by Prof. Fatos Xhafa, Technical University of Catalonia, Spain, and Dr. Juggapong Natwichai, Chiang Mai University, Thailand, who will present their recent work and will give new insights and ideas to the conference participants.

An international conference of this size requires the support and help of many people. A lot of people have helped and worked hard to produce a successful AINA-2024 technical program and conference proceedings. First, we would like to thank all authors for submitting their papers. We are indebted to Program Track Co-chairs, Program Committee Members and Reviewers, who carried out the most difficult work of carefully evaluating the submitted papers.

We would like to thank AINA-2024 General Co-chairs, PC Co-chairs, Workshops Organizers for their great efforts to make AINA-2024 a very successful event. We have special thanks to the Finance Chair and Web Administrator Co-chairs.

We do hope that you will enjoy the conference proceedings and readings.

AINA-2024 Organizing Committee

Honorary Chair

Makoto Takizawa   Hosei University, Japan

General Co-chairs

Minoru Uehara	Toyo University, Japan
Euripides G. M. Petrakis	Technical University of Crete (TUC), Greece
Isaac Woungang	Toronto Metropolitan University, Canada

Program Committee Co-chairs

Tomoya Enokido	Rissho University, Japan
Mario A. R. Dantas	Federal University of Juiz de Fora, Brazil
Leonardo Mostarda	University of Perugia, Italy

International Journals Special Issues Co-chairs

Fatos Xhafa	Technical University of Catalonia, Spain
David Taniar	Monash University, Australia
Farookh Hussain	University of Technology Sydney, Australia

Award Co-chairs

Arjan Durresi	Indiana University Purdue University in Indianapolis (IUPUI), USA
Fang-Yie Leu	Tunghai University, Taiwan
Marek Ogiela	AGH University of Science and Technology, Poland
Kin Fun Li	University of Victoria, Canada

Publicity Co-chairs

Markus Aleksy	ABB Corporate Research Center, Germany
Flora Amato	University of Naples "Federico II", Italy
Lidia Ogiela	AGH University of Science and Technology, Poland
Hsing-Chung Chen	Asia University, Taiwan

International Liaison Co-chairs

Wenny Rahayu	La Trobe University, Australia
Nadeem Javaid	COMSATS University Islamabad, Pakistan
Beniamino Di Martino	University of Campania "Luigi Vanvitelli", Italy

Local Arrangement Co-chairs

Keita Matsuo	Fukuoka Institute of Technology, Japan
Tomoyuki Ishida	Fukuoka Institute of Technology, Japan

Finance Chair

Makoto Ikeda	Fukuoka Institute of Technology, Japan

Web Co-chairs

Phudit Ampririt	Fukuoka Institute of Technology, Japan
Ermioni Qafzezi	Fukuoka Institute of Technology, Japan
Shunya Higashi	Fukuoka Institute of Technology, Japan

Steering Committee Chair

Leonard Barolli	Fukuoka Institute of Technology, Japan

Tracks Co-chairs and Program Committee Members

1. Network Architectures, Protocols and Algorithms

Track Co-chairs

Spyropoulos Thrasyvoulos	Technical University of Crete (TUC), Greece
Shigetomo Kimura	University of Tsukuba, Japan
Darshika Perera	University of Colorado at Colorado Springs, USA

TPC Members

Thomas Dreibholz	Simula Metropolitan Center for Digital Engineering, Norway
Angelos Antonopoulos	Nearby Computing SL, Spain
Hatim Chergui	i2CAT Foundation, Spain
Bhed Bahadur Bista	Iwate Prefectural University, Japan
Chotipat Pornavalai	King Mongkut's Institute of Technology Ladkrabang, Thailand
Kenichi Matsui	NTT Network Innovation Center, Japan
Sho Tsugawa	University of Tsukuba, Japan
Satoshi Ohzahata	University of Electro-Communications, Japan
Haytham El Miligi	Thompson Rivers University, Canada
Watheq El-Kharashi	Ain Shams University, Egypt
Ehsan Atoofian	Lakehead University, Canada
Fayez Gebali	University of Victoria, Canada
Kin Fun Li	University of Victoria, Canada
Luis Blanco	CTTC, Spain

2. Next Generation Mobile and Wireless Networks

Track Co-chairs

Purav Shah	School of Science and Technology, Middlesex University, UK
Enver Ever	Middle East Technical University, Northern Cyprus
Evjola Spaho	Polytechnic University of Tirana, Albania

TPC Members

Burak Kizilkaya	Glasgow University, UK
Muhammad Toaha	Middle East Technical University, Turkey
Ramona Trestian	Middlesex University, UK
Andrea Marotta	University of L'Aquila, Italy
Adnan Yazici	Nazarbayev University, Kazakhstan
Orhan Gemikonakli	Final International University, Cyprus
Hrishikesh Venkataraman	Indian Institute of Information Technology, Sri City, India
Zhengjia Xu	Cranfield University, UK
Mohsen Hejazi	University of Kashan, Iran
Sabyasachi Mukhopadhyay	IIT Kharagpur, India
Ali Khoshkholghi	Middlesex University, UK
Admir Barolli	Aleksander Moisiu University of Durres, Albania
Makoto Ikeda	Fukuoka Institute of Technology, Japan
Yi Liu	Oita National College of Technology, Japan
Testuya Oda	Okayama University of Science, Japan
Ermioni Qafzezi	Fukuoka Institute of Technology, Japan

3. Multimedia Networking and Applications

Track Co-chairs

Markus Aleksy	ABB Corporate Research Center, Germany
Francesco Orciuoli	University of Salerno, Italy
Tomoyuki Ishida	Fukuoka Institute of Technology, Japan

TPC Members

Hadil Abukwaik	ABB Corporate Research Center, Germany
Thomas Preuss	Brandenburg University of Applied Sciences, Germany
Peter M. Rost	Karlsruhe Institute of Technology (KIT), Germany
Lukasz Wisniewski	inIT, Germany
Angelo Gaeta	University of Salerno, Italy
Angela Peduto	University of Salerno, Italy
Antonella Pascuzzo	University of Salerno, Italy
Roberto Abbruzzese	University of Salerno, Italy
Tetsuro Ogi	Keio University, Japan

Yasuo Ebara	Osaka Electro-Communication University, Japan
Hideo Miyachi	Tokyo City University, Japan
Kaoru Sugita	Fukuoka Institute of Technology, Japan

4. Pervasive and Ubiquitous Computing

Track Co-chairs

Vamsi Paruchuri	University of Central Arkansas, USA
Hsing-Chung Chen	Asia University, Taiwan
Shinji Sakamoto	Kanazawa Institute of Technology, Japan

TPC Members

Sriram Chellappan	University of South Florida, USA
Yu Sun	University of Central Arkansas, USA
Qiang Duan	Penn State University, USA
Han-Chieh Wei	Dallas Baptist University, USA
Ahmad Alsharif	University of Alabama, USA
Vijayasarathi Balasubramanian	Microsoft, USA
Shyi-Shiun Kuo	Nan Kai University of Technology, Taiwan
Karisma Trinanda Putra	Universitas Muhammadiyah Yogyakarta, Indonesia
Cahya Damarjati	Universitas Muhammadiyah Yogyakarta, Indonesia
Agung Mulyo Widodo	Universitas Esa Unggul Jakarta, Indonesia
Bambang Irawan	Universitas Esa Unggul Jakarta, Indonesia
Eko Prasetyo	Universitas Muhammadiyah Yogyakarta, Indonesia
Sunardi S. T.	Universitas Muhammadiyah Yogyakarta, Indonesia
Andika Wisnujati	Universitas Muhammadiyah Yogyakarta, Indonesia
Makoto Ikeda	Fukuoka Institute of Technology, Japan
Tetsuya Oda	Okayama University of Science, Japan
Evjola Spaho	Polytechnic University of Tirana, Albania
Tetsuya Shigeyasu	Hiroshima Prefectural University, Japan
Keita Matsuo	Fukuoka Institute of Technology, Japan
Admir Barolli	Aleksander Moisiu University of Durres, Albania

5. Web-Based Systems and Content Distribution

Track Co-chairs

Chrisa Tsinaraki	Technical University of Crete (TUC), Greece
Yusuke Gotoh	Okayama University, Japan
Santi Caballe	Open University of Catalonia, Spain

TPC Members

Nikos Bikakis	Hellenic Mediterranean University, Greece
Ioannis Stavrakantonakis	Ververica GmbH, Germany
Sven Schade	European Commission, Joint Research Center, Italy
Christos Papatheodorou	National and Kapodistrian University of Athens, Greece
Sarantos Kapidakis	University of West Attica, Greece
Manato Fujimoto	Osaka Metropolitan University, Japan
Kiki Adhinugraha	La Trobe University, Australia
Tomoki Yoshihisa	Shiga University, Japan
Jordi Conesa	Open University of Catalonia, Spain
Thanasis Daradoumis	Open University of Catalonia, Spain
Nicola Capuano	University of Basilicata, Italy
Victor Ströele	Federal University of Juiz de Fora, Brazil

6. Distributed Ledger Technologies and Distributed-Parallel Computing

Track Co-chairs

Alfredo Navarra	University of Perugia, Italy
Naohiro Hayashibara	Kyoto Sangyo University, Japan

TPC Members

Serafino Cicerone	University of L'Aquila, Italy
Ralf Klasing	LaBRI Bordeaux, France
Giuseppe Prencipe	University of Pisa, Italy
Roberto Tonelli	University of Cagliari, Italy
Farhan Ullah	Northwestern Polytechnical University, China

Leonardo Mostarda	University of Perugia, Italy
Qiong Huang	South China Agricultural University, China
Tomoya Enokido	Rissho University, Japan
Minoru Uehara	Toyo University, Japan
Lucian Prodan	Polytechnic University of Timisoara, Romania
Md. Abdur Razzaque	University of Dhaka, Bangladesh

7. Data Mining, Big Data Analytics and Social Networks

Track Co-chairs

Pavel Krömer	Technical University of Ostrava, Czech Republic
Alex Thomo	University of Victoria, Canada
Eric Pardede	La Trobe University, Australia

TPC Members

Sebastián Basterrech	Technical University of Denmark, Denmark
Tibebe Beshah	University of Addis Ababa, Ethiopia
Nashwa El-Bendary	Arab Academy for Science, Egypt
Petr Musilek	University of Alberta, Canada
Varun Ojha	Newcastle University, UK
Alvaro Parres	ITESO, Mexico
Nizar Rokbani	ISSAT-University of Sousse, Tunisia
Farshid Hajati	Victoria University, Australia
Ji Zhang	University of Southern Queensland, Australia
Salimur Choudhury	Lakehead University, Canada
Carson Leung	University of Manitoba, Canada
Syed Mahbub	La Trobe University, Australia
Osama Mahdi	Melbourne Institute of Technology, Australia
Choiru Zain	La Trobe University, Australia
Rajalakshmi Rajasekaran	La Trobe University, Australia
Nawfal Ali	Monash University, Australia

8. Internet of Things and Cyber-Physical Systems

Track Co-chairs

Tomoki Yoshihisa   Shiga University, Japan
Winston Seah   Victoria University of Wellington, New Zealand
Luciana Pereira Oliveira   Instituto Federal da Paraiba (IFPB), Brazil

TPC Members

Akihiro Fujimoto   Wakayama University, Japan
Akimitsu Kanzaki   Shimane University, Japan
Kazuya Tsukamoto   Kyushu Institute of Technology, Japan
Lei Shu   Nanjing Agricultural University, China
Naoyuki Morimoto   Mie University, Japan
Teruhiro Mizumoto   Chiba Institute of Technology, Japan
Tomoya Kawakami   Fukui University, Japan
Adrian Pekar   Budapest University of Technology and
                           Economics, Hungary

Alvin Valera   Victoria University of Wellington, New Zealand
Chidchanok Choksuchat   Prince of Songkla University, Thailand
Jyoti Sahni   Victoria University of Wellington, New Zealand
Murugaraj Odiathevar   Sungkyunkwan University, South Korea
Normalia Samian   Universiti Putra Malaysia, Malaysia
Qing Gu   University of Science and Technology Beijing,
                           China

Tao Zheng   Beijing Jiaotong University, China
Wenbin Pei   Dalian University of Technology, China
William Liu   Unitec, New Zealand
Wuyungerile Li   Inner Mongolia University, China
Peng Huang   Sichuan Agricultural University, PR China
Ruan Delgado Gomes   Instituto Federal da Paraiba (IFPB), Brazil
Glauco Estacio Goncalves   Universidade Federal do Pará (UFPA), Brazil
Eduardo Luzeiro Feitosa   Universidade Federal do Amazonas (UFAM),
                           Brazil

Paulo Ribeiro Lins Júnior Instituto Federal da Paraiba (IFPB), Brazil

9. Intelligent Computing and Machine Learning

Track Co-chairs

Takahiro Uchiya	Nagoya Institute of Technology, Japan
Flavius Frasincar	Erasmus University Rotterdam, The Netherlands
Miltos Alamaniotis	University of Texas at San Antonio, USA

TPC Members

Kazuto Sasai	Ibaraki University, Japan
Shigeru Fujita	Chiba Institute of Technology, Japan
Yuki Kaeri	Mejiro University, Japan
Jolanta Mizera-Pietraszko	Military University of Land Forces, Poland
Ashwin Ittoo	University of Liège, Belgium
Marco Brambilla	Politecnico di Milano, Italy
Alfredo Cuzzocrea	University of Calabria, Italy
Le Minh Nguyen	JAIST, Japan
Akiko Aizawa	National Institute of Informatics, Japan
Natthawut Kertkeidkachorn	JAIST, Japan
Georgios Karagiannis	Durham University, UK
Leonidas Akritidis	International Hellenic University, Greece
Athanasios Fevgas	University of Thessaly, Greece
Yota Tsompanopoulou	University of Thessaly, Greece
Yuvaraj Munian	Texas A&M-San Antonio, USA

10. Cloud and Services Computing

Track Co-chairs

Salvatore Venticinque	University of Campania "Luigi Vanvitelli", Italy
Shigenari Nakamura	Tokyo Denki University, Japan
Sajal Mukhopadhyay	National Institute of Technology, Durgapur, India

TPC Members

Giancarlo Fortino	University of Calabria, Italy
Massimiliano Rak	University of Campania "Luigi Vanvitelli", Italy
Jason J. Jung	Chung-Ang University, Korea

Dimosthenis Kyriazis	University of Piraeus, Greece
Geir Horn	University of Oslo, Norway
Dario Branco	University of Campania "Luigi Vanvitelli", Italy
Dilawaer Duolikun	Cognizant Technology Solutions, Hungary
Naohiro Hayashibara	Kyoto Sangyo University, Japan
Tomoya Enokido	Rissho University, Japan
Sujoy Saha	NIT Durgapur, India
Animesh Dutta	NIT Durgapur, India
Pramod Mane	IIM Rohtak, India
Nanda Dulal Jana	NIT Durgapur, India
Banhi Sanyal	NIT Kurukshetra, India

11. Security, Privacy and Trust Computing

Track Co-chairs

Ioannidis Sotirios	Technical University of Crete (TUC), Greece
Michail Alexiou	Georgia Institute of Technology, USA
Hiroaki Kikuchi	Meiji University, Japan

TPC Members

George Vasiliadis	Hellenic Mediterranean University, Greece
Antreas Dionysiou	University of Cyprus, Cyprus
Apostolos Fouranaris	Athena Research Center, Greece
Panagiotis Ilia	Technical University of Crete, Greece
George Portokalidis	IMDEA, Spain
Nikolaos Gkorgkolis	University of Crete, Greece
Zeezoo Ryu	Georgia Institute of Technology, USA
Muhammad Faraz Karim	Georgia Institute of Technology, USA
Yunjie Deng	Georgia Institute of Technology, USA
Anna Raymaker	Georgia Institute of Technology, USA
Takamichi Saito	Meiji University, Japan
Kazumasa Omote	University of Tsukuba, Japan
Masakatsu Nishigaki	Shizuoka University, Japan
Mamoru Mimura	National Defense Academy of Japan, Japan
Chun-I Fan	National Sun Yat-sen University, Taiwan
Aida Ben Chehida Douss	National School of Engineers of Tunis, ENIT Tunis, Tunisia
Davinder Kaur	IUPUI, USA

12. Software-Defined Networking and Network Virtualization

Track Co-chairs

Flavio de Oliveira Silva	Federal University of Uberlândia, Brazil
Ashutosh Bhatia	Birla Institute of Technology and Science, Pilani, India

TPC Members

Rui Luís Andrade Aguiar	Universidade de Aveiro (UA), Portugal
Ivan Vidal	Universidad Carlos III de Madrid, Spain
Eduardo Coelho Cerqueira	Federal University of Pará (UFPA), Brazil
Christos Tranoris	University of Patras (UoP), Greece
Juliano Araújo Wickboldt	Federal University of Rio Grande do Sul (UFRGS), Brazil
Haribabu K.	BITS Pilani, India
Virendra Shekhavat	BITS Pilani, India
Makoto Ikeda	Fukuoka Institute of Technology, Japan
Farookh Hussain	University of Technology Sydney, Australia
Keita Matsuo	Fukuoka Institute of Technology, Japan

AINA-2024 Reviewers

Admir Barolli	Burak Kizilkaya
Aida ben Chehida Douss	Carson Leung
Akimitsu Kanzaki	Chidchanok Choksuchat
Alba Amato	Christos Tranoris
Alberto Postiglione	Chung-Ming Huang
Alex Thomo	Dario Branco
Alfredo Navarra	David Taniar
Amani Shatnawi	Elinda Mece
Anas AlSobeh	Enver Ever
Andrea Marotta	Eric Pardede
Angela Peduto	Euripides Petrakis
Anne Kayem	Evjola Spaho
Antreas Dionysiou	Fabrizio Messina
Arjan Durresi	Feilong Tang
Ashutosh Bhatia	Flavio Silva
Beniamino Di Martino	Francesco Orciuoli
Bhed Bista	George Portokalidis

Giancarlo Fortino
Giorgos Vasiliadis
Glauco Gonçalves
Hatim Chergui
Hiroaki Kikuchi
Hiroki Sakaji
Hiroshi Maeda
Hiroyuki Fujioka
Hyunhee Park
Isaac Woungang
Jana Nowaková
Jolanta Mizera-Pietraszko
Junichi Honda
Jyoti Sahni
Kazunori Uchida
Keita Matsuo
Kenichi Matsui
Kiki Adhinugraha
Kin Fun Li
Kiyotaka Fujisaki
Leonard Barolli
Leonardo Mostarda
Leonidas Akritidis
Lidia Ogiela
Lisandro Granville
Lucian Prodan
Luciana Oliveira
Mahmoud Elkhodr
Makoto Ikeda
Mamoru Mimura
Manato Fujimoto
Marco Antonio To
Marek Ogiela
Masaki Kohana
Minoru Uehara
Muhammad Karim
Muhammad Toaha Raza Khan
Murugaraj Odiathevar
Nadeem Javaid
Naohiro Hayashibara
Nobuo Funabiki
Nour El Madhoun
Omar Darwish

Panagiotis Ilia
Petr Musilek
Philip Moore
Purav Shah
R. Madhusudhan
Raffaele Guarasci
Ralf Klasing
Roberto Tonelli
Ronald Petrlic
Sabyasachi Mukhopadhyay
Sajal Mukhopadhyay
Salvatore D'Angelo
Salvatore D'Angelo
Salvatore D'Angelo
Salvatore Venticinque
Santi Caballé
Satoshi Ohzahata
Serafino Cicerone
Shigenari Nakamura
Shinji Sakamoto
Sho Tsugawa
Sriram Chellappan
Stephane Maag
Takayuki Kushida
Tetsuya Oda
Thomas Dreibholz
Tomoki Yoshihisa
Tomoya Enokido
Tomoya Kawakami
Tomoyuki Ishida
Vamsi Paruchuri
Victor Ströele
Vikram Singh
Wei Lu
Wenny Rahayu
Winston Seah
Yong Zheng
Yoshitaka Shibata
Yusuke Gotoh
Yuvaraj Munian
Zeezoo Ryu
Zhengjia Xu

AINA-2024 Keynote Talks

Agile Edge: Harnessing the Power of the Intelligent Edge by Agile Optimization

Fatos Xhafa

Technical University of Barcelona, Barcelona, Spain

Abstract. The digital cloud ecosystem comprises various degrees of computing granularity from large cloud servers and data centers to IoT devices, leading to the cloud-to-thing continuum computing paradigm. In this context, the intelligent edge aims at placing intelligence to the end devices, at the edges of the Internet. The premise is that collective intelligence from the IoT data deluge can be achieved and used at the edges of the Internet, offloading the computation burden from the cloud systems and leveraging real-time intelligence. This, however, comes with the challenges of processing and analyzing the IoT data streams in real time. In this talk, we will address how agile optimization can be useful for harnessing the power of the intelligent edge. Agile optimization is a powerful and promising solution, which differently from traditional optimization methods, is able to find optimized and scalable solutions under real-time requirements. We will bring real-life problems and case studies from Smart City Open Data Repositories to illustrate the approach. Finally, we will discuss the research challenges and emerging vision on the agile intelligent edge.

Challenges in Entity Matching in AI Era

Juggapong Natwichai

Chiang Mai University, Chiang Mai, Thailand

Abstract. Entity matching (EM) is to identify and link entities originating from various sources that correspond to identical real-world entities, thereby constituting a foundational component within the realm of data integration. For example, in order to counter-fraud detection, the datasets from sellers, financial services providers, or even IT infrastructure service providers might be in need for data integration, and hence, the EM is highly important here. This matching process is also recognized for its pivotal role in data augmenting to improve the precision and dependability of subsequent tasks within the domain of data analytics. Traditionally, the EM procedure composes of two integral phases, namely blocking and matching. The blocking phase associates with the generation of candidate pairs and could affect the size and complexity of the data. Meanwhile, the matching phase will need to trade-off between the accuracy and the efficiency. In this talk, the challenges of both components are thoroughly explored, particularly with the aid of AI techniques. In addition, the preliminary experiment results to explore some important factors which affect the performance will be presented.

Contents

Efficient Communication Protocol
for Programmable Matter

Jean-Paul A. Yaacoub[1], Benoit Piranda[1(✉)], Frederic Lassabe[2], and Hassan N. Noura[1]

[1] Univ. Franche-Comté, FEMTO-ST Inst., CNRS, Montbéliard, France
{jean.absyaacoub,benoit.piranda}@femto-st.fr,
Hassan.noura@univ-fcomte.fr
[2] UTBM, FEMTO-ST Inst., CNRS, Montbéliard, France
frederic.lassabe@utbm.fr

Abstract. Lattice-based modular robots are composed of modules arranged on a lattice and forming 3D shapes, if these robots are small enough and many enough, they form a programmable matter. This work proposes a method for optimising data communication times between modules by compressing the data. We have first analysed the communication delay between the end device modules, then a set of recent lossless compression algorithms was tested to select the optimal one to implement with *Blinky Block*. Based on the results obtained, we propose to add a lossless data compression scheme to reduce the communicated data size and consequently communication delay. We found that the "Brotli" compression algorithm is the most suitable one for modular robot communication as it achieved a good balance between computing and communication overhead. Then, based on the compression ratio and the communication delay interpolation, a significant gain is achieved by reducing the communication delay by a factor of 5.

1 Introduction

Programmable matter is made of small autonomous building blocks that can be programmed to achieve a wide range of geometric objects and structures with programmable capabilities to change their colour or shape, which leads to the creation of programmable matter [1].

Most of the algorithm use communication capabilities of robots to share local information in order to enlarge global knowledge of the set. These communications are the weak point of distributed algorithms, as they represent the longest processing time. We will show that communication time is mainly due to the size of the data embedded in the messages. Even if the computational capabilities of the robots used in the programmable subject are quite small, the idea developed here is to use these computational capabilities to process the data received in order to reduce the size of the data transmitted.

The context used in this article is central to the problem of self-reconfiguration of programmable matter. Self-reconfiguration consists in programming modular robots so that modules move relatively to each other to change the overall shape of the assembly [2–4].

© The Author(s), under exclusive license to Springer Nature Switzerland AG 2024
L. Barolli (Ed.): AINA 2024, LNDECT 201, pp. 1–11, 2024.
https://doi.org/10.1007/978-3-031-57870-0_1

The preliminary step in any self-reconfiguration algorithm is to give the modules a way of knowing the final shape to be made. In [5], Tucci et al. presented a very efficient 3D scene encoding model for the self-reconfiguration process that describes a 3D models in the form of a Constructive Solid Geometry tree (CSG tree) combining the simple geometric objects placed in the leaves. Combination can be union, intersection and difference of sub-trees. A string code may be generated from a depth search first traversing of the tree.

Blinky Blocks are small cubic modular robots that make up the key component of the Claytronics project to create highly adaptable and reconfigurable objects and environments (cf. Fig. 1). Each *Blinky Block* can be attached with its magnets to form complex geometric shapes, can exchange messages with directly connected neighbours, and react to noise by emitting sounds or/and changing colours. We use these real robots as a test bed to validate some parts of distributed algorithms for programmable matter. Despite the fact that they have no autonomous movement capability, their communication and computing capacity means that algorithms for programmable matter can be implemented on several hundred connected real robots.

The adoption of modular robots in the nature of the Internet of Things (IoT) [6] with the implementation of AI-empowered applications and services [7], can reshape the robotics concept [8,9] into a new modular self-reconfigurable swarm, capable of operating in large numbers synchronously and simultaneously.

Communication between similar modular robotic systems as IoT components is essential to perform the intended task. However, this can be delayed due to the message size, which the length of the message can prove to be challenging and result in communication delays. For that, several solutions for Wireless Multimedia Sensor Networks (WMSN) were presented such as in [10] to reduce this redundancy by discarding a certain number of data packets while guaranteeing its integrity (quality). Other solutions include low-overhead data compression techniques [11], Compressed Sensing (CS) algorithms for data compression [12], and data compression and transmission scheme for power reduction in IoT enabled wireless sensors [13].

However, they are prone to delays which can affect their ability to react in real-time which is often caused not only by the *Blinky Blocks* number but rather by the Message Length (ML), which the higher the message, the higher the delay will become. As a result, several experimental results were tested on different compression/decompression

Fig. 1. Left: *Blinky Block* Hardware. Right: a set of 768 *Blinky Blocks* running the same program to visualise a cutting plane of 3D Objects.

algorithms to verify which one is more suitable to be applied to *Blinky Blocks* to mitigate the issue of delay and ensure a higher real-time reaction to users' orders and commands.

The following section presents preliminary work on the study of robots to evaluate their communication and computation capabilities. The next part proposes a study of classical compression models compared to Huffman's method. Finally, our method is presented and completed by an experiment on a real problem applied to a large number of connected robots.

2   *Blinky Block* Benchmark and Compression Models

A preliminary study of *Blinky Blocks* has enabled to assess their communication and pure computing capabilities. *Blinky Blocks* use very standard communication systems (6 UARTs, one on each side of the *Blinky Block*) and a processor very common in embedded systems (ARM Cortex M0 from STMicroelectronics, the STM32F091CB with 32 KB RAM and 128 KB flash memory), which allows us to generalise this study to most distributed multi-robot systems used in the context of programmable matter, such as the *3D Catom* [14].

First, in order to analyse the communication delay on *Blinky Blocks* in terms of the Message Length (M_l) and number of *Blinky Blocks* (N_{BB}), we place N_{BB} *Blinky Blocks* forming a simple line and we compute the communication time of several messages with different size of embedded data using the configuration presented Fig. 2.

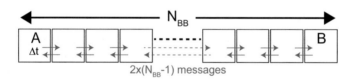

Fig. 2. Experimental network diagram used to measure message propagation times.

Measurement of the total time taken to transfer a message on all *Blinky Blocks* is carried out by a distributed program running on the robots. This program starts with the first extremity A sending a message to its only neighbour, at the local time t_0 stored in A. When the message reaches an internal module with two neighbours, the message received is sent back to the connected opposite port. When the message reaches the extremity B (which has only one neighbour connected), the message is sent back to the receiving port. When the back message reaches A at local time t_1, the time $\Delta_t = t_1 - t_0$ gives the average duration of $2 \times (N_{BB} - 1)$ message transfers where N_{BB} varies from 4 to 52 respectively.

We repeat this operation 1000 times to deduce the average duration of the transmission of messages (T_{ML}). The benchmark tests were performed on a series of 52 *Blinky Block* with each set being tested for a message of N bytes, with N taking 7 values in [2..227]. Based on the obtained results (see Fig. 3a and 3b), we found that only the message length has an effect on the communication time. Finally, we propose a linear approximation of the duration of the message depending on its length:

$$t = 0.08935 \times M_l + 1.516 \tag{1}$$

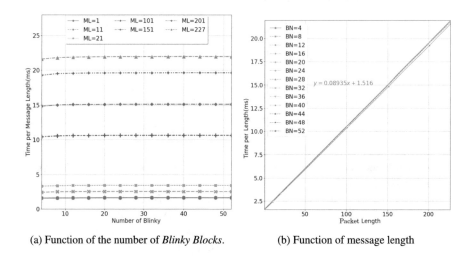

(a) Function of the number of *Blinky Blocks*.   (b) Function of message length

Fig. 3. Variation of communication delay

In a second study, we carried out a number of calculations on each *Blinky Block* set, such as mathematical operations and decompression using Huffman's method. This study led us to the conclusion that all the computations required in distributed algorithms for programmable matter were negligible compared with communication time. Here, for example, decompressing a Huffman code of 1061 bytes takes 15 ms, which is comparable to sending 150 bytes from a *Blinky Block* to a neighbour.

Thirdly, we studied the various compression algorithms available and compared them with the Huffman method implemented on our *Blinky Blocks*. Data compression algorithms can be divided into two classes: Lossless Compression which allows the original data to be fully reconstructed from the compressed data and with no information loss, and Lossy Compression which is especially used for multimedia data such as images and audio. It allows the original data to be reconstructed with a certain loss of information, but it can achieve better data reduction compared to lossless compression as it allows more space to be freed up.

In our case, the type of compression is message (textual data) compression and consequently, the required compression time should also be lossless, due to its ability to prevent the loss of any data during the compression/decompression process to avoid any modification to the original sent message. On the other hand, Fig. 4 represents a taxonomy of existing lossless data compression schemes. In this paper, a lossless set of these compression schemes was tested to confirm whether they are suitable to be implemented with lattice-based modular robots or not.

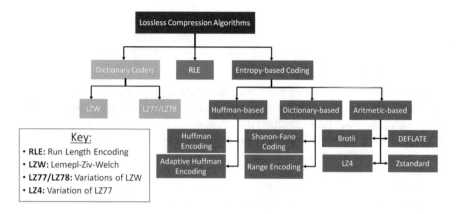

Fig. 4. Taxonomy of Existing Lossless Compression Algorithm Types.

A description of the most known and widely used lossless compression algorithms is presented in Fig. 4. We tried different kinds of compression methods to give a brief description of each of the widely selected lossless data compression algorithms [15]:

- **DEFLATE:** is a lossless compression algorithm widely used in many popular compression utilities like gzip, zip, and PNG. It uses a combination of Huffman coding and LZ77 sliding window compression to compress text data [16].
- **LZ77:** is a lossless compression algorithm that uses a sliding window technique to compress textual data. It works by identifying repeated patterns in the input text and replaces them with references to previous occurrences of the same pattern [17].
- **LZW:** stands for Lemepl-Ziv-Welch, is a dictionary-based lossless compression algorithm that is used in several popular file formats like GIF and TIFF. It works by building a dictionary of frequently occurring patterns in the input text and replaces them with shorter codes [18].
- **Brotli:** is a relatively new compression algorithm that was developed by Google. It uses a combination of a modern variant of the LZ77 algorithm, Huffman coding, and second-order context modelling to achieve higher compression ratios compared to other algorithms like DEFLATE [19].

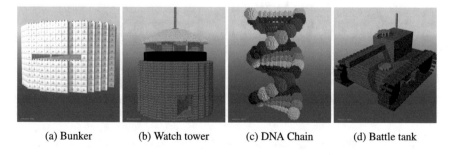

(a) Bunker   (b) Watch tower   (c) DNA Chain   (d) Battle tank

Fig. 5. VisibleSim view of 3D models used for experiments.

- **Zstd:** stands short for Zstandard, and is a compression algorithm developed by Facebook. It uses a combination of Huffman coding, Finite State Entropy (FSE) compression, and a fast dictionary search algorithm to achieve high compression ratios and fast decompression speeds [20].

To test the effectiveness of our proposed solution, we decided to create more or less complex shapes to get different message length. Therefore, we designed four 3D models on *OpenSCAD* [21] modeller: (a) a "Bunker", (b) a fortified "Watchtower", (c) an "ADN" and (d) a "Battle Tank" and integrated them on the VisibleSim simulator [22] to create a set of *Blinky Blocks* that fills the models as shown in Fig. 5.

Our vectorial description language alphabet being made of 32 different characters, the description models are encoded into a list of 5-bit codes building the CSG tree. In fact, Table 1 shows the code size for each one of them. this code can be compressed before sending it in the graph of modules and is locally decoded inside each module (without storing the model), before being integrated into our simulator and applied using Huffman decompression.

Table 1. Size of the Designed Data Models.

3D model	Brut Size	5-Bit Coded	Huffman Header	Huffman Body
Bunker	116 Bytes	580 bits	139 bits	429 bits
Watchtower	397 Bytes	249 bits	167 bits	1422 bits
DNA chain	3722 Bytes	18610 bits	139 bits	12147 bits
Tank	3986 Bytes	19930 bits	188 bits	14432 bits

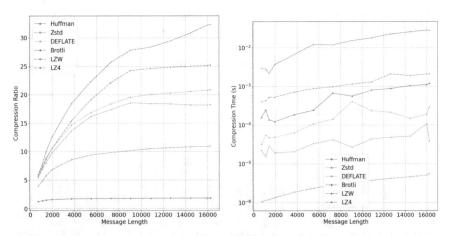

(a) Comparison of compression ratio versus message length.

(b) Variation of the compression time versus message length.

Fig. 6. Two efficiency comparisons for different lossless compression algorithms.

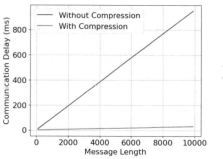 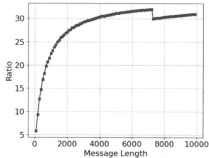

Fig. 7. Variation of Communication Delay (a) and Time Ratio (b) in Terms of Message Length with/without Brotli Compression.

3 Method and Experiments

To verify which compression algorithm is most suitable for both compression and decompression, a comparison was made in terms of the compression ratio of the data and the compression/decompression time.

The comparison was made between these lossless compression algorithms in terms of data compression ratio (see Fig. 6a), data compression, and data decompression times (see Fig. 6b). In fact, the testing was done on real messages that *Blinky Blocks* can use. Therefore, one can conclude that even though Brotli does not have the fastest compression and decompression times, except that it achieves the best result when it comes to message size compression by reducing the original message size as seen in Table 2, and data by 55%. Thus, it appears to be the best lossless compression algorithm for both data compression and decompression time, as seen in Fig. 7, and it is a suitable candidate for implementation with *Blinky Blocks*.

To be more precise, the compression process was performed only once on the master side to prepare data to be flooded into the network, while the decompression process was done on each *Blinky Block* at each computation of its colour.

Table 2. Numerical Example between different lossless compression algorithms using different message data sizes.

3D Model	Original Size (Bytes)	Huffman	Zstd	DEFLATE	Brotli	LZW	LZ4
Bunker	116	241	88	84	82	99	126
Tower	397	432	180	162	165	190	263
ADN Chain	3772	2252	288	266	229	253	443
Tank	3986	2577	569	491	444	507	1046

The data transmitted in our application is used to describe a 3D configuration (i.e. the shape to be occupied by the set of robots). These vector data are used to determine whether or not a grid position (occupied by a *Blinky Block*) is inside this shape. Compression is performed once by the external server, and then the compressed message (size N_{comp}) is sent to all robots via a module connected to the server. The decompression process, made in each *Blinky Block* in parallel, can be carried out by a single traversal of all received data. Then, data is 'decoded' on the fly, without storing a decompressed version of the message. This results in a complexity decompression algorithm $O(N_{comp})$.

Despite Brotli having a high compression time, it also has the lowest decompression time. However, since we only need to compress the message once and decompress it every single time per *Blinky Block*, we found that Brotli seems to be the most ideal solution for this. Based on the obtained experimental results, we have shown how Brotli outperforms the other lossless compression algorithms in terms of compression ratio (as shown in Fig. 6a).

As a result, one can clearly deduce the effectiveness of the Brotli lossless compression algorithm in terms of both data compression and decompression and the reduction of message length. Thus, it proves to be a very effective method to mitigate the delay problem and effectively reduce its computation and execution time. Its appliance on *Blinky Blocks* comes as a novel solution for, to our knowledge, we are the first to propose applying lossless compression algorithms to a set of modular robots *Blinky Block* in terms of "Programmable Matter" and select the best one. Regarding Fig. 6b, the experimental validation of the given remarks was applied. On the left side, the communication delay of messages with different lengths is compared in both compressed (using Brotli) and original (non-compressed) versions. These graphs clearly show the gains made from the use of Brotli compression to transmit messages. The second graph (right), shows the link between the compressed and non-compressed message lengths for different message sizes. This experience confirms that the gain is very important whenever we have a higher message length.

Therefore, Brotli is a versatile compression algorithm that offers excellent compression ratios, especially for *Blinky Blocks* messages, while still maintaining reasonable compression and decompression speeds. In fact, to further confirm the accuracy of our presented work, we tested it on *VisibleSim*, which is a software tool for simulating and programming modular robots (*Blinky Blocks*) and compared it with the already obtained results (see Fig. 6a and Fig. 6b) to show how close these results are and that the executed code remains the same wherever it is tested. Thus, it shows that the proposed algorithm has no compatibility or coding issues since it operates on the size of the message and not on the *Blinky Block* configuration.

After several tests on real data models and having it compared with other lossless compression algorithms in terms of compression/decompression time and compression ratio, Huffman will be replaced with the Brotli compression algorithm. Thus, offering the highest known compression ratio with far fewer compression and decompression times compared to Huffman.

Finally, we propose a more practical experiment to validate the complete process, consisting of compressing the 3D model, distributing the data code to a large set of connected *Blinky Blocks*, and decompressing many time the stored code in each *Blinky Block* to use the 3D data to set their colour. Figure 8 shows a picture of the setup of this experiment, also used to produce the video[1]. The setup shown on the left side of Fig. 8, includes a laptop connected to a grid of 768 *Blinky Blocks* (32×24). The laptop first sends the coordinates to each *Blinky Block* then sends the compressed model to the *Blinky Blocks*. At launch, the *Blinky Blocks* create a common coordinate system to obtain a position (cx, cy, cz) relative to the module in the lower left corner, by applying the algorithm proposed in [23]. The spanning tree created for this purpose will be used to distribute the code to all the blocks.

In this application we use a Huffman encoding algorithm which we ran on the laptop to create the code from the 3D model (the DNA model presented on the right side of Fig. 8), and sending it to one of the *Blinky Blocks*. To check that the data is well-received and uncompressed by each *Blinky Block*, after reception we repeat 60 rounds that compute the colour of an horizontal plane at level cz crossing the 3D scene.

At each stage, each *Blinky Block* analyses the encoding chain eight times to calculate the colour at eight different positions of the space inside the block. This method allows to create anti-aliasing effects. Positions are $(cx \pm 0.25 \times l, cy \pm 0.25 \times l, cz \pm 0.25 \times l)$ where l is the width the cubic *Blinky Block*. After half a second, each *Blinky Block* switches to the next stage by increasing its cz position by $0.25 \times l$ and recomputing a new colour.

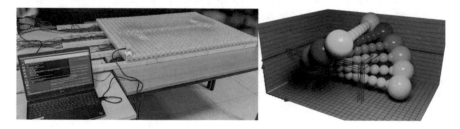

Fig. 8. An example of *Blinky Blocks* application with transmission and decompression of a 3D description model (the short DNA chain presented in the right picture).

4   Conclusion and Future Works

In this paper, we propose a study of the efficiency of a set of *Blinky Blocks* robots in terms of communication delay and computation time.

Based on the obtained results, we show that the communication delay linearly depends on the size of the message, and presented Huffman as a lossless compression algorithm as a novel method, which was substituted by Brotli as an ideal solution.

To reduce the communication delay, we propose to add a lossless compression algorithm. We compare a set of recent efficient algorithms with the Huffman coding method. We express the compression ratio and compression/decompression execution time for each of them. Moreover, the obtained results show that the Brotli algorithm requires the

---

[1] Video of Real time decompression on *Blinky Blocks*: https://youtu.be/xjAKxByAEll.

minimum overhead in terms of execution time and can achieve the maximum compression ratio. Therefore, this work indicates that the Brotli algorithm should be introduced at *Blinky Block* to reach a minimum communication delay.

In the future, this work will further extend to cover three main points:

- First, the adoption of Huffman as the first compression mechanism that can perform compression on *Blinky Blocks* proved to be a success. However, it cannot compress large messages within the accepted range of *Blinky Blocks'* message length, which varies from 1 to 227 bytes. Therefore, based on the presented results above, Brotli will be introduced as a successor to replace Huffman's compression.
- The constant integration of *Blinky Blocks* into the IoT domain [24, 25] and its interaction with different IoT devices will surely require not only textual data to be exchanged, but also audio, video, and even images. Therefore, other compression algorithms will be tested, depending on the changing nature of *Blinky Blocks* and the structure of the integrated data to reduce the communication delays between *Blinky Blocks*.
- Compression is surely an important mechanism to reduce communication delays. However, it is important to ensure that this communication is not intercepted by a malicious/non-malicious party. Therefore, a very lightweight cryptographic solution that takes into consideration the resource-constrained nature of *Blinky Blocks* is required and will be integrated with the compression mechanism to ensure the first crypto-compression solution for *Blinky Blocks* that reduces delays and secures the communication by preventing the interception of the compressed messages.

Acknowledgment. This work has been supported by the EIPHI Graduate School (contract "ANR-17-EURE-0002").

References

1. Bourgeois, J., et al.: Programmable matter as a cyber-physical conjugation. In: IEEE International Conference on Systems, Man, and Cybernetics (SMC 2016), Budapest, Hungary, pp. 002 942–002 947. IEEE (2016). https://publiweb.femto-st.fr/tntnet/entries/13257/documents/author/data
2. Rubenstein, M., Cornejo, A., Nagpal, R.: Programmable self-assembly in a thousand-robot swarm. Science **345**(6198), 795–799 (2014)
3. Stoy, K., Nagpal, R.: Self-reconfiguration using directed growth. In: Alami, R., Chatila, R., Asama, H. (eds.) Distributed Autonomous Robotic Systems 6, pp. 3–12. Springer, Tokyo (2007). https://doi.org/10.1007/978-4-431-35873-2_1
4. Bassil, J., Piranda, B., Makhoul, A., Bourgeois, J.: Repost: distributed self-reconfiguration algorithm for modular robots based on porous structure. In: IEEE RSJ International Conference on Intelligent Robots and Systems (IROS 2022), Kyoto, Japan (2022)
5. Knychala Tucci, T., Piranda, B., Bourgeois, J.: Efficient scene encoding for programmable matter self-reconfiguration algorithms. In: 32nd Annual Symposium on Applied Computing (SAC 2017), ser. ACM International Conference Proceedings, Morroco, Marrakesh, Marrakech, Morocco, pp. 256–261 (2017). https://publiweb.femto-st.fr/tntnet/entries/13417/documents/author/data
6. Yaacoub, J.-P.A., Noura, H.N., Piranda, B.: The internet of modular robotic things: issues, limitations, challenges, & solutions. Internet of Things 100886 (2023)

7. Yaacoub, J.-P.A., Noura, H.N., Salman, O.: Security of federated learning with IoT systems: issues, limitations, challenges, and solutions. Internet Things Cyber-Phys. Syst. **3**, 155–179 (2023)
8. Yaacoub, J.-P.A., Noura, H.N., Salman, O., Chehab, A.: Robotics cyber security: vulnerabilities, attacks, countermeasures, and recommendations. Int. J. Inf. Secur. **21**, 115–158 (2022)
9. Yaacoub, J.-P., Noura, H., Salman, O., Chehab, A.: Security analysis of drones systems: attacks, limitations, and recommendations. Internet Things **11**, 100218 (2020)
10. Tagne, E.F., Kamdjou, H.M., Amraoui, A.E., Nzeukou, A.: A lossless distributed data compression and aggregation methods for low resources wireless sensors platforms. Wireless Pers. Commun. **128**(1), 621–643 (2023)
11. Banerjee, R., Das Bit, S.: An energy efficient image compression scheme for wireless multimedia sensor network using curve fitting technique. Wirel. Netw. **25**, 167–183 (2019)
12. Chen, F., Chandrakasan, A.P., Stojanovic, V.M.: Design and analysis of a hardware-efficient compressed sensing architecture for data compression in wireless sensors. IEEE J. Solid-State Circuits **47**(3), 744–756 (2012)
13. Deepu, C.J., Heng, C.-H., Lian, Y.: A hybrid data compression scheme for power reduction in wireless sensors for IoT. IEEE Trans. Biomed. Circuits Syst. **11**(2), 245–254 (2016)
14. Peng, Y., et al.: A high-voltage generator and multiplexer for electrostatic actuation in programmable matter. IEEE J. Solid-State Circuits **58**(4), 915–928 (2023)
15. Alakuijala, J., Kliuchnikov, E., Szabadka, Z., Vandevenne, L.: Comparison of Brotli, Deflate, Zopfli, LZMA, LZHAM and Bzip2 compression algorithms. Google Inc. 1–6 (2015)
16. Oswal, S., Singh, A., Kumari, K.: Deflate compression algorithm. Int. J. Eng. Res. Gener. Sci. **4**(1), 430–436 (2016)
17. Ziv, J.: The universal LZ77 compression algorithm is essentially optimal for individual finite-length n-blocks. IEEE Trans. Inf. Theory **55**(5), 1941–1944 (2009)
18. Dheemanth, H.: LZW data compression. Am. J. Eng. Res. **3**(2), 22–26 (2014)
19. Alakuijala, J., et al.: Brotli: a general-purpose data compressor. ACM Trans. Inf. Syst. (TOIS) **37**(1), 1–30 (2018)
20. Collet, Y., Kucherawy, M.S.: Zstandard compression and the application/zstd media type. RFC, vol. 8878, pp. 1–45 (2018). https://api.semanticscholar.org/CorpusID:52962805
21. Büttrich, S.: 3D modeling with openscad-part 1. Low-Cost 3D Printing, p. 83 (2018)
22. Thalamy, P., Piranda, B., Naz, A., Bourgeois, J.: VisibleSim: a behavioral simulation framework for lattice modular robots. Robot. Auton. Syst. 103913 (2021). https://www.sciencedirect.com/science/article/pii/S0921889021001986
23. Piranda, B., Lassabe, F., Bourgeois, J.: Disco: a multiagent 3D coordinate system for lattice based modular self-reconfigurable robots. In: IEEE International Conference on Robotics and Automation (ICRA 2023), London, England (2023)
24. Yaacoub, J.-P.A., et al.: Securing internet of medical things systems: limitations, issues and recommendations. Futur. Gener. Comput. Syst. **105**, 581–606 (2020)
25. Yaacoub, J.-P.A., Salman, O., Noura, H.N., Kaaniche, N., Chehab, A., Malli, M.: Cyber-physical systems security: limitations, issues and future trends. Microprocess. Microsyst. **77**, 103201 (2020)

IoRT-Based Distributed Algorithm for Robust Team Formation and Its Application to Smart City Operation

Rajdeep Niyogi[1](✉) and Amar Nath[2]

[1] IIT Roorkee, Roorkee 247667, India
rajdeep.niyogi@cs.iitr.ac.in
[2] SLIET, Longowal 148106, India
amarnath@sliet.ac.in

Abstract. Robust team formation (RTF) is an important problem in multi-agent systems. Existing research addresses offline robust team formation. Our work addresses the online version of the problem where task requirements are gathered, team members are recruited, and the newly formed team executes the task with robustness guarantees. An Internet of Robotic Things (IoRT) based distributed robust team formation (DRTF) algorithm is developed to form a minimal-cost robust team for a given degree of robustness. We have implemented the proposed approach using ARGoS, a realistic multi-robot simulator, and validated the approach using a smart city operation of an emergency medical team attending an accident. The simulation results illustrate that team members are decided at runtime; cannot be determined *a priori*, since the robots' states, locations, and roles change with time.

1 Introduction

A team is considered to be k-robust (for a given non-negative integer k, referred to as the degree of robustness) if removing any k agents from it leads to a remaining team that can still accomplish the given task [1]. Given the parameters: a set of agents, a set of tasks, a description of each agent, a goal that is a subset of tasks, and the degree of robustness, the robust team formation problem seeks to find a minimum cost robust team. The computational complexity of this problem is known to be NP-hard [1]. The simplest formulation of such problems would require finding a subset of agents (referred to as a team) from a given set of agents such that the team satisfies some desirable property. A minimum-cost robust team has to be formed despite the absence of information on some relevant parameters; this paper deals with this issue.

In a smart city setting, none of the robotic agents, dispersed in the city doing some activities, has a global view; no agent knows the total number of agents present, their states, and their locations. The states and locations of agents may change over time; for example, they may be involved in some other task; agents may enter or exit the smart city for some reason. The arrival times and locations for tasks (any operation like rescue) are not known in advance.

L. Barolli (Ed.): AINA 2024, LNDECT 201, pp. 12–23, 2024.
https://doi.org/10.1007/978-3-031-57870-0_2

In a smart city, we may assume the robotic agents will continuously move across the city. Whenever it detects a task (for instance, the occurrence of an accident), it tries to communicate with other agents in its neighborhood to form a team. IoRT-enabled agents facilitate such communication [2]. In IoRT, robots have sensors, actuators, and communication capabilities that allow collaboration and automation. This enables robots to perform tasks more efficiently, make autonomous decisions based on real-time data, and collaborate with other robots and devices in various applications [3]. Thus, team formation could be done at runtime by exchanging messages with other agents without any centralized agent or human intervention. Explicit communication allows multi-agent coordination to be more effective [4]. So, a distributed algorithm [5] for team formation has to be devised. However, team formation may be unsuccessful due to a lack of suitable agents or the unpredictability of message delivery (i.e., the messages may be lost or delayed).

In this paper, we propose an IoRT-based distributed algorithm for forming a robust team with minimal cost in a smart city, where the number of agents that may fail during mission accomplishment is given. Still, the other relevant parameters have yet to be discovered in advance. The agents communicate with each other via sending and receiving messages for coordination.

We briefly illustrate the proposed approach with an operation in a smart city that involves a team of agents. An emergency medical team has to be formed to attend to an accident and thereafter transport the injured patient to the nearby hospital. Let us assume that two emergency medical technicians (EMTs) (Agent1, Agent5) and three paramedics (Agent2, Agent3, Agent4) are currently available and are dispersed at different locations (Fig. 1). For responding to an accident, a paramedic charges $150, and an EMT charges $100. These are the abilities of an EMT [6]: give sick or injured patients treatment, such as first aid or life support. A paramedic possesses all the same abilities as an EMT as well as the following abilities [6]: improved airflow management and ECG interpretation. Both an EMT and a paramedic can operate an ambulance. A team should have the following abilities to handle an accident: sophisticated airflow control, ECG interpretation, first aid or life support care, and ambulance driving. At least one EMT or paramedic should constantly observe the patient's vital signs. Additionally, the team should be robust enough for the remaining members to complete the assignment, even if one of the members becomes unwell.

Agent1 finds that an accident has occurred at location L1 and tries to form a 1-robust team by communicating with other agents in its neighborhood (Fig. 1(a)). Agent1 finds three possible 1-robust teams. Team 1 comprises Agents 1, 3, and 4. Team 2 comprises Agents 1, 2, and 4. Team 3 comprises Agents 1, 2, and 3. The cost of a team is determined by the costs of the team members and their distances from the task's location. Since Agent2 is located at a relatively longer distance from L1 compared to the locations of Agents 3 and 4, so Team 1 having minimum cost, is selected. The selected members of Team 1 reach location L1 (Fig. 1(b)). Finally, the patient is transported to location L6. If Agent3 falls ill during transportation, Agents 1 and 4 can still carry out the task (Fig. 1(c)).

The remainder of the paper is organized as follows. Related work is given in Sect. 2. Section 3 presents a formal framework for robust team formation. The proposed dis-

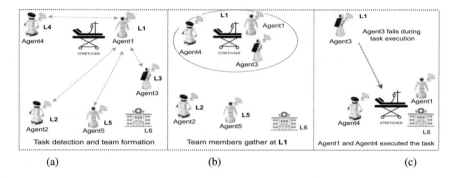

Fig. 1. An emergency medical team transporting an injured patient: Different phases

tributed robust team formation (DRTF) algorithm is given in Sect. 4. The implementation of the proposed approach using ARGoS [7] is presented in Sect. 5. The conclusions are drawn in Sect. 6.

2   Related Work

Robust team formation has been studied in [1, 8] by considering a bi-objective constraint optimization problem with a pareto-frontier of solutions trading off team cost and team robustness. These works assume complete knowledge of the environment and are not concerned with an agent's current state and location. A centralized algorithm based on branch and bound technique is suggested in [1]. The work [1] is extended to find multiple k-robust teams in [8], where a centralized algorithm based on branch and bound technique is suggested. The work [9] considers a robust team formation problem of finding a minimal cost k-robust team for a given desired robustness level k, assuming complete knowledge of the environment. [9] suggests approximation algorithms for the problem and its variants. A centralized algorithm to find a partially robust team is suggested in [10, 11]. We suggest a decentralized algorithm for finding a minimal-cost k-robust team.

A distributed paradigm for solving an optimization problem is DCOP (distributed constraint optimization problem). By exchanging their (local) subproblems, the agents collaborate to find a global optimal solution. Synchronized Branch-and-Bound (Syn-BB) is an algorithm for solving DCOPs that guarantees an optimal solution. An optimal solution is also guaranteed by the Synchronous Branch-and-Bound for Stable Teams (SynBB-ST) algorithm [12], which is based on Syn-BB. The following is how these algorithms work: A communication network topology, such as a tree or a graph, connects the agents by allowing them to send and/or receive messages from one another. This topology is chosen and independent of the DCOP and the algorithm considered. The communication network assumes that no messages are lost, and this network property is necessary for SynBB-based algorithms to guarantee an optimal solution.

Our work considers a general setting where messages might get lost. Our work does not rely on any particular network topology. A message propagation tree is constructed at runtime. Our approach typically yields a (local) optimal solution. A robust team identified by our algorithm has a global optimal cost if all environment agents receive the **Request** message and no other messages are lost. As a result, considering the communication network's unpredictability, we developed a distributed algorithm that leads to a locally optimal solution. Accordingly, we can't utilize the current DCOP algorithms (e.g., [12]) to tackle the issue thought about in this paper.

Swarm robotics [13] is inherently based on homogeneous robots that only interact with their nearest neighbors. MRS (multi-robot systems) comprise robots of different skills, so they are heterogeneous and interact with one or more robots that need not be their nearest neighbors. Whereas swarm robotics usually involves hundreds or thousands of robots, this number is relatively small in MRS [14]. If the number of robots is smaller, then much time would be spent on obstacle avoidance as each robot would act as an obstacle for the other. This time is orders of magnitude more than the time taken for agent communication. In swarm robotics, communication is implicit, based on stigmergy where information is shared by making modifications in the environment [14]. In MRS, the robots exchange messages via explicit communication. In this paper, we consider an MRS where the robots are IoRT-enabled and allow communication with one another by sending and receiving messages.

Auction-based approaches for team formation (task allocation) are suggested in [15, 16]. A non-initiator (initiator) in our work can be considered a bidder (auctioneer). Responding to a **Request** message by a non-initiator is similar to a bidder making a bid. In our work, two or more non-initiators may communicate with each other. This is not allowed in auction-based approaches [15, 16] since bidders do not communicate with each other.

Distributed team formation approaches have been suggested in [15, 17–19]. However, none of these works consider the robust team formation problem. In [18, 19], the assumptions about the environment and the types of messages are the same as those in this paper. In [18, 19] only the initiator and the non-initiators and in [17] only the leaders and the followers communicate with each other. So the nature of communication in [15, 17–19] is relatively much simpler than in this paper.

3   Formal Framework

In this work, agents refer to IoRT-enabled mobile robots that can send and receive messages. The agents work in a common environment with a set of locations. Let U be a universe of a set of skills.

Definition 1 (Agent). *An agent is represented by a tuple $r = \langle id, \psi, s, l, m, c, \theta, \theta' \rangle$, where id is a unique identifier, $\psi \subseteq U$ is the set of skills, $s \in \mathscr{S}$ is the current state from a set of states $\mathscr{S}$, l is the current location in the environment, $m \in \mathscr{M}$ is the last message sent/received from a set of messages $\mathscr{M}$, c denotes the cost that the agent charges for a task, $\theta, \theta' \in (0, 1]$ denote the remaining battery backup and battery consumption rate of the agent, respectively.*

An agent with a greater θ is less likely to fail. An agent with a smaller θ' is more likely to last for a longer time.

Definition 2 (Task). *A task τ is specified by a tuple $\tau = \langle v, \Psi, l, t \rangle$, where v is the name of a task (e.g., move glass container B to location l'), $\Psi \subseteq U$ is the set of skills required to execute the task, l is the location where the task arrived, and t is the time at which the task arrived.*

Definition 3 (Task allocation). *A task $\tau = \langle v, \Psi, l, t \rangle$ can be allocated to a team of agents $\Gamma = \{r_1, \ldots, r_n\}(n > 1)$, denoted by $\Gamma \leftarrow \tau$, if $\bigcup_{j=1}^{n} r_j.\psi \supseteq \tau.\Psi$.*

The minimum size of a team is 2 since only some agents can accomplish a task.

According to the taxonomy [20], the team formation problem discussed in this paper falls into the category ST-MR-IA where ST means a robot can only perform at most a single task at a time, MR means that some tasks may require multiple robots, and IA stands for instantaneous allocation (i.e., at runtime) of robots for the tasks.

Definition 4 (k-robust team). *Given $\Gamma = \{r_1, \ldots, r_n\}(n > 1)$, $\tau = \langle v, \Psi, l, t \rangle$, $k > 0$, $n > k$ such that $\Gamma \leftarrow \tau$. Suppose any k members are unavailable in the team Γ. Let Γ' be the new team with $(n - k)$ members such that $(n - k) > 1$. Γ is k-robust if $\Gamma' \leftarrow \tau$.*

For example, the minimum size of a 1-robust team is 3.

Definition 5 (Task execution). *A task $\tau = \langle v, \Psi, l, t \rangle$ can be executed by a team of agents $\Gamma = \{r_1, \ldots, r_n\}(n > 1)$, if $\Gamma \leftarrow \tau$ and all members of Γ reach location l at some time $t' > t$.*

Definition 6 (Cost of a k-robust team). *Given a task $\tau = \langle v, \Psi, l, t \rangle$ and $k > 0$. Let $\Gamma = \{r_1, \ldots, r_n\}$ be a k-robust team that can execute τ, where each member r_i of the team was located at l_i at some time $t' > t$. The cost of a team Γ for executing the task τ at t' is $\mathscr{C}_{\langle \Gamma, \tau, t' \rangle} = \sum_{i=1}^{n} c(r_i, \tau) \times \frac{1}{\theta_{r_i}} + d(l_i, l) \times \theta'_{r_i}$ where $d(l_1, l_2)$ is the Euclidean distance between l_1 and l_2.*

In the illustration of the robust team formation example (Fig. 1), it is assumed that the values of θ, θ' are identical for the same type of agents.

Definition 7 (Dominant team). *Given a task $\tau = \langle v, \Psi, l, t \rangle$ $k > 0$, and $t' > t$. Let $\Gamma_1, \ldots, \Gamma_m$ be the possible k-robust teams for executing τ. Γ_i is said to be a dominant team for τ if: $\mathscr{C}_{\langle \Gamma_i, \tau, t' \rangle} \leq \mathscr{C}_{\langle \Gamma_j, \tau, t' \rangle} \ \forall j \in \{1, \ldots, m\} \setminus \{i\}$*

Problem Statement: No task τ_i is known in advance. Design an algorithm to form a dominant team Γ_i at runtime for each task τ_i with degree of robustness $k_i > 0$ such that τ_i can be executed by Γ_i.

4   Proposed Approach: DRTF

We consider a wireless communication network where messages may be lost but are not corrupted during transmission. Message delay is nonzero but finite. Agents are IoRT-enabled mobile robots that can send and receive messages. An agent who is outside the environment does not receive any message. An agent can enter the environment at any time, but it can exit if it is not involved in any team formation. A finite number of agents can be present in the environment at any time.

Collaborative task execution is given in Algorithm 1. There can be only one initiator for a task. Distributed robust team formation (DRTF) comprises Algorithms 2 and 3. The agents communicate by sending different types of messages. **Request** message is used to advertise a task by the initiator to obtain potential team members. A non-initiator expresses its interest in becoming a team member for the task advertised by the initiator using the **Reply** message. **Declare** message (a notification) is used to convey the decision of the initiator about who are and who are not the team members of the advertised task. The messages can be modeled using agent communication languages [21]. However, modeling in such languages is beyond the scope of this paper.

Request and **Declare** messages are sent from the initiator to a non-initiator via zero or more non-initiators. A **Reply** message is sent opposite to a **Request** message. The meaning of the states is as follows. UNCOMMITTED state means an agent is available and not assigned to any task. COMMITTED state means that an agent has selected a task, but the team for the task still needs to form. ACTIVE state means a team for the task has been formed, an agent has become a team member, and task execution is ongoing or about to start. We consider a general situation where messages may get lost. However, we want the notifications to be received by the agents. The agents are cooperative and help in the team formation process even if they are not potential team members. Thus, we state these assumptions explicitly:

A1. **Request** and **Reply** messages may get lost in transit. A2. **Declare** messages are not lost in transit. A3. An agent who has propagated a **Request** message will remain in its communication range until it has propagated the **Declare** message.

Distributed Robust Team Formation Process: Algorithm 2 works as follows: An initiator A, who is trying to form a k-robust team for a task τ, changes its state from UNCOMMITTED to COMMITTED (Line 1), broadcasts a **Request** message to the agents that are within its communication range, and waits for some unit of time Δ for the **Reply** messages. $ReceivedReply^A$ is the set of agents whose **Reply** messages are received within Δ (Line 2). TABLE$_A$ is updated in Line 3. If a dominant team can be formed as per Definition 7, then A's state becomes ACTIVE, otherwise UNCOMMITTED (Lines 4–7). A intimates who its team members are (TM) and who are left out (NTM) by sending a **Declare** message to all the agents in $ReceivedReply^A$, and TABLE$_A$ is deleted (Line 9).

Algorithm 1: Collaborative task execution

Input : A mission M consists of a finite set of tasks where no task $\tau_i = \langle v_i, l_i, t_i, \Psi_i \rangle$ is known in advance
Output: M accomplished

1 state = UNCOMMITTED of r_j;
2 **repeat**
3   r_j detects τ_i at l_i and starts k_i-robust team formation using Algorithm 2 s.t. $r_j.\psi \cap \tau_i.\Psi_i \neq \emptyset$; // now r_j becomes initiator
4   **if** (*a dominant team* $\Gamma = \{x_1, \ldots, x_N\}$ *is formed for* τ_i) **then**
5          $\forall x_j \in \Gamma : x_j$ moves to l_i; Γ executes τ_i;
6   dissolve team Γ; $\forall x_j \in \Gamma$: state := UNCOMMITTED;
7   **end**
8 **until** (*all tasks of mission M are executed*);

Algorithm 2: k-robust team formation for initiator Y

---

1 state := COMMITTED;
2 broadcast **Request**($\langle Y, \tau \rangle$, Y, TABLE$_Y$) and wait for some time Δ for the **Reply** messages; let *ReceivedReply*^{Y} be
    the set of agents whose **Reply** messages are received within Δ;
3 update TABLE$_Y$;
4 **if** *(a dominant team is formed)* **then**
5 | state := ACTIVE;
6 **else**
7 | state := UNCOMMITTED;
8 **end**
9 send **Declare**($\langle Y, \tau \rangle$, Y, TM, NTM) to all the agents in *ReceivedReply*^{Y}; delete TABLE$_Y$; exit;

Algorithm 3: k-robust team formation for non-initiator X

---

1   **case** state = UNCOMMITTED **do**
2 | **if** *(Request is received first time from z)* **then**
3 | | PARENT := z;
4 | | **if** *(X.$\psi \cap \tau.\Psi \neq \emptyset$)* **then**
5 | | | construct TABLE$_X$ by appending the row of X to TABLE$_z$;
6 | | | state := COMMITTED;
7 | | **end**
8 | | relay($\langle Y, \tau \rangle$, X);
9 | | send **Reply**($\langle Y, \tau \rangle$, X, TABLE$_X$) to PARENT;
10 | **else**
11 | | **if** *(Declare is received)* **then**
12 | | | send **Declare** to all the agents in *ReceivedReply*^{X}; delete TABLE$_X$;
13 | | **else**
14 | | | skip;
15 | | **end**
16 | **end**
17 **end**
18 **case** state = COMMITTED **do**
19 | **if** *(Declare is received)* **then**
20 | | **if** $X \in TM$ **then**
21 | | | state := ACTIVE;
22 | | **else**
23 | | | state := UNCOMMITTED;
24 | | **end**
25 | | send **Declare** to all the agents in *ReceivedReply*^{X}; delete TABLE$_X$;
26 | **else**
27 | | skip;
28 | **end**
29 **end**

Algorithm 4: function **relay**($\langle I, \tau' \rangle$, S)

---

1 broadcast **Request**($\langle I, \tau' \rangle$, S, TABLE$_S$) and wait for some time Δ' for the **Reply**
   messages; let *ReceivedReply*^{S} be the set of agents whose **Reply** messages are received
   within Δ';
2 obtain TABLE$_S$ by merging the tables of the agents in *ReceivedReply*^{S};

Algorithm 3 works as follows. When a non-initiator B in UNCOMMITTED state receives the **Request** message for the first time (Line 2), say, from A, it makes A its parent (Line 3). If B receives the **Request** message again from another agent, say D, then it discards the message (Line 14, 27). If B's skill set matches with some skills of the task (Line 4), then it constructs a table by appending a row to the table received and changes its state to COMMITTED (Lines 5–6). TABLE$_X$ is a table that has the fields, *id*, skills, location, cost, and θ, θ', of an agent X. B then propagates the **Request** message with the table using the function *relay* (Line 8). If B's skill set does not match with some skills of the task, then it simply propagates the received table (Line 8). In the latter situation, where although B is not a potential team member, B helps in propagating the **Request** message to the extent possible without compromising on its resources. This reflects the cooperative behavior of the agents. Then B sends a **Reply** to its parent (Line 9).

Suppose that B is in COMMITTED state, and it receives a **Declare** message (Line 19). If B is a member of the team, then its state becomes ACTIVE, otherwise UNCOMMITTED (Line 20–23). B then sends the **Declare** message to all the agents in *ReceivedReply*^{B}, and TABLE$_B$ is deleted (Line 25).

The function **relay** works as follows. The **Request** message is broadcast by S to the agents that are within its communication range, and waits for the **Reply** messages for time Δ' (Line 1) as in Algorithm 2. *ReceivedReply*^{S} is the set of agents whose **Reply** messages are received within Δ' (Line 1). TABLE$_S$ is obtained by merging the tables of the agents in *ReceivedReply*^{S} (Line 2). Each agent has a timer that avoids indefinite wait. So Algorithms 2, 3 are non-blocking.

Salient features of the Algorithm: (i) Termination: Algorithms 2 and 3 terminate. (ii) Correctness condition/Invariant: Initiator's state becomes ACTIVE iff $\forall x \in TM$ state of x becomes ACTIVE. (iii) Communication complexity: The number of messages sent is $O(n^2)$ where n is the number of agents present in the environment. (iv) Optimality: if the **Request** message is received by all the other agents and no other messages are lost, then a k-robust team found by the algorithm has optimal cost. The proofs will be provided in an extended version of the paper.

5   Implementation

The working of the proposed approach is simulated and tested using the 3.0.0-beta50 version of ARGoS [7], a realistic multi-robot simulator, on Intel$^\oplus$ Core^{TM} i7 Processor with 8GB RAM and macOS Sierra operating system. We used foot-bot, a type of robot supported by ARGoS that has different types of sensors/actuators such as *Colored_blob_omnidirectional_camera, gripper, Positioning, proximity, Differential_steering, LED*, and *Range_and_bearing* [7]. An enclosed arena of dimension (10m × 10m) is created in ARGoS, where the borders are shown in black Fig. 2(a). Two types of movable objects of different shapes and sizes (cylindrical–shown in green and cubical–shown in red Fig. 2(a)) are taken. A task involves transporting such an object, placed somewhere in the arena (white space), to a border.

We consider an operation in a smart city where an emergency medical team has to respond to a major traffic accident. A robust team has to be formed at runtime that can safely transport an injured patient to the designated treatment site while constantly monitoring the patient's vital parameters and providing basic treatment. A task is determined by the severity of the injury caused by the accident. The accomplishment of a task depends on the combined skill set of a team. In our simulations, we abstract the different tasks by transporting objects of varying sizes and shapes.

Tasks τ_1, τ_2, τ_3 refer to transporting a small cube, a large cube, and a large cylinder respectively. The set of skills required for τ_1, τ_2, τ_3 is $\{s_1, s_2\}$, $\{s_2, s_3\}$, and $\{s_1, s_2, s_3\}$ respectively, where s_1 denotes vision, s_2 navigation, and s_3 terrain identification. All the robots are capable of lifting, grabbing, and pushing an object. The robots $r_1, \ldots, r_9$ (blue dots in Fig. 2(a)) used in the experiments have the attributes shown in Table 1.

Table 1. The attributes of the robots

Robot	ψ	c	θ'	θ
r_1, r_2, r_3	$\{s_1, s_3\}$	10	0.2	0.8
r_4, r_5, r_6	$\{s_2, s_3\}$	15	0.3	0.7
r_7, r_8, r_9	$\{s_1, s_2\}$	20	0.3	0.7

5.1   Simulation Results

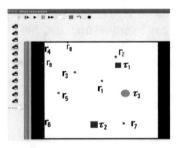

(a) Initial positions of tasks and robots

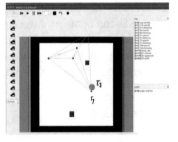

(b) r_7 begins team formation for task τ_3

(c) team is formed and the other members (r_1, r_5) have arrived at the task's location

(d) τ_3 is executed: the object is transported jointly

Fig. 2. 1-robust team formation and object transportation

For all the experimental runs the robots and objects are initially placed randomly in the arena. One such situation (Scenario 1) is shown in Fig. 2(a). Now r_7 (blue dot) detects the task τ_3 and begins the team formation for $k = 1$ (Fig. 2(b)). The other members r_1, r_5 arrive at the location of the task (Fig. 2(c)). Now the object is transported to the southern border (Fig. 2(d)). The team is then dissolved, and the other objects are transported similarly. For Scenario 1,2, we obtained the following results (Table 2): for τ_1 different teams are formed by different initiators, for τ_2 different teams are formed by the same initiator, for τ_3 the same initiator forms the same team, but the costs of the teams are different since the members' locations have changed.

The task execution time (TET) is the time elapsed in seconds from the beginning of team formation to the dissolution of the team; the communication time is very small compared to the time taken for the physical movement of the robots. When k is increased, (i) the team size increases, which increases the cost, and (ii) TET increases (Scenario 3, Table 2). In Fig. 2(c), for the transportation to begin, the robots re-orient themselves toward the destination, grab the object, and synchronize; then, the movement starts. The time taken for these activities increases with the size of the team. The experimental results validate that the team members are decided at runtime; cannot be determined *a priori*, since the robots' states, locations, and roles change with time. If the number of robots in the arena is greatly increased, most of the time is spent avoiding collisions as each robot acts as an obstacle and thus hinders movement.

Table 2. Results for the different scenarios

Task	Initiator	Team members	TET(sec)	Cost
Scenario 1 $k = 1$				
τ_1	r_1	$\{r_1, r_7, r_8\}$	172	72.04
τ_2	r_5	$\{r_3, r_5, r_9\}$	166	65.10
τ_3	r_7	$\{r_1, r_5, r_7\}$	160	64.30
Scenario 2 $k = 1$				
τ_1	r_6	$\{r_3, r_6, r_8\}$	172	95.45
τ_2	r_5	$\{r_2, r_5, r_7\}$	177	75.50
τ_3	r_7	$\{r_1, r_5, r_7\}$	182	72.45
Scenario 3 $k = 2$				
τ_1	r_1	$\{r_1, r_7, r_8, r_9\}$	232	101.81
τ_2	r_5	$\{r_3, r_4, r_5, r_6\}$	225	94.87
τ_3	r_8	$\{r_1, r_2, r_5, r_6, r_8\}$	244	100.63

Implementation Challenges: These include avoiding collisions, controlling the speed of a robot as it approaches other robots or obstacles, rotation, and alignment to reach the exact location, grabbing the object, and synchronization while executing. We addressed these concerns during the implementation of the algorithms using ARGoS. We have also handled a more nuanced situation in which two robots are present at the task location, and both simultaneously begin the team formation process. An omnidirectional camera sensor allows a robot to detect its surroundings. By using Beacon signals, the robots will exchange their identities and sets of skills. The initiator will be the one who has more skills. If their skills are identical, the robot with a lower identifier will take the lead.

6   Conclusions

In this paper, we addressed the challenging problem of minimal cost distributed robust team formation for a given degree of robustness. For this, a novel IoRT-based distributed algorithm DRTF with low communication overhead is presented. We developed a prototype model of DRTF in ARGoS and validated it using a smart city operation of an emergency medical team attending an accident. The simulation results demonstrate how team members are formed at runtime; cannot be determined *a priori*. Aside from team formation, the gathering of team members and task execution have all been successfully implemented. As a result, this paper attempted to achieve a complete realization of cooperative problem-solving. Formal verification of the proposed approach would be undertaken as part of future work.

Acknowledgement.   The first author was in part supported by a research grant from Google.

References

1. Tenda, O., Schwind, N., Clement, M., Ribeiro, T., Inoue, K., Marquis, P.: How to form a task-oriented robust team. In: AAMAS, pp. 395–403 (2015)
2. Vermesan, O., et al.: Internet of robotic things intelligent connectivity and platforms. Front. Robot. AI **7**, 104 (2020)
3. Ray, P.P.: Internet of robotic things: concept, technologies, and challenges. IEEE Access **4**, 9489–9500 (2016)
4. Zhi, Y., Jouandeau, N., Cherif, A.A.: A survey and analysis of multi-robot coordination. Int. J. Adv. Rob. Syst. **10**(399), 1–18 (2013)
5. Lynch, N.: Distributed Algorithms. Elsevier (1996)
6. U.S. Bureau of Labor Statistics. Occupational Outlook Handbook. https://www.bls.gov/ooh/healthcare/emts-and-paramedics.htm#tab-2
7. Carlo, P., et al.: ARGoS: a modular, parallel, multi-engine simulator for multi-robot systems. Swarm Intell. **6**(4), 271–295 (2012)
8. Tenda, O., Ribeiro, T., Bouchabou, D., Inoue, K.: Mission oriented robust multi-team formation and its application to robot rescue simulation. IJCAI (2016)
9. Crawford, C., Rahaman, Z., Sen, S.: Evaluating the efficiency of robust team formation algorithms. In: Osman, N., Sierra, C. (eds.) AAMAS 2016, vol. 10002, pp. 14–29. Springer, Cham (2016). https://doi.org/10.1007/978-3-319-46882-2_2
10. Nicolas, S., Demirovic, E., Inoue, K., Lagniez, J.M.: Partial robustness in team formation: bridging the gap between robustness and resilience. In: AAMAS, pp. 1154–1162 (2021)
11. Nicolas, S., Demirovic, E., Inoue, K., Lagniez, J.M.: Algorithms for partially robust team formation. J. Auton. Agents Multi-Agent Syst. **37**, 22 (2023)
12. Barambones, J., Richoux, F., Imbert, R., Inoue, K.: Resilient team formation with stabilisability of agent networks for task allocation. ACM Trans. Auton. Adapt. Syst. **15**(3), 24, Article no. 7 (2021)

13. Sahin, E.: Swarm robotics: from sources of inspiration to domains of application. In: International Workshop on Swarm Robotics, pp. 10–20 (2004)
14. Farinelli, A., Iocchi, L., Nardi, D.: Multi-robot systems: a classification focused on coordination. IEEE Trans. Syst. Man Cybern. Part B **34**, 2015–2028 (2004)
15. Yan, K., Zhang, M., Ye, D.: An auction-based approach for group task allocation in an open network environment. Comput. J. **59**(3), 403–422 (2016)
16. Gerkey, B.P., Mataric, M.J.: Sold!: auction methods for multirobot coordination. IEEE Trans. Robot. Autom. **18**(5), 758–768 (2002)
17. Lorenzo, C., Franceschetti, M.: Distributed team formation in multi-agent systems: stability and approximation. In: IEEE Conference on Decision and Control (CDC), pp. 2755–2760 (2012)
18. Nath, A., Arun, A.R., Niyogi, R.: DMTF: a distributed algorithm for multi-team formation. In: International Conference on Agents and Artificial Intelligence (ICAART), vol. 1, pp. 152–160 (2020)
19. Nath, A., Arun, A.R., Niyogi, R.: A distributed approach for autonomous cooperative transportation in a dynamic multi-robot environment. In: The 35th ACM/SIGAPP Symposium on Applied Computing (ACM-SAC), pp. 792–799 (2020)
20. Gerkey, B.P., Matarić, M.J.: A formal analysis and taxonomy of task allocation in multi-robot systems. Int. J. Robot. Res. **23**(9), 939–954 (2004)
21. Kumar, C.A., Singh, M.P.: An evaluation of communication protocol languages for engineering multiagent systems. J. Artif. Intell. Res. **69**, 1351–1393 (2020)

A WSN and Vision Based Energy Efficient and Smart Surveillance System Using Computer Vision and AI at Edge

Shreeram Hudda[✉], K. Haribabu, Rishabh Barnwal, and Abhishek Khurana

Birla Institute of Technology and Science, Pilani, Rajasthan, India
hudda.shhudda@gmail.com, khari@pilani.bits-pilani.ac.in,
rishabh.barnwal123@gmail.com, abhikhurana2003@gmail.com

Abstract. The current traditional surveillance systems frequently fall short in delivering satisfactory quality of service, leading to frustrated user experiences. Consequently, there is a growing demand for more efficient and intelligent surveillance solutions. This paper addresses this need by introducing a wireless sensor networking (WSN) and vision based approach that employs optical verification through computer vision and AI at the edge, specifically designed for resource constrained IoT nodes. To support the feasibility and effectiveness of the proposed system, the authors conducted experimental analyses using both simulation and a case study. The results of the study demonstrate that the suggested surveillance system is energy conservative and provides real time information, offering a promising solution to the limitations of traditional surveillance setups.

1 Introduction

The WSNs are widely used in different IoT applications such as medical, industry, battlefield, parking, and many more mostly for surveillance purposes [23]. A WSN consists of a large number of sensors or sensor nodes. The most widely used sensors for aforesaid applications are light, magnetic, infrared, optical, and ultrasonic sensors [6,10,18,25]. These sensor nodes have resource constraints such as battery powered energy source, less memory, less computing capabilities, shorter coverage areas thus there should be some energy efficient and less resource intensive algorithms due to which transfer of data would take place. Most of the time these sensor's collected data are suffering from noise and outliers which causes false positives, resulting in misinterpretations of such data. Several uncontrolled environmental factors such as high temperature, humidity, cloudy and rainy weather, radiation, data collection time (i.e. day or night) etc. can cause noise and outliers in the data. Thus, there should be mechanisms to deal with such faulty (i.e. noisy and outliers) data. Several proposed approaches in the literature survey are not power conservative since some of them require its sensors to be continuously operational. Some researchers proposed their own devices for energy efficiency, however these devices are passive since they need external commands from users or administrators for data collection and transmission, and sleep and wake ups [16,25,28]. Several other works have some other constraints such as they can not work at night time due to the shortage of energy sources.

L. Barolli (Ed.): AINA 2024, LNDECT 201, pp. 24–36, 2024.
https://doi.org/10.1007/978-3-031-57870-0_3

The Data sensing, collection, and transmissions are the main tasks of the sensor nodes [29]. So from a sensing point of view, the given surveillance solutions can be divided into three categories: static sensor solutions, moving sensors solution, and vision based solutions [1, 6, 16, 29]. Firstly, these static sensor solutions are nothing but WSN based solutions only. In WSN based solutions, a particular slot in each surveillance space needs a sensor, however there can be multiple slots in a surveillance space, resulting in multiple sensors being placed in a surveillance space to cover multiple slots. Such classical sensors are tiny in size, but still they require additional work for installation and maintenance. Such additional work is required for every sensor to be placed at every "single" slot in surveillance space. For these sensors, the data detection approaches have high false detection rate in certain circumstances such as the magnetic sensors are sensitive toward large nearby metals, and ultrasonic, optical, and infrared sensors are influenced by some of the uncontrolled environmental factors such as high temperature, humidity, cloudy and rainy weather, radiations, lightening conditions etc. One advantage of WSN based solutions is that even if some sensors stop working, the system can still work due to the huge number of sensors.

Secondly, the moving sensor solutions use mobile devices such as drones, and a hand held mobile phone to monitor surveillance space. If mobile phone is used as a mobile device, for that some apps will be installed into the mobile phones, later on these apps use the available sensors of the mobile phones. Such solutions are limited for several applications such as on street surveillance in urban areas, even these solutions can not work in rural areas where less number of moving sensors are available. Thirdly, vision based solutions are used increasingly since there are some recent advancements in the computer vision field. In WSN based solutions, there is a need for a particular dedicated sensor for each slot in the surveillance system, and also in moving sensor solutions a moving unit has only one sensor, whereas in vision based solutions one camera covers multiple slots in a single surveillance system, resulting in less installation and maintenance cost [1, 3, 6, 17, 18, 23]. The installation of cameras is an easy task compared to WSN based solutions where each slot requires the installation of one sensor. Also, the data collected through cameras is much richer than classical sensors i.e. WSN based solutions. A low frame rate and low resolution video data can generate more than 10Mb data per second and around 1Tb data per day. This requires huge computing resources and storage resources to handle such vast video data for a surveillance system. Thus, the vision based solutions generate a huge amount of data which requires high data transmission rate, storage, and processing, which results in more energy consumption [6, 9, 18].

For such energy consumption problem, one immediate and important solution can be use of edge servers instead of cloud servers for data processing [18]. The cloud server i.e. cloud computing is widely used for big data processing tasks. However, the cloud computing service not only requires high bandwidth but also gives high latency for such big data processing [8, 15, 18, 19]. Thus, it becomes challengeable for the services where limited bandwidth is available to do real time processing on the big data with low latency. The alternative solution can be to shift the processing and computing tasks at a centralized cloud server to near the network or at the edge of the network. Such demand can be fulfilled by the edge computing which can process the data near to the clients where it is generated rather than transfer it to the centralized cloud server. Finally, the authors can say that due to these sensor's collected data are suffering from noise and outliers, false positives, high false detection rate, high installation and maintenance cost, and less scalability, the use of vision based solutions are most favourable [16, 18]. However, vision based solutions generate vast amounts of video data that require real time processing for accurate decision making, and also need high storage and data transmission rate, resulting in more energy consumption. Thus, there is a need to combine both WSN based solutions and vision based solutions. Therefore, there is a need of an energy efficient mechanism for resource constrained IoT environments by utilizing the WSN, computer vision, and AI together at edge computing, and also permit the use of cloud computing service for making much better decisions in future.

In this paper, the authors are using parking space surveillance system as a case study. This paper proposes a smart surveillance system that integrates WSN with computer vision and AI at edge, offering a comprehensive and synergistic solution, to address the challenges available in traditional surveillance management systems. The WSN nodes, strategically placed within parking spaces, serve as the initial point of interaction, employing ultrasonic sensors for accurate vehicle detection. The system introduces optical verification through computer vision, ensuring precise determination of parking space occupancy. The integration of AI at edge empowers the WSNs with local processing capabilities, enabling real-time decision-making and reducing reliance on centralized servers.

This paper is organized as follows in subsequent sections: Sect. 2 mentions the background and related work related to WSN, vision based solutions and also how AI effect these solutions. Afterwards, two clustering algorithms, optical verification using computer vision and AI at edge, integration of WSN with optical verification, and AI at edge are described in Sect. 3. Section 4 shows the simulation results case study. The performance analysis of both R-CNN and Faster R-CNN models mentioned in Sect. 5. Finally, the result and future study are presented in Sect. 6.

2 Literature Review

The research [20] represents a pioneering effort in the utilization of WSN for real-time parking space occupancy detection, signifying a noteworthy advancement in the field of intelligent surveillance systems. Their methodology facilitated continuous monitoring, although confronted by challenges in accuracy attributable to the restricted granularity of sensors, particularly notable in complex parking layouts. Building upon this groundwork, the work [15] extended the research by specifically addressing energy efficient WSN designs with the aim of enhancing accuracy while concurrently minimizing power consumption. In spite of commendable efforts, the achievement of optimal accuracy and energy efficiency in WSNs remains a complex task, particularly in real-world environments where sensor performance is significantly influenced by environmental factors and layout complexities.

Concurrently, camera-only systems have emerged as robust alternatives, relying exclusively on image processing methodologies. The study [5] introduced camera-only systems, leveraging image processing for precise information retrieval. However, these systems encountered challenges in diverse and complex conditions, thereby impacting their energy efficiency and accuracy. Despite evolving into non-intrusive and scalable solutions with advanced features [26], concerns persist, notably related to privacy considerations. Addressing these concerns, the research [8] enhanced camera-based systems by integrating deep learning techniques, leading to a substantial improvement in accuracy but concurrently intensifying computational demands. The challenge of balancing high computational requirements with energy efficiency remains a significant important, particularly concerning real-time processing and varying environmental conditions.

In a paradigm shift, the study [18] introduced edge AI implemented on IoT devices, facilitating on-device real-time data analysis with the potential to minimize energy consumption by reducing reliance on centralized systems. This novel approach holds promise in making less severe challenges associated with data transmission and processing, particularly in large-scale surveillance deployments. Nevertheless, practical hurdles such as power constraints and network connectivity in diverse environments present substantial challenges to the widespread implementation and optimization of edge AI in smart surveillance systems. Exploring alternative technologies, the researchers in work [24] investigated Ultrasonic Sensor-based Parking Assistance Systems, exhibiting high accuracy. Nevertheless, concerns regarding energy consumption, especially during prolonged use, emerged as a potential limitation. The study [25] proposed wireless magnetic sensor nodes incorporating optical wake-up, emphasizing energy efficiency. Despite exhibiting promising efficiency, the optimization of power management for sustained operation in real-world conditions remains a domain requiring further improvement and research. Furthermore, the authors in study [22] established a benchmark for image-based space occupancy classification, making a significant contribution towards accuracy assessment. However, persistent concerns revolve around energy consumption and real-time applicability in practical systems. The complex challenge of achieving a delicate balance between accuracy, computational demands, and energy efficiency in image-based systems remains an important task of ongoing research.

The LEACH clustering mechanism [7, 13, 27] optimizes communication, minimizes power consumption, and manages energy resources within the WSN for efficient management. In the proposed smart surveillance system, LEACH-C [2, 12] enhances control and coordination, improving data aggregation and transmission efficiency. The integration of the HEED protocol [4, 30, 31] strategically organizes sensor nodes into clusters, emphasizing energy efficiency and minimizing redundant transmissions for effective energy resource management. In summary, each aforementioned approach contributes unique advancements to the domain of smart surveillance systems. Nevertheless, challenges persist in the optimization of energy efficiency without compromising accuracy, particularly in real-world deployment scenarios characterized by diverse environmental conditions. This necessitates ongoing research efforts to address and overcome these challenges in order to enhance the efficacy of intelligent surveillance systems.

3 Proposed Solution

This section introduces the proposed surveillance system, combining WSN nodes with computer vision and AI at edge. Initially, two clustering approaches [14] are presented, followed by a discussion on optical verification at the edge using computer vision and AI. The section concludes by detailing the integration of WSN and optical verification for the surveillance system.

3.1 Algorithm Without Degree

This approach [14] considers energy and distance in choosing a cluster head, eliminating probabilistic methods. The algorithm evaluates a candidate node's suitability based on the average distance to all nodes, calculating the sum of distances and determining the average. The node with the maximum energy and the minimum average distance is selected as the cluster head for the network.

3.2 Algorithm with Degree

This approach [14] enhances accuracy by avoiding the selection of idle cluster heads that waste energy. Eliminating the compulsory re-election in each round improves round speed and reduces fluctuations in energy consumption. After the initial election round, normal nodes are assigned to each cluster head, and the degree of each cluster head is calculated at the time of assignment. In subsequent rounds, cluster heads are verified based on two criteria: (i) Does the cluster head node have a degree >3% of operating nodes? (ii) Does the cluster head have remaining energy >20% of the initial energy of nodes? Failure to meet these criteria triggers re-election, ensuring energy preservation for the network's longevity without unnecessary re-elections.

3.3 Optical Verification Using Computer Vision and AI at Edge

In this work, the authors select a parking space surveillance system as a case study. They utilize the ACPDS dataset and employ two models, R-CNN and Faster R-CNN, as detailed in the study [22]. These models are trained with the ACPDS dataset and executed at the edge for local processing.

R-CNN
This model [22] employs ResNet50 [11] to analyze image patches corresponding to individual parking spaces directly from a high-resolution image (4000 × 3000). Unlike traditional R-CNN architectures, which use numerous region proposals, this approach processes a maximum of about 100 manageable image patches at reduced resolution (128 × 128). However, a drawback is the lack of communication between patches, potentially limiting the model's ability to accurately interpret occlusions or concealed elements within parking spaces. For more details about this model, the readers should refer to the study [22].

Faster R-CNN
In this model [22], a resized image is processed using ResNet50 and a feature pyramid. Extracted features for each parking space pass through a classification head to obtain final occupancy scores. The model adopts a heuristic approach [21] to determine pyramid layers, maintaining a consistent pooling resolution of (7 × 7). Unlike R-CNN,

it doesn't exploit the full dataset resolution but allows information exchange between parking spaces. This feature enhances the model's ability to deduce information about occlusions and interpret obstructed elements within the parking area, resulting in more accurate results. For more details about this model as well, the readers should refer to the study [22].

3.4 Integration of WSN with Optical Verification

In this section, the combined operation of WSN and optical verification using computer vision at the edge in the proposed system is explained. The integration of WSN for optical verification involves coordinating the WSN node's activity with the optical verification process, triggered by a zero (i.e. 0) to one (i.e. 1) transition indicating a change in parking space occupancy status. Upon detecting a transition, signifying a shift from unoccupied (0) to occupied (1) or vice versa, the entire WSN network enters a low-power or sleep mode. During this inactive state, the optical verification system is activated to accurately confirm the detected transition. The synchronization process between the WSN node's transition detection and optical verification involves a sequence of steps, as mentioned below:

1. The WSN node, equipped with sensors designed for the purpose of detecting the occupancy status of parking spaces, operates in a continuous monitoring mode to observe fluctuations in occupancy states. The sensors integrated into the node are configured to recognize alterations in the occupancy status, distinguishing between unoccupied (designated as 0) and occupied (designated as 1) states. Upon the identification of a transition from one state to another, whether from unoccupied to occupied or vice versa, the WSN node initiates the generation of a trigger signal. This trigger signal serves as a distinctive indication of the detected change in parking space occupancy, facilitating the dissemination of relevant information regarding the transitioning states within the monitored environment.

2. Following the detection of a state transition, the WSN node initiates a signaling mechanism designed to inform the entire WSN network of the occurrence of the transition event. Subsequent to this signaling process, the entire network undergoes activation of a sleep mode as part of an energy conservation strategy. This sleep mode is characterized by a substantial reduction in power consumption, achieved primarily through the end of continuous data transmission and the deactivation of sensor operations. The implementation of the sleep mode serves the purpose of enhancing the overall energy efficiency of the WSN, mitigating power consumption during periods of inactivity, and thereby contributing to prolonged network sustainability.

3. Concurrently, the trigger signal initiates the operational sequence of the optical verification system. This event prompts the activation of the gateway camera within the network, transitioning it from an inactive to an active state. The optical verification system, now engaged, undertakes visual monitoring tasks as dictated by its configured parameters or specific operational requirements. This activation of the optical verification system, facilitated by the trigger signal, highlights a synchronized response within the network to the detected event, aligning with the comprehensive objective of real-time surveillance facilitated by optical means.

4. The operationalized optical verification system captures an image of the network encompassing the transition node. Subsequently, this captured image undergoes processing through the R-CNN or the Faster R-CNN model. The primary objective of this computational analysis is to substantiate and authenticate the identified alteration in occupancy status, as signified by the WSN node. The R-CNN and Faster R-CNN models, well established within the domain of computer vision, are employed to meticulously examine and classify objects within the image, specifically focusing on the region associated with the WSN node and its immediate surroundings. This analytical process serves to validate the accuracy of the detected change in occupancy status through sophisticated pattern recognition and object localization techniques inherent to the utilized deep learning models.

5. Following the optical verification procedure and the substantiation of the detected transition, a wake-up signal is transmitted to the WSN network. This signal serves as a trigger for the network to cease its sleep mode and transition back to its regular operational state. Resuming its typical functionality, the WSN network recommences its continuous monitoring of parking space occupancy. The wake-up signal initiates the reactivation of network components that were previously in a in-active state during sleep mode, facilitating the seamless restoration of normal operations. This strategic synchronization of sleep and wake-up states optimizes energy conservation while ensuring the network's responsiveness to dynamic changes in the monitored environment, ultimately contributing to the overall efficiency and efficacy of the WSN infrastructure.

3.5   AI at Edge

AI at edge facilitates on-device inference, empowering sensors or cameras in the parking infrastructure (i.e. the parking surveillance management is the case study in this present study) to analyze data and make immediate decisions without relying on central servers. This capability ensures swift responses to changes in parking space occupancy detected by WSN nodes. AI at edge, seamlessly integrated into the optical verification system, efficiently process captured images. Computer vision techniques rapidly analyze visual data to validate changes in occupancy status identified by WSN nodes. By employing AI at edge for initial data processing and inference at edge devices, the system reduces the need to transmit large raw data volumes to central servers. This reduction in data transmission lowers bandwidth requirements and minimizes delays, thereby enhancing overall system responsiveness and energy efficiency. This solution is adaptable to changing conditions and can accommodate scalability needs. Both the clustering algorithms and computer vision techniques are optimized and updated locally, providing flexibility to adapt to new parking configurations or system enhancements without significant disruptions.

4   Experimental Results and Analysis

The authors implemented LEACH, LEACH-C, HEED, and two other clustering approaches using MATLAB environment for experimental analysis.

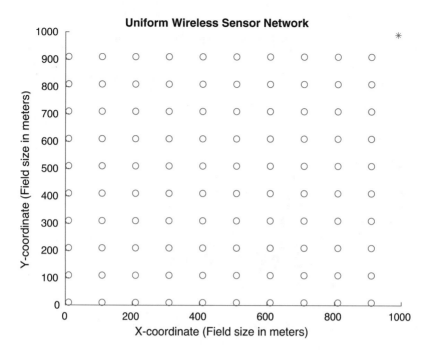

Fig. 1. WSN Node Positioning in a parking space. Each node is placed in a quadrilateral parking spot. The red star is the camera. A parking space contains multiple parking spots/lots/slots.

4.1 WSN Analysis

The study implemented and simulated clustering algorithms (LEACH, LEACH-C, HEED, Algorithm with Degree, and Algorithm without Degree) using MATLAB. Figure 1 illustrates a parking space with 100 WSN nodes (represented by circles) covering 100 parking spots or lots or slots, with a red star indicating the camera node at the gateway. These nodes collect parking spot status and communicate data to the gateway through various clustering algorithms. Figure 3 illustrates network statuses at intervals when the first node, half the network, and the entire network die, correlating with the number of rounds in the network. The most effective clustering algorithm observed is the Algorithm with Degree, as nodes require more rounds to become dead. This ultimately prolongs the network's lifespan, allowing it to achieve the maximum number of rounds when starting with the same initial energy for the combined network.

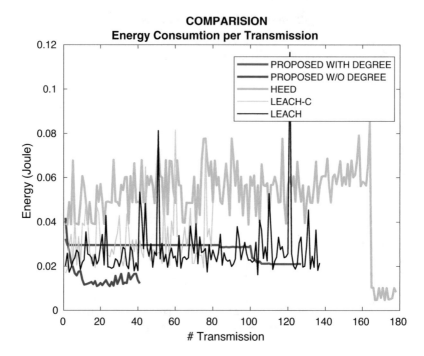

Fig. 2. The image shows the Energy Consumption per transmission of data through the various nodes.

4.2 Energy Analysis

Figure 2 compares the clustering algorithms in terms of network Energy Consumption per Transmission. LEACH and LEACH-C exhibit similar energy consumption until the first node dies. HEED consumes the highest energy due to its multiple rounds for stabilization. The Algorithm without Degree has the least energy consumption until the first node dies by selecting distance-biased nodes. The stability of the Algorithm with Degree is evident, avoiding the re-election of cluster head nodes in every round.

5 Performance Analysis

This research work also compares effectiveness of R-CNN and Faster R-CNN models for parking space detection [22]. R-CNN was tested at pooling resolutions of 64, 128, and 256, while Faster R-CNN was tested at resolutions of 800, 1100, and 1440 [22]. The results notably indicated a consistent superiority of the Faster R-CNN model over

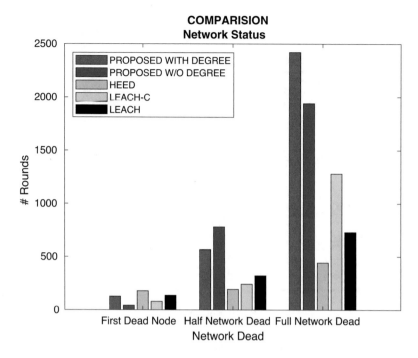

Fig. 3. The image shows the Network Status at crucial intervals (First node dead, Half network dead, and Full Network dead).

R-CNN, as depicted in Fig. 4. Square pooling emerged as the preferred method, particularly advantageous for R-CNN. The optimized resolution for R-CNN (128×128) was aligned with anticipated performance based on parking space dimensions, aiming to minimize patch-up sampling while retaining crucial information. Despite memory limitations, the resolution for Faster R-CNN was constrained to 1920×1440, still significantly lower than the original image size (4000×3000). Additionally, observations highlighted that Faster R-CNN exhibited superior generalization to the test dataset compared to R-CNN, likely due to systematic differences between the test and validation sets, including varying occlusion instances. These findings highlight the significance of model selection based on validation accuracy, with the emphasis on maintaining test accuracy as an unbiased estimator of generalization.

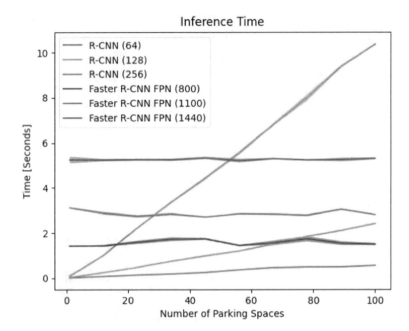

Fig. 4. Inference time for various models performed on i5-10210U CPU.

6 Conclusions

This work presents advancements in the domain of surveillance management, specifically focusing on the utilization of WSN with optical verification using computer vision and AI at the edge for IoT devices operating under resource constraints environment. The integration of WSN with optical verification introduces an approach for achieving precision in monitoring applications. Throughout this current study, the integration of clustering algorithms with optical verification is systematically demonstrated to explain the system's efficacy in terms of energy efficiency. This integration reduces energy consumption through the selectively activation of optical verification during transitions, thereby ensuring reliable validation of changes. It is acknowledged that the use of vision based solutions for surveillance management inherently raises privacy concerns, constituting a potential limitation of the present work. Therefore, in subsequent study, the authors will address these privacy related concerns while implementing surveillance management through the combined use of WSN and vision based solutions.

References

1. Abas, K., Porto, C., Obraczka, K.: Wireless smart camera networks for the surveillance of public spaces. Computer **47**(5), 37–44 (2014)
2. Al-Shaikh, A., Khattab, H., Al-Sharaeh, S.: Performance comparison of leach and leach-c protocols in wireless sensor networks. J. ICT Res. Appl. **12**(3), 219–236 (2018)

3. Amato, G., et al.: A wireless smart camera network for parking monitoring. In: 2018 IEEE Globecom Workshops (GC Wkshps), pp. 1–6. IEEE (2018)
4. Anitha, G., Vijayakumari, V., Thangavelu, S.: A comprehensive study and analysis of leach and heed routing protocols for wireless sensor networks-with suggestion for improvements. Indonesian J. Electr. Eng. Comput. Sci. **9**(3), 778–783 (2018)
5. Banerjee, S., Choudckar, P., Muju, M.: Real time car parking system using image processing. In: 2011 3rd International Conference on Electronics Computer Technology, vol. 2, pp. 99–103. IEEE (2011)
6. Baroffio, L., Bondi, L., Cesana, M., Redondi, A.E., Tagliasacchi, M.: A visual sensor network for parking lot occupancy detection in smart cities. In: 2015 IEEE 2nd World Forum on Internet of Things (WF-IoT), pp. 745–750. IEEE (2015)
7. Behera, T.M., Samal, U.C., Mohapatra, S.K.: Energy-efficient modified leach protocol for IoT application. IET Wirel. Sens. Syst. **8**(5), 223–228 (2018)
8. Bura, H., Lin, N., Kumar, N., Malekar, S., Nagaraj, S., Liu, K.: An edge based smart parking solution using camera networks and deep learning. In: 2018 IEEE International Conference on Cognitive Computing (ICCC), pp. 17–24. IEEE (2018)
9. Cheng, X., et al.: Camera sensor platform for high speed video data transmission using a wideband electro-optic polymer modulator. Opt. Express **27**(3), 1877–1883 (2019)
10. Chinrungrueng, J., Sunantachaikul, U., Triamlumlerd, S.: Smart parking: an application of optical wireless sensor network. In: 2007 International Symposium on Applications and the Internet Workshops, p. 66. IEEE (2007)
11. He, K., Zhang, X., Ren, S., Sun, J.: Deep residual learning for image recognition. In: Proceedings of the IEEE Conference on Computer Vision and Pattern Recognition, pp. 770–778 (2016)
12. Heinzelman, W.B., Chandrakasan, A.P., Balakrishnan, H.: An application-specific protocol architecture for wireless microsensor networks. IEEE Trans. Wirel. Commun. **1**(4), 660–670 (2002)
13. Heinzelman, W.R., Chandrakasan, A., Balakrishnan, H.: Energy-efficient communication protocol for wireless microsensor networks. In: Proceedings of the 33rd Annual Hawaii International Conference on System Sciences, pp. 10–pp. IEEE (2000)
14. Hudda, S., Haribabu, K., Barnwal, R.: A novel approach for energy-efficient communication in a constrained IoT environment. In: 2024 International Conference on Information Networking (ICOIN), pp. 702–707. IEEE (2024)
15. Joseph, J., Patil, R.G., Narahari, S.K.K., Didagi, Y., Bapat, J., Das, D.: Wireless sensor network based smart parking system. Sens. Transducers **162**(1), 5 (2014)
16. Joshi, A., Kanungo, D.P., Panigrahi, R.K.: WSN-based smart landslide monitoring device. IEEE Trans. Instrum. Measur. (2023)
17. Kamminga, J.W., Jones, M., Seppi, K., Meratnia, N., Havinga, P.J.: Synchronization between sensors and cameras in movement data labeling frameworks. In: Proceedings of the 2nd Workshop on Data Acquisition to Analysis, pp. 37–39 (2019)
18. Ke, R., Zhuang, Y., Pu, Z., Wang, Y.: A smart, efficient, and reliable parking surveillance system with edge artificial intelligence on IoT devices. IEEE Trans. Intell. Transp. Syst. **22**(8), 4962–4974 (2020)
19. Lee, C.P., Leng, F.T.J., Habeeb, R.A.A., Amanullah, M.A., ur Rehman, M.H.: Edge computing-enabled secure and energy-efficient smart parking: a review. Microprocess. Microsyst. 104612 (2022)
20. Lee, S., Yoon, D., Ghosh, A.: Intelligent parking lot application using wireless sensor networks. In: 2008 International Symposium on Collaborative Technologies and Systems, pp. 48–57. IEEE (2008)

21. Lin, T.Y., Dollár, P., Girshick, R., He, K., Hariharan, B., Belongie, S.: Feature pyramid networks for object detection. In: Proceedings of the IEEE Conference on Computer Vision and Pattern Recognition, pp. 2117–2125 (2017)
22. Marek, M.: Image-based parking space occupancy classification: dataset and baseline. arXiv preprint arXiv:2107.12207 (2021)
23. Nieto, R.M., Garcia-Martin, A., Hauptmann, A.G., Martinez, J.M.: Automatic vacant parking places management system using multicamera vehicle detection. IEEE Trans. Intell. Transp. Syst. **20**(3), 1069–1080 (2018)
24. Park, W.J., Kim, B.S., Seo, D.E., Kim, D.S., Lee, K.H.: Parking space detection using ultrasonic sensor in parking assistance system. In: 2008 IEEE Intelligent Vehicles Symposium, pp. 1039–1044. IEEE (2008)
25. Sifuentes, E., Casas, O., Pallas-Areny, R.: Wireless magnetic sensor node for vehicle detection with optical wake-up. IEEE Sens. J. **11**(8), 1669–1676 (2011)
26. Sirithinaphong, T., Chamnongthai, K.: The recognition of car license plate for automatic parking system. In: Proceedings of the Fifth International Symposium on Signal Processing and its Applications (IEEE Cat. No. 99EX359), ISSPA 1999, vol. 1, pp. 455–457. IEEE (1999)
27. Sun, H., Pan, D.: Research on optimisation of energy efficient routing protocol based on leach. Int. J. Ad Hoc Ubiquitous Comput. **41**(2), 92–107 (2022)
28. Vellela, S.S., Balamanigandan, R.: An intelligent sleep-awake energy management system for wireless sensor network. Peer-to-Peer Netw. Appl. **16**(6), 2714–2731 (2023)
29. Vítek, S., Melničuk, P.: A distributed wireless camera system for the management of parking spaces. Sensors **18**(1), 69 (2017)
30. Younis, O., Fahmy, S.: Distributed clustering in ad-hoc sensor networks: a hybrid, energy-efficient approach. In: IEEE INFOCOM 2004, vol. 1. IEEE (2004)
31. Younis, O., Fahmy, S.: Heed: a hybrid, energy-efficient, distributed clustering approach for ad hoc sensor networks. IEEE Trans. Mob. Comput. **3**(4), 366–379 (2004)

Mitigating Resource Depletion and Message Sequencing Attacks in SCADA Systems

Neminath Hubballi[(✉)] and Nisha Kumari Barsha

Department of Computer Science and Engineering, Indian Institute of Technology Indore,
Indore, India
{neminath,ms2204101007}@iiti.ac.in

Abstract. Electric grid networks bring measurement data to a control center for making safety critical decisions. These messages are sent using application layer protocols like Modbus and IEC-104, which operate over TCP/IP. Such communications are susceptible to different cyber attacks. In the direction of safeguarding grids, our contributions in this paper are threefold. First, we study two types of attacks known as malformed and message sequencing attacks. In the malformed message attack, an adversary injects a large number of malformed messages and have it sent to the control center to overwhelm it and depleting its computational resources. In the sequencing attacks a series of messages are sent to generate an attack. These messages if accepted by the control center can impact safety of the grid as it can lead to improper assessment of grid network state. Second, we describe a method to detect variants of malformed messages using first order logic statements. Third, we propose a method to filter the messages belonging to malformed and sequencing attacks using the Extended Berkeley Packet Filter. This is realized by implementing logical statements and screening the messages over a window period to show that such a filtering is effective and robust against attack variants and intensity.

1 Introduction

Electric grid networks are critical infrastructures as millions of users rely on them for their energy needs. Normally they use Supervisory Control and Data Acquisition (SCADA) systems for monitoring and controlling these critical infrastructures. With advent of smart-grids and small micro-grids, these networks heavily rely on ICT for communication. As such communications use TCP/IP networks, they inherit the vulnerabilities of these networks [1]. There are application layer protocols like IEC-104 and Modbus which are commonly used for carrying the measurement data from the sensors deployed at various places to the control center where control decisions are made. As these messages are sent over conventional networks, these can be manipulated or blocked to create different cyber attacks.

Traditional method for protecting the SCADA systems is to deploy firewall which will filter packets coming from known attackers, unknown sources, etc. However, this method of filtering falls short of the requirements. One of the issues with these firewalls is that they are implemented as software solutions, which themselves can be subjected to attacks. Moreover, such filtering will not be reliable unless they inspect the

L. Barolli (Ed.): AINA 2024, LNDECT 201, pp. 37–47, 2024.
https://doi.org/10.1007/978-3-031-57870-0_4

content of these packets. There are attack detection methods [10, 12, 17] proposed for detecting different attacks. These detection techniques usually run as applications. The monitoring code implemented in software runs in user space which has overheads of bringing the packets through kernel space involving several context switches and data copy operations. This will place significant overhead on the resources. Taking motivation from this, we propose to offload filtering operation to the network interface card. In this direction, we make following contributions in this paper.

(i) We study different types of malformed attacks that can be generated by manipulating the application layer messages of IEC-104 and Modbus.
(ii) We propose a method to identify all malformed messages using formal constructs generated from protocol specification.
(iii) We study the impact of malformed packet attacks on the system resources advocating the need for mitigation technique.
(iii) We propose a method to offload screening and filtering of malformed packets to the programmable network interface cards saving significant system resources.

2   Background

Here we provide a brief overview of the two application layer protocols which are commonly used in SCADA communications.

(i) IEC-104 Protocol Overview: The IEC 60870-5-104 protocol stack is an International Electrotechnical Commission (IEC) standard designed for telecontrol of equipment/systems using TCP/IP. This standard outlines a message format called the Application Protocol Data Unit (APDU) as shown in Fig. 1 facilitating communication between a master station (SCADA master) and remote devices (slaves). The APDU consists of mandatory Application Protocol Control Information (APCI) or an APCI combined with an Application Service Data Unit (ASDU). APCI has attributes for executing functions like start, stop, test, reset, and data transfer, ensuring session establishment, termination, reliable data transmission, and link integrity checks. ASDU within the APDU, comprises a data identifier field and the data itself.

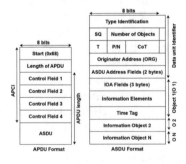

Fig. 1. APDU Message Format

The data identifier field encompasses type Identification (ASDUType), the number of information objects (NUMIX) in the APDU, Cause of Transmission (CoT), Originator Address (ORG), and ASDU address field (ADDR). ASDUType signifies the transmitted information type, while CoT elucidates the purpose or event triggering the transmission. Data, consisting of one or more information objects, uses IOA to identify specific data within a defined station. Analyzing values in APDU fields, the number of APDU messages, and their sequence aids in detecting anomalies.

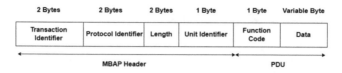

2 Bytes	2 Bytes	2 Bytes	1 Byte	1 Byte	Variable Byte
Transaction Identifier	Protocol Identifier	Length	Unit Identifier	Function Code	Data

MBAP Header ◄──────────────────────────────────────► ◄► PDU

Fig. 2. Modbus Messaging over TCP/IP

(ii) Modbus Protocol Overview: Modbus over TCP/IP is a widely used industrial communication protocol, leveraging TCP/IP networks for seamless device interaction. The message structure of a Modbus communication is shown in Fig. 2. Key features include encapsulation for reliable message delivery, IP-based addressing, various function codes for control and monitoring, and support for diverse data types. Different fields of the message frame include (a) Transaction Identifier (TID) which identifies message transactions, (b) Protocol Identifier (PID) which designates Modbus over TCP/IP, (c) Length field indicating remaining bytes, (d) Unit Identifier (UID) identifying the remote device, (e) Function Code which specifies the action type and (f) Data pertaining to function-specific information.

3 Malformed and Message Sequencing Attacks

In this section, we discuss how an adversary can generate malformed and message sequencing attacks taking a reference architecture. A typical grid has power generation sources at the one end and consumers on the other end as shown in Fig. 3. A control center at the center of these two coordinates to estimate the demand and supply and also to detect faults in the network. This is important for the safety of the grid and enables identification, localization, and quick recovery from faults. The control center receives measurement data from different locations of the grid network to assess the stability.

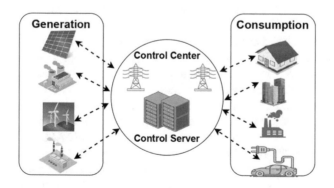

Fig. 3. Grid Network Communication Architecture

3.1 Malformed Messages and Detection

In the setup shown in Fig. 3, we consider an attack scenario where an adversary over-loads the control center with a random set of application messages. We assume that the attacker is an insider who can generate the required set of packets or has compromised systems inside the network which she can leverage to mount the attack. These messages can have two types of implications as (i) misinterpretation of status and executing unwanted/undesirable steps if the messages are honored. (ii) server resource depletion as these messages have to be processed even if they are not honored. Such messages can be generated with ease using packet generation tools and with little programming. Following is a non-exhaustive list of such messages and also its implications.

Modbus Malformed Messages

(i) Protocol Identifier Field Error: Should always be set to 0x00. If any other value is used, the message is termed as malformed, leading to communication errors and potential misinterpretation of the message by the devices in the network.

(ii) Incompatible Value Message: This category of attacks can be generated by placing contradicting values in the message. For e.g., using a non-standard Protocol Identifier results in an incompatible value message. Another example is setting the function code for a query as 'x' where '$x\&0x80 == 1$' indicates an exception range, causing a conflict in the values of the function code and leading to an incompatible message. These inconsistencies can hinder proper communication and data exchange in SCADA systems.

IEC-104 Malformed Messages

(i) Wrong Start Character (other than 0x68): Using other values can lead to invalid or missing start characters, causing packet synchronization issues.

(ii) Incompatible ASDUType and CoT Values: This can lead to misinterpretation of commands, impacting the proper functioning of SCADA systems. For e.g., when ASDUType is set to 2, only specific CoT values such as 3, 5, 11, and 12 are valid. Deviating from these specified combinations, like using an incompatible CoT value such as 8 can result in misinterpretation of commands, potentially leading to disruptions in the proper functioning of the system. In total, there are 127 ASDUType and each ASDUType has a set of valid CoT values associated with it leading to a large number of combinations.

(iii) Incompatible Length and Number of Objects Fields: These messages create packet parsing and interpretation errors as there will be a mismatch in the number mentioned and data available within APDU.

(iv) Incorrect Test Bit Setting (T Bit): This bit is set for testing purposes only. Setting and unsetting this bit can lead to misinterpretations, causing test messages to be mistaken for operational commands or vice versa.

(v) Mismatch in P/N Bit Setting: This indicates success or failure of an execution. A mismatch in the Positive/Negative (P/N) bit (P/N = 0 or P/N = 1) can result in incorrect assumptions about the success or failure of command execution, impacting the reliability and accuracy of the grid system.

The malformed message/packet examples mentioned above are broadly of two types namely (i) having illegal settings and (ii) dependent attributes value violations. Here, we provide a generic method to identify such messages. We use the first order logic framework to represent constraints as per the specification in order to detect all variants of malformed packets.

We consider the Modbus and IEC-104 packet as a message $M = \{a_1, a_2, \cdots a_n\}$ with $a_i \in M$ is an attribute of M. Each a_i takes values from a well defined domain D_i. The two types of malformed messages are formally identified with logic statements constructed as follows.

(i) Identifying illegal settings: Let $L(a_i)$ indicate whether the attribute a_i takes values from the respective domain D_i in which case it is evaluated to *true* otherwise it will be evaluated to *false*. A message M is said to be malformed if any one of the attribute a_i is taking value not in its domain D_i. This can be represented as a logical statement $\exists a_i \neg L(a_i)$.

(ii) Dependent attribute violations: Let a_i and a_j be two attributes of M with a_i having domain $D_i = \{x_1, x_2, \cdots, x_m\}$ and $a_j = \{y_1, y_2, \cdots, y_k\}$ with $m \geq 1$ and $k \geq 1$. Here we call a_i as the primary attribute and a_j as the dependent attribute to capture the dependency relationship. Suppose a_i takes a value x_i then the dependent attribute a_j can take subset of values $D_j^{'} = \{y_1, y_2, \cdots, y_p\}$ with $D_j^{'} \subseteq D_j$. In order to represent such dependencies as logical statements, we first extract such dependency pairs (a_i, a_j) from the specification of the protocols. Subsequently, we expand the attribute set and partition the domain set. An attribute a_i is expanded into p attributes with each a_{it} corresponding to a domain $D_{it} \subseteq D_i$ (with possible overlapping) as follows. $a_i = \{a_{i1}, a_{i2}, \cdots, a_{ip}\}$ and $D_i = \{D_{i1}, D_{i2}, \cdots, D_{ip}\}$ with a_{it} corresponding to D_{it} for $1 \leq t \leq p$. $a_j = \{a_{j1}, a_{j2}, \cdots, a_{jp}\}$ and $D_j = \{D_{j1}, D_{j2}, \cdots, D_{jp}\}$ with a_{jt} corresponding to D_{jt} for $1 \leq t \leq p$. In this case, if predicate $L(a_{it})$ is *true*, meaning the attribute a_i takes one of the values from the set D_{it}, then the attribute a_{jt} should take values from domain D_{jt}. This constraint can be expressed as a logical statement $\nexists (a_{it}, a_{jt}) \neg (L(a_{it}) \wedge L(a_{jt}))$.

3.2 Message Sequencing Attacks

In this case, an attacker generates a sequence of application layer messages such that all of them put together causes stability issues at the control center. For example, if an attacker sends a sequence of messages by setting specific values for CoT and ASDU-Type attributes in IEC-104, this can lead to the target device turn on and off for each such message [2]. As the attribute values are legal; but how they are used is an issue, this requires a different approach for detecting these sequencing attack.

Let $M_1, M_2, \cdots, M_n$ be a sequence of messages received within a time period W with each $M_i = \{a_1, a_2, \cdots a_m\}$ with $1 \leq i \leq n$ and $1 \leq t \leq m$. We log the attribute values of each M_i in an auxiliary storage along with the timestamp as follows. $Log(W_q) = (t_1, M_1), (t_2, M_2), (t_3, M_3), \cdots, (t_n, M_n)$ for the qth window period. For every window period, the logs are screened and values of attributes are verified if the pattern required to detect an attack is found.

4   Mitigating Attack

We propose an attack mitigation technique with filtering employed at the OS kernel/NIC
of the server located at control center. The proposed technique screens the packets (IEC-
104 and Modbus) to check if they are well-formed and carrying legal combination of
values inside. In order to filter malformed packets, we use the Extended Berkeley Packet
Filter (eBPF). It allows to execute custom (user written) code in the Linux kernel by
attaching programs to kernel hooks or can be completely offloaded to programmable
network interface cards as shown in Fig. 4. These hooks can be any events for e.g.,
system calls or arrival of a new packet, etc. This feature allows developers to run safe,
custom programs in the kernel space without modifying the kernel itself. eBPF programs are written in restricted C language and is verified by the verifier for safety of
execution so that kernel is not affected (e.g., crashed).

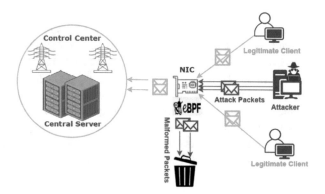

Fig. 4. Proposed Packet Filtering with eBPF

We propose to exploit the feature offered by eBPF and write validation code in
the eBPF to check the incoming Modbus and IEC-104 messages. As this code can be
executed either in the kernel or completely offloaded to network interface card there are
two scenarios. If it is running in kernel space, it will help improving performance by
avoiding several context switch events and data copy operations required for copying
the content from kernel buffers to user space buffers. On the other hand, if the code is
offloaded to hardware NIC, the packets will be directly sent to user space application
without involving the kernel further improving performance. There are utilities like
XDP [3] which allow eBPF programs to be offloaded to SmartNICs. The eBPF code
performs a two step validation, where in the first step it detects if the packet is a Modbus
or IEC-104 protocol by looking at the packet header details and destination port number
details. Subsequently, it executes a series of rules formed from the logic statements
outlined in Sect. 3 for detecting malformed message attacks. It uses maps feature offered
by the eBPF to store the timestamped messages along with attribute values to detect and
filter sequencing attacks.

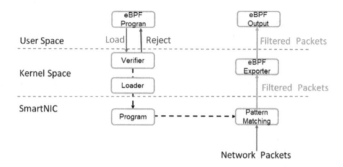

Fig. 5. eBPF based Filtering of Packets at NIC

The eBPF screening program matches or search for certain patterns and combination of attribute values in real-time. It sends only packets which are well-formed to the user space application directly as shown in Fig. 5.

5   Experiments

In this section, we describe the evaluations done with the proposed attack detection and mitigation methods. We provide the details of testbed setup, details of implication study of malformed messages, and how eBPF based filtering help mitigating the two types of attacks in the next four subsections respectively.

(i) Testbed Setup: In order to assess the implication of malformed message attacks and also to implement the filtering methods (for both the attack types), we created a testbed consisting of 8 computers. One of these machines represents the control center and also performs the duty of master device. One more machine acts as legitimate client and send/respond to the queries sent by the master. Both master and clients are simulated with python programs generating appropriate messages of Modbus and IEC-104. The server will invoke requests and clients receive them and respond accordingly with reply messages having hypothetical readings of voltage, current, angle values, etc. Remaining set of machines are used for generating different malformed and message sequencing attacks to the server. All these machines use scapy library [4] to craft appropriate and custom types of messages.

(ii) Malformed Message Implication: We first study the implications of malformed messages on the server. As mentioned in Sect. 1 and 3, at the least they have an implication on the server resources where resources are utilized to process the malformed messages. Here, we investigate the CPU utilization on the server machine with and without malformed messages. For this study, there is a single legitimate client sending measurements periodically to the control center/server at the rate of 200 messages per minute. Another 7 malicious clients send malformed messages at the rate of 2000 messages per minute cumulatively. The CPU utilization with and without attack in this case is shown in Fig. 6a. We also studied the utilization by varying the intensity of attack by sending 500 to 2500 malformed packets per minute in the step size of 500. Figure 6b shows the utilization variation against the intensity. We can notice that the CPU utilization increases significantly as the number of malformed messages increases.

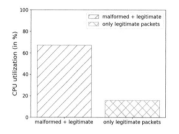 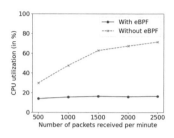

(a) CPU Utilization with and without Attack (b) CPU Utilization with Varying Message Intensity

Fig. 6. Number of Messages v/s CPU Utilization

(iii) Filtering Malformed Messages: Here we show the implementation details of eBPF based filtering and how it helps to protect the server resources. We used the logical representation presented in Sect. 3 to define attributes of interest and corresponding partitions. We translated these statements into *if-then-else* rules in eBPF C code for the purpose of filtering. eBPF code also requires a corresponding user space program to interact with it directly. Here, the server program is extended to interact with the eBPF program loaded at the NIC. The domains of corresponding attributes were stored using the maps feature available in eBPF.

In order to assess the filtering ability of eBPF code, we performed evaluations by generating malformed messages at different rates between 500 to 2500. Figure 6b shows the CPU utilization on the server machine with and without filtering. We can notice that as the intensity of the attack increases the CPU utilization increases sharply and with filtering enabled all malformed messages get dropped and hence will not be processed by the user space server program retaining a constant CPU utilization around 16%.

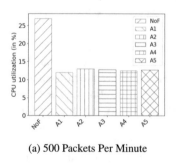 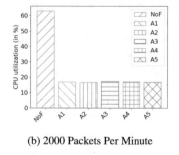

(a) 500 Packets Per Minute   (b) 2000 Packets Per Minute

Fig. 7. CPU Utilization with eBPF Filtering for Different Attacks using IEC-104

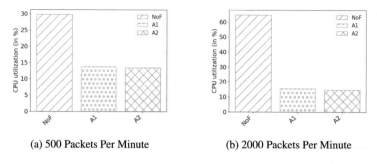

(a) 500 Packets Per Minute   (b) 2000 Packets Per Minute

Fig. 8. CPU Utilization with eBPF Filtering for Different Attacks using Modbus

We also show the CPU utilization for individual attack cases as well. Figure 7a and Fig. 7b show these graphs for attack intensities of 500 and 2000 respectively for IEC-104. In these figures, NoF denotes the utilization without filtering as a baseline without eBPF, and A1-A5 represents different attacks. A1 is an attack with incompatible length and object fields, A2 is with incorrect test bit setting, A3 is with P/N setting mismatch, A4 with wrong start character, and A5 is incompatible ASDUType and CoT values. The bars representing attacks show the CPU utilization when an attack of intensity as indicated in the caption is generated and those messages are filtered by eBPF. We can notice that a nearly consistent utilization here indicating that filtering is attack type agnostic. Similarly, Fig. 8 shows the CPU utilization for the Modbus protocol with two attacks (A1 with protocol identifier error and A2 with incompatible value message). We can make similar observations here also indicating the filtering technique is robust and can scale well.

(iv) Filtering Message Sequencing Attack Messages: Here we study an example sequencing attack and implement the eBPF code to detect and filter messages pertaining to this attack. We consider an attack generated with one adversary

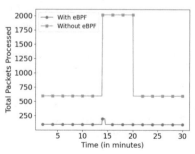

Fig. 9. Packets Processed in Message Sequencing Attack

which is to toggle the machine state by setting ASDUType value to 46 and changing the CoT values to 6, 7 and 10 respectively in that order. For detecting the attack we used the timestamp and message attribute values as outlined in Sect. 3.2. We study the filtering ability by collecting traffic for a duration of 30 min each; one time without filtering and second time with filtering implemented. In this 30 min period, the toggling attack was generated for a duration of 6 min starting at 14 min. Our eBPF code filters all messages coming from the adversary, once the attack is detected. Figure 9 shows the number of messages processed by the control center without and with filtering cases. We can notice that with filtering the number of messages processed is considerably less except the period in which the attack is detected in the beginning.

6   Prior Works

Existing works related to the security of industrial control systems (ICS) fall into the following two categories.

(i) Attack Vectors: Several prior works [5–7] cover cyber attacks against grid and SCADA systems and associated protocols. Most studied attacks include false data injections [7] and denial of service attacks [8]. False data injection attacks generate false measurement data and have it sent to control center to mislead the state estimation or assessment. On the other hand DoS attacks prevent the normal operation of either control center or measurement units/clients from performing its operations. We study a different class of attacks which target computational resources of the control center. Similar attacks are studied in other domains like web servers [9].

(ii) Detecting the Attacks: These methods develop techniques to detect different types of attacks. There are several machine learning algorithms [10, 11], state transition modeling approaches [2, 12, 13] and commercial IDS solutions [14] available for this purpose. Many of these techniques rely on Deep Packet Inspection (DPI) [15] for detecting these attacks. However, these methods detect conventional attacks like scanning, injection, connection loss, etc. Lin et al. [16] adopted Bro [18] (now called zeek) intrusion detection system for SCADA networks. They also propose to filter malformed messages in SCADA systems. However, their study is limited to a limited number of attribute types and that to of only DNP3 protocol. Further, Bro rules were manually written for detection. Unlike this, we describe a systematic way to identify all types of malformed messages.

7   Conclusion

Electric grid networks use SCADA communication systems to exchange safety critical information. As these messages are sent over conventional TCP/IP networks, they are susceptible to different cyber attacks. In this paper, we studied two types of attacks namely resource depletion attacks and message sequencing attacks. In a resource depletion attack, an adversary sends a large number of malformed messages with the intention of consuming resources and in message sequencing, she sends a sequence of messages to disturb the operations of the SCADA system. We described a method to identify such malformed messages using rules generated with first order logic statements. We also proposed an attack mitigation method by filtering malformed messages using the Extended Berkeley Packet Filter. Our simulation based study showed that such filtering is effective and robust against variants of attacks and also to intensity of attacks.

Acknowledgement. Work reported in this paper is financially supported by funding from IHUB NTIHAC foundation IIT Kanpur through grant number IHUB-NTIHAC/2021/01/24 and also by Science and Engineering Research Board via grant CRG/2022/005198-G. Authors thankfully acknowledge the funding received.

References

1. Acarali, D., Rajarajan, M., Chema, D., Ginzburg, M.: Modelling DoS attacks & interoperability in the smart grid. In: Proceedings of the 29th International Conference on Computer Communications and Networks, ICCC 2020, pp. 1–6. IEEE (2020)
2. Matoušck, P., Havlena, V., Holík, L.: Efficient modelling of ICS communication for anomaly detection using probabilistic automata. In: 2021 IFIP/IEEE International Symposium on Integrated Network Management (IM), pp. 81–89 (2021)
3. Jørgensen, T.H., et al.: The EXpress data path: fast programmable packet processing in the operating system kernel. In: Proceedings of the 14th International Conference on Emerging Networking Experiments and Technologies, CoNEXT 2018, pp. 54–66 (2018)
4. SCAPY. https://scapy.net/
5. Beasley, C., Zhong, X., Deng, J., Brooks, R., Venayagamoorthy, G.K.: A survey of electric power synchrophasor network cyber security. In: IEEE PES Innovative Smart Grid Technologies Conference Europe, pp. 1–5 (2014)
6. Evangeliou, I.E.: Vulnerabilities of the Modbus protocol. Ph.D. thesis, University of Piraeus, Greece (2018)
7. Aoufi, S., Derhab, A., Guerroumi, M.: Survey of false data injection in smart power grid: attacks, countermeasures and challenges. J. Inf. Secur. Appl. **54** (2020)
8. Zhu, B., Joseph, A., Sastry, S.: A taxonomy of cyber attacks on SCADA systems. In: CPSCom: Proceedings of the 4th IEEE International Conference on Cyber, Physical and Social Computing, pp. 380–388 (2011)
9. Tripathi, N., Hubballi, N.: Application layer denial-of-service attacks and defense mechanisms: a survey. ACM Comput. Surv. **54**(4), 1–30 (2021)
10. Phillips, B., Gamess, E., Krishnaprasad, S.: An evaluation of machine learning-based anomaly detection in a SCADA system using the Modbus protocol. In: Proceedings of the 2020 ACM Southeast Conference, ASM-SE 2020, pp. 188–196 (2020)
11. Anwar, M., Lundberg, L., Borg, A.: Improving anomaly detection in SCADA network communication with attribute extension. Energy Inform. **5**(1), 1–22 (2022)
12. Goldenberg, N., Wool, A.: Accurate modeling of Modbus/TCP for intrusion detection in SCADA systems. Int. J. Crit. Infrastruct. Prot. **6**(2), 63–75 (2013)
13. Kleinmann, A., Wool, A.: Automatic construction of statechart-based anomaly detection models for multi-threaded industrial control systems. ACM Trans. Intell. Syst. Technol. **8**(4), 1–21 (2017)
14. StationGuard. https://www.omicronenergy.com/en/solution/intrusion-detection-system-ids-for-the-power-grid/#
15. Nyasore, O.N., Zavarsky, P., Swar, B., Naiyeju, R., Dabra, S.: Deep packet inspection in industrial automation control system to mitigate attacks exploiting Modbus/TCP vulnerabilities. In: 2020 IEEE International Conference on Intelligent Data and Security (IDS), pp. 241–245 (2020)
16. Lin, H., Slagell, A., Di Martino, C., Kalbarczyk, Z., Iyer, R.K.: Adapting Bro into SCADA: building a specification-based intrusion detection system for the DNP3 protocol. In: Proceedings of the Eighth Annual Cyber Security and Information Intelligence Research Workshop, pp. 1–4 (2013)
17. Chan, A.C., Zhou, J.: Non-intrusive protection for legacy SCADA systems. IEEE Commun. Mag. (2023)
18. ZEEK IDS. https://old.zeek.org/manual/2.5.5/broids/index.html

API Descriptions for the Web of Things

Aimilios Tzavaras[ID], Chrisa Tsinaraki[ID], and Euripides G. M. Petrakis[✉][ID]

School of Electrical and Computer Engineering, Technical University of Crete (TUC),
73100 Chania, Crete, Greece
{atzavaras,ctsinaraki}@tuc.gr, petrakis@intelligence.tuc.gr

Abstract. The Web of Things (WoT) aims to integrate the Internet of Things with the Web. One requirement is that all things (i.e., devices, sensors, etc.) must be described to be machine discoverable on the web so that they can be searched for and reused in applications. The W3C defines a Thing Description (TD) reference model, but users and industry do not widely accept it. The proposed work revolves around the idea that things should be described similarly to services using OpenAPI or AsyncAPI for Things implementing synchronous or asynchronous communication respectively. Both are becoming the de facto industry standards for service descriptions and can be used to define widely accepted TDs equally well. The emerging TD description model should be based on the combination of both.

Keywords: Web of Things · Thing Description · OpenAPI · AsyncAPI

1 Introduction

The Web of Things (WoT) initiative [3] suggests that Things should advertise their identity and properties on the Web so that they can be discovered by Web search engines and reused in applications. Thing Descriptions (TD) are the entry points for discovering services and resources related to Things. Hosted in a directory service on a gateway or in the cloud, they offer a Web interface to register and search for Things.

The WoT architecture recommendation of W3C [5] sets the requirements for interacting with Things on the Web using REST [7]. The Thing Description [4] of the W3C architecture is a representation of Thing properties (e.g., data types and operations) based on JSON-LD [8]. It describes Things exposing metadata and shows how a client can interact with Things (e.g., retrieve their Properties, request Actions, or trigger Events). It is not tied to a specific application, protocol, or implementation, and exposes Thing metadata on the Web so that other Things or clients can interact with them.

Despite W3C's suggestion for REST Things, synchronous communication is not always suitable for real-world IoT scenarios. A way around this is to deploy a Web Proxy, that runs on the cloud (or on a gateway), translates IoT-specific

L. Barolli (Ed.): AINA 2024, LNDECT 201, pp. 48–58, 2024.
https://doi.org/10.1007/978-3-031-57870-0_5

communication to REST HTTP(S), and keeps the virtual image of each Thing (e.g., a JSON representation). Then, Things can be accessed via Web Proxies and become part of the Web, so they can be published, consumed, aggregated, updated, and searched for. Essential parts of the WoT proxy are, an API interface to allow connections with the outside world and a directory of available services (e.g., a database or an ontology). Nonetheless, this is a hard constraint (i.e., lack of flexibility and introduces delays) in device communication.

IoT devices can be either synchronous or asynchronous. They can communicate using any of a wide range of IoT or application-specific protocols (e.g., Bluetooth, MQTT, ZigBee, LoRa, CoAP, HTTP, etc.) either synchronously or asynchronously. Accordingly, TDs must be capable of describing all common IoT protocols, in addition to HTTP(S). Asynchronous APIs, also known as *async APIs* or *event-driven APIs*, are preferred in resource-constrained scenarios where a triggering request or immediate response is not required, or when processing time is critical. This is typical when an IoT device (e.g., traffic light or sensor) can operate asynchronously and independently of requests.

In our previous work [10], Things are described similar to RESTful services using OpenAPI [6]. OpenAPI is a description format for REST APIs. It is based on JSON (or YAML) and comprises a large set of properties for composing service descriptions. OpenAPI Thing Descriptions (TDs) form a lightweight version of the general-purpose description of REST services specialized for Web Things. W3C TD and OpenAPI TD share common features and serve the same purpose (i.e., discovering Things in the WoT).

Similar to OpenAPI for REST devices, AsyncAPI [1] is a relatively new approach that describes asynchronous, event-based service communication. Unlike OpenAPI, it targets Streaming APIs (i.e., not REST APIs) and supports additional schemas, parameters, and protocols such as AMQP, MQTT, Kafka, STOMP, etc. AsyncAPI is a promising choice for describing (especially) asynchronous APIs in the WoT context. However, it is not clear to what extent AsyncAPI competes or complements OpenAPI or W3C TD for the description of Things.

W3C TD is the reference approach. Although a W3C recommendation, it has limited user and industry acceptance. The proposed work supports the hypothesis that both OpenAPI and AsyncAPI conform to the W3C principles and can be used in place of W3C TD. OpenAPI is the de-facto industry standard, and AsyncAPI is the emerging standard for synchronous and asynchronous communication respectively. We argue that there is no need for yet another standard like W3C TD to describe Things in WoT.

The discussion that follows suggests that the choice of TD must be based on the scenario and communication pattern of the application. For instance, an IoT smart lamp API using MQTT (i.e., asynchronous communication) can be described using AsyncAPI or W3C TD but, a smart lock (i.e., synchronous) can be described using OpenAPI or W3C TD.

The three TD approaches (i.e., W3C TD, OpenAPI, AsyncAPI) are presented in Sect. 2. Their comparison is discussed in Sect. 3 followed by, conclusions in Sect. 4.

2   Thing Description Approaches for the Web of Things

The fundamentals of W3C, OpenAPI, and AsyncAPI TD are discussed next. In a previous paper [9] we show how Things are described using OpenAPI and this approach is compared with W3C TD (comparison with AsyncAPI was left for future work).

2.1   W3C Thing Description (TD)

The W3C TD defines the architectural aspects of a thing, including interaction capabilities, data schemas, and security configurations. It describes Properties (e.g., sensor pressure), Actions (e.g., smart window opening and closing), and Events (e.g., smart window state transitioning to open). Events associated with event sources, push data (e.g., notifications) to registered clients, typically implemented using functions such as subscribing to a Webhook callback URI. Additionally, it defines the Thing Model as a generic template for a type of thing with common properties (as opposed to OpenAPI TD [10] which focuses on Thing instances). The TD document includes the Thing's name, unique identifier, security requirements, title, optional human-readable description, and interactions. It can be enriched with a context field for converting the JSON format to JSON-LD.

Listing 1.1 is an example TD for a Smart Light Switch Actuator with (a) a context attribute that extends the definition with additional vocabulary terms, (b) a device identifier, (c) an indicative title, (d) the security configuration (Basic Authentication in this example), (e) supported interactions; the state property, turn on/off actions, the turn on event (i.e., the light turning on), and, (f) the forms field that describes how each interaction can be performed, by specifying the protocol that should be used (i.e., HTTPS) and the operation endpoint. The Properties Object specifies the endpoint for retrieving the (last) light switch state value. The Actions Object specifies protocols and endpoints for the turn-on/off actions. The Events Object specifies the protocol, endpoint, and sub-protocol (e.g., the exact mechanism used for asynchronous notifications) for the light turning-on event. This example TD provides key details for a Smart Light Switch Actuator in a concise format.

Listing 1.1. Thing Description for the Smart Light Switch.

```
1   {
2     "@context": "http://www.w3.org/ns/td",
3     "id": "urn:dev:ops:32473-WoTSmartLightSwitch-1234",
4     "title": "MySmartLightSwitch",
5     "securityDefinitions": {
6       "basic_sc": {"scheme": "basic", "in": "header"}
7     },
8     "security": "basic_sc",
9     "properties": {
10      "state": {
11        "type": "string",
12        "forms": [{"href": "https://mysmartlightswitch.example
            .com/state"}]
```

```
13      }
14    },
15    "actions": {
16      "turn-on": {
17        "forms": [{"href": "https://mysmartlightswitch.example
              .com/turn-on"}]
18      },
19      "turn-off": {
20        "forms": [{"href": "https://mysmartlightswitch.example
              .com/turn-off"}]
21      }
22    },
23    "events":{
24      "turning-on":{
25        "description": "Smart Light Switch turns on",
26        "data": {"type": "string"},
27        "forms": [{
28          "href": "https://mysmartlightswitch.example.com/
              turning-on",
29          "subprotocol": "longpoll"
30        }]
31      }
32    }
33  }
```

2.2 OpenAPI Thing Description (TD)

OpenAPI Thing Descriptions (TDs) [10] conform to the principles of W3C. The OpenAPI format uses JSON (or YAML) and comprises a large set of properties for composing Thing descriptions similar to Web services. It is a lightweight version of the general-purpose OpenAPI description for REST services specialized for Web Things. It is an alternative to the TD of the W3C and provides an informative mechanism for exposing the functionality of Things on the Web similar to RESTful services. Both representations share common features and serve the same purpose (i.e., discovering Things in the WoT).

Webhooks[1] (e.g., a Webhook notification for a sensor's temperature value change) is the only asynchronous operation supported by OpenAPI v3.1. OpenAPI can also describe subscriptions to Things. A subscription is the result of subscribing to a specific resource of a Thing (e.g., a particular property or action) to get notified of changes in the Thing's state information (e.g., new humidity value). The subscriptions are stored in databases so that they can be retrieved by a subscription identifier.

Listing 1.2 showcases (in YAML format) an operation that creates a new subscription to a specific resource associated with the Thing. Subscriptions are ideally implemented through custom callbacks (i.e., Webhooks) which are naturally offered by the WebSocket protocol. An HTTP POST request is required to create and store a new subscription. The particular endpoint (i.e., /subscriptions) and operation are specified in the Paths object. Although the subscription information is defined in the request body by the subscriber (i.e., client or service), the schema object that describes the payload can be a reusable object

---

[1] https://www.asyncapi.com/blog/openapi-vs-asyncapi-burning-questions.

which is defined in the Components object. The object contains an indicative name, a description, the subscription type (e.g., Webhook), the callback URL, an object containing (as object properties) the type and the name of the resource to which the subscription is made, the expiration date of the subscription and, a throttling parameter which is used to specify a minimum inter-notification arrival time for the subscription. A response header containing the subscription identifier is returned as long as the operation is successful (i.e., a 200 OK response is returned).

Listing 1.2. The operation for creating a subscription.

```
1  /subscriptions:
2    post:
3      tags:
4        - Subscriptions
5      summary: Create a subscription
6      description: A Web Thing should support subscriptions
                   for its resources.
7      operationId: createSubscription
8      x-operationType: 'https://schema.org/CreateAction'
9      requestBody:
10       description: Create a new subscription
11       content:
12         application/json:
13           schema:
14             $ref: '#/components/schemas/SubscriptionRequestBody'
15       required: true
16     responses:
17       '200':
18         description: OK
19       '404':
20         description: Not found
```

2.3   AsyncAPI Thing Description (TD)

AsyncAPI maintains some compatibility with OpenAPI [2] but differs from OpenAPI in its support for additional schemas and payload (i.e., the request or response body can be any value other than a schema), parameters, and protocols (e.g., AMQP, MQTT, WebSockets, Kafka, STOMP, Mercure, etc.). AsyncAPI handles all security mechanisms of OpenAPI (i.e., password, certificates, API keys, and OAuth2.0) and also mechanisms for asynchronous communication (e.g., SASL authentication[2], which is commonly used with Apache Kafka and LDAP). The Publish-Subscribe pattern is a major concept of distributed and messaging architectures, and a notable use case for AsyncAPI. It allows the description of topics, events, and message payloads in publication-subscription architectures, providing a standardized way to communicate the structure and format of messages.

AsyncAPI is similar to OpenAPI in terms of core objects such as Info, Server, Operation, and Components. Differences are related to Channels in the place of

---

[2] https://www.iana.org/assignments/sasl-mechanisms/sasl-mechanisms.xhtml.

OpenAPI Paths and structural differences regarding the Operations. The Channel object represents a communication channel, a topic or an event. The Message object defines the data structure and format of messages exchanged on channels, including payload details and associated metadata. The Operation object defines publish or subscribe operations within channels, specifying the operation action (i.e., sending or receiving a message), a summary, and optional parameters. The Components object holds reusable elements like schemas and security schemes, enhancing consistency throughout the AsyncAPI document.

Similar to W3C TD, AsyncAPI is protocol-agnostic (i.e., it is not tied to any specific communication protocol). In contrast to OpenAPI, which is limited to HTTP(S) and REST, AsyncAPI can be used with various lightweight, low-overhead messaging protocols that are often used in IoT (e.g., MQTT). AsyncAPI accommodates different communication patterns, including message queues (e.g., RabbitMQ), publish-subscribe systems (e.g., Apache Kafka), and other asynchronous messaging mechanisms (e.g., long polling, Server-Sent Events, etc.).

In summary, AsyncAPI is designed to better describe publish-subscribe channels and operations rather than REST API endpoints and operations. However, it can be used to describe HTTP(S) streaming APIs that allow real-time, two-way communication over HTTP(S) in cases where continuous or real-time updates are necessary. For example, AsyncAPI can describe devices that communicate using Server-Sent Events (SSE) or WebSockets after establishing an SSE or WebSocket connection to a server. This is especially beneficial in situations where Things need to exchange information in a timely and efficient manner. Overall, the OpenAPI TD and AsyncAPI TD approaches are complementary and can coexist in the asynchronous or synchronous WoT realm.

Listing 1.3 is a YAML excerpt of the AsyncAPI Thing Description[3] for an API exposed by the smart streetlights of a city, which utilizes Apache Kafka and allows users to remotely manage the city lights (i.e. by sending commands to turn the lights on or off). It describes a publish operation that produces a streetlight measurement to Kafka, and a subscribe operation that consumes the turn-on event of a streetlight (i.e. triggered after a user command). After receiving information about environmental lighting conditions, the first operation can publish a measurement in a Kafka topic. The second operation can be used to subscribe to a Kafka topic and accept user commands for turning a specific streetlight on.

Two Operation objects are utilized in this example to represent the specific operations within a Channels object. The example incorporates references (i.e., *ref*) to Message objects describing the messages exchanged by the operations: a) messages that represent light measurements (named as *lightMeasured*) and messages that represent commands to turn the streetlights on or off (named as *turnOnOff*). In addition, the example includes references to an Operation Trait object (named as *kafka*). Operation Trait objects are generally used in AsyncAPI to define reusable sets of properties that may apply to different operations.

---

[3] https://editor-next.swagger.io/.

They can include various properties (like, for example, descriptions, bindings, headers, etc.), depending on the specific needs of the API. In this example, a Kafka operation trait (i.e. including protocol bindings for Kafka) is specified in the Components Object. Finally, the example includes references to Parameter objects used to denote the identifiers (IDs) of particular streetlights (using the parameter *streetlightId*).

Listing 1.3. Two operations for smart city streetlights.

```
1  channels:
2    smartylighting.streetlights.1.0.event.{streetlightId}.
       lighting.measured:
3      description: The topic on which measured values may be
         produced and consumed.
4      parameters:
5        streetlightId:
6          $ref: '#/components/parameters/streetlightId'
7      publish:
8        summary: Inform about environmental lighting
           conditions of a particular streetlight.
9        operationId: receiveLightMeasurement
10       traits:
11         - $ref: '#/components/operationTraits/kafka'
12       message:
13         $ref: '#/components/messages/lightMeasured'
14
15   smartylighting.streetlights.1.0.action.{streetlightId}.
       turn.on:
16     parameters:
17       streetlightId:
18         $ref: '#/components/parameters/streetlightId'
19     subscribe:
20       operationId: turnOn
21       traits:
22         - $ref: '#/components/operationTraits/kafka'
23       message:
24         $ref: '#/components/messages/turnOnOff'
```

3   Comparison

W3C TD and AsyncAPI TD both distinguish themselves as protocol-agnostic approaches. They can describe HTTP(S) and other IoT or messaging (e.g., publish-subscribe) protocols such as MQTT, CoAP, AMQP, WebSocket, and Apache Kafka. They are particularly well suited for event-driven communication which is categorized as asynchronous. Asynchronous APIs emphasize the exchange of messages, events, or notifications between components in a system. OpenAPI TD is not protocol-agnostic and is designed primarily for synchronous HTTP(S) communication. RESTful APIs revolve around resources identified by URIs and resource-centric interactions, prioritizing the request-response model, and statelessness. Hence, a fundamental consideration in selecting the appropriate method for each scenario revolves around the communication pattern (i.e., synchronous or asynchronous) and the communication protocol. However, Webhooks and HTTP Live Streaming (HLS) communication are exceptions. Webhooks, although asynchronous, are also supported in OpenAPI v3.1, and consequently, by all three approaches. HLS on the other hand, is used to deliver real-time multimedia (e.g. audio, video) over the Internet. Despite being an HTTP-based protocol, it is typically described by AsyncAPI or W3C TD.

3.1 Protocol Support

AsyncAPI TD can describe HTTP(S) Streaming APIs, but not RESTful APIs. A W3C TD can describe extra IoT protocols (e.g., CoAP, MQTT, Modbus, OPC UA). Additional protocols can be described using the Context Extension mechanism of JSON-LD. For example, the W3C TD of a sensor may specify an MQTT Protocol Binding similar to AsyncAPI. W3C TD and AsyncAPI TD are protocol-agnostic approaches, while OpenAPI lags in terms of IoT protocol support and is limited to HTTP(S). This might not be considered vital, though, considering that the WoT Proxy (i.e., a gateway) resumes all IoT protocol conversion to HTTP(S). In that case, the end-user communicates with Things using HTTP(S), and the conversion to IoT protocol resorts to the WoT proxy. Table 1 summarizes the above comparison.

Table 1. Communication protocol support.

Protocols	OpenAPI TD	W3C TD	AsyncAPI TD
HTTP/HTTPS and Webhook mechanism	Yes	Yes	Yes
MQTT/Secure-MQTT	No	Yes	Yes
CoAP	No	Yes	Yes
AMQP/AMQPS	No	Possible using semantics	Yes
Kafka/Kafka-Secure	No	Possible using semantics	Yes
STOMP/STOMPS	No	Possible using semantics	Yes
WebSocket/WSS	No	Possible using semantics	Yes
JMS	No	Possible using semantics	Yes

3.2 Hypermedia Controls Support

OpenAPI is a promising technology for understanding and constructing Web services that meet the HATEOAS[4] requirement of the REST architectural style. Links is a new feature introduced in the latest OpenAPI v3.1 (along with Callbacks). OpenAPI Links are defined in service responses to allow values returned by a service call to be used as input for the next call. More specifically, Links are defined as Link objects in the responses of API operations. This is an attempt to incorporate HATEOAS functionality into the specification. A truly RESTful service sends a client the information on how to form the next requests

---

[4] https://neurosys.com/blog/hateoas-links.

according to the requested resources. OpenAPI Links differ from HATEOAS, as they do not always come from the service (i.e., they can just be present in the service description). Therefore, OpenAPI Links do not formally meet the HATEOAS requirement of REST. OpenAPI TD can also describe operations that clients may perform on Things. OpenAPI operations are similar to W3C TD Web forms. Hence, OpenAPI supports hypermedia controls that refer to Things and their functionality.

Hypermedia controls in W3C TD describe client interaction with Things using Web links and Web forms. Although a TD can describe links, it does not refer to links in the same way as OpenAPI. In W3C TD, links are used in a broader sense (e.g., to include a documentation resource). Web links offer navigation affordances for resource discovery. Web forms enable clients to describe operations (e.g., read a property) that can even change a Thing's state (e.g., turn on a device), and not just discover resources using URIs. W3C TDs use forms to describe how to perform specific operations on Things (e.g., to retrieve information about a property, invoke an action, etc.). Overall, both W3C TD and OpenAPI TD support hypermedia controls such as Web links and Web forms. Links are used by both, but they represent a different concept in each approach. W3C TD uses Web forms to describe operations on Things. Web forms do not exist in the OpenAPI (operations are described by Paths objects).

HATEOAS is commonly linked with RESTful APIs, employing hypermedia controls for navigating an application's state. In contrast, AsyncAPI is applied to describe asynchronous (not RESTful) APIs centered around messages or events. Therefore, HATEOAS is not supported by AsyncAPI. AsyncAPI TD can support Web forms using AsyncAPI operations (but these are not RESTful operations) in contrast to W3C TD and OpenAPI TD. Table 2 summarizes this comparison.

Table 2. Hypermedia control support.

Hypermedia Controls	OpenAPI TD	W3C TD	AsyncAPI TD
Links	Yes (OpenAPI Links)	Yes (TD Links)	No
Forms	Yes (OpenAPI Operations)	Yes (TD Forms)	No

3.3   Security Support

The HTTP(S) security schemes, vocabulary, and syntax of W3C TD share many similarities with OpenAPI v3.0.1 (and later). AsyncAPI supports all security schemes of OpenAPI. All TD methods support security configuration for different HTTP(S) authentication schemes (e.g., Basic Authentication), for API key authentication, and OAuth 2.0 common flows. Pre-Shared key authentication (PSK) using pre-shared keys (e.g., TLS-PSK) is supported by W3C TD. PSK authentication is used for client authentication on wireless networks for WPA and WPA2 encryption. It is not supported in the OpenAPI TD and does not seem to be supported in the AsyncAPI TD. OpenID Connect authentication can be supported in all the approaches.

AsyncAPI describes security schemes not supported by OpenAPI (i.e., SASL Authentication[5] and X.509 Certificate[6]). These can be described by W3C TD using contextual definitions (in some namespace) that integrate additional semantics (this mechanism is referred to as "possible using semantics"). Hence, W3C TD allows for the description of a security scheme, such as SASL Authentication, that does not belong to the default security schemes outlined in the W3C Thing Description specification[7]. This is possible by using an external semantic model (in JSON-LD) to define the particular security scheme. Similar to the ComboSecurityScheme of the W3C TD security definition, the combination of security requirements is possible in OpenAPI and AsyncAPI. If no authentication is required to access the resources of a Thing, this can be declared in the TD security definition (NoSecurityScheme) of W3C TD, and also in OpenAPI using an empty array. In AsyncAPI, the *security* field can simply be omitted. Table 3 summarizes this comparison.

Table 3. Security schemes support.

Security Schemes	OpenAPI TD	W3C TD	AsyncAPI TD
Basic Authentication	Yes	Yes	Yes
Bearer Authentication	Yes	Yes	Yes
API Key Authentication	Yes	Yes	Yes
Digest Authentication	Yes	Yes	Yes
OAuth2.0 Authentication	Yes	Yes	Yes
OpenID Connect Authentication	Yes	Possible using semantics	Yes
PSK Authentication	No	Yes	No
User/Password Authentication	Yes	Yes	Yes
SASL Authentication	No	Possible using semantics	Yes
X.509 Certificate	No	Possible using semantics	Yes
Authentication	Yes	Yes	Yes
Combo Authentication (Combination of security schemes)	Yes	Yes	Yes
No Security	Yes	Yes	Yes

3.4 Discussion

W3C TD focuses on IoT and does not provide the same level of detail and specificity for RESTful API documentation as OpenAPI does. Likewise, AsyncAPI is particularly beneficial for systems that follow messaging patterns (e.g., event-driven architectures) and offers a level of detail and specificity regarding the characteristics of asynchronous communication. While W3C TD may describe the asynchronous Thing interactions in the IoT, it is not specifically designed to

---

[5] https://www.iana.org/assignments/sasl-mechanisms/sasl-mechanisms.xhtml.
[6] https://www.ietf.org/rfc/rfc2459.txt.
[7] https://www.w3.org/TR/wot-thing-description11/.

address the detailed requirements of asynchronous APIs in the same way that AsyncAPI does. Another observation has to do with the standardization status or adoption of the three approaches: OpenAPI and AsyncAPI are de facto standards, with wide adoption in the industry, while the W3C TD is a W3C recommendation, without strong support from the industry. Moreover, the OpenAPI and AsyncAPI specifications are compatible at the schema level [2], thus allowing, when combined, full support of both synchronous and asynchronous communication requirements.

4   Conclusions

Thing Description is a central building block of any Web of Things architecture. This work suggests that OpenAPI and AsyncAPI conform to the W3C Architecture model principles and can co-exist in WoT to describe devices implementing synchronous or asynchronous communication respectively. An advantage of the OpenAPI and AsyncAPI Thing Descriptions interplay is the uniformity of the representation of both Things and services that also co-exist and interact on the Web. It then becomes easier to compose them in new applications on the WoT. This is an interesting issue for future research.

References

1. AsyncAPI (2023). https://www.asyncapi.com
2. AsyncAPI: Coming from OpenAPI (2024). https://www.asyncapi.com/docs/tutorials/getting-started/coming-from-openapi
3. Guinard, D., Trifa, V.: Building the Web of Things. Manning Publications Co., Greenwich (2016). https://www.manning.com/books/building-the-web-of-things/
4. Kaebisch, S., Kamiya, T., McCool, M., Charpenay, V., Kovatsch, M.: Web of Things (WoT) Thing Description (2020). https://www.w3.org/TR/wot-thing-description/, w3C Recommendation
5. Kovatsch, M., Matsukura, R., Lagally, M., Kawaguchi, T., Toumura, K., Kajimoto, K.: Web of Things (WoT) Architecture (2020). https://www.w3.org/TR/wot-architecture/, w3C Recommendation
6. Miller, D., Whitlocak, J., Gartiner, M., Ralphson, M., Ratovsky, R., Sarid, U.: OpenAPI Specification v3.1.0 (2021). https://spec.openapis.org/oas/latest.html, openAPI Initiative, The Linux Foundation
7. Richardson, L., Amundsen, M., Ruby, S.: RESTful Web APIs. O'Reilly (2013). https://www.oreilly.com/library/view/restful-web-apis/9781449359713/
8. Sporny, M., Longley, D., Kellogg, G., Lanthaler, M., Champin, P., Lindström, N.: JSON-LD 1.1: A JSON-based Serialization for Linked Data (2021). https://w3c.github.io/json-ld-syntax/, w3C Editor's Draft
9. Tzavaras, A., Mainas, N., Bouraimis, F., Petrakis, E.: OpenAPI thing descriptions for the web of things. In: IEEE International Conference on Tools with Artificial Intelligence (ICTAI 2021), pp. 1384–1391 (2021). https://ieeexplore.ieee.org/document/9643304
10. Tzavaras, A., Mainas, N., Petrakis, E.G.: OpenAPI framework for the Web of Things. Internet of Things **21**, 100675 (2023). https://doi.org/10.1016/j.iot.2022.100675

SmartDriveAuth: Enhancing Vehicle Security with Continuous Driver Authentication via Wearable PPG Sensors and Deep Learning

Laxmi Divya Chhibbar[✉], Sujay Patni, Siddarth Todi, Ashutosh Bhatia, and Kamlesh Tiwari

Birla Institute of Technology and Science, Pilani, Pilani, Rajasthan, India
{p20210023,f20190575,f20190991,ashutosh.bhatia, kamlesh.tiwari}@pilani.bits-pilani.ac.in

Abstract. The paper introduces a novel approach for continuous driver authentication in vehicle security, utilizing wearable photoplethysmography (PPG) sensors and Long Short-Term Memory (LSTM)–based deep learning. This study aims to overcome the limitations of traditional one-time authentication (OTA) methods, which typically involve passwords, PINs, or physical keys. While effective for initial identity verification, these conventional methods do not continuously validate the driver's identity during vehicle operation. The proposed system leverages an LSTM-based prediction model to efficiently predict the subsequent PPG values using the raw PPG signals from wrist-worn devices. The predicted values are continuously compared with actual real-time data (received from the sensors) for authentication. The proposed system eliminates the need to permanently store user biometrics in a database. Motion artifacts and momentary disruptions have minimal impact on system performance. Experimental validation was conducted with 15 participants driving in varied conditions to simulate real-life driving conditions. The study evaluated the system's accuracy, achieving an Equal Error Rate (EER) of 4.8%, demonstrating its potential as a viable solution for continuous driver authentication in dynamic environments.

1 Introduction

Technology has been a cornerstone in addressing safety and security threats since last few decades. One of the most significant of these threats is impersonation attacks, which pose serious risks to both personal safety and property. In our digital era, such attacks are commonly countered with authentication methods. These methods require users to provide credentials to prove their legitimacy and rights to access specific resources. This process, known as One-time Authentication (OTA), grants access for an entire session after initial verification. However, OTA has limitations in critical sectors such as healthcare, transportation, aviation, and corporate environments. Take the transportation sector, for example.

© The Author(s), under exclusive license to Springer Nature Switzerland AG 2024
L. Barolli (Ed.): AINA 2024, LNDECT 201, pp. 59–72, 2024.
https://doi.org/10.1007/978-3-031-57870-0_6

Companies must ensure the safety of passengers and goods, which involves managing a large fleet and assigning verified drivers. Here, the use of OTA poses a potential risk as a driver, once authenticated, may transfer control of the vehicle to an unverified or malicious individual. This scenario raises concerns, particularly in contexts involving transportation of critical military equipment, valuable items, or the vulnerable individuals such as children, women, and senior citizens. Allowing an unverified person with malicious intent access to the vehicle may result in severe and potentially harmful incidents. Thus, there is a pressing need for a method that continuously or frequently verifies the legitimacy of a driver, known as Continuous Authentication (CA). CA is a security mechanism that constantly or periodically validates a user's identity based on behavioral or physiological data, collected transparently without requiring active user participation. Notably, in the context of transportation sector, CA also ensures accountability as the drivers cannot repudiate the actions taken by them as they are getting authenticated continuously.

Biometrics are the cornerstone of CA systems due to their inherent transparency (as they can be captured without explicit user interaction). This characteristic is essential for CA, where continuous verification should occur unobtrusively. Biometric data types include both behavioral and physiological biometrics. Behavioral biometrics are derived from patterns in device interaction and motion dynamics, while physiological biometrics involve physical or biological traits such as facial recognition, ECG, EEG, and PPG signals [4,7]. For driver authentication, non-intrusive biometric data collection during driving is imperative. Following a thorough evaluation, photoplethysmography (PPG) signals from wrist-worn wearables emerged as the optimal choice, attributed to their heightened stability compared to behavioral-related biometrics. PPG sensors are commonplace in commercial smartwatches and fitness bands, offering excellent discriminability for authentication purposes. However, PPG signals from wrist-worn devices face challenges due to noise and motion artifacts, posing significant hurdles for CA, especially given CA systems' general avoidance of extensive preprocessing to recover or enhance signal quality. In contrast to prior studies employing diverse filtering and feature extraction methods, our research introduces a novel approach utilizing a Long Short-Term Memory (LSTM)-based prediction model for CA. This 'lightweight' method leverages the capabilities of LSTM to handle raw data inputs, eliminating the necessity for complex preprocessing or feature extraction. Our innovative approach aims to enhance CA systems' efficiency. The contributions of this work are pivotal to the advancement of Continuous Authentication (CA) systems using wearable technology. Firstly, we have developed a novel framework for a CA system based on wearables, utilizing LSTM-based prediction method for the authentication process. This framework represents a significant step forward in leveraging biometric data for authentication purposes. Secondly, the implementation of this framework is unique as it utilizes raw PPG signals obtained from wrist-worn devices. Despite the inherent challenges of motion artifacts [12] in such data, our work explores the efficacy of the LSTM model in this context, showcasing its potential to overcome these obstacles. Lastly, we have conducted a thorough evaluation of our proposed method using real PPG data. The results of this evaluation are crucial in demonstrating the practical applicability and effectiveness of our LSTM-based authentication method in real-world scenarios, setting a new benchmark in the field of biometric authentication.

The remainder of the paper is organized to provide a comprehensive understanding of our approach and findings. Section 2 reviews related work on LSTM-based CA and CA using PPG signals. Section 3 delves into the design challenges of a CA system and is followed by Sect. 4 detailing our proposed system, SmartDriveAuth. The section includes specifics of our LSTM-prediction-based authentication approach. The subsequent experimentation section (Sect. 5) provides dataset details and results highlighting our proposed system's accuracy and overall performance. The paper concludes with Sect. 6, which addresses broader implications and limitations and suggests future research directions in wearable-based CA for transportation and related sectors.

2   Related Work

2.1   Long Short-Term Memory (LSTM) for CA

Long Short-Term Memory (LSTM) networks, a specialized form of deep learning models, are particularly suited for processing sequential data. As a variant of Recurrent Neural Networks (RNNs), LSTMs are designed to overcome the challenges associated with long sequences, such as the vanishing or exploding gradient problem. In the context of authentication models, LSTMs are adept at binary classification tasks, distinguishing between legitimate users and impostors. The application of LSTM in Continuous Authentication (CA) systems has been explored in various studies. For instance, in [1], the authors proposed a deep learning-based CA approach using built-in smartphone sensors to capture users' behavioral patterns. Their LSTM model, trained on data from 84 subjects, achieved an impressive F1-Score of 98% and an Equal Error Rate (EER) of 0.41%. This study demonstrates the efficacy of LSTM in accurately analyzing behavioral data for authentication purposes. Another significant contribution in this field is DeepAuth [3], an LSTM-based CA method. DeepAuth utilizes accelerometer and gyroscope data to capture behavioral patterns, indicative of user identity. Their experimental results, based on data collected from 47 participants, showed a high authentication accuracy of 96.7%. In [10] also, the authors proposed a CA scheme that used a LSTM based classification model to authenticate the users using their behavioral features. The model achieved a high authentication accuracy of 97%. These studies further reinforce the capability of LSTM models in handling complex, time-based biometric data, making them highly suitable for CA systems that require rapid and accurate user verification.

2.2   Photoplethysmography (PPG) for CA

Photoplethysmography (PPG) signals, harnessed from specialized sensors, represent cardiac activity by quantifying changes in blood volume through light reflection off the skin. PPG sensors necessitate direct contact with the skin for optimal signal accuracy, commonly integrated into wrist-worn wearables such as smartwatches and fitness bands. The ubiquity and affordability of these devices make PPG an ideal focus for our research investigations. Several studies have explored the effectiveness of PPG signals in CA. In a preliminary study [5], researchers examined the suitability of PPG signals for continuous authentication (CA) systems. The findings underscored the discriminative potential of PPG signals, endorsing their use in biometric-based CA systems. A subsequent study [6] explored the efficacy of bio-signals captured from PPG sensors in various smart bands for CA. PPG signals, derived from real-life data and subjected to systematic preprocessing, were input into five classification algorithms (k-Nearest Neighbour, Random Forest (RF), Multi-Layer Perceptron, Logistic Regression, N-Naive Bayes). The experimental setup achieved an impressive Equal Error Rate (EER) of 4.4%, particularly notable when the RF algorithm processed data from the Empatica E4 smartwatch. Another noteworthy work on a PPG-based CA system is Trueheart [12], which utilized Gradient Boosting Tree (GBT) as a matching algorithm, achieving accuracy surpassing 90% and a low False Detection Rate (FDR) of 4%. Another innovative approach [11] combined body movements (captured via motion sensors - accelerometer and gyroscope) with PPG signals for CA. The system achieved an average accuracy rate of 98.5% and an F1-score of 86.67%. Another study [8] proposed a PPG-based authentication leveraging Convolutional Networks to learn features from raw PPG signals directly. The results indicated Area Under the Curve (AUC) values ranging from 78.2% to 86.4% on the PulseId dataset and 73.8% to 83.2% on Troika datasets, with the proposed system exhibiting promising low complexity for real-world CA scenarios.

Notably, our literature review revealed that none of the existing research on continuous authentication has explored the use of an LSTM-based prediction model for CA using PPG signals, presenting a novel direction for our research endeavors.

3   Continuous Driver Authentication: Design Challenges and Assumptions

Every CA system presents a few challenges that must be addressed for practical application in real-world scenarios. In this section, we identify the key design challenges specific to a CA system for driver authentication and the assumptions required to be made by our system.

3.1   CA System Requirements

Continuous Authentication systems, especially in the context of vehicular security, necessitate adherence to specific criteria to ensure efficacy and reliability.

Transparency: Paramount in CA systems is the requirement of transparency or implicitness. An ideal CA system operates unobtrusively, capturing required authentication data without necessitating active user participation. This feature is vital in vehicular environments where driver distraction must be minimized.

Efficiency: Efficiency in a CA system is measured by its response time. The system must process and authenticate data swiftly to ensure security and user convenience. A delay in authentication could compromise the system's usability and security integrity.

Accuracy: The cornerstone of any authentication system, CA systems included, is accuracy. These systems must effectively differentiate between legitimate users and impostors, which is particularly challenging given the typically low quality of biometric signals. Balancing high accuracy with the constraint of minimal preprocessing time is a significant challenge for CA systems in dynamic environments.

Availability: CA systems must authenticate users continuously without interrupting or denying access to legitimate users. Minimizing false negatives is crucial to ensure that legitimate users maintain uninterrupted access, even in the face of technical challenges or signal quality issues.

Lightweight: Given the processing limitations of wearable devices that are used for capturing biometric data, it's imperative that the CA system be lightweight, ensuring it doesn't overburden the computational capabilities of these devices.

Security and Privacy: The CA system must safeguard sensitive data, including biometric and credential information, from unauthorized access and data tampering, upholding the highest data security standards.

3.2   Assumptions

Our system is predicated on several critical assumptions:

- We make the assumption that the driver is wearing a smartwatch, securely attached to the vehicle, which is equipped with Photoplethysmography (PPG) sensors and has the capability to connect via Bluetooth or WiFi. Additionally, we presume that a dedicated application for collecting PPG data is pre-installed on the smartwatch and that it is successfully paired with the device responsible for authentication.
- The system is considered to have an initial layer of authentication, which could be based on passwords, tokens, or biometrics. Also, we assume that the vehicle's ignition is contingent upon the driver wearing the device and successfully passing initial authentication. This first factor of authentication acts as the preliminary security layer.
- Following the initial authentication, a 5-minute stability assumption is made. We assume that the driver cannot change in this duration.
- A secure communication channel, established through the pairing of devices, is assumed to exist between the authentication device and the smartwatch. This channel is essential for safeguarding sensor data during transmission and when stored.

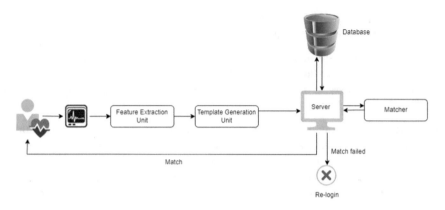

Fig. 1. Continuous Authentication System

- It is assumed that there is no time gap or breaks in the PPG signals received from the sensors. We also assume that there is no delay in detecting an imposter if the smartwatch is transferred from the legitimate driver to an unauthorized individual.

4   SmartDriveAuth: Continuous Driver Authentication Using PPG Wearable Sensors

A typical CA system matches the fresh input templates (query templates consisting of features extracted from biometrics) with the enrolled biometric templates stored in the database, as shown in Fig. 1. The process iterates upon successful matches. Conversely, if a match fails, the user is prompted to perform the initial authentication again and re-login. In this section, we discuss how our SmartDriveAuth is different from a typical CA system. This section outlines the unique features of SmartDriveAuth in contrast to a typical Continuous Authentication (CA) system and elucidates its underlying architecture.

SmartDriveAuth introduces a novel architecture in the domain of Continuous Authentication, specifically for automotive drivers, by harnessing wearable sensor technology, notably smartwatches with PPG sensors. Tailored for the dynamic automotive environment, it non-intrusively gathers PPG signals in real-time to perform continuous driver identity verification.

SmartDriveAuth's versatile architecture seamlessly integrates wearable sensors with a sophisticated Authentication Processing System. This employs Long Short-Term Memory (LSTM) networks for precise analysis of PPG signals (and other physiological data), matcher for comparing predicted and actual values; and a Confidence Computing Unit to calculate the Confidence level of the current system, as shown in Fig. 2. The intricately designed system unifies hardware components, data communication protocols, and advanced computational models to deliver seamless, reliable, and real-time authentication specifically crafted for the automotive realm.

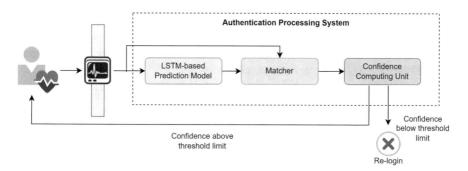

Fig. 2. Architecture of SmartDriveAuth

4.1 Wearable Technology Integration

Central to SmartDriveAuth's architecture are wearable devices equipped with Photoplethysmography (PPG) sensors, commonly found in contemporary smartwatches. These devices are instrumental in capturing continuous physiological data, particularly heart rate signals, vital for the authentication process. The utilization of wearables ensures non-intrusiveness and continuous data collection without any active effort from the driver.

4.2 Data Transmission and Security

The architecture ensures secure transmission of data from the wearable devices to the authentication system, facilitated through wireless communication protocols such as Bluetooth or WiFi. A pivotal aspect of the system is the establishment of a secure channel, which is essential for maintaining the integrity and confidentiality of the biometric data in transit.

4.3 Authentication Processing System

SmartDriveAuth's core architecture revolves around an advanced authentication processing system, shown in Fig. 2. It is comprising of the following components:

1. **LSTM-Based Prediction Model:** Long Short-Term Memory (LSTM) networks are proficient in handling sequential data for time-series prediction, making them particularly suitable for processing the PPG signals from the wearables for predicting the next values with high accuracy. Furthermore, the distinctive capability of LSTM to learn and retain long-term dependencies further ensures the precision of predictions. The PPG data from the wearable is directly fed to the LSTM model. We train the LSTM model using the data received during the period of initial 5 min. The model begins predicting future values after the initial 5 min of training.

2. **Matcher and Confidence Computing Unit:** The predicted values are continuously compared with real-time data from the PPG sensor. The system employs this comparison, considering both current and historical differences, to dynamically calculate the confidence level.

Authentication decisions hinge on the computed confidence level. If the confidence level drops below a predefined threshold, the user is logged out and prompted to undergo initial authentication again. Conversely, if the confidence level is above the threshold, the system continues its operations seamlessly, maintaining the established loop for continuous user validation.

5   Experimentation

5.1   Dataset Selection

In our SmartDriveAuth system, continuous time series Photoplethysmography (PPG) data is utilized to authenticate the identity of a person with a certain level of confidence. PPG readings from smartwatches often face unique challenge of motion artifacts. Existing PPG datasets often lack representation of a variety of activities in real-life settings. To address this, Attila Reiss created a comprehensive dataset under more natural conditions, involving 15 subjects (seven male and eight female, aged 30.60 ± 9.59 years) who performed eight different activities, including driving. Each subject drove for about 15 min following a predefined route, incorporating varied driving conditions. Data was collected using the Empatica E4 wrist-worn device on the non-dominant wrist, sampling PPG data at 64 Hz, a standard rate for commercial smartwatches [9]. The focus for our CA experimentation is on the task of driving. Our experimentation is structured in sets of 225 sample test cases, where the LSTM model, trained on one individual's data, is tested against the data of all 15 participants. The LSTM training and prediction considered a 'lookback parameter' of 64.

5.2   Results and Analysis

In this section, we undertake an analysis and discussion to assess the performance of the proposed system. The experiments run on a laptop and all the functional modules of the proposed system are implemented using Python.

Key Observations on System Performance
Figure 3 shows that our LSTM-based prediction model is performing accurately as after time = 64 (when the system starts predicting subsequent values), the PPG values being predicted align closely with the actual PPG readings from the authorized user. This emphasizes the model's ability to maintain accuracy even with limited available PPG values. Figure 3 also shows the unwavering confidence level at 1 during the authentic user's operation of the vehicle, underscoring the system's reliability. This steadfast confidence provides assurance of consistent

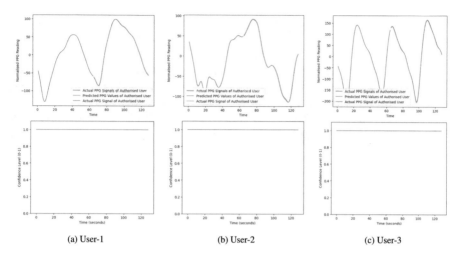

Fig. 3. The precision of predicted PPG values as the authentic user continues to drive the vehicle and the corresponding confidence level

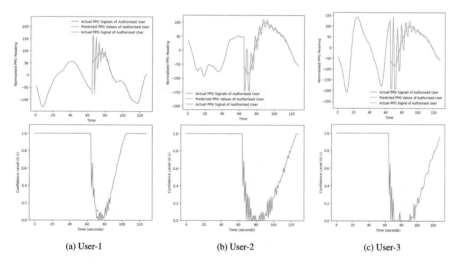

Fig. 4. Disruption in predicted values and the corresponding decline in confidence level when an imposter takes control of the vehicle

performance when operated by the authorized user. Conversely, Fig. 4 vividly illustrates the anomaly in the otherwise accurate prediction of PPG values when an imposter takes control of the system. This emphasizes the pivotal capability of our proposed system to promptly detect and respond to unauthorized access, showcasing its effectiveness in identifying instances where an imposter's data is introduced. Furthermore, The adjoining graph demonstrates a notable drop in confidence levels when the switching of users occur, offering a clear visual representation of the system's response to imposter data. This decline serves as a reliable indicator, allowing for timely detection and alerting when unauthorized input is detected, thereby enhancing the overall security and integrity of the system. It is imperative to promptly generate an alert once the confidence level falls below an acceptable threshold. This quick response is crucial as, over time, the model starts predicting future values based on imposter input data and consequently, the Confidence level rises back to 1. This emphasizes the importance of timely alerts and reporting of impersonation.

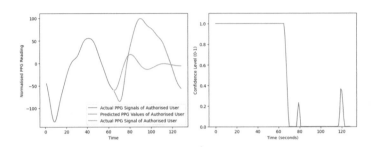

Fig. 5. System performance when previously predicted values are used for prediction of next PPG value instead of actual PPG readings

It's crucial to emphasize that the stated prediction accuracy holds true only when the system receives continuous, uninterrupted data. If there are long breaks or gaps in the data stream, various issues may arise. This is attributed to the practice of utilizing actual, not predicted, values when forecasting the $(n + 1)$th value based on previous inputs. For instance, if the model uses the first 10 values to predict x as the next value, the authentication process involves comparing this prediction with the actual data value y obtained from PPG sensors. However, when predicting the 12th value, the model relies on the actual value y, not the previously predicted one. The system's performance is adversely affected if predicted values are considered (as shown in Fig. 5).

In Fig. 6, 7 the system's performance is evaluated under scenarios where there are gaps in data signals, with 2 and 10 missing values respectively. The results indicate a decline in confidence levels. This is because the system is designed to take the previously predicted values for predicting the next value when actual values are not available. The findings illustrate that broader gaps, extending over multiple seconds or minutes, may lead to a substantial decrease in confidence values, resembling the system's response when confronted with an imposter. This highlights a critical limitation in our approach.

Authentication Performance and Discussion

To evaluate the authentication performance of the proposed CA system, we have utilized the False Positive Rate (FPR), False Negative Rate (FNR), True Positive Rate (TPR), and Equal Error Rate (EER) and plotted the ROC curve. In the

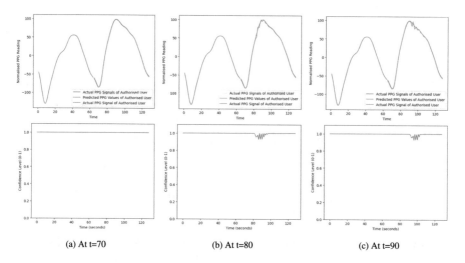

Fig. 6. System performance when 2 values are missing at different time intervals

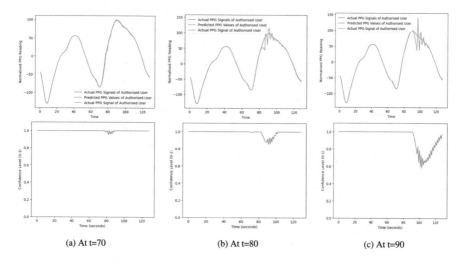

Fig. 7. System performance when 10 values are missing at different time intervals

context of SmartDriveAuth, we adopt the following definitions of True positives, False positives, True negatives and False negatives:

- **False Positives:** Cases where an imposter's data is falsely recognized as genuine. That is, the authentication system is still indicating that an authorized person is driving the car when an imposter has actually taken over.
- **False Negatives:** Cases where the genuine person's data lead to a drop in confidence levels. That is, the authentication system is falsely suggesting that an imposter has taken over even when the authorized user is only driving the car.

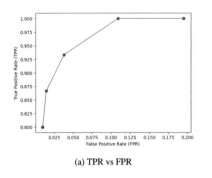
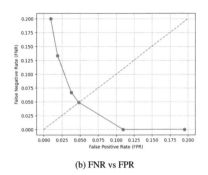

(a) TPR vs FPR   (b) FNR vs FPR

Fig. 8. Receiver Operating Characteristic (ROC) Curve

- **True Positives and True Negatives:** These cases correctly reflects the change or constancy in identity. True positives maintain maximum confidence levels when the individual remained the same, while true negatives showed a significant drop in confidence levels when an imposter has taken over.

Figure 8a and Fig. 8b shows the ROC curve (The relation between the FPR and FNR by varying the acceptable threshold). When setting a threshold, a trade-off requires to be done between security and user acceptability (in terms of availability). In the standard ROC curve, a crossover point emerges where FPR and FNR are equal. That is, false acceptance and false rejection errors hold equal likelihood [2]. A lower crossover point signifies enhanced accuracy in the biometric authentication system. In the depicted ROC curve, this intersection aligns with a threshold of 0.72 (72%). At this threshold, the FPR and FNR for the test across 15 users stand at 0.048285 (4.8285%) and 0.048814 (4.8814%), respectively. That is, our LSTM-based SmartDriveAuth achieved an EER of 4.8% with a lookback of 64.

6   Conclusion and Future Work

Our study demonstrates that the LSTM-based prediction models can be used for CA using PPG signals obtained from wrist-worn wearables. We proposed a CA system wherein predicted values are continuously compared with actual real-time data (received from the sensors), and the calculated differences, thus obtained over time, are used for computing the current confidence level. This confidence level represents the likelihood that the current user is the genuine, authorized user of system resources. Implementing the proposed system with PPG data collected from 15 users (while they were driving), we conducted experiments to establish its accuracy, achieving an EER of 4.8%. Therefore, the system effectively and efficiently detects impersonation attacks. Notably, our proposed system excels in its approach by eliminating the need to permanently store user biometrics and avoiding the need to employ protection against potential attacks or data breaches targeting databases (storing biometrics intended for use as authentication credentials). Moreover, momentary spikes or disruptions in incoming PPG signals have minimal impact on user availability or acceptability, as our confidence computation is not solely dependent on differences in current values of actual and predicted data. Therefore, the system adeptly addresses key challenges of classical driver authentication systems and also remains adaptable to gradual changes in user PPG data. Additionally, the raw PPG data is directly fed to the LSTM model without filtering or pre-processing.

However, it is crucial to acknowledge significant limitations in our system- the assumption that the authorized user will not hand over the vehicle to an imposter or be replaced by one for at least the first 5 min after the initial authentication. Moreover, the system shows a significant decline in confidence levels when there is a prolonged interruption in the continuity of data signals, closely resembling its response in situations involving an imposter. This presents another significant constraint in the real-world applicability of our approach, urging further consideration and refinement for enhanced security and practicality. Also, we had a small dataset (of 15 data subjects) available for our experiments, which limited us in precisely determining the performance of our system. In our future work, we commit to addressing these limitations, striving for further enhancements in the accuracy of our system (reducing the EER to 0 or negligible), and actively working towards making our system more privacy-preserving. Our continuous efforts aim to refine and extend the capabilities of our proposed CA system, ensuring its resilience in diverse and dynamic real-world scenarios.

References

1. Abuhamad, M., Abuhmed, T., Mohaisen, D., Nyang, D.H.: AUToSen: deep-learning-based implicit continuous authentication using smartphone sensors. IEEE Internet Things J. **7**(6), 5008–5020 (2020)
2. Ahmed, A.A.E., Traore, I.: A new biometric technology based on mouse dynamics. IEEE Trans. Dependable Secure Comput. **4**(3), 165–179 (2007)
3. Amini, S., Noroozi, V., Pande, A., Gupte, S., Yu, P.S., Kanich, C.: DeepAuth: a framework for continuous user re-authentication in mobile apps. In: Proceedings of the 27th ACM International Conference on Information and Knowledge Management, pp. 2027–2035 (2018)
4. Baig, A.F., Eskeland, S.: Security, privacy, and usability in continuous authentication: a survey. Sensors **21**(17), 5967 (2021)
5. Bonissi, A., Labati, R.D., Perico, L., Sassi, R., Scotti, F., Sparagino, L.: A preliminary study on continuous authentication methods for photoplethysmographic biometrics. In: 2013 IEEE Workshop on Biometric Measurements and Systems for Security and Medical Applications, pp. 28–33. IEEE (2013)
6. Ekiz, D., Can, Y.S., Dardagan, Y.C., Ersoy, C.: Can a smartband be used for continuous implicit authentication in real life. IEEE Access **8**, 59402–59411 (2020)
7. Hernández-Álvarez, L., de Fuentes, J.M., González-Manzano, L., Hernández Encinas, L.: Privacy-preserving sensor-based continuous authentication and user profiling: a review. Sensors **21**(1), 92 (2020)
8. Luque, J., Cortes, G., Segura, C., Maravilla, A., Esteban, J., Fabregat, J.: End-to-end photopleth YsmographY (PPG) based biometric authentication by using

convolutional neural networks. In: 2018 26th European Signal Processing Conference (EUSIPCO), pp. 538–542. IEEE (2018)

9. Reiss, A., Indlekofer, I., Schmidt, P., Van Laerhoven, K.: Deep PPG: large-scale heart rate estimation with convolutional neural networks. Sensors **19**(14), 3079 (2019)

10. Sahu, A.K., Sharma, S., Raja, R.: Deep learning-based continuous authentication for an IoT-enabled healthcare service. Comput. Electr. Eng. **99**, 107817 (2022)

11. Guannan, W., Wang, J., Zhang, Y., Jiang, S.: A continuous identity authentication scheme based on physiological and behavioral characteristics. Sensors **18**(1), 179 (2018)

12. Zhao, T., Wang, Y., Liu, J., Chen, Y., Cheng, J., Yu, J.: TrueHeart: continuous authentication on wrist-worn wearables using PPG-based biometrics. In: IEEE INFOCOM 2020-IEEE Conference on Computer Communications, pp. 30–39. IEEE (2020)

A Vision Transformer Based Indoor Localization Using CSI Signals in IoT Networks

Gaurav Prasad$^{(\boxtimes)}$, Aditya Gupta, Avnish Aryan, and Sudhir Kumar

Indian Institute of Technology, Patna, Patna, India
{gaurav_2221ee21,aditya_2101ee87,avnish_2101ee85,
sudhir}@iitp.ac.in

Abstract. In recent years, the Channel State Information (CSI) based fingerprint localization method has shown promising growth in locating users indoors. However, deep-learning-based mapping of CSI signals into location remains a challenge due to the signal's complex nature. The existing Convolutional Neural Network (CNN) algorithms are limited in capturing long-range CSI sub-carrier dependency, which represents location-specific information. In this paper, CNN-aided Vision Transformer is considered, which utilizes both local and global structures present in CSI for improved learning. CNN's receptive field captures local structure among CSI sub-carriers aiding Transformer to learn global dependency by utilizing its self-attention mechanism. The proposed method outperforms baseline deep learning models such as CNN and Long-Short Term Memory (LSTM) on public CSI fingerprint testbeds.

1 Introduction

The last decade shows extensive research and development in the area of indoor localization to provide location-based services in Internet of Things (IoT) networks. With the demand for locating users precisely in environments such as Museums and Shopping Malls [1], indoor localization technology needs to advance further over time for precise positioning tasks. Indoor localization approaches are mainly divided into range-based and fingerprint-based methods. The latter shows promising results in handling complex indoor environments such as Non-Line of Sight (NLOS) cases [2] and includes two modes: offline and online. In offline mode, signals at known reference locations are collected to create a radio map and during the online mode, features from unknown locations are matched with the collected radio map to locate the user.

Out of all available mainstream wireless signal measurements, Received Signal Strength (RSS) and CSI are widely adopted for fingerprint-based indoor positioning systems. However, RSS contains coarse channel information and is inherited to multi-path effect and temporal changes, which limits its accuracy [3]. Whereas, CSI is more stable and provides fine-grained channel information across sub-carriers [2–4]. The CSI-based methods carry sub-carrier amplitude and phase information of the propagation environment, thus leading to precise location-based positioning. Thus, feature learning methods accurately modeling the relationship between multiple diverse CSI sub-carriers and distances is the need to enhance localization performance [3,4].

© The Author(s), under exclusive license to Springer Nature Switzerland AG 2024
L. Barolli (Ed.): AINA 2024, LNDECT 201, pp. 73–83, 2024.
https://doi.org/10.1007/978-3-031-57870-0_7

The structure of the paper is as follows: Sect. 1.1 and 1.2, represent the related work in CSI-based fingerprint methods and our contributions respectively, Sect. 2 explained the proposed model structure followed by self-attention mechanism, Sect. 3 represents the experiment performed on a public testbed and Sect. 4 concludes our work with future aspects.

1.1 Related Work and Motivation

In recent years, deep learning-based indoor localization algorithm [5–8] have shown promising results with CNN, its variants [5,6], and Recurrent Neural Network (RNN) models such as LSTM [7], Bi-directional (Bi-LSTM) [8] to extract useful features from CSI. However, the existing CNN-based model shows below-par performance due to its limited receptive field [9], whereas the RNN model only extracts global features and neglects the local critical location information. Also, LSTM-based models [7,8] learn long-range dependency but are bounded by input sequence length [12]. However, a few recent works [10,11] consider the attention mechanism to capture local sub-carrier features for improved performance.

With Intel 5300 NIC cards, Channel Frequency Response (CFR) is collected within Wi-Fi bandwidth as CSI [13]. It includes both amplitude and phase response as fingerprints to be utilized for indoor localization. However, in [10,22], only amplitude [10] and phase [22] responses are separately considered as features, and on capturing non-neighboring sub-carrier relationships, these methods show better performance. This motivates us to model both local and global dependencies in CSI fingerprints by considering amplitude and phase responses for accuracy improvement.

Recently, the Vision Transformer (ViT) deep learning model has shown effective performance in Computer Vision applications due to its self-attention mechanism. In ViT, self-attention directly captures the connection between each part of the sequence to learn long-term dependency. However, the absence of inductive bias, such as locality and translation equivariance in comparison to CNN's, limits its performance [14]. Earlier, transformer-based localization approaches [15] utilize CSI as a sequence of observations similar to word sequence in a sentence. This increases its complexity and requires past sequence CSI values to accurately locate a user in a fixed-defined path only.

1.2 Contributions

In this paper, we proposed CViTLoc, a CNN-aided ViT-based localization method utilizing both the local feature extraction property of CNN and the global preference of ViT. As per the author's knowledge, Vision Transformer is not considered for CSI-based localization tasks, adding a novelty to our work. We consider the Convolutional Token Embedding approach borrowed from [21] for the combined model. The major contributions of our paper are as follows:

1. The combined CNN and Vision Transformer effectively learn both local and global location-specific information present in CSI signals. The Linear projection layer of the Vision Transformer is replaced with the Convolution layer to capture the local structure between neighboring sub-carriers which are highly correlated.

2. Our model, in comparison with other baseline algorithms, shows better performance on public CSI testbed with localization error within the centimeter level.

2   Combined CNN and Vision Transformer Model

This section provides a brief theory on Channel State Information and an overview of the proposed CViTLoc model. With CNN and ViT blocks operation, we also explain the self-attention mechanism considered.

2.1   System Model

We consider 'L' links between transmitter and receiver antennas, which receive OFDM signal from the user device to nearby Access Points (APs) as base-station [16]. Orthogonal Frequency Division Multiplexing (OFDM) provides CSI as channel responses over 'K' sub-carriers, which includes amplitude and phase response for each sub-carrier. The channel response (in frequency domain) H is given in matrix form as:

$$H = \begin{bmatrix} h_0^1 & h_1^1 & h_2^1 & \dots & h_{K-1}^1 \\ \vdots & \vdots & \vdots & \ddots & \vdots \\ h_0^L & h_1^L & h_2^L & \dots & h_{K-1}^L \end{bmatrix} \in \mathbb{C}^{L \times K} \tag{1}$$

where, $h_k^l = |h_k|e^{j\theta_k}$ represents complex frequency response for each kth sub-carrier in l^{th} link, and $|h_k|$ and θ_k represents amplitude and phase response of sub-carrier respectively. The Channel Frequency Response as CSI is our training dataset collected over defined reference locations. The goal is to learn an efficient mapping between CSI samples and location information.

Also, CSI data are pre-processed as adapted by [17], which includes phase unwrapping and outlier removal from phase information. In addition, de-noising is performed to remove high-frequency noise in amplitude and phase of each sub-carrier by a moving average window.

2.2   Structure of Proposed CNN-Vision Transformer Model

This sub-section provides an overview of the proposed CViTLoc model, with CNN and ViT blocks followed by a fully connected prediction layer as shown in Fig. 1.

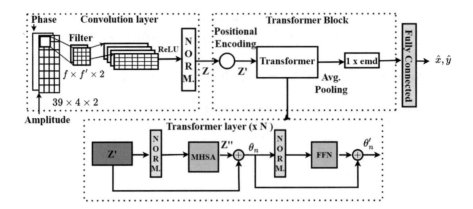

Fig. 1. The combined Convolutional Neural Network and Vision Transformer model.

Convolution Neural Network: The pre-processed CSI data with size $[K \times L \times 2]$ are considered as two channels representing amplitude and phase. The CSI are transformed using convolution operation into $[\frac{K-f}{s}+1, \frac{L-f'}{s}+1, n_f]$ with filter size $[f \times f' \times 2]$ where n_f and s are number of filters and stride respectively. The CNN layer captures local features between neighboring sub-carriers by utilizing local receptive fields and shared weights [18]. To further improve CNN feature representation, Rectified Linear Unit (ReLU) is considered, which adds sparsity and non-linearity in learning, whereas layer normalization (Norm.) alleviates internal co-variate shift issues [19].

Vision Transformer: In Vision Transformer, the two-dimensional input is flattened into patches and then linear projected into a non-overlapping one-dimensional sequence called token embeddings [14]. In this paper, we are substituting Linear Projection with Convolution projection of filter size $f' = L$. This transformed 2-D CSI matrix $[K \times L \times 2]$ into size $[\frac{K-f}{s}+1, 1, n_f]$, which is directly squeezed into non-overlapping sequence as $\mathbf{Z} = [z_0, z_1 \ldots z_{[\frac{K-f}{s}+1]}]^T \in \mathbb{R}^{[\frac{K-f}{s}+1] \times n_f}$, where each 1-D vector $\mathbf{z_i} \in \mathbb{R}^{1 \times n_f}$ is our token embedding, here n_f is also represented as embedding size (emd). The CSI tokens are further encoded by positional embedding only without class token as in [14], this provides spatial identity to each convoluted CSI fine-grained element 'z_i'. Each z_i is mapped into a unique embedding vector in embedding matrix $\mathbf{E}_{pos}$ which is updated during model training as in [14]:

$$\mathbf{Z}' = \mathbf{Z} + \mathbf{E}_{pos}, \; \mathbf{E}_{pos} \in \mathbb{R}^{(\frac{K-f}{s}+1) \times emd} \tag{2}$$

The encoded CSI sequence $\mathbf{Z}'$ is followed by 'N' Transformer blocks, with each block consisting of one Multi-Head Self-Attention (MHSA) layer and a feed-forward layer to utilize the transformer's global context fusion. The MHSA block includes M self-attention blocks, with each block having its learnable weight matrices $[W^{Q_m}, W^{K_m}, W^{V_m}]$, where $m = 0, 1, \ldots (M-1)$. The encoded CSI sequence is projected onto these weight matrices into $Q = \mathbf{Z}'W^{Q_m}$, $K = \mathbf{Z}'W^{K_m}$ and $V = \mathbf{Z}'W^{V_m}$. The output of each self-attention layer is given by $\mathbf{Z}'' = Softmax(\frac{QK^T}{\sqrt{d_k}}V)$ which provides the attention scores [20]. Further, each self-attention block output is concatenated into the single matrix with projected weight matrix W_o. The MHSA operation is represented as [20]:

$$\theta_n = \text{MHSA}\left(\text{LN}(Z''_{n-1})\right) + Z''_{n-1}, \ n = 1, ..., N \tag{3}$$

where LN represents layer normalization. Further, θ_n is followed by layer normalization (LN) into a feed-forward Neural Network in every n^{th} Transformer block:

$$\theta'_n = \text{MLP}\left(\text{LN}(\theta_n)\right) + \theta_n, \ n = 1, ..., N \tag{4}$$

The Transformer layer output is average pooled into $[1 \times emd]$ and at last, a fully connected feed-forward (FFN) layer makes the final location prediction $[\hat{x}, \hat{y}]$ utilizing Transformer block output as:

$$[\hat{x}, \hat{y}] = \text{FFN}(\theta') \tag{5}$$

We consider the location prediction task in the regression setting with Mean Squared Error (MSE) as our loss function and update all the model weights by the backpropagation algorithm.

3 Experiment and Results

In this section, we consider a public OpenCSI fingerprint testbed [17] for localization in an indoor environment of area 5×3.5 square meters (m^2). CSI fingerprints are collected from 3983 reference locations, with roughly 1000 samples collected from locations spaced one centimeter (cm) apart. Figure 2 shows CSI amplitude variation at two locations, A and B. The amplitude response for 4 links (channels) varies at each location, which is exploited by the model for learning.

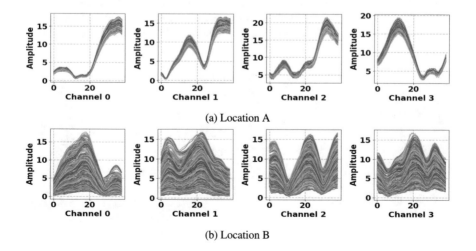

(a) Location A

(b) Location B

Fig. 2. CSI Amplitude variation

The long-range dependency is also analyzed by considering amplitude variation between sub-carriers for all the samples collected at one location. In Fig. 3, the nearby sub-carriers amplitude are correlated and represent a unique relationship with non-neighboring sub-carriers for different locations. To evaluate the localization performance of our model, Median Localization Error (MLE) is considered, defined as the median of differences between actual and estimated locations and is less sensitive to outliers.

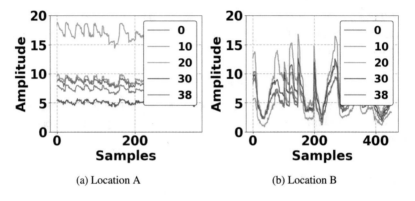

(a) Location A   (b) Location B

Fig. 3. Long-range dependency among sub-carriers.

3.1   Model Training

The pre-processed CSI samples collected from 3983 locations are considered for training the proposed network. In total, we used 3585 locations for training, 179 locations for validation and another 179 locations for testing. We train the proposed model using Adam optimizer with a batch size of 512, learning rate of 0.001, and hyper-parameters

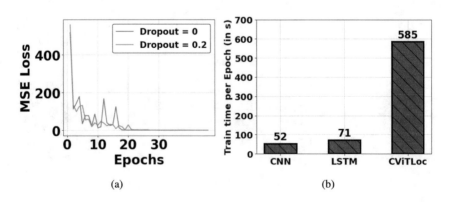

(a)   (b)

Fig. 4. (a) Model training plot (b) Training Time Comparison

value $\beta_1 = 0.9$, $\beta_2 = 0.99$. On considering reduced learning rate on plateau scheduler for training, with a patience level of 5 epoch results in optimum performance. We train the model till the mean squared loss converges with increasing epochs, as shown in Fig. 4a. The figure also represents dropout variation as a regularizer to prevent over-fitting of the model, however, we observe the network converges better when not considered. We consider hyper-parameter selection by fixing all other parameters except the one and consider the same approach for others to find the optimum value.

We also compute the training time required to complete one epoch, with our model consuming longer training time due to the self-attention mechanism which requires high computational cost on capturing complex relationships as shown in Fig. 4b.

3.2 Model Training Parameters

On training, we consider batch size, filter size, number of filters, transformer blocks, and size of feed-forward network as our parameters for tuning. Batch size selection leads to better generalization and achieves global optimum convergence. We observe the best model performance with 512 batch of samples, kernel size of 5×4, 4 transformer blocks layers, 8 number of multi-head, and 128 neurons in the Transformer feed-forward layer with least median localization error of 2.46 cm. In addition, the number of CNN feature embeddings was fixed to 512, however, increasing embedding size further improves accuracy but leads to increased complexity of the model and may lead to over-fitting.

3.3 Ablation Study

CNN Kernel Variation: On considering different kernel sizes $f \times 4$ for CNN model as shown in Fig. 6c where f represents the width of filter and captures consecutive sub-carriers. By considering, $f = 5$ shows the best performance among all other combinations with improved training and localization error. By considering Convolution projection instead of linear, had also showed improved performance for image-based classification tasks [21].

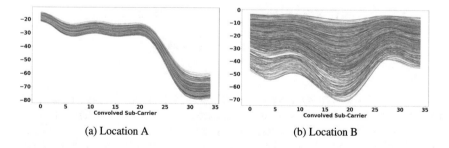

(a) Location A   (b) Location B

Fig. 5. CNN embedding with the captured local structure for each location A and B.

Also, we represent the embedding from the CNN block end which shows that the local structure captured leads to improved representation in comparison to raw CSI values (Fig. 2). Here, the kernel size is 5 which records a better representation as shown in Fig. 5.

Number of Filters Variation (n_f): Figure 6d represents the effect of the number of filters on localization accuracy. By varying the number of filters, the model captures different features from input CSI which results in improved learning. For smaller n_f the network is unstable and does not learn meaningful CSI representation. We vary n_f between 128 to 512 yielding good convergence with the least error for $n_f = 512$.

Number of Transformer Encoder Layers: We also consider the number of transformer encoder layers variation which depends on the size of data and complexity at hand. For CSI data, we start with a single transformer layer and increase further as shown in Fig. 6a. We achieve the best performance by stacking 4 transformer blocks.

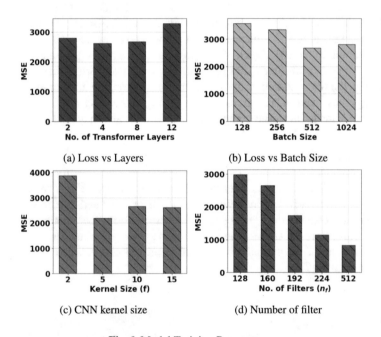

Fig. 6. Model Training Parameters.

t-SNE Plot for Transformer Embedding: To visualize the model performance on the prediction task, we plot a t-distributed stochastic neighboring (t-SNE) plot for Transformer layer embedding. t-SNE used as a dimension reduction method reduces high dimensional 512 embedding into two dimensions. We consider five different location CSI samples to generate embeddings from the trained model. The t-SNE plot shows our model was able to separate five locations in Fig. 7a.

3.4 Results

Figure 7b represents a comparison with baseline deep learning models such as CNN and LSTM using the same test samples. The CNN architecture is the same as OpenCSI [15] with two convolutions and a fully connected layer, which results in MLE of 65 cm for test samples. The network is trained using Adam optimizer with a drop-out value 0.5 and early stopping is considered to avoid over-fitting. As CNN model ignores the global structure present in CSI samples resulting in coarse learning of features and only exploits nearby sub-carrier information.

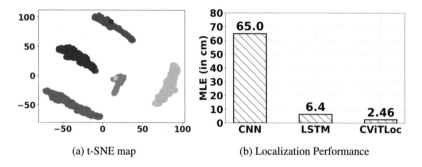

(a) t-SNE map   (b) Localization Performance

Fig. 7. (a) represents t-SNE map for five location embeddings, (b) localization error comparison with baseline model.

For, LSTM we consider the similar architecture as in [7] after taking the Inverse Fast Fourier Transform (IFFT) of frequency domain CSI, which is converted into time domain CSI features. We consider two LSTM layers followed by dense layers to provide the estimated locations in the last layer. The model is trained considering the Adam optimizer with a learning rate 0.001 which provides improved performance. We also consider reducing the learning rate on the plateau as learning rate scheduler for training, on testing we achieve MLE of 6.41 cm. LSTMs are Recurrent Neural Networks (RNN) that compress the past and future sub-carrier information into a vector to reduce the vanishing gradient problem in RNN using a gating mechanism. Also, LSTM uses a forget gate in managing the information through the network by selectively remembering and forgetting information from previous time steps. However, this results in the loss of long-range CSI information as representation is restricted to a single vector.

In comparison, our proposed method effectively utilizes each CSI fine-grained information by learning local as well as long-range dependency and adaptively determines their importance with each sub-carrier information. We also achieve an improved localization error of 2.46 cm using the CViTLoc model, the best among all the results. Also, a similar approach is used for Received Signal Strength (RSS) based localization, with limited accuracy due to the challenge associated with AP sequence selection for spatial representation [23].

4   Conclusion

This paper presents a novel deep-learning method to enhance CSI feature learning for indoor localization. The proposed method accurately detects location in a lab environment within cm level accuracy. Additionally, it shows Vision Transformers is effective in learning long-range dependencies with an improved localization error of 2.46 cm in comparison to other baseline deep learning methods. Further, a detailed ablation study shows the effectiveness of our method supported by a t-SNE embedding plot. The future work includes reducing the model size as well as computational complexity for end-device applications.

Acknowledgment. This work is supported in part by the Department of Science and Technology, Government of India, under Grant NGP/GTD/Sudhir/IITPatna/BR/05/2022 and in part by the Science and Engineering Research Board (SERB), Government of India, Project MTR/2021/000380.

References

1. Sen, S., Radunovic, B., Choudhury, R.R., Minka, T.: You are facing the Mona Lisa: spot localization using PHY layer information. In: Proceedings of the 10th International Conference on Mobile Systems, Applications, and Services, pp. 183–196 (2012)
2. Liu, W., et al.: Survey on CSI-based indoor positioning systems and recent advances. In: International Conference on Indoor Positioning and Indoor Navigation (IPIN) 2019, pp. 1–8 (2019)
3. Liu, M., Liao, X., Gao, Z., Li, Q.: FT-Loc: a fine-grained temporal features based fusion network for indoor localization. IEEE Internet Things J. **11**(3), 4324–4334 (2023)
4. Yang, Z., Zhou, Z., Liu, Y.: From RSSI to CSI: indoor localization via channel response. ACM Comput. Surv. (CSUR) **46**(2), 1–32 (2013)
5. Wang, X., Gao, L., Mao, S., Pandey, S.: CSI-based fingerprinting for indoor localization: a deep learning approach. IEEE Trans. Veh. Technol. **66**(1), 763–776 (2017)
6. Hsieh, C.-H., Chen, J.-Y., Nien, B.-H.: Deep learning-based indoor localization using received signal strength and channel state information. IEEE Access **7**, 33256–33267 (2019)
7. Zhang, Y., Qu, C., Wang, Y.: An indoor positioning method based on CSI by using features optimization mechanism with LSTM. IEEE Sens. J. **20**(9), 4868–4878 (2020)
8. Panda, A.K., Manikandan, M.S., Ramkumar, B.: Bidirectional LSTM based sequence classification approach for human presence detection using channel state information. In: 2023 International Conference on Microwave, Optical, and Communication Engineering (ICMOCE), pp. 1–6 (2023)
9. Luo, W., Li, Y., Urtasun, R., Zemel, R.: Understanding the effective receptive field in deep convolutional neural networks. In: Advances in Neural Information Processing Systems, vol. 29 (2016)
10. Xiao, J., Wu, K., Yi, Y., Ni, L.M.: FIFS: fine-grained indoor fingerprinting system. In: 2012 21st International Conference on Computer Communications and Networks (ICCCN), pp. 1–7 (2012)
11. Zhang, B., Sifaou, H., Li, G.Y.: CSI-fingerprinting indoor localization via attention-augmented residual convolutional neural network. IEEE Trans. Wirel. Commun. **22**(8), 5583–5597 (2023)
12. Dai, Z., Yang, Z., Yang, Y., Carbonell, J., Le, Q.V., Salakhutdinov, R.: Transformer-XL: attentive language models beyond a fixed-length context. arXiv preprint arXiv:1901.02860 (2019)

13. Halperin, D., Hu, W., Sheth, A., Wetherall, D.: Predictable 802.11 packet delivery from wireless channel measurements. ACM SIGCOMM Comput. Commun. Rev. **40**(4), 159–170 (2010)
14. Dosovitskiy, A., et al.: An image is worth 16×16 words: transformers for image recognition at scale. arXiv preprint arXiv:2010.11929 (2020)
15. Zhang, Z., Du, H., Choi, S., Cho, S.H.: TIPS: transformer based indoor positioning system using both CSI and DoA of WiFi signal. IEEE Access **10**, 111363–111376 (2022)
16. Yang, Z., Zhou, Z., Liu, Y.: From RSSI to CSI: indoor localization via channel response. ACM Comput. Surv. (CSUR) **46**(2), 1–32 (2013)
17. Gassner, A., Musat, C., Rusu, A., Burg, A.: OpenCSI: an open-source dataset for indoor localization using CSI-based fingerprinting. arXiv preprint arXiv:2104.07963 (2021)
18. LeCun, Y., Haffner, P., Bottou, L., Bengio, Y.: Object recognition with gradient-based learning. In: Forsyth, D.A., Mundy, J.L., di Gesu, V., Cipolla, R. (eds.) Shape, Contour and Grouping in Computer Vision, vol. 1681, pp. 319–345. Springer, Heidelberg (1999). https://doi.org/10.1007/3-540-46805-6_19
19. Zhao, G., Zhang, Z., Guan, H., Tang, P., Wang, J.: Rethinking ReLU to train better CNNs. In: 2018 24th International Conference on Pattern Recognition (ICPR), pp. 603–608 (2018)
20. Vaswani, A., et al.: Attention is all you need. In: Advances in Neural Information Processing Systems, vol. 30 (2017)
21. Wu, H., et al.: CVT: introducing convolutions to vision transformers. In: Proceedings of the IEEE/CVF International Conference on Computer Vision, pp. 22–31 (2021)
22. Foliadis, A., Garcia, M.H.C., Stirling-Gallacher, R.A., Thomä, R.S.: CSI-based localization with CNNs exploiting phase information. In: IEEE Wireless Communications and Networking Conference (WCNC), Nanjing, China, pp. 1–6 (2021)
23. Savin, N.: Exploring indoor localization with transformer-based models: a CNN-transformer hybrid approach for WiFi fingerprinting. BS thesis, University of Twente (2023)

Deep Reinforcement Learning for VNF Placement and Chaining of Cloud Network Services

Wided Khemili[1,2], Jalel Eddine Hajlaoui[1], Mohand Yazid Saidi[2(✉)],
Mohamed Nazih Omri[3], and Ken Chen[2]

[1] University of Sousse, Street of Khalifa Karwi, Sahloul 4, 4002 Sousse, Tunisia
[2] L2TI - Institut Galilée, Université Sorbonne Paris Nord, 93430 Villetaneuse, France
{saidi,ken.chen}@univ-paris13.fr
[3] MARS Laboratory, University of Sousse, Sousse, Tunisia
mohamednazih.omri@eniso.u-sousse.tn

Abstract. With Network Function Virtualization (NFV), network services can be swiftly and efficiently constructed by instantiating Virtual Network Functions (VNFs) on host servers. In order to optimize resource utilization (such as bandwidth) and meet Service-Level Agreements (end-to-end delay for instance), efficient solutions are required for VNF placement, consolidation, and chaining. To tackle these challenges, we propose two approaches: (1) a novel Formal Concept Analysis (FCA)-based method focused on placing VNFs on appropriate virtual machines, and (2) a Deep Reinforcement Learning-based approach that integrates two parallel modules for VNF placement and chaining: Markov Decision Process (MDP) and Long Short-Term Memory (LSTM).

The parallel operation of these modules enables the extraction and capture of the current NFV environment and historical transitions. In our proposal, we employ Policy Gradient for agent training, aiming to identify suitable hosts for each VNF in the Service Function Chain (SFC), thereby enhancing various quality metrics such as latency, resource cost, and throughput. Simulation results show the effectiveness of our approach, achieving a 47% improvement in rewards compared to the deep VNF approach and a 52% improvement compared to the First Fit approach.

Keywords: Network Function Virtualization · VNF Placement and Chaining · VNF Consolidation · resource allocation · Service-Level Agreement

1 Introduction

Cloud computing represented a big leap in technological evolution. Where virtualization is considered the main driver of cloud computing, which serves to separate physical resources to meet the demands of a gigantic data center. Network Function Virtualization (NFV) creates virtual copies of network services (Firewall, router, Deep Packet Inspection (DPI)) and run them on virtual machines.

L. Barolli (Ed.): AINA 2024, LNDECT 201, pp. 84–96, 2024.
https://doi.org/10.1007/978-3-031-57870-0_8

In addition, NFV works to facilitate the execution and distribution of virtualized network functions (VNFs) on different servers or move and migrate them dynamically from server to server according to request, ie anywhere on the network. In another sense, it allows to create hybrid constructor where the network functions and resources can coexist and consolidate from one host machine to another as needed. This flexibility and dynamism contributes to the possibility of controlling and managing virtual resources easily to optimize cost and energy consumption. These features provide a suitable environment to exploit and apply placement and consolidation algorithms which is mainly a migration [2] of virtualized network functions or resources from one host or source machine to another according to their requirements and capacity. Indeed, the emergence of new technologies that require a very high network throughput, reduced delay and latency (particularly for real-time applications like Voice over IP, telemedicine, etc.) presents a new challenge and requires the finding of suitable solutions and harmonics with the problem of consolidation and placement of VNF. To confront this challenge, several placement and consolidation approaches based on different resources. But, most of this research has focused on gain and minimizing cost before quality of service QoS [10,17]. Despite the numerous lines of research dedicated to addressing the consolidation and placement of Virtual Network Functions (VNFs), certain issues remain either inadequately addressed or overlooked. One such overlooked aspect is the VNF chaining problem, which intricately ties in with the challenges of VNF placement and consolidation.

These observations highlight the need for more comprehensive research endeavors that are cognizant of the intricacies of this problem. Further exploration is justified, marked by increased experience and heightened awareness to comprehensively address the multifaceted nature of the VNF consolidation and placement challenges.

As the Service Function Chain (SFC) structure consists of a set of VNFs linked in a chain according to a predefined order, it poses another problem of chaining in parallel with placement and consolidation. This chaining imposes new rules for the placement of VNFs which require compliance with a predefined order of a set of VNFs to deploy the expected network service and to respond to an SFC request.

In this context, the paper's remaining sections are organized as follows: In Sect. 2, we present significant works related to the VNF placement and chaining problem. We define and formulate the VNF placement problem in Sect. 3. Then, we formulate and present the chaining issue as extension of VNF placement problem in Sect 4. Section 5 describes FCA methodology and deep learning-based solution we've suggested and its strategy. Section 6 illustrates and supports the effectiveness of our FCA and deep learning based technique compared to the current approaches. Section 8 serves as the paper's conclusion.

2 Related Work

In fact, the growing demand on telecommunication networks represent the first reason for the advent of virtualization. NFV creates virtual copies of network services and run them on virtual machines. In addition, NFV works to facilitate the execution and distribution of virtualized network functions on different servers or move and migrate them dynamically from server to server according to request [19]. This characteristics provides a suitable environment of the consolidation which contributes to reduce the number of active physical devices and therefore the number of virtual devices, to improve the exploitation of the resources and to minimize the energy consumed. [22] tackles the problem of consolidation under interference constraint. We notice that the dimensions or the members of consolidation are different from one approach to another. In [18], the consolidation is performed between the VM and the server, [15, 23] between VM and tasks, [16] between VNFs and servers, [9] between VNFs and VMs, in [5, 6] between the Software Licenses. The proposed approaches are also characterized by the use of different classes of algorithms, for example combined algorithm VCMM(Virtual consolidation method) proposed in [16], heuristics [3, 4], genetic algorithms [8, 20], "Gossiping" [11], etc. Some approaches indicate the use and exploiting of resources by tackling only one or some resources. However, some works do not consider the importance of this factor which also requires a large amount of energy. As an extension of the VNF placement problem, it cannot solve the VNF placement problem without taking into account the chaining characteristics of the VNFs. A flow of VNFs is routed in chains according to a predefined order, which constitutes a network service chain called SFC [13, 14]. Furthermore, the flow on the router and switch is a challenge at the level of the control unit (SDN) [12, 16]. Thus, the growing traffic requires a high-performance routing strategy to minimize the routing computation time. Lately, a multiplicity of academic research that has addressed the problem of VNF selection and chaining. In [23], Coa et al developed a log-competitive online COATS algorithm to direct traffic in an SDN network. This algorithm aims to control the traffic by considering the time of arrival and departure of traffic. However, it does not address the bandwidth availability issue considering the VNF instance multiplicity. In [7, 21] a new strategy based on deep learning has been proposed to solve the VNF selection and chaining problem to minimize the end-to-end delay and get the best SFC routing path. Moreover, these two studies [21, 23] focus on the role of SDN to control traffic in multi-instance VNF network. [1] focused on improving shared resource utilization of edge servers and physical links within latency bounds. The variety of this research proves that the VNF placement and chaining problem is a complex optimization problem that requires in-depth study.

3   Formulation and Description of VNF Placement and Chaining Problem

3.1   Description

To understand the issue of placement and chaining of VNF in SFC of requested service, we divide it into two unseparated problems, one is the extension of the other. The first problem consists in the placement and consolidation of VNF in the most appropriate VM to minimize the number of active virtual machines and to put the machines which are emptied in standby in order to minimize the energy consumed at the cloud data centers and to minimize the latency and the cost of resources. The second problem consists in placing the VNFs in a chain according to a predefined order to meet the requested service requirements (SFC). The traffic and direction of the VNFs is determined by taking into consideration the multiplicity of VNF instances, the congestion at the link level (according to the bandwidth) and at the VM node level (according to the memory and storage capacity). Our SFC is involved four VNFs used for network security which are: Firewall (FW) which monitors traffic to ensure network security, deep packet inspection (DPI) which is an application used to process the packet deep in the load of IP data, intrusion prevention system (IPS) is a framework that provides a security system checking errors, and deception system (DS) is a system that defends hacker attacks based on the identification of the signature of hackers.

3.2 Problem Formulation

A. Placement: To formulate the VNF placement problem, suppose our system is a context $K = (VM, VNF, I)$. $VM = \{vm_1, vm_2, ...vm_i\}$ is the set of objects, $VNF = \{vnf1, vnf2, ...vnf_j\}$, and $I \in (VM, VNF)$ is the binary relationship between VM objects and VNF attributes. A concept of context (VM, VNF, I) is a couple of object and attribute (vm1, vnf1) where, $vm1 \subseteq VM$ and $vnf1 \subseteq VNF$ which verifies the application $vm' = vnf$ and $vnf' = vm$. For a concept (vm, vnf), vm is said to be its extension and vnf its intention. Let app be the function of placing a VNF partition in virtual machines with VM × VNF → app. We define VNF placement problem formally by the next triplet:

$$PPr_{VNF} \equiv \prec VM, VNF, app \succ \tag{1}$$

The app application represents the operation of placing a set of VNFs into a set of VMs. Each vm_i requires a vnf_j to run. After the identification and formulation of the problem, we will apply the approach based on the FCA methodology to extract the concept (vm_i, vnf_j) which indicates the best placement of VNF in VM.

B. Chaining: After solving the VNF placement problem, we extract the candidate VNFs to use them to deploy the requested network service according to a predefined order. We formulate the chaining problem as follows.

Let be a model of an undirect network graph $G = (V, E)$, V is the node set VM, E is the edge set. Each $VM_i \in V$ is equipped with $R = 1, 2, 3$ resources (memory, storage, cpu). F represents the set of virtualized network functions (VNFs) where $F = f1, f2, f_i..., f_t$ which is characterized by latency $d_{i,j}$ between node i and j, where $E \in (i, j)$. Let S be the service function chain (SFC) which contains a series of VNFs f_i which are placed according to a predefined order, where f is the i_th VNF in S and t is the total number of VNFs in S. Service function chain request S is defined by R, this request has a source node v and destination node v. We consider N_f as the set of candidate VM nodes for hosting VNF. The binary value $x_{i,j}$ shows if the edge e on the path from ingress node v to egress node v. Also, the binary value y identifies if VNF f is mapped to node v.

4   FCA Based Approach

4.1   Placement and Consolidation Strategy

Generally, the strategy proposed is based on three main steps of FCA: (i) the first step consists on the organization of the VNF according to a hierarchy, described by a lattice of concepts, in order to group the VNF and the dependent NS in the same concept. (ii) the second step is the selection of the candidate concepts (cc.int) of FCA and to attach them and use as VNF to the second process of PBDRL and (iii) the third step is to construct the lattice of Galois, where the circle represents the concept, and the arcs between the rectangles express the relationship from the more general (above) to the more specific (below). The Galois lattice aims to eliminate unnecessary and redundant concepts without losing information.

4.2   FCA (Formal Concept Analysis)

We apply FCA to place VNFs in the minimum and most suitable VM. We use the context $K = (VM, VNF, I)$ which is defined in Sect. 3.2 according to Table 1. Given a concept $C = (VNF, VM)$ of context K, we define the placement problem by the triplet in Eq. (1) and solve the rest of the placement problem as follows.

1. Determine the dependency between two VNFs jointly dependent on one or more VMs:

$$Dependency_{vnfi,vnfj}^{VNF} = Count(vm_{vnf_i} \cap vm_{vnf_j}) \qquad (2)$$

This dependency is explained by the number of network service VM that use both vnf_i and vnf_j. Conversely, the dependency between two VM jointly dependent on one or more VNF which can formulate as follows:

$$Dependency_{vm_i,vm_j}^{VM} = Count(vnf_{vm_i} \cap vnf_{vm_j}) \qquad (3)$$

To extract the candidate concept, we first define the concept weight and then the maximum coverage. Let c_i be any concept, where $ci.Int$ is the intention of the concept c_i. The weight of a concept c_i, denoted by $W(c_i)$, is written as follows:

$$W(c_i) = \frac{|c_i.Int|}{|vnf_i|} \qquad (4)$$

And the maximum coverage that covers the set of V M objects denoted by MC is defined as follows:

$$MC = \begin{cases} 1 & if \cup_{i \in I} c_{i.Int=1} \\ 0 & else \end{cases} \quad (5)$$

Let $C = cc_1, ..., cc_p$ be a set of concept. We define the candidate concept, that has a maximum weight and cover all VNF, noted by $Cand(CC)$, as follows:

$$Cand(cc) = \begin{cases} 1 & if \sum_{k=0} W(cc_p) = 1 and MC(cc) = 1 \\ 0 & else \end{cases} \quad (6)$$

Table 1. Table of context

O/A	vnf_1	vnf_2	vnf_3	vnf_4
vm_1	X		X	
vm_2	X	X		
vm_3			X	X
vm_4	X			

5 Deep Reinforcement Based Approach: PBDRL

In this section, we describe our deep reinforcement learning-based proposal which is based on two parallel modules LSTM and MDP (see Fig. 1). This approach is presented as an extension of the FCA approach using candidate VNFs as the input for the deep neural network used in the NFV environment. DRL is a combination of Deep Learning and Reinforcement Learning (RL). This combination

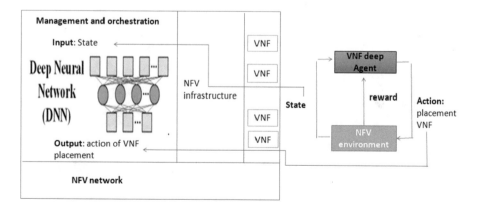

Fig. 1. Deep reinforcement learning as a combination of deep neural network (DNN) and reinforcement learning (RL).

makes it possible to adapt the proposed environment (NFV) of Reinforcement Learning with the characteristics of neural network to transform the set of service input function chain requested (set of VNF in predefined order) into a set of outputs for deploy SFC while respecting the constraints of available capacities and resources. Reinforcement Learning allows control of inputs through the agent receiving the VNF of SFC in which an agent learns to make decisions through trial and error. MDP is used to extract motion state from included NFV environment transitions (state, action, and reward) to be learned. LSTM is used to extract the motion state of historical transitions. We combine MDP and LSTM to extract the S_{new} state. At each timestep, the agent receives a $s_n ew$ state, takes an action a, receives a reward and transitions to next state according to the dynamic characteristics of the environment. DRL algorithms are based on deep learning for the high-dimensional MDP problem where $\pi(a/s_{new})$ represents the policy or function acquired as a neural network.

6   VNF Placement and Chaining Approach for Requested SFC Based on Different Timeslots

NFV network dynamic transitions and movements are expressed in different time slots. At each time slot (i.e. time step), the agent and the environment exchange information and learn it. To reduce the discrete action space, the deep reinforcement learning strategy in our approach is based on serialization and backtracking methods to VNFs in SFCs. Only one VNF is treated in each state transition. If a VNF cannot be placed owing to insufficient resources, delay, or bandwidth constraints, the request can not be fulfilled. Thus, our system backtracks to the previous network state. Note that between every two-time slots, there are two cases:

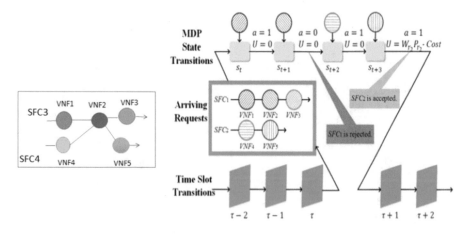

Fig. 2. Placement of VNFs for each SFC request at different time slots.

(1) Intra time slot: our system takes one VNF of an SFC after another to conserve the sequence order. MDP state is changed when a VNF is rejected or accepted knowing that MDP model is formally defined as: $\prec S, A, P, R, \gamma \succ$ where S is the collection of discrete states, A is the collection of discrete action, $P : S \times A \times S$ is the distribution of the transition probability. $R : S \times A$ is the reward function and $\gamma \in [0, 1]$is a discount factor of future rewards. As shown in Fig. 2, when two requests come (such as $SFC1$ and $SFC2$) in the time slot, $VNF1$ is processed by the NFV agent which tries to place it in a node satisfying the bandwidth and resource constraints. An action is then passed to set the $VNF1$ on the appropriate node. No rewards are returned to our system agent as $SFC1$ is not fully deployed. Therefore, the system enters state s_{t+1}, where the reward $U(s_t, a) = 0$. In state S_{t+1}, if no action is taken then $SFC1$ is rejected. This is either because there are no VM node satisfying all the constraints related to $VNF2$. By applying the same method, $SFC2$ is successfully deployed at time slot s_{t+3} with a reward $U(s_{t+3}, a)$ that is the throughput of $SFC2$ minus the cost of the resource consumption: $U(s_{t+3}, a) = SFC_i Bl_i \tau Cost(S_{t+3}, a)$. To avoid multiple VNF instances, in order to improve the resource utilization and choose the nearest node minimizing the traffic and the delay, many SFCs can share a same VNF that is deployed on the same node. In this case, the same procedures are put in place to place the VNFs in the chain in the intra and inter slots as shown in Fig. 2.

(2) Inter time slot: In the last time slot $\tau_k (K \in N+)$, SFC requests stop arriving at the agent. In this case, the state does not change and no action occurs. For every two time slots, the system removes the timeout and releases the resources. In the next time slot, a new request arrives and the agent takes state transition and according to constraints gives an action or rejected the SFC. Accordingly, a reward $U(s_t, a)$ is calculated as follows:

$$U(s_t, a) = \begin{cases} -\sigma_t latence(s_t, a) + w_{r_i}, \text{ si } r_i \text{ is accepted} \\ 0 \qquad\qquad\qquad r_i \text{ is rejected or not fully deployed.} \end{cases}$$

$$(7)$$

7   Experiments and Results Analysis

To evaluate our approach of two algorithms FCA and PBDRL, we first prepare the necessary data for the framework of each algorithm. For FCA, the dataset is made up of a set of VMs whose number varies from 100 to 1000 and a set of VNFs between 150 and 1100 which are distributed as illustrated in Table 2. The number of VMs chosen by expert is always lower than the number of VNFs. For PBDRL, the dataset is mainly made up of a set of SFC request inputs and a set of VNFs which are grouped and distributed into evaluation, training and result datasets. The values of the resources requested for each VNF in terms of memory, CPU and I/ O are chosen randomly in the range [1000, 5000]. Whereas, the values of resource capacity in terms of memory for each VM, CPU, and I/

O are chosen by chance from the range [10, 000, 50, 000]. In implementation, we applied our solution of placement of VNF by the principle of FCA, where each VNF is automatically migrated to the most appropriate VM on the basis of the received stimulus and its internal response. After, the VNF candidate extracted which will use in the following step.

7.1   Evaluation and Control Parameters

The different evaluation and control parameters for the results of this work are:

- The first criterion is packaging efficiency which is shown from the quantity of reduction in number of active virtual machines and reflects on the total energy consumption.
- The second criterion is energy consumption for server and network.
- The performance evaluation is based on the comparison of computational time of different solutions for a set of test problems of different sizes and complexities.
- Number of VM: the number of nodes that directly affects the cost of resources (CPU, memory), and therefore controls the reward. Obviously, when the number of nodes decreases, the reward is increases.
- Number of VNFs: the number of VNFs affects the number of trials and mainly the number of episodes. Indeed, when the number of VNFs increases in each request, the number of episodes also increases, as well as the requirements of each episode in bandwidth, latency, throughput, and resources.
- Resources (CPU, memory): the resources also present themselves as a setting parameter and control the reward and the cost of our network.

7.2   Results Analysis

From the FCA algorithm database, we were able to extract and determine the evolution of the average latency and reduction in the number of active virtual machines by comparing it with the MultiSwarm algorithm. For the FCA hybridization result with PBDRL approach, we were able to extract the reward results obtained, the average throughput and the error rate by comparing it with the First Fit and NFV deep algorithm.

Table 2 shows the number of active virtual machines (VM) versus the total number of incoming network services to Cloud for each time period (test). The results clearly show the effectiveness of the clustering which is directed by FCA to reduce the number of active virtual machines (Table 2) as well as to minimize the unused resources (UR) as shown in Table 3. The percentage of unused resources exhibits a slight increase with the growing number of virtual machines (VMs), impacting the Packaging Efficiency (PE) rate, as depicted in Table 3. These findings affirm the effectiveness of our approach in minimizing the number of active virtual machines, thereby demonstrating efficiency in reducing energy consumption.

Table 2. Number of active VMs compared to the total incoming VM number and Unused resources for our FCA algorithm.

Sets of tests	Number of VM	Number of active VM
S_1	100	60
S_2	200	78
S_3	300	116
S_4	400	212
S_5	500	204
S_6	600	182
S_7	700	322
S_8	800	406
S_9	900	448
S_{10}	1000	449

Compared to the MultiSwarm algorithm, our approach achieved lower average latency. As illustrated in Table 4, the average latency of our approach at level 150 VNF is 0.016 ms while it exceeds 0.050 ms for Multiswarm algorithm which confirms the effectiveness of our approach.

Table 4 shows the SFC request acceptance rate in different time steps for our hybrid approach (FCA and PBDRL) compared to First Fit algorithm. So, our approach ensures a very high acceptance rate compared to the First Fit algorithm which reflects the efficiency of our approach to pre-estimate for appropriate SFC. Table 5 compares the reward of our approach with the NFV deep and First Fit approach. This comparison shows that our approach is 60.5% more effective than

Table 3. Packaging efficiency (PE) evolution according to Unused Resource (RU) evolution for our approach FCA.

Sets of tests	Number of VM	PE%	RU%
S_1	150	2.5	77
S_2	250	3.20	71.5
S_3	350	3.01	71.65
S_4	450	2.12	80.29
S_5	600	2.94	74.23
S_6	700	3.84	64.63
S_7	800	2.48	77.35
S_8	900	2.21	80.004
S_9	100	2.23	79.96
S_{10}	1100	2.44	77.45

Table 4. Efficiency of our approach in terms of average latency and acceptance rate

Test	VNF number	Average latency of our approach	Average latency of MultiSwarm	Our approach acceptance rate	First Fit acceptance rate
S_1	150	0.016	0.052	0.85	0.69
S_2	250	0.012	0.05	0.65	0.6
S_3	350	0.0089	0.02	0.7	0.58
S_4	450	0.0047	0.009	0.6	0.55
S_5	600	0.0049	0.0084	0.71	0.53
S_6	700	0.0054	0.0081	0.57	0.5
S_7	800	0.0031	0.007	0.48	0.48
S_8	900	0.0024	0.0066	0.46	0.46
S_9	1000	0.0022	0.005	0.44	0.43
S_{10}	1100	0.002	0.0045	0.40	0.41

Table 5. Reward of each request

Reward/Time slot per episode	1000	2000	3000	4000	5000	6000
Our approach	2.92	3.03	3.09	3.16	3.15	3.17
NFV deep	1.4	1.38	1.41	1.37	1.42	1.42
First Fit	1.25	1.27	1.3	1.1.31	1.35	1.39

the NFV deep approach and 62% more efficient than the First Fit algorithm. So, for episodes going from 1000 time-slots to 6000-time slots, the reward of our approach increases more quickly compared to than that obtained with the NFV deep and First Fit approach.

As shown in Table 6, the average throughput of our approach is higher than NFV deep and First Fit at each time slot per episode. So, the average throughput of our approach reaches 1.55 for 2048-time slots per episode, 1.4 for NFV deep approach, and 1.12 for First Fit approach.

Table 6. Average throughput

Average throughput/Time slot per episode	2048	4096	6144	8192	10240	12288
Our approach	1.55	1.56	1.58	1.59	1.57	1.59
NFV deep	1.4	1.41	1.42	1.41	1.43	1.43
First Fit	1.12	1.15	1.25	1.3	1.36	1.37

8   Conclusion

In this paper, our primary emphasis was on resolving the VNF placement problem without taking into account the order of VNFs. The aim was to minimize the number of active virtual machines, reduce resource costs, and optimize energy consumption. We applied the principle of Formal Concept Analysis (FCA) grouping by incorporating the concept of labor division inspired by swarm intelligence within a cloud environment. This approach was utilized to extract candidate Virtual Network Functions (VNFs).

To thoroughly assess our approach, we conducted a comparative simulation with the MultiSwarm algorithm, which we implemented using the EdgeCloudSim tool. A series of test scenarios were conducted to assess the performance of our proposal and its ability to meet the specified constraints. The experimental results demonstrate that our VNF placement algorithm outperformed the MultiSwarm algorithm used for comparison. It ensures the minimization of the number of active virtual machines for VNF allocation, reduces unused resources, achieves high packing efficiency, and provides reliable computation time.

In the subsequent stage, our efforts extended to addressing the problem of both placement and chaining of VNFs, considering the predefined order of VNFs in Service Function Chain (SFC) series for the deployment of the requested network service. In this study, we delved into the feasibility of applying a reinforcement deep learning strategy to improve the Virtual Network Function chaining and the deployment of Service Function Chain requests in a cloud network. Our Parallel Bi-state Deep Reinforcement Learning approach yielded superior outcomes, demonstrating a reduction in delay through enhanced network throughput and an improved SFC request acceptance rate.

In summary, our methodology involves problem formulation using Formal Concept Analysis (FCA) and Deep Reinforcement Learning (DRL), placement prediction, and SFC request deployment through deep reinforcement learning based on policy gradient (PG) for action decision-making (VNF placement for an SFC request). This decision-making process takes into account the environmental state and its history, captured through Markov Decision Process (MDP) and Long Short-Term Memory (LSTM).

References

1. Abdelhamid, A.: Service function placement and chaining in network function virtualization environments. Ph.D. thesis, Bordeaux (2019)
2. Cho, D., Taheri, J., Zomaya, A.Y., Bouvry, P.: Real-time virtual network function (VNF) migration toward low network latency in cloud environments. In: 2017 IEEE 10th International Conference on Cloud Computing (CLOUD), pp. 798–801. IEEE (2017)
3. Ferdaus, M.H., Murshed, M., Calheiros, R.N., Buyya, R.: Virtual machine consolidation in cloud data centers using ACO metaheuristic. In: Silva, F., Dutra, I., Santos Costa, V. (eds.) Euro-Par 2014. LNTCS, vol. 8632, pp. 306–317. Springer, Cham (2014). https://doi.org/10.1007/978-3-319-09873-9_26
4. Gao, Y., Guan, H., Qi, Z., Hou, Y., Liu, L.: A multi-objective ant colony system algorithm for virtual machine placement in cloud computing. J. Comput. Syst. Sci. **79**(8), 1230–1242 (2013)
5. Helali, L., Omri, M.N.: A survey of data center consolidation in cloud computing systems. Comput. Sci. Rev. **39**, 100366 (2021)
6. Helali, L., Omri, M.N.: Software license consolidation and resource optimization in container-based virtualized data centers. J. Grid Comput. **20**(2), 13 (2022)

7. Helali, L., Omri, M.N.: Machine learning compliance-aware dynamic software allocation for energy, cost and resource-efficient cloud environment. Sustain. Comput. Inform. Syst. **41**, 100938 (2024)
8. Joseph, C.T., Chandrasekaran, K., Cyriac, R.: A novel family genetic approach for virtual machine allocation. Procedia Comput. Sci. **46**, 558–565 (2015)
9. Khemili, W., Hajlaoui, J.E., Omri, M.N.: Energy aware fuzzy approach for placement and consolidation in cloud data centers. J. Parallel Distrib. Comput. **161**, 130–142 (2022)
10. Lal, S., Taleb, T., Dutta, A.: NFV: security threats and best practices. IEEE Commun. Mag. **55**(8), 211–217 (2017)
11. Marzolla, M., Babaoglu, O., Panzieri, F.: Server consolidation in clouds through gossiping. In: 2011 IEEE International Symposium on a World of Wireless, Mobile and Multimedia Networks, pp. 1–6. IEEE (2011)
12. Mavridis, I., Karatza, H.: Combining containers and virtual machines to enhance isolation and extend functionality on cloud computing. Future Gener. Comput. Syst. **94**, 674–696 (2019)
13. Mijumbi, R., Serrat, J., Gorricho, J.-L., Bouten, N., De Turck, F., Boutaba, R.: Network function virtualization: state-of-the-art and research challenges. IEEE Commun. Surv. Tutor. **18**(1), 236–262 (2015)
14. Nikolai, J., Wang, Y.: Hypervisor-based cloud intrusion detection system. In: 2014 International Conference on Computing, Networking and Communications (ICNC), pp. 989–993. IEEE (2014)
15. Pei, J., Hong, P., Li, D.: Virtual network function selection and chaining based on deep learning in SDN and NFV-enabled networks. In: 2018 IEEE International Conference on Communications Workshops (ICC Workshops), pp. 1–6. IEEE (2018)
16. Qi, D., Shen, S., Wang, G.: Virtualized network function consolidation based on multiple status characteristics. IEEE Access **7**, 59665–59679 (2019)
17. Shojafar, M., Canali, C., Lancellotti, R., Baccarelli, E.: Minimizing computing-plus-communication energy consumptions in virtualized networked data centers. In: 2016 IEEE Symposium on Computers and Communication (ISCC), pp. 1137–1144. IEEE (2016)
18. Soualah, O., Mechtri, M., Ghribi, C., Zeghlache, D.: Energy efficient algorithm for VNF placement and chaining. In: 2017 17th IEEE/ACM International Symposium on Cluster, Cloud and Grid Computing (CCGRID), pp. 579–588. IEEE (2017)
19. Szabo, R., Kind, M., Westphal, F.-J., Woesner, H., Jocha, D., Csaszar, A.: Elastic network functions: opportunities and challenges. IEEE Netw. **29**(3), 15–21 (2015)
20. Tang, M., Pan, S.: A hybrid genetic algorithm for the energy-efficient virtual machine placement problem in data centers. Neural Process. Lett. **41**, 211–221 (2015)
21. Watada, J., Roy, A., Kadikar, R., Pham, H., Bing, X.: Emerging trends, techniques and open issues of containerization: a review. IEEE Access **7**, 152443–152472 (2019)
22. Zhang, Q., Liu, F., Zeng, C.: Adaptive interference-aware VNF placement for service-customized 5G network slices. In: IEEE INFOCOM 2019-IEEE Conference on Computer Communications, pp. 2449–2457. IEEE (2019)
23. Zhou, J., Hong, P., Pei, J., Li, D.: Multi-task deep learning based dynamic service function chains routing in SDN/NFV-enabled networks. In: ICC 2019-2019 IEEE International Conference on Communications (ICC), pp. 1–6. IEEE (2019)

Optimising Water Quality Classification in Aquaculture Using a New Parameter Pre-selection Approach

Mahdi Hamzaoui[1(✉)], Mohamed Ould-Elhassen Aoueileyine[1], Lamia Romdhani[2], and Ridha Bouallegue[1]

[1] Innov'COM Laboratory, Higher School of Communication of Tunis, Technopark Elghazala, Raoued, 2083 Ariana, Tunisia
{mahdi.hamzaoui,mohamed.ouldelhassen,ridha.bouallegue}@supcom.tn
[2] University of Qatar, Doha, Qatar
lamia.romdhani@qu.edu.qa

Abstract. Water stands as a pivotal element in aquaculture, and its quality plays a crucial role in the management of fish farming. The inherent non-linearity, dynamics, and instability of its parameters render it a highly complex system to oversee. Conventional methods for assessing water quality in relation to fish farming often prove inadequate. The integration of technology becomes imperative for successful outcomes. In this context, the utilization of artificial intelligence and machine learning techniques emerges as a promising solution. This paper presents a comparison between several methods of pre-selecting water parameters to optimise the classification of water quality for the culture of marine species. The SFI method selects the parameters Temperature, DO, and pH. The results showed that the SFI-KNN combination is the best. It achieved an accuracy rate equal to 99.87%.

1 Introduction

Aquaculture, an essential practice for cultivating aquatic organisms, plays a vital role in ensuring human food security. The significance of fishing in addressing food crises, securing the food supply, enhancing quality of life, and boosting export activities cannot be overstated. In 2016, the global fishery production achieved an unprecedented milestone, totaling 171 million tons. A substantial portion, amounting to 88%, is directly consumed by humans [1]. Nevertheless, with the ongoing expansion of the global population, the demand on the world's fisheries is poised to escalate further [2,3].

Water is a critical component in the aquaculture industry. Its quality has a considerable impact on fish farming practices. Optimal water quality is critical for promoting maximum fish development, assuring high-quality goods output, and reducing the danger of disease and mortality. The combination of these variables contributes to increased fish production, which benefits both national and international economic growth [4].

Numerous parameters within water serve as indicators of its quality. In the field of aquaculture, specific standard value ranges exist [5]. If any parameter surpasses these established limits, it will impact the overall quality of the water.

L. Barolli (Ed.): AINA 2024, LNDECT 201, pp. 97–107, 2024.
https://doi.org/10.1007/978-3-031-57870-0_9

Dissolved oxygen (DO), ammonia (NH3, NH4+), nitrite (NO2−), nitrate, turbidity, pH, and temperature are the key indicators of water quality in aquaculture. Water quality is influenced by a wide range of factors, including biological, physical, and human activity, making it a complex, non-linear, and dynamic system. In intensive aquaculture, accurate classification of water quality is critical. It acts as an early warning system for changes in water quality, which helps to reduce aquaculture losses. Because of the complicated composition of water, traditional techniques of water quality classification are less effective. Outdated categorization models do not produce optimal results. As a result, incorporating cutting-edge technology like artificial intelligence and machine learning emerges as a realistic and effective alternative.

This paper introduces an innovative methodology for the pre-selection of water parameters before their integration into a machine learning model for water quality classification. To offer context, the "Related Work" section reviews and discusses previous studies on the same topic, along with their outcomes. The "Background" section provides theoretical definitions of the various methods employed in this research. Moving to the "Methodology" section, it outlines the data acquisition phase and details the data pre-processing methods applied in the study. This section critically compares diverse approaches to pre-selecting water parameters to achieve a precise classification of water quality. Subsequently, the "Results and Discussion" section presents the study's findings and assesses the performance of our novel approach in comparison with previous methodologies. This section also includes the presentation of results from other studies akin to the current research. Finally, the paper concludes with the "Conclusion" section, offering a forward-looking perspective and outlining avenues for future research.

2   Related Work

This section provides an overview of pertinent studies in the domain of water quality classification and prediction that leverage machine learning techniques. In the last few years, several models utilizing Artificial Neural Networks (ANN) and Deep Learning (DL) have been proposed to predict water quality indicators in the context of aquaculture [6,7]. To develop a forecasting model for water quality in aquaculture, certain approaches employed a Backpropagation (BP) neural network technique, incorporating various activation functions such as tansig, logsig, and purelin. The study conducted by Donya Dezfooli et al. focused on the Probabilistic Neural Network (PNN) in the context of classifying drinking water. The findings revealed that PNN outperforms K-nearest neighbors and support vector machine in terms of classification accuracy [8]. Tingting Li et al. applied various methods, including BPNN, RBFNN, SVM, and LSSVM, in the field of aquaculture to forecast water parameter values such as dissolved oxygen (DO), pH, ammonium-nitrogen (NH3 - N), nitrate nitrogen (NO3 -N), and nitrite-nitrogen (NO2 -N). Their study revealed that SVM yielded the most accurate predictions, demonstrating high stability in the results [9]. Common approaches for predicting water quality parameters encompass the time-series method, the Markov method, the support vector regression machine method, and the grey system theory method [10–12]. Nevertheless, these widely employed methodologies have associated drawbacks, including challenges with weak generalization, low processing efficiency, and limited, uncertain forecasting accuracy. Consequently, these conventional methods often struggle to meet the growing requirements for precision in aquaculture. Numerous other studies have suggested hybrid models for the prediction of water quality. Shuangyin Liu et al. introduced a hybrid methodology to forecast dissolved oxygen (DO) and water temperature values within a crab rearing environment. This approach combines two algorithms, namely Real-Valued Genetic Algorithms (RGA) and Support Vector Regression (SVR). The method involves utilizing RGA to search for the optimal SVR parameters and subsequently employing these optimal parameters to construct the SVR models [13]. Elias Eze et al. introduced a hybrid model that integrates Empirical Mode Decomposition (EEMD), Deep Learning (DL), and Long Short-Term Memory (LSTM). EEMD plays a role in decomposing water parameters, while both EEMD and DL contribute to predicting the values of these parameters [14]. Applying a decomposition method to the original signal has been shown to enhance the characteristics of predicted signals through multi-scale forecast techniques, as compared to relying solely on single-scale features [15]. This approach involves breaking down each segment of the original signal, thereby unveiling its distinct intrinsic characteristics. Smail Dilmi et al. employed a novel approach that integrates deep learning and feature extraction techniques for the monitoring and classification of water quality [16]. Neha Radhakrishnan et al. conducted a comparison of various classification methods, ultimately determining that the decision tree is the most effective classification approach [5].

3   Methodology

In this section, we delineate various methodologies employed to select aquaculture water parameters before their integration into a machine learning model. Each of these approaches is geared towards optimizing the classification phase, with the ultimate goal of ensuring enhanced and conducive freshwater quality for the well-being of fish life. Subsequently, we introduce our own approach, a fusion of multiple techniques aimed at identifying the most impactful parameters during the classification phase. The methodology can be succinctly summarized in the following steps.

3.1   Data Gathering and Preprocessing

The precision of system performance hinges on the accuracy of collected data, making data collection a pivotal undertaking in experimental processes. To compile the dataset, twelve fishponds were utilized for this study. The data collection occurred in Nsukka, Nigeria, spanning from June 18th to July 24th, 2021, employing six sensors temperature (TEMP), turbidity (TURB), dissolved oxygen (DO), pH, ammonia (AMMO), and nitrate (NITR), managed by an ESP 32 microcontroller. Data points were recorded at five-second intervals throughout this period. It's noteworthy that this project is undertaken by the HiPIC Research Group within the Department of Computer Science at the University of Nigeria, Nsukka, Nigeria. Table 1 shows a set of data captured on 2021-07-07 between 15:15:34 and 15:20:12 from the fishpond 10. The total catch data is 345,000 catches spread over the 12 fishponds as shown in Table 2. Data losses have been reported, stemming from a combination of the substantial volume of records and the sensitivity of data sensors to environmental conditions, particularly evident in the case of wireless sensors. Employing data preprocessing mechanisms has been instrumental in addressing this issue by populating the vacant fields. Additionally, the imperative to normalize the data is underscored, as it plays a pivotal role in enhancing the performance of algorithms. In this work, we use Min-Max Normalization to put all values between 0 and 1. This technique makes it easier to train our model afterwards.

Table 1. Table representing a sample of data.

created_at	entry_id	TEMP	TURB	DO	pH	AMMO	NITR
2021-07-07 15:15:34	121	25.8125	100	0	6.33187	129.85301	828
2021-07-07 15:15:53	122	25.9375	100	0	6.33641	133.92024	845
2021-07-07 15:16:13	123	25.9375	100	0	6.33641	254.43057	868
2021-07-07 15:16:32	124	25.9375	100	1.561	6.34549	127.87219	919
2021-07-07 15:16:52	125	25.9375	100	0	6.34095	158.22548	922
2021-07-07 15:17:11	126	25.875	100	2.306	6.34095	339.83801	907
2021-07-07 15:17:31	127	26	100	1.142	6.34549	226.16016	934
2021-07-07 15:17:50	128	25.9375	100	1.465	6.34095	131.86873	933
2021-07-07 15:18:14	129	-127	100	1.543	6.32733	231.25386	873
2021-07-07 15:18:33	130	25.9375	100	30.476	6.34095	161.60397	816
2021-07-07 15:18:53	131	25.875	100	0	6.34095	303.96719	934
2021-07-07 15:19:13	132	26	100	2.132	6.34549	263.24252	922
2021-07-07 15:19:33	133	26	100	1.401	6.35003	116.0966	866
2021-07-07 15:19:52	134	25.9375	100	1.874	6.32733	150.91507	912
2021-07-07 15:20:12	135	25.9375	100	1.537	6.35003	141.02589	866

3.2   Water Parameters Values in Aquaculture

Numerous characteristics in water can be used to determine its quality; in aquaculture, these parameters have defined value ranges, and if a parameter's value exceeds these ranges, the water's quality may be impacted [5]. The following Table 3 shows the suitable ranges of water parameters in aquaculture:

Table 2. The distribution of data on fishponds.

FishPond	The total number of recordings
FishPond 1	83127
FishPond 2	172250
FishPond 3	169186
FishPond 4	89845
FishPond 5	8024
FishPond 6	91051
FishPond 7	279613
FishPond 8	70745
FishPond 9	151786
FishPond 10	621
FishPond 11	3166
FishPond 12	3591

Table 3. Table of water parameter suitable ranges in aquaculture.

Parameters	Ranges	Units
PH	7–8.5	–
Temperature	27–30	c
Dissolved Oxygen	4–9	g/ml
Nitrate	0–350	g/ml

3.3   Different Approaches to Parameter Pre-selection

3.3.1   Variance Inflation Factor Method (VIF)

A statistical tool commonly employed to assess the degree of collinearity among independent variables within a linear regression model is the Variance Inflation Factor (VIF). The VIF gauges the increase in variance observed in a regression coefficient estimate due to collinearity between an independent variable and others. To compute the VIF for a particular variable, one compares the variance of the regression estimator in the full model with the variance of the regression estimate for that variable in a basic model, which includes only that variable. A high VIF, in accordance with conventional wisdom, indicates a notable correlation between variables and may signify a collinearity concern within the model. To mitigate collinearity issues and enhance the overall goodness of fit of the model, it is customary to consider eliminating variables with elevated VIFs from the model.

3.3.2   Pearson's Correlation Coefficient Method (PCC)

Correlation analysis serves as a valuable tool for evaluating the extent of association between two sets of items. The strength of this association is determined by considering factors such as direction, form, and dispersion strength. Frequently, numerical expression of this relationship is achieved through the utilization of a correlation coefficient. The correlation coefficient is computed within a predefined range, the specifics of which depend on the chosen algorithm. By assessing the coefficient's value within this range, one can discern both its strength and direction. A positive coefficient signifies a positive correlation between the two variables, while a negative value indicates a negative correlation.

In the context of Pearson's correlation coefficient approach, a statistical examination of the collinear relationship between two variables is conducted. This involves the consideration of the covariance ratio as well as the standard deviation of the data values associated with the two variables in question. To illustrate, let's consider variables A and B. The calculation of Pearson's correlation coefficient for these variables follows the formula below:

$$C_{A,B} = \frac{covariance(A,B)}{\sigma_A \sigma_B} \tag{1}$$

where $C_{A,B}$ is the correlation coefficient, $covariance(A,B)$ is the covariance, and σ_A and σ_B are the standard deviations of A and B, respectively.

3.3.3   The Approach Presented in the DTKNN+ Model

DTKNN+ represents a hybrid methodology that integrates two models, namely the decision tree and K-Nearest Neighbors (KNN), for the purpose of classifying water quality in aquaculture. The process commences with the acquisition of raw data pertaining to water parameters. Following initial dataset preprocessing, the data is fed into the decision tree algorithm for training. This algorithm possesses the capability to assess the significance of each feature in the model by utilizing the Feature Importance function. Within this study, a dedicated Python module is developed to selectively retain only the most crucial features during the training phase. These identified important features are then exclusively passed on to the KNN algorithm for further training [17].

In essence, the presented approach involves pre-training a Decision Tree that incorporates a specific function known as the Feature Importance function. This function is instrumental in revealing the importance of each parameter during the classification phase. The DTKNN+ methodology employs a Decision Tree to discern and retain relevant parameters, guided by the equation: DB = 1/NBF, where DB represents the decision boundary and NBF denotes the number of features.

3.3.4   Select Features Importance Method (SFI)

SFI is a feature selection method devoloped by ourselves in another study in which we pre-select the most important parameters in the fish weight prediction phase. The process involves employing two multicollinearity analysis methods, namely VIF and PPC, to eliminate less important features in predicting fish weight. Specifically, it opts for features with low multicollinearity values with the "Weight" feature according to both VIF and PCC. Any features demonstrating high collinearity with the "Weight" feature, as determined by both VIF and PCC methods, are automatically excluded from the prediction process. This innovative method will be used to select the most important water parameters for judging whether water quality is good or bad [18].

3.3.5   The General Workflow of Our Work

As already illustrated in Fig. 1, a preprocessing operation is applied to the dataset in order to normalise the data and fill in the empty fields. Secondly, the water-related parameters are extracted. Next, a comparison is made between several parameter pre-selection methods to determine the most important parameters for judging water quality. After this comparison, the selected parameters are passed to KNN for training, validation and to output the final water quality classification result.

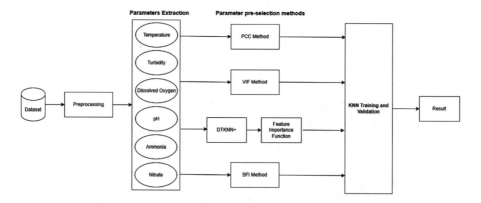

Fig. 1. The workflow for the different parameter selection approaches

4   Result and Discussion

4.1   Analysis Using the VIF Method

This method shows in Table 4 that the inflation factor is high for the Nitrate and Turbidity parameters. It is low for the Ammonia, Temperature, DO and pH parameters. Based on this method, the removal of the Nitrate and Turbidity parameters is useful to obtain a favourable atmosphere for model training.

Table 4. The variance inflation factor VIF of the water parameters.

Features	VIF
Nitrate	3201.23
Turbidity	2418.59
Ammonia	144.73
Temperature	122.65
DO	64.46
pH	47.50

4.2   Analysis Using the PCC Method

This method reveals the correlations among various parameters, with the identification of strong correlations during the training process potentially hindering model efficiency. By employing the PCC method, we systematically eliminate parameters exhibiting pronounced correlations with others. In Fig. 2, the correlation values between different parameters are depicted. Notably, a substantial correlation is observed between pH and Nitrate, as well as pH and Ammonia. Consequently, Nitrate and Ammonia are excluded from the set of parameters utilized in the model training process.

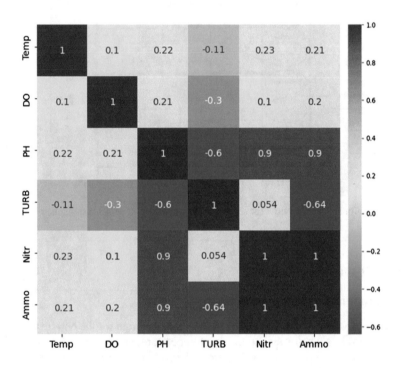

Fig. 2. Correlations between water parameters

4.3   Analysis Using the DTKNN+ Approach

The DTKNN+ approach pre-train using the Decision Tree algorithm. Thanks to the specific function of this algorithm called Feature Importance, DTKNN+ selects the most important parameters in KNN training. As shown in Fig. 3, DTKNN+ selects DO and pH as the most important parameters.

4.4   Analysis Using the SFI Approach

The SFI approach represents a convergence between the parameter lists generated by PCC and those returned by VIF. By adopting this method, the SFI approach simultaneously incorporates the parameters identified by both PCC and VIF. This integration

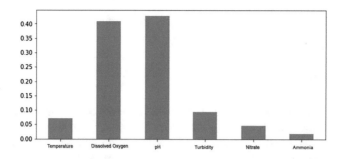

Fig. 3. Feature importance visualization

ensures a comprehensive inclusion of relevant factors, leveraging the strengths of both methodologies to provide a more robust and encompassing set of parameters. The result is a unified approach that captures the shared insights of PCC and VIF, enhancing the overall effectiveness of the analysis. As a result, the SFI Approach identifies DO, pH, and temperature as the most crucial parameters during the training phase of KNN. These findings underscore the significance of these specific variables in influencing the predictive performance of the KNN algorithm.

4.5   Evaluation of Different Parameter Selection Approaches with KNN

Table 5 illustrates the parameters selected by various approaches and the effectiveness of KNN in the water quality classification phase following its integration with each approach. KNN training with the full set of parameters including pH, DO, Temperature, Ammonia, Turbidity, and Nitrate achieved an accuracy of 93.56%. VIF selects only the parameters Ammonia, Temperature, pH, and DO. KNN's accuracy with this data set is 96.56%. PCC removes Ammonia and Nitrate from the parameter set. It keeps Temperature, DO, pH, and Turbidity. With this preselection of parameters, KNN achieves a water quality classification rate of 97.24%. DTKNN+ returns only DO and pH to KNN for training. The accuracy rate of this combination is 99.28%. Finally the SFI-KNN approach gives the best result by achieving an accuracy of 99.87%. We consider that SFI is the most appropriate approach for our context of water parameter selection in

Table 5. Results obtained using different approaches.

Approach	Selected parameters	Mean Absolute Error	Root Mean Squared Error	Evaluation of Accuracy %
KNN	All parameters	0.0643	0.2535	93.56
VIF-KNN	Ammonia, Temperature, DO, pH	0.0481	0.1287	96.56
PCC-KNN	Temperature, DO, pH, Turbidity	0.0413	0.0945	97.24
DTKNN+	DO, pH	0.0071	0.0847	99.28
SFI-KNN	Temperature, DO, pH	0.0052	0.0412	99.87

aquaculture. Temperature, DO and pH are the key parameters for ensuring comfortable water quality for the survival of marine species.

5 Conclusion

This study attempts to monitor and classify water quality in response to several issues with water quality. It starts by comparing several approaches to selecting water parameters. The PCC and VIF methods already exist, while SFI and DTKNN+ are two other approaches developed by ourselves in previous work. By combining the different approaches separately with a KNN model, we find that the SFI-KNN combination gives more efficient results than the other combinations. An evaluation showed that the accuracy rate improved with SFI-KNN, reaching 99.87%. In the future, an Internet of Things solution will be put into place to record real-time data on water parameters, choose the pertinent parameter set using the SFI approach, train the selected data using KNN, and then visualize the outcome using a mobile or online application.

References

1. Nakai, J.: Food and agriculture organization of the united nations and the sustainable development goals. Sustain. Dev. **22**, 1–450 (2018)
2. Delgado, C.L.: Fish to 2020: Supply and Demand in Changing Global Markets, vol. 62. WorldFish (2003)
3. Caddy, J.F., Cochrane, K.L.: A review of fisheries management past and present and some future perspectives for the third millennium. Ocean Coast. Manag. **44**(9–10), 653–682 (2001)
4. Ahmad, A., Abdullah, S.R.S., Hasan, H.A., Othman, A.R., Ismail, N.I.: Aquaculture industry: supply and demand, best practices, effluent and its current issues and treatment technology. J. Environ. Manag. **287**, 112271 (2021)
5. Kassem, T., Shahrour, I., El Khattabi, J., Raslan, A.: Smart and sustainable aquaculture farms. Sustainability **13**(19), 10685 (2021)
6. Xiao, Z., Peng, L., Chen, Y., Liu, H., Wang, J., Nie, Y.: The dissolved oxygen prediction method based on neural network. Complexity **2017**, 4967870 (2017)
7. Liu, J., et al.: Accurate prediction scheme of water quality in smart mariculture with deep Bi-S-SRU learning network. IEEE Access **8**, 24784–24798 (2020)
8. Dezfooli, D., Hosseini-Moghari, S.M., Ebrahimi, K., Araghinejad, S.: Classification of water quality status based on minimum quality parameters: application of machine learning techniques. Model. Earth Syst. Environ. **4**, 311–324 (2018)
9. Li, T., Lu, J., Wu, J., Zhang, Z., Chen, L.: Predicting aquaculture water quality using machine learning approaches. Water **14**(18), 2836 (2022)
10. Rozario, A.P., Devarajan, N.: Monitoring the quality of water in shrimp ponds and forecasting of dissolved oxygen using Fuzzy C means clustering based radial basis function neural networks. J. Ambient. Intell. Humaniz. Comput. **12**(5), 4855–4862 (2021)
11. Liu, S., Xu, L., Jiang, Y., Li, D., Chen, Y., Li, Z.: A hybrid WA-CPSO-LSSVR model for dissolved oxygen content prediction in crab culture. Eng. Appl. Artif. Intell. **29**, 114–124 (2014)
12. Li, Z., Jiang, Y., Yue, J., Zhang, L., Li, D.: An improved gray model for aquaculture water quality prediction. Intell. Autom. Soft Comput. **18**(5), 557–567 (2012)

13. Liu, S., Tai, H., Ding, Q., Li, D., Xu, L., Wei, Y.: A hybrid approach of support vector regression with genetic algorithm optimization for aquaculture water quality prediction. Math. Comput. Model. **58**(3–4), 458–465 (2013)
14. Eze, E., Halse, S., Ajmal, T.: Developing a novel water quality prediction model for a South African aquaculture farm. Water **13**(13), 1782 (2021)
15. Li, C., Li, Z., Wu, J., Zhu, L., Yue, J.: A hybrid model for dissolved oxygen prediction in aquaculture based on multi-scale features. Inf. Process. Agric. **5**(1), 11–20 (2018)
16. Dilmi, S., Ladjal, M.: A novel approach for water quality classification based on the integration of deep learning and feature extraction techniques. Chemom. Intell. Lab. Syst. **214**, 104329 (2021)
17. Hamzaoui, M., Aoueileyine, M.O.E., Bouallegue, R.: A hybrid method of K-nearest neighbors with decision tree for water quality classification in aquaculture. In: Nguyen, N.T., et al. (eds.) ICCCI 2023. CCIS, vol. 1864, pp. 287–299. Springer, Cham (2023). https://doi.org/10.1007/978-3-031-41774-0_23
18. Hamzaoui, M., Aoueileyine, M.O.E., Romdhani, L., Bouallegue, R.: Optimizing XGBoost performance for fish weight prediction through parameter pre-selection. Fishes **8**(10), 505 (2023)

Topic Analysis of Japanese Sentences Using Sentence Embeddings

Kenshin Tsumuraya[1], Huang Yonghui[1], Minoru Uehara[1(✉)], and Yoshihiro Adachi[2]

[1] Graduate School of Information Sciences and Arts, Toyo University, Kawagoe, Japan
`{s3B102200130,s3B102200084,uehara}@toyo.jp`
[2] RIIT, Toyo University, Kawagoe, Japan
`adachi@toyo.jp`

Abstract. Previously, we proposed a topic analysis method that clusters sentence embeddings generated by Japanese Sentence-BERT and assigns a topic-word list to each cluster according to the values of a topic-word evaluation function. For a word in each cluster, the function is defined as a linear combination of the word's cluster-based TFIDF and the cosine similarity between the word embedding and cluster centroid with hyperparameter α. In this study, we propose a method to automatically determine an appropriate value for α and evaluate the topic-word list for each cluster given by the determined α using topic coherence and topic diversity. Furthermore, we define a new metric called cluster-recall to evaluate whether a topic-word list adequately expresses the content of the corresponding cluster and evaluate the topic-word list attached to each cluster using cluster-recall. Automatically attached topic-word lists make it easy to understand the topic analysis results of a large-scale sentence dataset.

1 Introduction

In recent years, an increasing amount of digital text data, such as office documents, news, and reviews on the Internet, has been created and accumulated everywhere. It is difficult to manually understand and analyze this large amount of digital text data. Topic analysis (also called topic modeling) is a technique for automatically detecting topics (or themes) contained in text. The resulting topical information can be used in many areas, such as text mining, classification, and summarization.

As a conventional topic analysis model, latent Dirichlet allocation (LDA) [1], which is a statistical model for identifying latent topics by describing text as a bag of words, has been applied in various fields as a powerful unsupervised topic analysis tool. As extensions of the LDA model, in [2], the proposed approach attempted to model the content of documents and the author's interests simultaneously, and in [3, 4], the proposed model analyzed short documents on social media platforms, such as Twitter.

Non-negative matrix factorization (NMF) is an algorithm that decomposes a non-negative matrix into two non-negative matrices, which makes it easier to examine the latent features of the original matrix. NMF has also become widely used as a topic analysis method since Lee et al. developed a simple and efficient algorithm [5].

In recent studies on topic analysis, researchers have focused on using sentence embeddings generated by machine learning models. In [6], the authors clustered the embeddings generated by machine learning models, such as Word2vec [7] and bidirectional encoder representation from transformers (BERT) [8], and obtained topic words by adjusting and re-ranking words close to the center of each cluster by term frequency. Grootendorst [9] performed topic analysis by clustering sentence embeddings generated by Sentence-BERT [10]. In [9], the author proposed a topic representation model that assigns topics to each cluster using a cluster-based TFIDF procedure.

To date, in topic analysis studies, researchers have focused on relatively long documents (coarse-grained topic analysis in terms of document length), such as news articles, which generally consist of multiple sentences. As the length of the document increases, the number of topic words increases. Document embeddings are calculated by averaging these word embeddings. Therefore, the influence of feature information from an individual topic word included in a document on the document embedding is relatively small.

In the topic analysis of, for example, questionnaire responses, reviews, and literary works, topic analysis techniques that use a sentence as a unit (fine-grained topic analysis in terms of document length) are very important. Using a small-scale corpus of Japanese sentences with ground truth labels (GTLs) [11] and Japanese literary works [12] as the target datasets for analysis, we conducted fine-grained topic analysis. In [11], using Japanese Sentence-BERT (JSBERT), we generated an embedding from each sentence in the Japanese sentence dataset and developed a method to classify sentences according to topic features by clustering these sentence embeddings. Then, we devised a new method for generating a topic-word list (i.e., topics) that explains the content of each cluster obtained by clustering a large-scale sentence dataset in an easy-to-understand manner. This method associates the values of a topic-word evaluation function with the words in each cluster, sorts the words in descending order according to these values, and selects the top N words as the topic-word list.

The topic-word evaluation function is a function that linearly combines the cluster-based TFIDF and the proximity (cosine similarity) of the word embedding to the cluster centroid for each word in the cluster using hyperparameter α. In [11, 12], we experimentally demonstrated that an appropriate topic-word list was attached to each cluster when hyperparameter α was set appropriately. In particular, in the topic analysis of Japanese literary works reported in [12], the topic-word list attached to each cluster obtained by clustering the sentences that make up a literary work was very useful for analyzing the topic characteristics of the work.

In the topic analysis of large-scale text datasets, it is very important that each cluster is labeled with an appropriate topic-word list so that the content of the cluster can be easily understood. The appropriate value for hyperparameter α depends on the dataset being analyzed and the clustering results. Therefore, it is desirable to develop a method to automatically determine the value of α that attaches an appropriate topic-word list to each cluster.

Topic coherence [13] and topic diversity (TD) [14] have been widely used as evaluation methods for the topic-word lists obtained for each cluster. Recently, methods based on measuring topic relevance [15] and topic coverage [16] have been proposed to evaluate topic models. These methods do not evaluate the coherence or diversity of the topic words generated by the topic model, but rather evaluate whether the topic words generated by the topic model cover the topic content in the text that is the subject of topic analysis. However, these methods require the creation of large-scale datasets of reference topics, such as topic words or topic documents, by topic analysis experts, which requires great effort.

In this study, as a method used to evaluate whether the topic-word list covers the content features of each cluster appropriately, we devised a new metric called cluster-recall to evaluate the extent to which the corresponding cluster can be reconstructed from the original dataset as a result of searching for sentences using a topic-word list.

The results of this study are as follows:

- We developed a method to automatically determine the optimal value of hyperparameter α under the condition of maximizing the product of topic consistency and TD.
- Using the optimal value of α, we evaluated the topic-word list attached to each cluster using topic coherence and TD metrics.
- We defined a new metric called mean average cluster-recall (MACR), which evaluates whether a topic-word list appropriately expresses the corresponding cluster content, and evaluated the topic analysis results using MACR.

The structure of the remainder of this paper is as follows: In Sect. 2, we describe the techniques used in this study: the Japanese sentence embedding generation model JSBERT, two clustering methods, a topic-word evaluation function, and topic-word list evaluation methods. In Sect. 3, we present the details of sentence clustering using embeddings and its performance evaluation results. In Sect. 4, we describe the main results of this study, a method for automatically determining hyperparameter α of the topic word evaluation function, and the performance evaluation results of the topic-word lists generated using the determined α using topic consistency and TD metrics. In Sect. 5, we explain MACR and describe the results of evaluating topic-word lists using MACR. In Sect. 6, we summarize the study.

2   Related Materials

In this section, we describe the techniques used in this study: sentence embedding generation, clustering, cluster performance evaluation, and topic-word selection.

2.1 Sentence Embedding Generation

A Sentence-BERT model is a modification of the pre-trained BERT network that embeds very rich contextual information to generate sentence embeddings that can successfully evaluate the semantic proximity of sentences using cosine similarity [10]. Therefore, sentence embeddings generated by Sentence-BERT are suitable for sentence clustering and similar-sentence retrieval based on topics using cosine similarity.

In this study, we used JSBERT, which we developed, as a machine learning model to generate sentence embeddings [12]. To implement JSBERT, we used the National Institute of Information and Communications Technology (NICT) BERT pre-trained model [17] as a pre-trained model with a mode that does not split words by "wordpiece" so that we could label the resulting clusters with appropriate topic words. We fine-tuned JSBERT using the Japanese translation of the Stanford Natural Language Inference corpus [18].

JSBERT generates sentence embeddings, that is, 768-dimensional real vectors, from an input sequence of 512 tokens or less. Two types of JSBERT exist according to the sentence embedding generation method: JSBERT-normal generates a sentence embedding from all the words that make up an input sentence and JSBERT-nouns generates a sentence embedding only from the noun words that make up the input sentence.

2.2 Clustering Methods

In this study, we used well-known clustering methods; k-means++ (KM) and the Gaussian mixture model (GMM).

K-means is a non-hierarchical and hard clustering algorithm. This method clusters data groups into k groups based on the idea that data that are close to each other belong to the same cluster. KM improves the selection of the initial value of the k-means method [19].

GMM is a clustering method that considers the data to be generated using the linear superposition of multiple normal distribution (Gaussian distribution) data. It not only divides the data into clusters but also provides the probability of belonging to each cluster for each sample.

In this study, we used the Python machine learning library scikit-learn [20] to perform clustering using KM and GMM. The performance evaluation results for clustering and topic analysis were almost the same regardless of whether we used KM or GMM. Therefore, in this paper, we only describe the experimental results using KM.

2.3 Evaluation of the Clustering Results

Adjusted mutual information (AMI) is an evaluation metric that compares two clustering results. It is a method for evaluating the accuracy of clustering results that is corrected by subtracting the expected value of mutual information (MI) when a random prediction is made from the MI between clustering results [21]. AMI returns a value of 1 if the two clustering results are identical (i.e., a perfect match) and a value closer to 0 if the clustering results are different. In this study, we used the AMI to evaluate the clustering accuracy between a sentence dataset with GTLs and the results obtained by clustering the sentences that belong to the dataset.

2.4 Topic-Word Evaluation Function

For each cluster obtained by clustering, we defined a topic-word evaluation function to select a topic-word list that well reflects the content of the cluster from the words in the cluster [11, 12].

For word w in cluster c, the topic-word evaluation function $f(w, c)$ is defined as

$$f(w, c) = \alpha \times \textit{tfidf}(w, c) + (1 - \alpha) \times \textit{cosim}(w, c), \tag{1}$$

where $\textit{tfidf}(w, c)$ is the cluster-based TFIDF of word w in cluster c, and $\textit{cosim}(w, c)$ is the cosine similarity between the distributed representations of word w and the centroid of cluster c. Hyperparameter α is greater than or equal to 0 and less than or equal to 1. The list of unique words that belong to each cluster is sorted in descending order according to the value of the topical word evaluation function and the top N words in the list are selected as the topic-word list of the cluster. Hereafter, unless otherwise specified, we use the topic-word list for $N = 10$.

We verified that a relatively appropriate topic-word list was selected when $\alpha = 0.5$ in topic analysis experiments on a small Japanese corpus [11] and Japanese literary works [12]. Generally, the appropriate value of α depends on the text dataset to be analyzed and the clustering results. In this study, we developed a method to automatically determine the value of α that can obtain a topic-word list that appropriately expresses the content in each cluster.

2.5 Topic-Word List Evaluation Method

It is desirable that the topic-word list attached to each cluster of the clustering result accurately reflects the content of the cluster and has a high interpretability for humans. We used topic consistency and TD, which are widely used in conventional topic analysis, to evaluate the quality of topic-word lists.

Various evaluation methods for topic coherence have been defined and their characteristics have been evaluated [13]. For fine-grained topic analysis using sentence-based clustering, because the number of words included in the elements (i.e., sentences) in a cluster is small, coherence evaluation methods that use the co-occurrence of words in the elements, such as UMass coherence, were not suitable in this study. Thus, we used Word2Vec coherence (W2V) and C_V coherence (C_V).

We calculated the W2V score of a topic-word list by computing word similarities using word embeddings generated by a Word2Vec model [22] trained on Japanese Wikipedia. The W2V formula definition is

$$W2V = \frac{2}{N(N-1)} \sum_{i=1}^{N-1} \sum_{j=i+1}^{N} cosim(v_i, v_j) \tag{2}$$

where N ($N \geq 2$) is the length of the topic-word list, and $cosim(v_i, v_j)$ is the cosine similarity between the i-th word embedding and the j-th word embedding in the topic-word list. W2V is a value between 0 and 1; the closer the value to 1, the higher the W2V.

In this study, we used the Python natural language processing library Gensim [23, 24] to compute C_V. In calculating C_V, we obtained word occurrence probabilities and co-occurrence probabilities using Japanese Wikipedia. C_V is a value between 0 and 1; the closer the value to 1, the higher the coherence.

We evaluated the diversity of topic-word lists attached to all clusters resulting from clustering using TD [14]. TD is the number of unique words in the topic-word lists divided by the number of all words in the topic-word lists. TD is a value between 0 and 1; the closer the value to 1, the higher the diversity.

2.6 Semantic Search with Queries Containing Logical Operations

Embeddings generated by JSBERT contain the rich topical information in sentences. We conducted a study on semantic retrieval from sentence datasets with queries containing logical operations {AND, OR, NOT} [25]. In the present study, we propose an evaluation method for a topic-word list using semantic retrieval including the AND operation.

Semantic retrieval using the AND operation for queries x and y from a dataset is executed as follows: For embedding e of each sentence in the dataset, compute the cosine similarity between the embedding of x and e, and the cosine similarity between the embedding of y and e, and store them in vectors v_x and v_y, respectively, corresponding to the sentence order. After scaling vectors v_x and v_y to the interval [0, 1], compute vector v_z that takes the minimum value of each element of v_x and v_y. Then, sort v_z in descending order and return sentences corresponding to the top N of v_z as the search results.

3 Sentence Clustering and Evaluation

In this section, we describe the target Japanese sentence datasets for topic analysis, clustering, and the performance evaluation of clustering results.

3.1 Analysis Target Corpus

Test2400 Corpus [25]. The Test2400 corpus (Test2400) is a corpus of Japanese sentences that consists of 2,400 sentences on 12 topic categories that we collected for topical sentence classification and retrieval experiments. Table 1 shows the 12 labels of the Test2400 corpus and the number of sentences with each label.

Table 1. Twelve labels and the number of sentences with each label in the Test2400 corpus.

label	English translation	number of sentences
政治	politics	170
傷病	illness/injury	188
衣服	clothes	207
料理	cooking	240
スポーツ	sports	227
災害	disasters	118
天気	weather	162
ゲーム	games	193
動物	animals	180
小説	novels	135
勉強	study	388
IT	IT	192

Part of the Test2400 is shown in Table 2. The English translation of each Japanese sentence in Table 2(b) is presented using the same line numbers as those in Table 2(a).

Japanese Daily Dialogue [26]. Japanese Daily Dialogue (JDD) is a dataset that contains daily conversations on five topics, that is, "daily life," "school," "travel," "health," and "entertainment," and consists of approximately 30,000 sentences.

Livedoor News Corpus [27]. The Livedoor News Corpus (LNC) consists of approximately 150,000 sentences categorized into nine categories: "single woman," "information technology," "home appliances," "male," "movies," "female," "mobile," "sports," and "news."

Table 2. (a) Part of the Test2400 and (b) its English translation.

(a)

line number	sentence	label
1	今日の国会では原子力事業所に関する法律案について議論していた	政治
2	両側の唾液腺に腫脹が見られ，流行性耳下腺炎の疑いを認める。	傷病
3	今日は厚手の靴下を履いていこう。	服
4	麻婆豆腐をご飯にかけて食べるのが好きだ。	料理
5	明日はオリンピックのゴルフの予選を見に行きます。	スポーツ
6	去年オーストラリアで発生した森林火災は，甚大な被害をもたらした。	災害
7	高気圧の影響で、日中は暑い	天気
8	スーパーマリオは楽しいです。	ゲーム
9	ワニは、淡水域の生態系において生態ピラミッドの最高次の地位を占めています。	動物
10	川端康成の作品は心動かされる	小説
11	いとこが九州の国公立大学に通っています。	勉強
12	持ち運びやすく高性能な MacBook シリーズは、携帯性を重視するクリエイターの方におすすめです。	IT

(b)

line number	sentence	label
1	Today's parliament was discussing a bill about nuclear power plants	politics
2	Swelling is seen in the salivary glands on both sides, suggesting mumps.	illness/injury
3	Let's wear thick socks today.	clothes
4	I like to eat mapo tofu over rice.	cooking
5	Tomorrow I'm going to see the Olympic golf qualifying.	sports
6	A forest fire in Australia last year caused enormous damage.	disasters
7	It is hot during the day owing to the effect of high pressure	weather
8	Super Mario is fun.	games
9	Crocodiles occupy the highest position in the ecological pyramid of freshwater ecosystems.	animals
10	Yasunari Kawabata's work is moving	novels
11	My cousin attends a national public university in Kyushu.	study
12	The easy-to-carry, high-performance MacBook series is recommended for creators who value portability.	IT

3.2 Clustering Results

We evaluated the AMI of clusters obtained by clustering Test2400 sentence embeddings generated by JSBERT, NICT-BERT [17], and Word2Vec [7] into 12 clusters using KM.

These results are shown in Table 3 together with the clustering results using LDA and NMF for comparison. The row "JSBERT-normal + KM" in Table 3 shows the AMI value of the result of clustering the set of embeddings generated by JSBERT-normal from each sentence in Test2400 using KM. The substrings "-normal + KM" and "-nouns + KM" have a similar meaning to that presented above for NICT-BERT and Word2Vec. Each AMI in Table 3 is the average of 30 trials.

Table 3. AMIs of the clustering results for Test2400.

Clustering method	AMI
JSBERT-normal + KM	0.788
JSBERT-nouns + KM	**0.792**
NICT-BERT-normal + KM	0.420
NICT-BERT-nouns + KM	0.619
Word2Vec-normal + KM	0.391
Word2Vec-nouns + KM	0.604
LDA	0.037
NMF	0.041

Table 3 shows that clustering by topic achieved the highest performance when using embeddings generated by JSBERT, particularly the JSBERT-nouns model.

Because the JDD and LNC datasets did not have GTLs, we attempted to estimate the appropriate number of clusters required to cluster them and determine the topic word list for each cluster. We tried various methods, such as the elbow method, silhouette method, Calinski–Harabasz method, and Davies–Bouldin method; however, it was not possible to clearly estimate the optimal number of clusters.

Based on the above experimental results, we performed topic analysis on the data clustered into 12 clusters using sentence embeddings generated by JSBERT-nouns for Test2400. For JDD and LNC, we performed topic analysis on the data clustered into 20, 30, and 40 clusters using the sentence embedding generated by JSBERT-nouns, and used the average of the results as the performance evaluation result.

4 Topic-Word List Generation and Evaluation

In this section, we describe the automatic generation method of topic-word lists, which is the main result of this study, and the evaluation results of the obtained topic-word lists.

4.1 Automatic Determination of Hyperparameter α

The appropriate value of hyperparameter α in the topic-word evaluation function (Eq. (1) in Sect. 2.4) for selecting a topic-word list generally depends on the topic analysis target data and clustering results.

We formulated the problem of determining an appropriate α ($0 \leq \alpha \leq 1$) as the problem of maximizing the value of the following function:

$$g(\alpha, N) = W2V(\alpha, N) \times TD(\alpha, N), \tag{3}$$

where $W2V(\alpha, N)$ and $TD(\alpha, N)$ are the W2V score and TD diversity score, respectively, when the length of the topic word list is N with hyperparameter α. We used the Brent method from Python's SciPy package [28] to determine the value of α that maximizes $g(\alpha, N)$.

4.2 Examples of a Topic-Word List

In Table 4, we show the topic-word list when $N = 5$ obtained by clustering Test2400 into 12 clusters using KM and using the α value that maximizes Eq. (5) for each cluster obtained.

The topic of each cluster can be clearly understood from the topic-word lists in Table 4. Increasing the length of the topic-word list N makes the topic of the cluster easier to understand. However, if the topic-word list is too long, topic words that are not the subject of the cluster appear, and it becomes difficult to understand the content of all the clusters by looking at the topic-word lists. Therefore, in this study, we proceeded with the evaluation experiment for topic analysis using $N = 10$.

Table 4. (a) Topic-word list when $N = 5$ in Test2400 and (b) its English translation.

(a)

cluster	word 1	word 2	word 3	word 4	word 5
1	素材	コーディネート	トップス	服装	ファッション
2	地震	被害	災害	台風	豪雨
3	ゲーム	プレイ	キャラクター	カード	プレイヤー
4	情報	画面	データ	ネットワーク	ＩＴ
5	料理	ラーメン	調理	食事	調味料
6	動物	パンダ	ペット	イルカ	飼育
7	小説	作品	文学	物語	作家
8	選手	競技	スポーツ	オリンピック	野球
9	症状	病気	治療	感染	接種
10	気温	気圧	天気	前線	季節
11	大学	勉強	学部	講義	授業
12	大統領	大臣	首相	選挙	総理

(b)

cluster	word 1	word 2	word 3	word 4	word 5
1	material	coordination	tops	clothing	fashion
2	earthquake	damage	disaster	typhoon	heavy rain
3	game	play	character	card	player
4	information	screen	data	network	IT
5	cooking	ramen	cooking	meal	seasoning
6	animal	panda	pet	dolphin	breeding
7	novel	the work	literature	story	writer
8	player	competition	sports	Olympic	baseball
9	symptoms	disease	treatment	infection	vaccination
10	temperature	atmospheric pressure	weather	front	season
11	university	study	faculty	lecture	class
12	president	cabinet minister	prime minister	election	prime minister

4.3 Performance Evaluation of the Topic-Word List

In Table 5, we show the results of evaluating W2V, C_V, and TD for the topic-word list when $N = 10$ assigned to each cluster with the optimal value α that maximizes the function defined in Eq. (5) for Test2400, JDD, and LNC. For Test2400, we clustered the data into 12 clusters and evaluated W2V, C_V, and TD for the topic-word list assigned to each cluster. For JDD and LNC, we calculated W2V, C_V, and TD for the topic-word lists for 20, 30, and 40 clusters, and then averaged them.

Table 5. W2V, C_V, and TD for an automatically selected optimal value of α when $N = 10$.

	Test2400			JDD			LNC		
	W2V	C_V	TD	W2V	C_V	TD	W2V	C_V	TD
$\alpha = 0$	0.660	0.568	0.975	0.637	0.391	0.825	0.625	0.380	0.668
$\alpha = $ optimal	**0.675**	0.633	**1.000**	**0.656**	**0.472**	0.936	**0.641**	0.494	0.811
$\alpha = 1$	0.664	**0.639**	0.992	0.653	0.469	0.941	0.639	**0.498**	0.814
LDA	0.584	0.336	0.900	0.561	0.277	**1.000**	0.563	0.353	**0.990**
NMF	0.610	0.499	0.908	0.585	0.374	0.671	0.608	0.429	0.739

In Table 5, we show W2V, C_V, and TD for the row with $\alpha = 0$, which resulted from sorting the topic-word lists in descending order by cosine similarity alone; the row with $\alpha = 1$, which resulted from sorting the topic-word lists in descending order by cluster-based TFIDF alone; and the row with $\alpha = $ optimal, which resulted from sorting the topic-word lists in descending order by the α value that maximized Eq. (5). From this table, we observe that, when $\alpha = $ optimal, W2V is the maximum value for the Test2400, JDD, and LNC datasets, and C_V is also close to the maximum value.

5   MACR

5.1   Definition of MACR

The topic coherence and TD discussed thus far only evaluate the characteristics of topic-word lists, not how well the topic-word lists cover the content characteristics of each cluster. Therefore, we devised the MACR metric, which evaluates how well a topic-word list expresses the content features of the corresponding clusters.

The average cluster-recall $ACR(C, N)$ for cluster C obtained by clustering the original dataset is a metric that evaluates how well the cluster can be recalled using semantic search with queries combined by AND operations for the words in the topic-word list attached to the cluster. We define $ACR(C, N)$ as

$$ACR(C, N) = \frac{1}{N} \sum_{n=1}^{N} CR(C, n). \tag{4}$$

Let $|C|$ be the number of elements in cluster C. A semantic search is performed on the original dataset using a query that combines the top n words in the topic-word list corresponding to cluster C using the AND operation described in Sect. 2.6. $CR(C, n)$ is the number of sentences that belong to cluster C (i.e., correct answers) in the sentence set obtained by the above semantic search divided by $|C|$. Therefore, when a two-class classification problem is considered using a set of sentences that belong to cluster C as a correct example, $CR(C, n)$ represents the precision, recall, and F1-score of the semantic search results using the top n topic words.

$MACR(N)$ is the average of the $ACR(C, N)$ for each cluster C. Let the cluster set obtained by clustering the original dataset be $\{C_1, C_2, \cdots, C_M\}$. $MACR(N)$ is defined as follows: $MACR(N)$ is a value between 0 and 1; the closer the value to 1, the higher the topic coverage:

$$MACR(N) = \frac{1}{M} \sum_{m=1}^{M} ACR(C_m, N).$$ (5)

5.2 MACR Evaluation Results for Test2400, JDD, and LDN

Table 6 shows the $MACR(10)$ evaluation results for Test2400, JDD, and LNC. From this table, we observe that when α = optimal, the value of $MACR(10)$, that is, the cluster-recall rate, is higher than in other cases.

Table 6. $MACR(10)$ for Test2400, JDD, and LNC.

	Test2400	JDD	LNC
$\alpha = 0$	0.573	0.366	0.231
α = optimal	**0.586**	**0.403**	**0.319**
$\alpha = 1$	0.570	0.396	0.299
LDA	0.468	0.259	0.251
NMF	0.409	0.272	0.241

6   Conclusion

In this paper, we reported results for a fine-grained (sentence-based) topic analysis method and its evaluation method. The topic analysis method uses JSBERT to generate sentence embeddings from the data to be analyzed and clusters them. Then, a topic-word list with high topic coherence and diversity is automatically selected and added to each cluster using a topic-word evaluation function. We proposed a new MACR metric to evaluate how well topic-word lists reflect the content features of each cluster. Then, we evaluated the automatically selected topic-word list using coherence, diversity, and MACR.

In future work, we will consider topic analysis using dimensionality-reduced sentence embeddings. Similar to semantic search using sentence embeddings [29], we expect the reduction of the dimensionality of the sentence embeddings to increase the accuracy of the topic analysis by making it robust against noise. We also expect the reduction of the dimensionality to increase the speed of the topic analysis.

The fine-grained topic analysis method in this study is very useful for the topic analysis of questionnaire responses, reviews, and literary works.

References

1. Blei, D.M., Ng, A.Y., Jordan, M.I.: Latent Dirichlet allocation. J. Mach. Learn. Res. **3**, 993–1022 (2003)
2. Rosen-Zvi, M., Griffiths, T., Steyvers, M., Smyth, P.: The author-topic model for authors and documents. In: Proceedings of the Twentieth Conference on Uncertainty in Artificial Intelligence (UAI2004), pp. 487–494 (2004)
3. Zhao, W.X., Jiang, J., Weng, J., He, J., Lim, E.P.: Comparing Twitter and traditional media using topic models, In: Clough, P., et al. (eds.) Advances in Information Retrieval. ECIR 2011. LNCS, vol. 6611, pp. 338–349. Springer, Cham (2011). https://doi.org/10.1007/978-3-642-20161-5_34
4. SMU Text Mining Group: Twitter-LDA. https://github.com/minghui/Twitter-LDA. Accessed 8 Jan 2023
5. Lee, D.D., Seung, H.S.: Algorithms for non-negative matrix factorization. In: Advances in Neural Information Processing Systems 13: Proceedings of the 2000 Conference, pp. 556–562. MIT Press (2001)
6. Sia, S., Dalmia, A., Mielke, S.J.: Tired of topic models? Clusters of pretrained word embeddings make for fast and good topics too! In: Proceedings of the 2020 Conference on Empirical Methods in Natural Language Processing, pp. 1728–1736 (2020)
7. Mikolov, T., Chen, K., Corrado, G., Dean, J.: Efficient estimation of word representations in vector space. arXiv:1301.3781 (2013)
8. Devlin, J., Chang, M.W., Lee, K., Toutanova, K.: BERT: pre-training of deep bidirectional transformers for language understanding. arXiv:1810.04805 (2018)
9. Grootendorst, M.: BERTopic: Neural topic modeling with a class-based TF-IDF procedure. arXiv:2203.05794 (2022)
10. Reimers, N., Gurevych, I.: Sentence-BERT: sentence embeddings using Siamese BERT-Networks. In: Proceedings of the 2019 Conference on Empirical Methods in Natural Language Processing. Association for Computational Linguistics, (Online) (2019). https://arxiv.org/abs/1908.10084
11. Tsumuraya, K., Amano, M., Uehara, M., Adachi, Y.: Topic-based clustering of Japanese sentences using sentence-BERT. In: CANDARW 2022 (2022)
12. Amano, M., Tsumuraya, K., Uehara, M., Adachi, Y.: An analysis of representative works of Japanese literature based on emotions and topics. In: Barolli, L. (eds.) Advanced Information Networking and Applications. AINA 2023. LNNS, vol. 654, pp. 99–112. Springer, Cham (2023). https://doi.org/10.1007/978-3-031-28451-9_9
13. Röder, M., Both, A., Hinneburg, A.: Exploring the space of topic coherence measures. In: WSDM 2015: Proceedings of the Eighth ACM International Conference on Web Search and Data Mining, pp. 399–408 (2015)
14. Terragni, S., Fersini, E., Galuzzi, B.G., Tropeano, P., Candelieri, A.: OCTIS: comparing and optimizing topic models is simple! In: Proceedings of the 16th Conference of the European Chapter of the Association for Computational Linguistics: System Demonstrations, pp. 263–270 (2021)
15. Chuang, J., Gupta, S., Manning, C.D., Heer, J.: Topic model diagnostics: assessing domain relevance via topical alignment. In: Proceedings of the 30th International Conference on Machine Learning, JMLR: W&CP, vol. 28 (2013)
16. Korenčić, D., Ristov, S., Repar, J., Šnajder, J.: A topic coverage approach to evaluation of topic models. arXiv:2012.06274 (2021)
17. National Institute of Information and Communications Technology: NICT BERT Japanese Pre-trained models. https://alaginrc.nict.go.jp/nict-bert/index.html. Accessed 29 Oct 2022

18. Yoshikoshi, T., Kawahara, D., Kurohashi, S.: Multilingualization of a natural language inference dataset using machine translation, SIG Technical reports, vol. 2020-NL-244, no. 6 (2020). (in Japanese)
19. Arthur, D., Vassilvitskii, S.: K-means++: The Advantages of Careful Seeding. https://theory.stanford.edu/~sergei/papers/kMeansPP-soda.pdf. Accessed 16 Aug 2023
20. Pedregosa, F., et al.: Scikit-learn: machine learning in Python. J. Mach. Learn. Res. **12**, 2825–2830 (2011)
21. Vinh, N.X., Epps, J., Bailey, J.: Information theoretic measures for clusterings comparison: variants, properties, normalization and correction for chance. J. Mach. Learn. Res. **11**, 2837–2854 (2010)
22. Mikolov, T., Chen, K., Corrado, G.S., Dean, J.: Efficient estimation of word representations in vector space. In: International Conference on Learning Representations, Workshop Track Proceedings (2013)
23. Řehůřek, R., Sojka, P.: Software framework for topic modelling with large corpora. In: Proceedings of the LREC 2010 Workshop on New Challenges for NLP (2010)
24. Řehůřek, R.: GENSIM: Topic modelling for humans. https://radimrehurek.com/gensim/index.html. Accessed 28 Aug 2023
25. Tsumuraya, K., Uehara, M., Adachi, Y.: Semantic search of Japanese sentences using distributed representations. In: The 8th International Workshop on GPU Computing and AI (2023)
26. Akama, R., Isobe, J., Suzuki, J., Inui, K.: Construction of a Japanese Daily dialogue corpus [nihongo nichijyou taiwa corpus no kouchiku]. In: Proceedings of the 29th Annual Meeting of the Association for Natural Language Processing, pp. 108–113 (2023). (in Japanese)
27. RONDHUIT: Livedoor News Corpus. https://www.rondhuit.com/download.html. Accessed 5 Sept 2023
28. The SciPy community: SciPy. https://scipy.org/. Accessed 8 Sept 2023
29. Tsumuraya, K., Uehara, M., Adachi, Y.: Performance Improvement of Semantic Search Using Sentence Embeddings by Dimensionality Reduction, AINA2024 (submitted)

Performance Improvement of Semantic Search Using Sentence Embeddings by Dimensionality Reduction

Kenshin Tsumuraya[1], Minoru Uehara[1(✉)], and Yoshihiro Adachi[2]

[1] Graduate School of Information Sciences and Arts, Toyo University, Kawagoe, Japan
{s3B102200130,uehara}@toyo.jp
[2] RIIT, Toyo University, Kawagoe, Japan
adachi@toyo.jp

Abstract. Semantic search, which searches for sentences with a high similarity in meaning to that of queries, allows a user to search for the desired sentences even when they cannot think of the appropriate keywords for a lexical search. Moreover, the search function can appropriately handle synonyms and spelling variations. We previously reported a semantic search method for Japanese sentences using sentence embeddings that appropriately processed queries in which sentences were combined using the logical operators AND, OR, and NOT. Reducing the dimensionality of sentence embeddings is expected to make semantic search more robust to noise in the embeddings, resulting in improved search accuracy and faster semantic search computation. In this study, we experimentally verified the improvement in semantic search performance by reducing the dimensionality of sentence embeddings generated by Japanese SimCSE. We also evaluated the runtimes for generating sentence embeddings and reducing dimensionality with PCA.

1 Introduction

As digital transformation progresses in various areas of society, vast numbers of documents are being stored as digital data. Techniques to search for desired documents from a vast number of electronic documents are also becoming increasingly important. Traditional methods of full-text search or index search based on words or phrases [1–3] have been used for document search, but it is difficult to implement appropriate search processing for synonyms and spelling variations in these methods.

In recent years, classification and retrieval techniques for digital documents using sentence embeddings have been actively researched and reported (e.g., [4, 5]). In particular, a simple contrastive sentence embedding (SimCSE) framework can produce superior sentence embeddings, and SimCSE yields better performance on standard semantic text similarity (STS) tasks compared to the best previous results [6]. Japanese SimCSE (JSimCSE) was also developed, and JSimCSE showed the highest performance for Japanese STS tasks compared to previous sentence embedding generation models [7, 8].

In semantic search, it is a difficult task to write appropriate queries that are logically connected using the logical operators AND, OR, and NOT to retrieve the desired results.

L. Barolli (Ed.): AINA 2024, LNDECT 201, pp. 123–132, 2024.
https://doi.org/10.1007/978-3-031-57870-0_11

Furthermore, semantic search using long queries containing many topics is generally inaccurate. Therefore, it is desired to develop a semantic search technique using queries composed of logical operations on sentences containing a small number of topics.

In searches based on semantic similarity using sentence embedding, information on semantic similarity accompanies each sentence in the results. It is not appropriate to implement the search process for queries in which sentences (or words) are combined by the logical operators by simply performing the corresponding set operation on the search results for each sentence (or word).

We previously proposed a method for semantic search that contained logical operators in the queries, using computations such as minimum and maximum values for semantic similarities, with reference to fuzzy logic computation methods [9]. We then conducted an experiment using a dataset (Test2400 corpus) consisting of 2,400 sentences with supervised labels, and we verified that the method can return appropriate search results for queries that contain logical operations.

By reducing the dimensionality of the sentence embeddings, the semantic search process should become robust against noise and the search accuracy should improve. Moreover, dimensionality reduction increases the speed of semantic search. In this study, we used principal component analysis (PCA) to reduce the dimensionality of sentence embeddings and verified improvements in the accuracy and speed of the semantic search proposed in [9].

The results of this study are as follows:

- Using PCA, we analyzed the 768-dimensional sentence embedding vectors generated by JSimCSE from Test2400 and confirmed that the cumulative contribution rate (CCR) is over 60% when they exceed 49 dimensions and is over 80% when they exceed 119 dimensions.
- We reduced the sentence embedding vectors of Test2400 to 49 and 119 dimensions using PCA and verified the improvements in semantic search accuracy and computation speed.

This paper is structured as follows. In Sect. 2, we explain the Japanese sentence embedding generation model JSimCSE, the accuracy evaluation methods for the search results, and PCA. In Sect. 3, we provide a detailed explanation of semantic search using sentence embeddings for queries containing logical operators. In Sect. 4, we describe the improvement in search accuracy obtained by reducing the dimensionality of sentence embeddings using PCA. In Sect. 5, we describe the experimental results regarding the improvement in semantic search speed caused by dimensionality reduction. In Sect. 6, we summarize the study.

2   Related Materials

In this section, we describe sentence embedding generation, the evaluation of search result accuracy, and PCA.

2.1 Sentence Embedding Generation

SimCSE is a sentence embedding generation method that uses contrastive learning [6]. A supervised SimCSE is a SimCSE model that uses entailment pairs as positive examples and contradictory pairs as strict negative examples in natural language inference (NLI) datasets. The supervised SimCSE significantly outperforms other sentence embedding methods on STS tasks.

In this study, we used the Japanese Supervised SimCSE base model (JSimCSE) [7, 8] as a Japanese sentence embedding generation model. This JSimCSE uses cl-tohoku/bert-base-japanese-v3 [10] as the base model and is fine-tuned using the supervised SimCSE method using the Japanese translation [11] of the Stanford Natural Language Inference corpus (JSNLI) [12, 13] as the training dataset.

2.2 Evaluation Methods for Search Result Accuracy

Accuracy of semantic search is evaluated using well-known metrics for classification: precision (P), recall (R), and F1-score (F). Metrics P, R, and F take values between 0 and 1, and the closer they are to 1, the better each performance is. However, there is a trade-off relationship between P and R with respect to the F value.

Semantic search results are obtained in a ranked form according to similarity to the query. Therefore, we also evaluate semantic search results using MAP@K, a commonly used evaluation index for search tasks that yield ranking results. MAP@K takes a value between 0 and 1, and the closer it is to 1, the higher the correct items are ranked.

2.3 Dimensionality Reduction of Sentence Embeddings

The sentence embedding generated by JSimCSE is a 768-dimensional vector. We expected that the reduction of the dimensionality of the embeddings would increase the accuracy of the semantic search by making it robust against search noise. We also expected it to increase the speed of search.

PCA is a linear dimensionality reduction method [14]. It identifies important elements in the data, that is, principal components, and expresses the data as a set of principal components. The principal components are obtained by sequentially projecting the data in the directions in which the variance is large. PCA can reduce the dimensionality of newly input samples by mapping them into a reduced-dimensional space of the original dataset. This function is important when searching large-scale document databases with newly entered queries.

3 Semantic Search Using Sentence Embeddings

We explain a semantic search procedure using sentence embeddings.

3.1 Semantic Search Procedure

We provide an overview of the query processing procedure for semantic search proposed in [9]. The cosine similarities between sentence embeddings in a search target dataset and an embedding of a query are regarded as the degrees of membership of a fuzzy set.

For a basic query that does not include logical operations, the cosine similarity between the embedding of the query and the embedding of each sentence in the dataset is obtained. For queries consisting of an AND operation of sentences, the minimum values of the similarities obtained from each sentence are calculated, and for a query consisting of an OR operation of sentences, the maximum values of the similarities obtained from each sentence are calculated. A NOT operation on a query is calculated using $1 - e$ for each element e of the similarities obtained for the query. Sentences with high similarity to the query are returned as semantic search results.

3.2 Dataset for the Demonstration Experiment

Test2400 is a corpus consisting of 2,400 Japanese sentences labeled with 12 topic categories {politics, illness/injury, clothes, cooking, sports, disasters, weather, games, animals, novels, study, IT} that we collected for our sentence classification and search experiments using sentence embedding [9].

4 Improvement in Search Accuracy Obtained by the Dimensionality Reduction of Sentence Embeddings

In this section, we report experimental results that show reducing the dimensionality of sentence embeddings using PCA improves the accuracy of the semantic search proposed in [9]. We used the cuML library [14–16] on Google Colaboratory [17] for the PCA computations.

4.1 Cumulative Contribution Rate and MAP Value

When reducing the dimensionality of sentence embedding vectors, it is important that the number of dimensions after reduction be as small as possible while retaining sufficient information for classification and search. In machine learning, when the dimensionality reduction of feature vectors is performed using PCA, the principal components are usually extracted so that the CCR is 80% or more. In some cases, a CCR of 60% may be sufficient depending on the analysis target.

Figure 1 shows a graph of the CCR using PCA for Test2400. In Test2400, the CCR value exceeds 60% for 49 dimensions or more and exceeds 80% for 119 dimensions or more.

Table 1 shows the MAP@2400 values of semantic search results using sentence embeddings with dimensions reduced by PCA. This table shows that reducing the dimensions to 49 or 119 dimensions improves MAP@2400 by about 10% compared with the original 768-dimensional sentence embeddings.

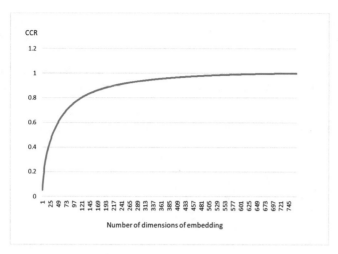

Fig. 1. CCR using PCA for Test2400.

Table 1. MAP@2400 values of semantic search results for dimensionality-reduced embeddings.

Dimension	MAP@2400 with PCA
49 (CCR $\cong$ 60%)	0.894
119 (CCR $\cong$ 80%)	0.877
768 (original)	0.797

4.2 Effect of Dimensionality Reduction on Search Accuracy

Table 2 presents the search accuracy at $N = 100, 200$, and 300 when the sentence embeddings of the Test2400 and queries are reduced to 49 dimensions by PCA. Furthermore, Table 3 shows the search accuracy at $N = 100, 200$, and 300 when the embeddings are reduced to 119 dimensions.

Table 2. Search accuracy using 49-dimensional embeddings reduced by PCA.

Query	$N = 100$			$N = 200$			$N = 300$		
	P	R	F	P	R	F	P	R	F
politics	0.990	0.582	0.733	0.735	0.865	0.795	0.520	0.918	0.664
illness/injury	0.980	0.521	0.681	0.760	0.809	0.784	0.560	0.894	0.689
⋮	⋮	⋮	⋮	⋮	⋮	⋮	⋮	⋮	⋮
IT	0.930	0.484	0.637	0.750	0.781	0.765	0.563	0.880	0.687
average	0.948	0.514	0.656	0.785	0.825	0.790	0.589	0.909	0.701

Comparing the results in Table 2 with the corresponding results in [9], it can be seen that reducing the original 768-dimensional sentence embedding to 49 dimensions improves the average F of the semantic search by about 3.6% to 8.1%. Comparing the results in Table 3 with the corresponding results in [9], it can be seen that reducing to 119 dimensions improves the average F by about 2.9% to 6.7%.

Table 3. Search accuracy using 119-dimensional embeddings reduced by PCA.

Query	$N = 100$			$N = 200$			$N = 300$		
	P	R	F	P	R	F	P	R	F
politics	0.960	0.565	0.711	0.690	0.812	0.746	0.507	0.894	0.647
illness/injury	0.970	0.516	0.674	0.750	0.798	0.773	0.567	0.904	0.697
⋮	⋮	⋮	⋮	⋮	⋮	⋮	⋮	⋮	⋮
IT	0.900	0.469	0.616	0.710	0.740	0.724	0.543	0.849	0.663
average	0.939	0.509	0.649	0.771	0.812	0.776	0.587	0.906	0.699

4.3 Effect of Dimensionality Reduction on Search Accuracy for Queries Containing Logical Operators

Tables 4 and 5 show the changes in search accuracy for queries containing logical operators by reducing the dimensionality of sentence embeddings. Comparing the results in Table 4 with the corresponding results in [9], the dimensionality reduction of sentence embedding increases F for OR search accuracy by about 5% or more. Table 5 shows the dimensionality reduction of sentence embeddings improves the search performance for queries containing AND, OR, and NOT.

Table 4. Search accuracy using dimension-reduced sentence embeddings for queries containing OR operator.

Query	Accuracy (49 dim., $N = 200$)			Accuracy (119 dims., $N = 200$)		
	P	R	F	P	R	F
"illness/injury"	0.760	0.809	0.784	0.740	0.787	0.763
"illness" OR "injury"	0.855	0.910	0.881	0.845	0.899	0.871

Table 5. Search results using dimension-reduced sentence embeddings for the queries "illness OR injury" and "(illness OR injury) AND NOT cancer."

	"illness OR injury"		"(illness OR injury) AND NOT cancer" min = 0.38, max = 1	
	768 dim., $N = 200$	49 dim., $N = 200$	768 dim., $N = 200$	49 dim., $N = 200$
illness/injury	159	171	155	156
cancer	10	10	3	2

5 Improvements in Search Speed Obtained by the Dimensionality Reduction of Sentence Embeddings

We show the measurement results of sentence embedding generation time, dimensionality reduction time, and semantic search time. This measurement was performed using GPU A100 on Google Colaboratory [17] by duplicating sentences to obtain the required number of sentences from the Test2400 corpus.

5.1 Speed of Sentence Embeddings Generation

Figure 2 shows the time needed to generate the 768-dimensional embedding using JSim-CSE on the Test2400 corpus. The horizontal axis indicates the number of sentences in the sentence data to be processed. It took approximately 6.2 s to generate embeddings for 10,000 sentences and approximately 59.7 s to generate embeddings for 100,000 sentences.

5.2 Speed of Dimensionality Reduction

Table 6 shows the computation speed of dimensionality reduction from 768 to 49 and 119 using PCA. The experimental results in Table 6 reveal that it takes almost the same number of seconds to reduce the embeddings to 49 or 119 dimensions. The dimension reduction time increases linearly with the number of sentences.

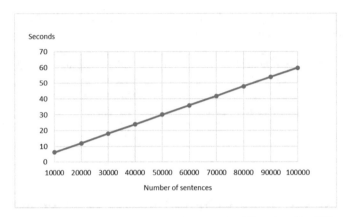

Fig. 2. Time needed to generate sentence embeddings by JSimCSE.

Table 6. Dimensionality reduction time in seconds by PCA.

Number of sentences	49-dimensional embedding	119-dimensional embedding
10000	0.550	0.553
20000	1.093	1.093
50000	2.701	2.703
100000	5.360	5.392

5.3   Effect of Dimensionality Reduction on Search Speed

Figure 3 compares the speeds of semantic search without logical operations when using the original 768-dimensional embeddings, reduced 49-dimensional embeddings, and reduced 119-dimensional embeddings. Using the dimensionality-reduced embeddings was eight times faster than using the original embeddings. The speed of semantic search using 49- and 119-dimensional embeddings was almost the same.

The execution times in seconds for semantic searches containing logical operations are shown in Table 7. The "AND," "OR," and "NOT" columns in this table are the execution times for queries containing one AND, OR, or NOT operation, respectively.

The execution times of the NOT operation in Table 7 are almost the same as the times in Fig. 3, which are for queries without logical operations. The execution times for the AND and OR operations of the two sentences in Table 7 are approximately twice the execution times for the cases shown in Fig. 3. This reveals that most of the query processing time was spent computing the cosine similarity between the embedding of the query and the embeddings of the dataset sentences under search. The computations of AND implemented using the mathematical "min" operation and OR implemented using the "max" operation are fast and take almost no time.

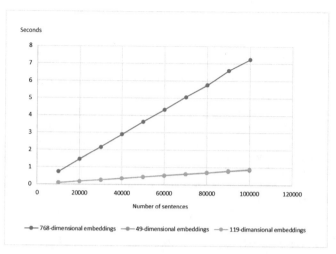

Fig. 3. Comparison of the semantic search speeds for 768-dimensional, 49-dimensional, and 119-dimensional embeddings.

Table 7. Execution time for queries containing a logical operator.

Number of sentences	768-dimensional embeddings			49-dimensional embeddings			119-dimensional embeddings		
	AND	OR	NOT	AND	OR	NOT	AND	OR	NOT
10000	1.415	1.464	0.636	0.194	0.192	0.104	0.197	0.198	0.097
20000	2.842	2.903	1.257	0.363	0.357	0.182	0.368	0.374	0.195
50000	7.143	7.275	3.644	0.863	0.868	0.445	0.899	0.895	0.499
100000	14.114	14.429	7.334	1.790	1.718	0.867	1.777	1.761	0.901

6 Conclusion

Semantic search allows users to search for desired sentences by entering queries with similar meanings without knowing the exact words or phrases needed to perform a lexical search. Additionally, semantic search provides a search function that appropriately handles synonyms and spelling variations.

We investigated the effect of dimensionality reduction of sentence embedding vectors on semantic search performance for queries that contain logical operators. As expected, we verified that reducing the dimensionality of sentence embeddings generated by JSimCSE increases semantic search accuracy and speeds up the search.

Future work includes evaluation experiments on improving the performance of semantic search by reducing the dimensionality of sentence embedding using standard datasets of Japanese sentences (e.g., [18]).

References

1. Lewis, D.D., Jones, K.S.: Natural language processing for information retrieval. Commun. ACM **39**(1), 92–101 (1996)
2. Namazu Project. Namazu: a Full-Text Search Engine. http://www.namazu.org/index.html.en. Accessed 21 June 2023
3. Groonga Project. Groonga. https://groonga.org/. Accessed 21 June 2023
4. Hugging Face. Using Sentence Transformers for semantic search. https://huggingface.co/spaces/sentence-transformers/embeddings-semantic-search. Accessed 21 June 2023
5. Elastic. Accelerate time to insight with Elasticsearch and AI. https://www.elastic.co/. Accessed 21 June 2023
6. Gao, T., Yao, X., Chen, D.: SimCSE: simple contrastive learning of sentence embeddings. arXiv:2104.08821v4 (2022)
7. Tsukagoshi, H., Sasano, R., Takeda, K.: Japanese SimCSE technical report. arXiv:2310.19349v1 (2023)
8. Tsukagoshi, H.: Japanese Simple-SimCSE. https://github.com/hppRC/simple-simcse-ja. Accessed 15 Dec 2023
9. Tsumuraya, K., Uehara, M., Adachi, Y.: Semantic search of Japanese sentences using distributed representations. In: CANDAR2023 GCA (2023)
10. Tohoku NLP Group. BERT base Japanese (unidic-lite with whole word masking, CC-100 and jawiki-20230102). https://huggingface.co/cl-tohoku/bert-base-japanese-v3. Accessed 15 Dec 2023
11. Yoshikoshi, T., Kawahara, D., Kurohashi, S.: Multilingualization of a natural language inference dataset using machine translation. SIG Technical reports, vol. 2020-NL-244, no. 6 (2020). (in Japanese)
12. Bowman, S.R., Angeli, G., Potts, C., Manning, C.D.: A large annotated corpus for learning natural language inference. In: Proceedings of the 2015 Conference on Empirical Methods in Natural Language Processing (EMNLP) (2015)
13. Stanford NLP Group. The Stanford Natural Language Inference (SNLI) Corpus. https://nlp.stanford.edu/projects/snli/. Accessed 1 June 2022
14. Jolliffe, I.T.: Principal Component Analysis. Springer Series in Statistics. Springer, New York (2002). https://doi.org/10.1007/b98835, ISBN 978-0-387-95442-4
15. Raschka, S., Patterson, J., Nolet, C.: Machine learning in Python: main developments and technology trends in data science, machine learning, and artificial intelligence. arXiv:2002.04803 (2020)
16. cuML - GPU Machine Learning Algorithms. https://github.com/rapidsai/cuml. Accessed 31 May 2023
17. Google Colaboratory. https://colab.research.google.com/. Accessed 31 May 2023
18. NII, NTCIR (NII Testbeds and Community for Information access Research) project. Accessed 5 Sept 2023

A Fuzzy-Based System for Assessment of Performance Error in VANETs

Ermioni Qafzezi[1(✉)], Kevin Bylykbashi[1], Shunya Higashi[2], Phudit Ampririt[2], Keita Matsuo[1], and Leonard Barolli[1]

[1] Department of Information and Communication Engineering, Fukuoka Institute of Technology (FIT), 3-30-1 Wajiro-Higashi, Higashi-Ku, Fukuoka 811-0295, Japan
qafzezi@bene.fit.ac.jp, {kt-matsuo,barolli}@fit.ac.jp
[2] Graduate School of Engineering, Fukuoka Institute of Technology (FIT),
3-30-1 Wajiro-Higashi, Higashi-Ku, Fukuoka 811-0295, Japan
bd21201@bene.fit.ac.jp

Abstract. Acknowledging the critical role of human factors in vehicular safety, our research addresses the existing gap in understanding and quantifying these factors in Vehicular Ad Hoc Networks (VANETs) domain. This paper explores the application of Fuzzy Logic (FL) in the context of VANETs, focusing on the assessment of driver performance. The three input parameters (driver experience, driver skills and mental and physical condition of the driver) are employed to determine the performance error of drivers. We present a methodological framework that leverages FL to model and analyze the intricate relationships between the specified input parameters and the resultant performance errors. Our findings not only provide valuable insights into the nuanced dynamics of driver behavior in VANETs but also contribute to enhancing the overall safety and efficiency of vehicular communication systems. This work underscores the significance of incorporating human-centric considerations in the design and optimization of intelligent transportation systems, paving the way for advancements in the field.

1 Introduction

In the pursuit of creating intelligent transportation systems, understanding and mitigating the influence of human factors on vehicular safety remains paramount. Among these factors, driver behavior emerges as a pivotal element that significantly impacts the occurrence of road accidents. The intricate interplay of drivers' cognitive abilities, experiences, and physical conditions introduces a complex dynamic to the road environment. As technological advancements propel the automotive industry towards the era of connected and autonomous vehicles, the role of drivers in the overall safety necessitates comprehensive examination.

A key focus within this domain is the quantification and assessment of performance errors exhibited by drivers. Performance errors encompass deviations from optimal driving behavior and may result from a myriad of factors including insufficient experience, inadequate skills, or compromised mental and physical states. Recognizing these

© The Author(s), under exclusive license to Springer Nature Switzerland AG 2024
L. Barolli (Ed.): AINA 2024, LNDECT 201, pp. 133–142, 2024.
https://doi.org/10.1007/978-3-031-57870-0_12

errors is imperative for preemptive intervention and the implementation of strategies to enhance road safety.

The integration of Fuzzy Logic (FL) into this context offers a sophisticated approach to discerning and categorizing performance errors. By encapsulating the inherent vagueness and imprecision associated with human behavior, FL provides a robust foundation for modeling and analyzing the nuanced relationships between driver attributes and potential risks. Ascertaining the critical threshold beyond which a driver's performance is deemed hazardous allows a real-time evaluation and, subsequently, the implementation of preventive measures.

In our previous work [10], we proposed a FL-based system that primarily considers internal factors, such as recognition errors, decision errors, and performance errors, as the key drivers of driver mistakes. In this work, we aim to contribute to the understanding of accident causation and prevention. This paper delves into the realm of driver behavior analysis within the framework of VANETs. By elucidating the concept of performance errors and proposing a FL-based methodology, our research seeks not only to understand the intricacies of human factors in driving but also to contribute actionable insights towards fortifying the safety landscape of modern roads. Emphasizing the amalgamation of technology and human-centric considerations, our work propels the discourse on making roads safer in the era of evolving transportation paradigms.

The paper's organization is as follows. Section 2 provides an overview of VANETs, tracing their evolution from traditional systems to cloud-fog-edge SDN-VANETs. In Sect. 3, we delve into the emerging technologies within VANETs, specifically focusing on SDN, Network Slicing, and enumerating their respective advantages. Section 4 is dedicated to the propsed and implemented FL-based system. The simulation results are deliberated in Sect. 5. Lastly, Sect. 6 concludes the paper.

2   Overview of VANETs

VANETs are a subset of Mobile Ad Hoc Networks (MANETs), designed to facilitate seamless communication among vehicles and between vehicles and roadside infrastructure. The establishment of network connectivity within the vehicular environment is achieved through the utilization of wireless communication technologies. Vehicles function as mobile nodes, engendering the real-time exchange of information amongst themselves and with stationary infrastructure components, such as traffic signals and road signs. This communication framework serves as the foundation for an array of applications, wielding considerable implications for road safety, traffic efficiency, and the overall driving experience [2,3,9,11].

The applications of VANETs encompass a myriad of domains, with a pronounced emphasis on enhancing road safety. Cooperative collision warning systems leverage VANETs to promptly alert drivers about nearby vehicles, thereby mitigating the inherent risk of accidents. Furthermore, VANETs contribute significantly to traffic management by enabling dynamic control of traffic signals and facilitating real-time data exchange to circumvent congestion. The resultant improvement in traffic flow translates into reduced travel times and decreased fuel consumption. Additionally, VANETs support infotainment and navigation applications by furnishing drivers with up-to-date information concerning points of interest, parking availability, and optimal route selection.

However, the integration of VANETs introduces challenges. The high mobility of vehicles and the dynamic nature of network topologies may lead to intermittent connectivity, impacting the reliability of communication. Ensuring the security and privacy of communication within VANETs is of paramount importance, given the potential vulnerability to malicious attacks that could compromise safety and data integrity. The deployment of VANET infrastructure, including roadside units, poses logistical and cost-related challenges. Effectively managing the substantial volume of real-time data generated by vehicles necessitates the implementation of efficient data dissemination strategies and scalable communication protocols.

Despite these challenges, ongoing advancements persist in VANET technology. The convergence of VANETs with emerging technologies, such as autonomous vehicles, holds considerable promise. Autonomous vehicles can leverage VANET data to augment decision-making processes, thereby enhancing navigation and coordination for safer operations. Additionally, the advent of 5G and advanced communication technologies presents opportunities for improved VANET connectivity, reduced latency, and enhanced data rates. Concurrent research endeavors focus on the development of machine learning and Artificial Intelligence (AI) algorithms to analyze and predict traffic patterns, facilitating more proactive traffic management strategies.

3 Technological Advancement in VANETs

The evolution of VANETs is intricately tied to technological advancements that extend beyond traditional communication paradigms. The integration of autonomous vehicles marks a significant stride in VANET development. While autonomous vehicles leverage onboard sensors and sophisticated algorithms for decision-making, they benefit immensely from the connectivity provided by VANETs. These networks enhance the real-time awareness of autonomous vehicles, enabling them to receive crucial information about the surrounding environment from other connected vehicles and infrastructure elements. The symbiosis between autonomous vehicles and VANETs is pivotal for achieving a comprehensive and context-aware approach to navigation and coordination.

The AI plays a pivotal role in VANETs by enhancing the intelligence of vehicles within the network. AI algorithms can analyze vast amounts of data generated by connected vehicles, enabling predictive analytics for traffic patterns and potential hazards. This predictive capability is crucial for autonomous vehicles to anticipate and adapt to dynamic road conditions. Moreover, AI facilitates efficient decision-making, such as route optimization and adaptive traffic management, contributing to the overall efficiency and safety of VANETs.

Despite the advancements, the full potential of autonomous vehicles and AI in VANETs relies on the continued presence of connected vehicles. Connected vehicles act as data sources, providing real-time information that enriches the capabilities of autonomous vehicles and AI algorithms. The synergy between autonomous, AI-enabled vehicles, and their connected counterparts forms a robust ecosystem, ensuring the seamless functioning and effectiveness of VANETs.

The integration of Edge-Fog-Cloud into VANETs represents another leap in network technological advancement. This architecture leverages distributed computing resources at the network edge, enabling faster data processing and reducing latency. In VANETs, the Edge-Fog-Cloud enhances the efficiency of real-time data exchange, contributing to quicker decision-making and response times. This decentralized approach aligns with the dynamic nature of vehicular environments, where low-latency communication is crucial for safety-critical applications.

Moreover, the adoption of Software-Defined Networking (SDN) and network slicing further refines VANETs. SDN allows the centralized control and programmability of the network, facilitating efficient resource allocation and dynamic adaptation to changing traffic conditions. Network slicing, on the other hand, enables the creation of virtualized, dedicated network segments tailored to specific applications or services. In VANETs, this integration ensures optimized communication pathways, meeting the diverse requirements of different vehicular applications.

The ongoing integration of autonomous vehicles, AI, Edge-Fog-Cloud, SDN, and network slicing into VANETs underscores a dynamic landscape where the synergy of these technologies is essential for realizing the full potential of intelligent vehicular communication systems. The collaborative interplay between autonomous, AI-enabled, and connected vehicles, coupled with advanced network architectures, heralds a new era in vehicular communication, promising heightened safety, efficiency, and adaptability on the roadways of the future.

4   Proposed Fuzzy-Based System

The proposed system is based on FL, which is used to assess driver performance considering the multifaceted nature of human behavior and the dynamic intricacies of real-world driving scenarios. Unlike traditional binary logic systems, the FL accommodates the inherent imprecision and vagueness inherent in human decision-making and responses. In the context of driver assessment, where factors such as experience, skill, and mental and physical conditions are inherently fuzzy and subjective, the flexibility of FL enables a more realistic modeling of the complex relationships between these variables. The ability of FL to handle uncertainty and variability aligns seamlessly with the nuanced and context-dependent nature of driver behavior, allowing for a comprehensive and adaptable evaluation of performance errors. Moreover, FL capacity to incorporate linguistic variables and rules facilitates a more interpretable and user-friendly system, essential for practical implementation in the domain of intelligent driving support systems [1,4–8,12–14].

In this paper, we introduce an intelligent driving support systems called Fuzzy System for Evaluation of Performance Errors (FS-APE). With an overarching commitment to advancing driving safety, FS-APE represents a novel and advanced tool designed to assess performance errors in real-time. Unlike conventional systems, FS-APE adopts a comprehensive approach, taking into account a diverse array of factors that contribute to performance errors. By examining variables such as driver experience, driver skills and mental and physical condition of the driver, FS-APE aims to provide a nuanced evaluation of driver behavior. The unique strength of FS-APE lies in its adaptability and precision, offering a sophisticated framework to unravel potential risks embedded in the dynamic context of real-world driving scenarios. This fuzzy-based system, therefore, emerges as a robust and flexible solution poised to contribute significantly to the enhancement of driving safety and the evolution of intelligent transportation systems.

Table 1. FS-APE parameters and their term sets.

Parameters	Term Sets
Experience (EX)	Inexperienced (InE), Moderate (MoE), Experienced (Exp)
Skill (SK)	Low (Lo), Moderate (Mo), High (Hi)
Mental and Physical Condition (MPC)	Good (G), Moderate (M), Bad (B)
Performance Error (PE)	RE1, RE2, RE3, RE4, RE5, RE6, RE7

We explain in detail the input and output parameters in following.

Experience (EX): The experience of the driver, a crucial input parameter in the proposed fuzzy-based system, encapsulates the cumulative exposure and familiarity gained by driver over time in diverse driving scenarios. This parameter considers factors such as years of driving, types of roads navigated, and encountered traffic conditions. The aim is to discern the impact of experiential knowledge on the driver's ability to navigate complex situations, contributing insights into the correlation between experience levels and potential performance errors.

Skill (SK): The skill of the driver, another pivotal input parameter, delves into the proficiency and dexterity with which the driver maneuvers the vehicle. This parameter encompasses aspects like handling, reaction times, and adherence to driving regulations. Skill is assessed through an intricate evaluation of driving maneuvers, response to dynamic road conditions, and the utilization of advanced driving techniques. The analysis of driver skill aids in understanding the nuanced relationship between driving proficiency and the likelihood of performance errors.

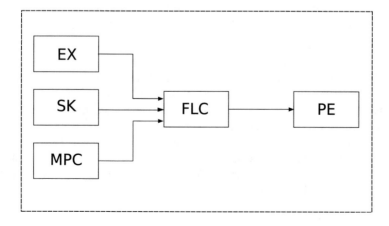

Fig. 1. Proposed system structure.

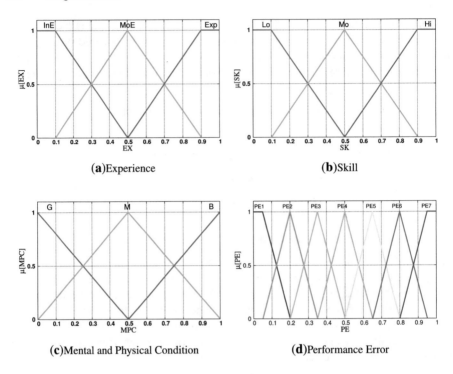

(a)Experience   (b)Skill

(c)Mental and Physical Condition   (d)Performance Error

Fig. 2. Membership functions of FS-APE.

Mental and Physical Condition (MPC): The mental and physical condition of the driver introduces a multifaceted dimension to the proposed fuzzy-based system, considering both cognitive and physical aspects that may influence driving performance. Mental factors encompass cognitive load, attention levels, and decision-making capabilities, while physical factors include fatigue, health conditions, and reflexes. This parameter aims to capture the holistic state of the driver, providing insights into how the intricate interplay of mental and physical well-being contributes to potential performance errors.

Performance Error (PE): The PE is the output parameter. It encapsulates deviations from optimal driving behavior, encompassing instances of erratic maneuvers, delayed responses, or failure to adhere to traffic regulations. The FS-APE system processes the input parameters to generate a nuanced evaluation of the driver's performance error. This output parameter serves as a critical metric, offering a quantifiable measure of the potential risks associated with the driver's behavior in real-time driving scenarios.

The design of the proposed FS-APE system is shown in Fig. 1. Details regarding the term sets for the input and output parameters are shown in Table 1. The membership functions for the input and output parameters are visually represented in Fig. 2, while the Fuzzy Rule Base (FRB) is shown in Table 2.

Table 2. FRB of FS-APE.

No	EX	SK	MPC	PE	No	EX	SK	MPC	PE
1	InE	Lo	G	PE5	15	MoE	Mo	B	PE5
2	InE	Lo	M	PE6	16	MoE	Hi	G	PE2
3	InE	Lo	B	PE7	17	MoE	Hi	M	PE3
4	InE	Mo	G	PE4	18	MoE	Hi	B	PE4
5	InE	Mo	M	PE5	19	Exp	Lo	G	PE3
6	InE	Mo	B	PE6	20	Exp	Lo	M	PE4
7	InE	Hi	G	PE3	21	Exp	Lo	B	PE5
8	InE	Hi	M	PE4	22	Exp	Mo	G	PE2
9	InE	Hi	B	PE5	23	Exp	Mo	M	PE3
10	MoE	Lo	G	PE4	24	Exp	Mo	B	PE4
11	MoE	Lo	M	PE5	25	Exp	Hi	G	PE1
12	MoE	Lo	B	PE6	26	Exp	Hi	M	PE2
13	MoE	Mo	G	PE3	27	Exp	Hi	B	PE3
14	MoE	Mo	M	PE4					

5   Simulation Results

The simulations for the proposed FS-APE system were executed using FuzzyC, presenting outcomes within three distinct scenarios characterized by varying levels of driver experience: Inexperienced, Moderate, and Experienced, shown in Fig. 3(a), Fig. 3(b), and Fig. 3(c), respectively. The results depict the relationship between PE and MPC for different SK values.

For scenarios characterized by low experience (EX = 0.1), as shown in Fig. 3(a), the results exhibit the highest values of PE, even for drivers characterized as highly skilled. This trend persists particularly when the mental and physical condition of drivers is compromised. This phenomenon suggests that the impact of limited driving experience may outweigh the mitigating effects of high skill levels, emphasizing the critical role of experience in navigating adverse conditions and maintaining optimal performance.

In contrast, for scenarios involving moderate experience (EX = 0.5), as shown in Fig. 3(a), the output values of PE are decreasing. Notably, PE consistently surpasses 0.5 for drivers with low skill levels, irrespective of the corresponding MPC values. This trend signifies a heightened susceptibility to performance errors among moderately experienced drivers with lower skill proficiency, underscoring the significance of skill development in mitigating errors even within moderately experienced cohorts.

The results reveal that for scenarios featuring experienced drivers (EX = 0.9), shown in Fig. 3(c), the PE values are the lowest. This suggests that drivers with extensive experience demonstrate a reduced likelihood of performance errors, emphasizing the positive impact of accumulated knowledge and familiarity with diverse driving situations.

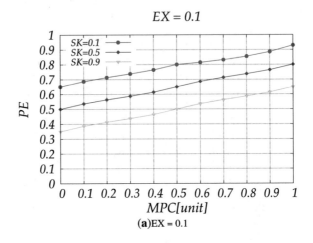

(a)EX = 0.1

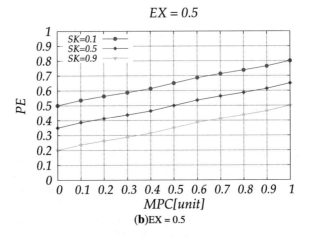

(b)EX = 0.5

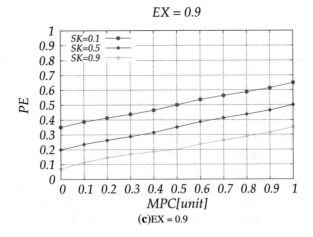

(c)EX = 0.9

Fig. 3. Simulation results for FS-APE.

The findings underscore the nuanced interplay between experience, skill, and mental and physical condition in influencing the likelihood of performance errors, providing valuable insights for the optimization of intelligent driving support systems.

6   Conclusions

In this paper, we presented FS-APE system for the assessment of driver performance error in VANETs. Through comprehensive simulations conducted using FuzzyC, we explored three distinct scenarios characterized by varying levels of driver experience: low, moderate, and high experience. We presented the relation between PE with MPC and SK. For experienced drivers (EX = 0.9), the observed low values of PE affirm the positive correlation between extensive experience and reduced performance errors. Most notably, in scenarios featuring inexperience drivers (EX = 0.1), the findings reveal that even highly skilled drivers exhibit heightened PE values, particularly under compromised mental and physical conditions.

These outcomes contribute to a deeper understanding of the complex relationships among experience, mental and physical condition, skill of drivers with performance errors during driving. The nuanced insights gleaned from this study have implications for the refinement of intelligent driving support systems, shedding light on the importance of tailoring interventions based on varying levels of driver experience and skill.

As we continue to advance in the realm of intelligent transportation, these findings pave the way for more effective strategies aimed at enhancing road safety and optimizing the performance of autonomous and connected vehicles in diverse driving scenarios.

References

1. Bojadziev, G., Bojadziev, M., Zadeh, L.A.: Fuzzy Logic for Business, Finance, and Management. Advances in Fuzzy Systems - Applications and Theory, vol. 12. World Scientific (1997)
2. Hartenstein, H., Laberteaux, K.P. (eds.): VANET: Vehicular Applications and Inter-Networking Technologies. Intelligent Transportation Systems, Wiley, Hoboken (2010). https://doi.org/10.1002/9780470740637
3. Hartenstein, H., Laberteaux, L.: A tutorial survey on vehicular ad hoc networks. IEEE Commun. Mag. 46(6), 164–171 (2008)
4. Kandel, A.: Fuzzy Expert Systems. CRC Press Inc., Boca Raton (1992)
5. Klir, G.J., Folger, T.A.: Fuzzy Sets, Uncertainty, and Information. Prentice Hall, Upper Saddle River (1988)
6. Klir, G.J., Yuan, B.: Fuzzy Sets and Fuzzy Logic - Theory and Applications. Prentice Hall (1995)
7. McNeill, F.M., Thro, E.: Fuzzy Logic: A Practical Approach. Academic Press Professional Inc., San Diego (1994)
8. Munakata, T., Jani, Y.: Fuzzy systems: an overview. Commun. ACM 37(3), 69–77 (1994)
9. Peixoto, M.L.M., et al.: FogJam: a fog service for detecting traffic congestion in a continuous data stream VANET. Ad Hoc Netw. 140, 103046 (2023). https://doi.org/10.1016/j.adhoc.2022.103046

10. Qafzezi, E., Bylykbashi, K., Higashi, S., Ampririt, P., Matsuo, K., Barolli, L.: A fuzzy-based error driving system for improving driving performance in VANETs. In: Barolli, L. (ed.) Complex, Intelligent and Software Intensive Systems, CISIS 2023, vol. 176, pp. 161–169. Springer, Cham (2023). https://doi.org/10.1007/978-3-031-35734-3_16

11. Schünemann, B., Massow, K., Radusch, I.: Realistic simulation of vehicular communication and vehicle-2-X applications. In: Proceedings of the 1st International Conference on Simulation Tools and Techniques for Communications, Networks and Systems & Workshops, SimuTools 2008, Marseille, France, 3–7 March 2008, p. 62. ICST/ACM (2008). https://doi.org/10.4108/ICST.SIMUTOOLS2008.2949

12. Zadeh, L.A., Kacprzyk, J.: Fuzzy Logic for the Management of Uncertainty. Wiley, New York (1992)

13. Zadeh, L.A., Klir, G.J., Yuan, B.: Fuzzy Sets, Fuzzy Logic, and Fuzzy Systems - Selected Papers by Lotfi A Zadeh. Advances in Fuzzy Systems - Applications and Theory, vol. 6. World Scientific (1996). https://doi.org/10.1142/2895

14. Zimmermann, H.J.: Fuzzy control. In: Zimmermann, H.J. (ed.) Fuzzy Set Theory and Its Applications, pp. 203–240. Springer, Dordrecht (1996). https://doi.org/10.1007/978-94-015-8702-0_11

Carbon Credits Price Prediction Model (CCPPM)

Inam Alanazi[✉], Firas AL-Doghman, Abdulrahman Alsubhi,
and Farookh Hussain

The University of Sydney Technology, 15 Broadway, Ultimo, NSW 2007, Australia
{Inam.alanazi,Abdulrahmannafeyea.alsubhi}@student.uts.edu.au,
{Firas.al-doghman,Farookh.Hussain}@uts.edu.au

Abstract. The valuation of carbon credits is a multifaceted task and is influenced by a wide range of factors encompassing economic activity, energy prices, weather conditions, policy adjustments, and market expectations. The accurate prediction of carbon credit prices is essential for traders, investors, regulators, and policymakers. To address the various facets of the carbon price prediction challenge, this paper contributes to this evolving field by proposing a solution that employs a machine learning methodology to enhance the accuracy and reliability of carbon credit price predictions. The proposed model is strong, precise and has the potential to guide decision making in the carbon market domain, with its proven accuracy and reliability highlighting its advantages as a valuable tool for stakeholders dealing with the intricacies of the carbon credit landscape.

1 Introduction

Carbon credits play a pivotal role in combating climate change by enabling industries to offset their carbon emissions. The accurate pricing of these credits is essential for fostering effective sustainable practices. This paper examines the realm of carbon credit pricing, with a specific focus on predicting prices through the utilisation of a machine learning model. The objective is to navigate the complexities inherent in this burgeoning market. There are several challenges associated with the accurate prediction of carbon credit prices. A fundamental obstacle is determining the optimal price for carbon credits. The work in [1] notes that research faces formidable hurdles due to the dynamic nature of global markets and complex environmental policies. The work in [2] further underscores the need for market-based instruments, such as carbon pricing mechanisms, to tackle the global scale of climate change and the diversity of emissions.

Additionally, [3] emphasises the pivotal role of precision in predicting carbon prices, asserting its significance in developing a reliable and beneficial carbon pricing mechanism while providing valuable insights for strategic business decisions. These studies collectively emphasise the urgent need for innovative and reliable methodologies in predicting carbon credit prices. To address these challenges, our proposed solution involves leveraging machine learning. This entails

L. Barolli (Ed.): AINA 2024, LNDECT 201, pp. 143–150, 2024.
https://doi.org/10.1007/978-3-031-57870-0_13

deploying advanced algorithms to analyse real-time data and environmental policies. The goal is to construct a predictive model that comprehensively understands patterns in the carbon credit market, ultimately enhancing pricing accuracy. This approach seeks to bring clarity to market participants. The precise forecasting of carbon prices is indispensable in establishing a robust and sustainable carbon pricing system, offering valuable support for informed business decision making. With the implementation of this proposed solution, we anticipate more stable and universally accepted carbon credit pricing mechanisms.

This paper is structured as follows: Sect. 2 presents the related work on carbon price prediction, Sect. 3 details the methodology employed in this research, Sect. 4 presents the results and provides a discussion and Sect. 5 concludes the paper.

2   Related Work

The valuation of carbon credits is subject to a myriad of influencing factors, encom- passing economic activity, energy prices, weather conditions, policy adjustments, and market anticipations.

Consequently, the prediction of carbon credit prices constitutes a formidable and pivotal endeavour for traders, investors, regulators, and policymakers alike. Diverse methodologies have been employed to forecast carbon credit prices, broadly classifiable into three categories: traditional econometric methods, artificial intelligence (AI) algorithms, and hybrid models.

2.1   Statistical Models

Traditional econometric approaches leverage statistical models to scrutinise histor- ical data and discern the variables impacting the supply and demand dynamics of carbon credits. Exemplifying this, [1] and [4] both applied multiple linear regression analysis, with the work in [1] additionally incorporating an auto-regressive integrated moving average model.

2.2   Machine Learning Models

AI algorithms, on the other hand, harness machine learning techniques to glean insights from data, uncovering nonlinear patterns and dependencies in the carbon market. For instance, [5] employed a PSO-RBF model to predict China's carbon trading market prices, showcasing its superior efficacy compared to other neural network models.

The work in [6] introduced an innovative combinatorial optimisation prediction method based on unstructured data, demonstrating its effectiveness in forecasting carbon trading prices in China. The work in [7] developed an ensemble prediction system integrating advanced data feature extraction technology and three sub-models, resulting in enhanced ac- curacy and stability in carbon price forecasting.

The work in [8] utilized six machine learning models, including extreme gradient boosting and support vector machines, to predict daily carbon prices and trading volumes in China, with the CEEMDAN-GWO-KNEA and CEEMDAN-RBFNN models exhibit- ing superior performance. The work in [9] employed a variety of algorithms, including linear regression, decision tree, random forest, extreme gradient boosting, and support vector machines, to predict the trading volume of carbon emissions, with random forest emerging as the most effective.

2.3 Hybrid Models

Hybrid models amalgamate the strengths of both econometric and AI methods, employing diverse techniques to address different facets of the carbon price prediction challenge. Noteworthy examples include the work in [10] which used a hybrid model comprising extreme point symmetric mode decomposition, an extreme learning machine, and a grey wolf optimiser algorithm, which surpassed the benchmark methods in predicting carbon prices in Hubei, Beijing, Shanghai, and Guangdong.

The work in [11] introduced a hybrid model integrating multi-resolution singular value decomposition and an extreme learning machine optimised by the adaptive whale optimisation algorithm, demonstrating superiority over benchmark methods in predicting carbon prices in China and the EU. The work in [12] proposed a hybrid model combining the ICEEMDAN decomposition-reconstruction method with the Sparrow search algorithm-optimised extreme learning machine model, resulting in heightened prediction accuracy, speed, and stability.

Additionally, the work in [13] advocated for a hybrid ARIMA and least squares support vector machine methodology, which proved effective in forecasting carbon prices. The work in [14] identified that a hybrid ARIMA and LSTM deep learning model exhibited the most favourable predictive performance for carbon trading prices in Shenzhen.

3   Method

This section explores the approach employed for carrying out our research. This section overviews the suggested model and the experimental process undertaken to derive results from this model. We use mean absolute error (MAE) which represents the average absolute difference between the predicted values and the actual values [15] to measure the efficiency of our proposed solution.

The level of accuracy impacts the utility or reliability of a model or predictions. A low MAE means that the predicted values are close to the actual values. This precision is crucial, especially in scenarios where accurate predictions are essential for decision making.

3.1 Prepossessing

In the realm of contemporary data-driven research, the significance of meticulous data preprocessing cannot be overstated. High-quality data forms the bedrock upon which accurate and reliable results are built, making the preprocessing phase an indispensable precursor to any analytical endeavour.

The integrity of research outcomes hinges on the quality of the input data, and as such, the process of refining and cleansing raw data emerges as a critical facet in ensuring the robustness and validity of subsequent analyses. Without diligent preprocessing, the potential for distorted or biased results increases, compromising the overall credibility and applicability of the research findings.

In light of these considerations, this section elucidates the various steps under- taken to refine, clean, and prepare the dataset for meaningful analysis, underscoring the pivotal role of data preprocessing in the pursuit of rigorous and dependable research outcomes.

3.1.1 The Selected Database

For the purpose of training and testing, we selected a dataset called "RCPI-data-public-aug2.xlsx," which is accessible via this link here Our choice of this dataset has been driven by its relevance to our intended objectives. We focus only on three columns in the Excel spreadsheet, namely the price, countries (with a specific focus on the USA), and the GHG level (carbon emissions) in line with global policy. These factors play a crucial role in influencing the decision-making process related to the carbon price in our proposed solution. Moreover, once the initial price is determined based on these variables, we further consider the dynamics of supply and demand to establish the ultimate price.

3.2 Algorithm

We used machine learning technology to build the solution and used the dataset to test its performance. Then, we represent our result as MAE.

The pseudo-code:

Function to compute real-time data based on multiple factors function computeRealTimeData(currentPrice, co2EmissionLevel, newRegulation, selectedCountry):

- Step 1: Calculate the impact of current price priceImpact = calculatePriceImpact (currentPrice)
- Step 2: Assess the influence of CO2 emission level emissionImpact = calculateEmissionImpact (co2EmissionLevel)
- Step 3: Evaluate the effect of new regulations regulationImpact = calculateRegulationImpact (newRegulation)
- Step 4: Consider the impact of the selected country countryImpact = calculateCountryImpact (selectedCountry)

- Step 5: Combine individual impacts to derive real-time data realTimeData = combineImpacts (priceImpact, emissionImpact, regulationImpact, countryImpact)
- return realTimeData

Our predictive model for global carbon credit prices is a robust tool that considers a range of influencing factors, ensuring a nuanced and accurate forecast. By incorporating the current price of carbon credits, the model establishes a baseline and identifies trends that can shape future valuations. The analysis of CO2 emission levels allows for an understanding of demand dynamics, reflecting the market's response to environmental commitments. New regulations are meticulously tracked, providing insights into the evolving landscape of carbon trading. Additionally, the model's consideration of selected countries acknowledges the diverse regulatory environments, enabling more precise regional predictions. The real-time data integration further enhances the model's adaptability, ensuring that predictions remain relevant in the face of dynamic market conditions. In essence, our model offers a comprehensive and dynamic approach to predicting carbon credit prices, empowering stakeholders with actionable insights for strategic decision making.

3.3   Implementation

In the development of our proposed solution, we used the Python programming language to craft the solution's code due to its versatility, extensive libraries, and ease of integration. Leveraging Python facilitates efficient coding practices and supports seamless integration with various data processing and analysis tools.

The core of our solution involves the utilisation of the RandomForestRegressor (RFR) machine learning algorithm. This algorithm, a part of the scikit-learn library, is well-suited for regression tasks and achieves robust performance in predicting outcomes based on complex datasets. Random forest is a robust and highly effective tool, demonstrating a level of performance that ranks among the most accurate methodologies to date, as highlighted by [16] The processed database, which forms the foundation of our model, undergoes thorough preprocessing to ensure data quality and relevance. This implementation choice aligns with our goal of creating a robust, adaptable, and accurate solution to address the challenges presented by our problem domain.

4   Result and Discussion

The exceptional MAE result of 0.011507294045069475 obtained from the carbon credits price prediction model underscores its excellence and reliability. This low MAE value demonstrates that, on average, the model's predictions are remarkably close to the actual market prices of carbon credits. The precision achieved by the model is particularly noteworthy in the context of carbon credit markets, where accurate forecasts are pivotal for informed decision making.

Fig. 1. The performance of the pricing prediction model

The model's ability to consistently provide predictions with such a small margin of error reflects its proficiency in capturing the complex and dynamic factors influencing carbon credit prices, as shown in Fig. 1. This outstanding performance enhances the credibility of the model and positions it as a valuable tool for various stakeholders, including businesses, investors, and policymakers. The reliability of the model, as evidenced by its low MAE, instils confidence in its predictions, making it an excellent resource for strategic planning, financial decision support, and sustainability initiatives. Overall, the noteworthy accuracy demonstrated by the model, as reflected in its low MAE, substantiates its status as an important and dependable tool in the domain of carbon credit price prediction.

5   Conclusion

In conclusion, this research presents a comprehensive exploration of the methodologies employed in predicting carbon credit prices. The proposed solution, underpinned by machine learning, demonstrates exceptional performance, as evidenced by a remarkably low MAE of 0.0115. This outstanding accuracy positions the model as a reliable tool for various stakeholders, offering valuable insights for strategic planning, financial decision support, and sustainability initiatives. The preprocessing phase, highlighted in the methodology section, underscores the significance of high-quality data in ensuring the robustness and validity of predictions. The selected dataset, focusing on price, countries (with a specific emphasis on the USA), and carbon emissions, aligns with global policy considerations, enhancing the relevance of the proposed solution. The implementation of the solution involves a step-by-step algorithm, including real-time data computation and the consideration of factors such as price, CO2 emission levels, new regulations, and selected countries. Overall, the research contributes to the evolving landscape of carbon credit price prediction, providing a robust and accurate model that can inform decision-making in carbon markets. The demonstrated precision and reliability underscore the potential of the proposed solution as a valuable asset for stakeholders navigating the complexities of the carbon credit landscape.

References

1. Kim, H., Kim, Y., Ko, Y., Han, S.: Performance comparison of predictive methodologies for carbon emission credit price in the Korea emission trading system. Sustainability **14**(13), 8177 (2022)
2. Aldy, J.E., Stavins, R.N.: The promise and problems of pricing carbon: theory and experience. J. Environ. Dev. **21**(2), 152–180 (2012)
3. Niu, X., Wang, J., Zhang, L.: Carbon price forecasting system based on error correction and divide-conquer strategies. Appl. Soft Comput. **118**, 107935 (2022)
4. Guðbrandsdóttir, H.N., Haraldsson, H.Ó.: Predicting the price of EU ETS carbon credits. Syst. Eng. Proc. **1**, 481–489 (2011)
5. Huang, Y., Hu, J., Liu, H., Liu, S.: Research on price forecasting method of China's carbon trading market based on PSO-RBF algorithm. Syst. Sci. Control Eng. **7**(2), 40–47 (2019)
6. Huang, Y., He, Z.: Carbon price forecasting with optimization prediction method based on unstructured combination. Sci. Total Environ. **725**, 138350 (2020)
7. Yang, Y., Guo, H., Jin, Y., Song, A.: An ensemble prediction system based on artificial neural networks and deep learning methods for deterministic and probabilistic carbon price forecasting. Front. Environ. Sci. **9**, 740093 (2021)
8. Lu, H., Ma, X., Huang, K., Azimi, M.: Carbon trading volume and price forecasting in China using multiple machine learning models. J. Cleaner Prod. **249**, 119386 (2020)
9. Wong, F.: Carbon emissions allowances trade amount dynamic prediction based on machine learning. In: 2022 International Conference on Machine Learning and Knowledge Engineering (MLKE), pp. 115–120. IEEE (2022)
10. Zhou, J., Huo, X., Xu, X., Li, Y.: Forecasting the carbon price using extreme-point symmetric mode decomposition and extreme learning machine optimized by the grey wolf optimizer algorithm. Energies **12**(5), 950 (2019)
11. Sun, W., Zhang, C.: Analysis and forecasting of the carbon price using multiresolution singular value decomposition and extreme learning machine optimized by adaptive whale optimization algorithm. Appl. Energy **231**, 1354–1371 (2018)
12. Zhou, J., Chen, D.: Carbon price forecasting based on improved CEEMDAN and extreme learning machine optimized by sparrow search algorithm. Sustainability **13**(9), 4896 (2021)
13. Zhu, B., Chevallier, J.: Carbon price forecasting with a hybrid ARIMA and least squares support vector machines methodology. In: Zhu, B., Chevallier, J. (eds.) Pricing and Forecasting Carbon Markets: Models and Empirical Analyses, pp. 87–107. Springer, Cham (2017). https://doi.org/10.1007/978-3-319-57618-3
14. Hu, Y., Xiao, W., He, B., Tang, X.: Carbon trading price forecasting with a hybrid arima and lstm deep learning methodology. In: Proceedings of the 2022 6th International Conference on E-Business and Internet, pp. 289–293 (2022)

15. Hyndman, R., Koehler, A.: Another look at measures of forecast accuracy. Int. J. Forecast. **22**, 679–688 (2006)
16. Svetnik, V., Liaw, A., Tong, C., Culberson, J.C., Sheridan, R.P., Feuston, B.P.: Random forest: a classification and regression tool for compound classification and QSAR modeling. J. Chem. Inf. Comput. Sci. **43**(6), 1947–1958 (2003)

Experimental Exploration of the Power of Conditional GAN in Image Reconstruction-Based Adversarial Attack Defense Strategies

Haibo Zhang[1]([⊠]) and Kouichi Sakurai[2]

[1] Department of Information Science and Technology, Graduate School of Information Science and Electrical Engineering, Kyushu University, Fukuoka, Japan
zhang.haibo.892@s.kyushu-u.ac.jp
[2] Department of Information Science and Technology, Faculty of Information Science and Electrical Engineering, Kyushu University, Fukuoka, Japan
sakurai@inf.kyushu-u.ac.jp

Abstract. Adversarial attacks pose a significant threat to the reliability and security of deep learning models, particularly in image processing applications. Defending against these sophisticated manipulations requires innovative strategies, with Generative Adversarial Networks (GANs) emerging as a promising solution. This paper presents an experimental exploration of the power of conditional Generative Adversarial Networks (cGANs) in image reconstruction-based strategies for defending against adversarial attacks. Our study involves a comparative analysis of four distinct image reconstruction models: the traditional GAN-based Defense-GAN, the cGAN-based method exemplified by pix2pix, a hybrid approach combining pix2pix with perceptual loss, and a generator model centered around residual blocks. The results of our experiments demonstrate that cGAN models exhibit significantly enhanced efficacy in defending against adversarial attacks compared to other image reconstruction methods. This superiority is attributed to the inherent characteristics of cGANs, which we delve into in detail. The findings provide crucial insights for developing more robust defense strategies against adversarial attacks in diverse image processing and machine learning applications.

1 Introduction

Adversarial attacks present a significant threat to the integrity of deep neural network (DNN) models, manipulating inputs to elicit erroneous outputs and bypassing conventional detection mechanisms. These attacks, exploiting the inherent vulnerabilities of DNNs, pose risks across a broad spectrum of applications, from image recognition to autonomous systems. Traditional countermeasures against these adversarial incursions have exhibited diversity; however, contemporary trajectories in machine learning scholarship are increasingly orienting towards adopting Generative Adversarial Networks (GANs) to bolster resilience against such adversarial threats.

© The Author(s), under exclusive license to Springer Nature Switzerland AG 2024
L. Barolli (Ed.): AINA 2024, LNDECT 201, pp. 151–162, 2024.
https://doi.org/10.1007/978-3-031-57870-0_14

Conventional methodologies to mitigate adversarial assaults encompass strategies such as input preprocessing, fortification of model robustness, and the implementation of ensemble techniques. However, the dynamic and evolving nature of adversarial attacks necessitates more adaptive and sophisticated approaches. GANs have surfaced within this paradigm as a formidable instrument. Specifically, within the GAN paradigm, Conditional Generative Adversarial Networks (cGANs) are distinguished by their distinctive proficiency in integrating conditional variables within the generative mechanism. This attribute confers a strategic advantage in devising defense architectures with enhanced resilience against adversarial threats.

This paper undertakes a comprehensive empirical investigation to evaluate the effectiveness of four divergent GAN-centric architectures as countermeasures against adversarial attacks. The models under scrutiny include the traditional GAN-based Defense-GAN [1], the cGAN-based method exemplified by pix2pix [2,3], a hybrid approach amalgamating pix2pix with perceptual loss [4], and a generator model focusing on residual blocks [5]. These models are rigorously tested across four diverse datasets: MNIST, Fashion-MNIST (F-MNIST), CIFAR-10, and ImageNet. The evaluation encompasses various aspects, from model stability and image fidelity to robustness against sophisticated adversarial attacks.

Our findings reveal a multifaceted landscape where each GAN variant contributes uniquely to defense strategies. The traditional Defense-GAN shows promise in simpler, grayscale datasets but falters in complex, high-resolution environments. In contrast, the cGAN-based models, particularly pix2pix, demonstrate versatility and robustness across a broader range of datasets. By comprehensively analyzing the performance of each model under different scenarios, this study aims to shed light on the strengths and limitations of each approach, providing a nuanced understanding of their roles in fortifying DNNs against adversarial threats.

2 Background and Related Works

2.1 Adversarial Attacks

In deep learning and machine learning, adversarial attacks represent a critical challenge to the integrity and reliability of predictive models. These attacks are engineered to bamboozle models through minute alterations to the input data, inducing fallacious outputs while eluding detection by human scrutiny. This vulnerability is particularly prevalent in high-stakes domains such as autonomous driving, healthcare, and cybersecurity [6]. The nature of adversarial attacks varies widely, ranging from white-box attacks, where the attacker has complete knowledge of the model, including its architecture and parameters [7], to black-box attacks, where the attacker has no direct access to the model's internals [8]. The complexity of these attacks and their potential to exploit subtle model weaknesses were further elaborated by Kurakin et al. [9].

The evolution of adversarial attacks has spurred a parallel advancement in defense strategies. A significant strand of research has focused on enhancing model robustness through training with adversarial examples, known as adversarial training [10]. This approach, however, is not a panacea, as it often leads to a trade-off between accuracy and robustness. The search for more effective defenses continues, with recent studies exploring using Generative Adversarial Networks (GANs) for detecting and mitigating adversarial perturbations. The dynamic nature of GANs, capable of generating realistic data samples, offers a promising avenue for countering the evolving sophistication of adversarial attacks, as exemplified in recent research named Defense-GAN [1].

2.2 Defense Models

2.2.1 Defense-GAN

Defense-GAN introduces a novel defense mechanism leveraging the generative capabilities of a Wasserstein Generative Adversarial Network (WGAN). As shown in Fig. 1, the primary objective is to use the generator to "denoise" or cleanse adversarial examples, ensuring that they closely resemble natural images, thereby making the classifier robust to adversarial noise.

At its core, Defense-GAN operates on the understanding that adversarial examples might fool deep learning models primarily because they often exist outside the natural data manifold of genuine images. The adversarial noise is mitigated by redirecting these adversarial samples back to the natural data manifold through a GAN.

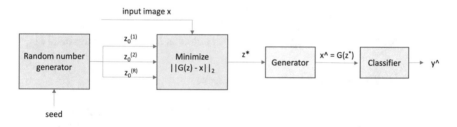

Fig. 1. The architecture of Defense-GAN illustrates the data flow from a random number generator to a generator, with optimization via minimization of loss, leading to a classifier.

2.2.2 Pix2pix-Based Method

Pix2pix is an image-to-image translation method that employs cGANs [2]. The primary objective of pix2pix is to map input images to corresponding output images, often used for tasks such as colorization, segmentation, and image synthesis. When applied for defense against adversarial attacks, Pix2pix can reconstruct images to their natural, unperturbed form.

Pix2pix operates on learning a mapping function from a source domain to a target domain. In defense against adversarial attacks, the adversarially perturbed images serve as the source domain, while the natural, unperturbed images represent the target domain. The goal is to learn this mapping in a way that removes the adversarial perturbations.

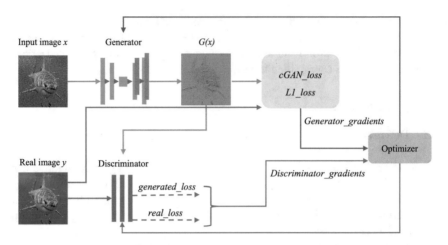

Fig. 2. The architecture of the pix2pix-based image reconstruction method.

Figure 2 introduces the architecture of a pix2pix-based image reconstruction method. Pix2pix's approach to image reconstruction offers an alternative method to Defense-GAN for mitigating adversarial noise. The key advantage of pix2pix lies in its ability to leverage paired data during training, ensuring accurate mappings from perturbed to natural images. However, like all defense mechanisms, its efficacy is contingent on the nature and strength of adversarial attacks it's defending against [11].

2.2.3   Hybrid Approach with Perceptual Loss

As Fig. 3 illustrates, The core augmentation in this methodology [4] is the integration of a feature extractor, which computes the perceptual distance between the feature maps of the generated image, $G(x)$, and the target image, y. By minimizing this perceptual distance, the generator is encouraged to reproduce pixel-wise images close to the target

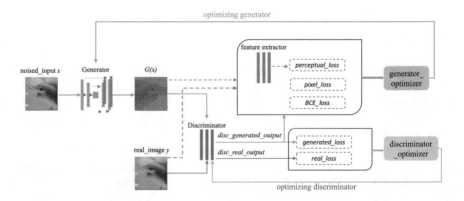

Fig. 3. The architecture of the hybrid image reconstruction method.

and to ensure that they reside in a similar feature space, enhancing their perceptual quality.

Furthermore, traditional losses such as pixel loss and binary cross-entropy (BCE) loss are retained to ensure the integrity of the translation process. The pixel loss ensures the generated image remains close to the target pixel-wise. In contrast, the BCE loss maintains the adversarial nature of the training, driving the discriminator to distinguish between real and generated images and, in turn, pushing the generator toward producing more realistic translations.

2.2.4   Residual Block-Based Generator

In defending against adversarial attacks by reconstructing images, a notable piece of work has been presented by Zhang et al. [5]. Their research is particularly significant due to the innovative use of perceptual loss in their image reconstruction network to bolster defenses against adversarial attacks.

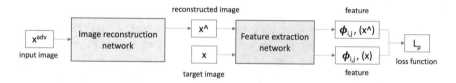

Fig. 4. The architecture of the residual block-based generator method.

Central to their approach is perceptual loss, which distinguishes their model from conventional reconstruction networks, as Fig. 4 shows. The incorporation of perceptual loss greatly suppresses the error amplification effect, enabling the creation of a more robust model capable of effectively counteracting adversarial perturbations. In essence, their method works two-fold: firstly, by converting adversarial images into non-adversarial representations, and secondly, by utilizing the perceptual difference between the adversarial image and the resultant clean reconstruction, ensuring the mitigation of error propagation.

3   Experimental Exploration

In this section, we compare four defense strategies on four image datasets: MNIST, FMNIST, CIFAR-10, and ImageNet. The defenses are evaluated under three adversarial attacks: FGSM ($\varepsilon = 0.3$), PGD ($\varepsilon = 0.3$), and the C&W attack. Our comprehensive analysis aims to understand the robustness and effectiveness of each defense in mitigating the adversarial perturbations introduced by these attacks.

3.1   Target Datasets

1. **MNIST:** The Modified National Institute of Standards and Technology (MNIST)[12] database is one of the foundational datasets in the machine learning community, predominantly used for handwritten digit classification. It consists

of 60,000 training and 10,000 test images, each of which is a grayscale image of 28×28 pixels.

2. **Fashion-MNIST:** Fashion-MNIST [13] is an alternative to the traditional MNIST dataset for benchmarking machine learning algorithms. It contains 60,000 training and 10,000 testing grayscale images, each 28×28 pixels in size, representing ten different categories of fashion items such as shirts, sandals, and bags.

3. **CIFAR-10:** The CIFAR-10 dataset [14] comprises 60,000 32×32 color images spread across ten classes, with 6,000 images per class. The dataset is divided into 50,000 training images and 10,000 test images. The classes encompass vehicles and animals, such as airplanes, automobiles, birds, and horses.

4. **ImageNet 2012:** The ImageNet Large Scale Visual Recognition Challenge (ILSVRC) 2012 dataset [15] is a benchmark in image classification and object detection. The dataset contains more than one million high-resolution images distributed over 1,000 categories. It has been the touchstone for evaluating the performance of deep neural architectures over the years.

3.2 Target Models

For the four datasets in question, we employed distinct image recognition models tailored to each dataset's unique characteristics. Specifically, for both the MNIST and Fashion-MNIST datasets, we adopted the LeNet model architecture [16]. We trained this architecture separately on each dataset, resulting in two pre-trained image recognition models, denoted as LeNet models. Regarding the CIFAR-10 dataset, we chose the ResNet architecture [17]. After training on the CIFAR-10 data, we obtained a pre-trained image recognition model, which we refer to as the ResNet model. As for the ImageNet dataset, we leveraged the pre-trained Inception_V3 model [18] available on the TensorFlow platform. This approach ensured that we utilized the most fitting models for each dataset, optimizing our chances for accurate image recognition.

3.3 Four Metrics Evaluation

Table 1 provides a comparative performance evaluation of four different models under FGSM attack with $\varepsilon = 0.3$. The models under consideration are the defense-GAN, pix2pix-based method, hybrid method with perceptual loss, and residual block-based generator method across datasets such as MNIST, F-MNIST, CIFAR-10, and ImageNet. The metrics evaluated include Accuracy (%), where higher values indicate better performance; PSNR, where higher values are better; Processing Time (ms), where lower values indicate faster processing; and Training Epochs, representing the number of epochs required for model convergence, where lower values are preferable. The best-performing values for each metric on each dataset have been highlighted in bold.

1. **Accuracy Analysis:** The hybrid method on the MNIST dataset showcases the highest accuracy at 98.65%, marginally higher than the next best, Res_generator method, at 96.4%. Defense-GAN and Pix2pix-based methods have similar performances, but the hybrid method outperforms them. On the F-MNIST dataset, all models showcase a decreased accuracy compared to MNIST. However, the hybrid method remains the

Table 1. Comparative Performance Metrics for Various Models on Different Datasets under FGSM ($\varepsilon = 0.3$) Attack.

Model	Accuracy (%)	PSNR (dB)	Processing Time (ms)	Training Epochs
MNIST (Before Attack: 98.73; After Attack: 27.48)				
Defense-GAN	96.6	34.48	2450	30,000
Pix2pix-based	94.3	34.09	**2.07**	2,000
Hybrid	**98.65**	34.27	2.1	**1000**
Res_generator	96.4	**34.73**	6.7	**1000**
F-MNIST (Before Attack: 88.8; After Attack: 10.32)				
Defense-GAN	87.5	30.69	2630	30,000
Pix2pix-based	84.8	32.61	**2.1**	2,000
Hybrid	**88.3**	**31.98**	**2.1**	**1000**
Res_generator	87.1	31.93	6.6	**1000**
CIFAR-10 (Before Attack: 89.63; After Attack: 8.31)				
Defense-GAN	9.69	29.46	3750	50,000
Pix2pix-based	80.54	33.28	**3.4**	500
Hybrid	**94.64**	**34.14**	3.5	400
Res_generator	90.32	34.07	28	**350**
ImageNet (Before Attack: 100; After Attack: 15.63)				
Defense-GAN	N.A.	N.A.	N.A.	N.A.
Pix2pix-based	77.54	77.86	11	200
Hybrid	**90.3**	**81.73**	**10.3**	150
Res_generator	88.4	79.54	82	**100**

top performer with 88.3%, followed closely by the Res_generator method. For the CIFAR-10 dataset, the hybrid method outperforms the others with an accuracy of 94.64%. The pix2pix-based method lags considerably behind the other models in this dataset. In the ImageNet dataset, the hybrid method maintains its superiority with an accuracy of 90.3%, followed by the Res_generator method.

2. **PSNR Analysis:** PSNR is an essential measure for evaluating the reconstruction quality, with higher values indicating better performance. Across all datasets, the hybrid method consistently delivers competitive or superior PSNR values, indicating a high image reconstruction quality, especially under adversarial attacks.

3. **Processing Time Analysis:** Across all datasets, the hybrid method boasts impressive processing speeds, with the lowest time being 2.1 ms for both MNIST and F-MNIST. This showcases the efficiency and optimization embedded within the hybrid method. Despite its commendable accuracy, Defense-GAN has a notably high processing time across datasets, which might be a concern for real-time applications.

4. **Training Epochs Analysis:** Training epochs indicate how long the model took to converge during training. Lower values suggest faster convergence, making the

model potentially more efficient in training. The hybrid method consistently converges with 1,000 epochs for MNIST, F-MNIST, and CIFAR-10, suggesting a consistent training behavior across varied datasets. Defense-GAN requires a notably higher number of epochs for convergence, especially on MNIST and F-MNIST, which may indicate a more intricate model architecture or perhaps a need for optimization.

3.4   Deep Analysis

The hybrid method consistently demonstrates superior or competitive performance across all metrics and datasets. Its high accuracy, impressive PSNR values, swift processing times, and consistent training epochs underscore its robustness and efficiency, especially under FGSM adversarial attacks. However, a deeper dive into the specific architectural differences, optimization techniques, and potential trade-offs among the models would provide a more comprehensive understanding of their strengths and weaknesses.

1. **Trade-off between Model Complexity and Efficiency:**
   Observing the processing times, even though our method demonstrates the highest accuracy across datasets, it achieves this with impressive efficiency. Such a combination is desirable but often challenging, as increased accuracy might come at the cost of greater complexity, leading to longer processing times. "Our Model" efficiency could be attributed to a well-optimized architecture or effective pruning techniques.
2. **Robustness under Adversarial Attacks:**
   The table assesses model performance under the FGSM attack. Adversarial attacks aim to perturb the input data minimally, making the model misclassified. The consistently high performance of our method across datasets suggests that it has a robust defense mechanism against such attacks, possibly utilizing techniques like adversarial training or leveraging architectures inherently resistant to adversarial perturbations.
3. **Training Efficiency:**
   The number of training epochs required for convergence can indicate the model's training efficiency. Our method consistently converges around 1,000 epochs for three datasets, implying a stable and efficient training behavior. This could result from superior optimization techniques, such as advanced gradient descent variants, regularization, or an effective learning rate strategy.
4. **Implications for Real-world Applications:**
   For real-world deployment, both accuracy and processing time are crucial. Despite its decent accuracy, Defense-GAN showcases a higher processing time, which might render it less suitable for real-time applications. In contrast, our method strikes a balance, making it a prime candidate for real-time, high-stakes applications, such as medical imaging or autonomous driving.

3.5   Results of PGD and C&W Attacks

Table 2 provides information on the resilience of different models to two adversarial attacks: PGD and C&W. It compares four models across four datasets: MNIST,

Table 2. Comparative Performance Metrics for Various Models on Different Datasets under PGD ($\varepsilon = 0.3$) and C&W ($L_2 norm$) Attacks.

Model	Accuracy (%)		PSNR (dB)	
	PGD	C&W	PGD	C&W
MNIST (Before Attack: 98.73)				
Attack	9.35	12.68	30.87	29.39
Defense-GAN	95.87	96.58	34.16	**34.93**
Pix2pix-based	93.24	94.52	33.96	33.78
Hybrid	**97.84**	**98.01**	**34.87**	34.02
Res_generator	95.47	94.03	34.65	34.42
F-MNIST (Before Attack: 88.8)				
Attack	2.09	1.06	29.18	27.38
Defense-GAN	82.36	74.04	30.53	30.25
Pix2pix-based	83.94	77.57	**32.18**	**31.92**
Hybrid	**87.26**	79.05	31.53	31.24
Res_generator	76.03	**84.72**	31.85	31.57
CIFAR-10 (Before Attack: 89.63)				
Attack	1.39	0.19	27.89	27.46
Defense-GAN	9.54	8.58	29.82	29.57
Pix2pix-based	79.83	78.07	33.05	32.76
Hybrid	**92.97**	**91.57**	**33.82**	33.37
Res_generator	88.94	87.54	33.97	**33.58**
ImageNet (Before Attack: 100)				
Attack	2.42	0	51.98	52.59
Defense-GAN	N.A.	N.A.	N.A.	N.A.
Pix2pix-based	76.76	80.4	77.58	77.08
Hybrid	**81.04**	**85.1**	**81.37**	**81.86**
Res_generator	79.59	82.08	80.75	79.58

F-MNIST, CIFAR-10, and ImageNet. The table presents accuracy (%) and PSNR (dB) metrics for each combination. The standout results for each metric across the datasets are underscored in bold.

Across all datasets, the hybrid method consistently emerges as the most resilient model against PGD and CW attacks, showcasing superior accuracy metrics. Notably, Defense-GAN also demonstrates commendable robustness, especially on the MNIST dataset. Different datasets present unique challenges. For instance, while Defense-GAN exhibits strong resilience on MNIST, its performance drops noticeably on CIFAR-10 and is absent for ImageNet, suggesting potential scalability issues. Conversely, the pix2pix-based method and the Res_generator method maintain relatively stable performances across datasets, indicating their general applicability.

Regarding PSNR, which gauges the perceptual quality of the defended images, models generally maintain consistent scores across attacks. The hybrid method ranks high, particularly on CIFAR-10, implying effective defense and good preservation of image quality. However, Defense-GAN's PSNR scores on F-MNIST under PGD attacks highlight its ability to defend while ensuring minimal distortions. The ImageNet dataset, representing high-resolution images, poses challenges for several models.

3.6   Visual Effects

Figure 5 illustrates the comparative results of four different image reconstruction-based adversarial defense strategies. The traditional Defense-GAN reconstructions exhibit notable distortions and a loss of detail, especially evident in the complexity of the natural images. In contrast, the cGAN-based methods show marked improvement in image quality and fidelity. The reconstructed images are more explicit, retain original features, and exhibit fewer artifacts. This comparison underscores the superior effectiveness of cGAN-based models in maintaining image integrity against adversarial attacks, which is particularly critical in real-world machine learning applications where accuracy and detail are paramount.

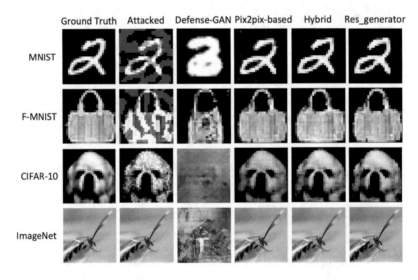

Fig. 5. A comprehensive matrix display illustrating the progressive refinement in image processing.

4   Conclusion

In summary, this research has demonstrated that cGANs, particularly when enhanced with perceptual loss and residual blocks, significantly outperform traditional GAN-based defense mechanisms against adversarial attacks. Our rigorous testing across multiple datasets reveals that cGANs offer a robust defense, especially in complex image scenarios, thereby establishing a new benchmark for adversarial defense strategies. These findings mark a substantial step forward in developing secure machine-learning models poised to withstand the evolving challenges presented by adversarial threats in diverse real-world applications. Future work will aim to refine these strategies, scaling to broader applications and further improving the resilience of neural networks.

Acknowledgement. This research is funded by JSPS international scientific exchanges between Japan and India, Bilateral Program DTS-JSP, grant number JPJSBP120227718.

References

1. Samangouei, P., Kabkab, M., Chellappa, R.: Defense-gan: protecting classifiers against adversarial attacks using generative models. arXiv preprint arXiv:1805.06605 (2018)
2. Isola, P., Zhu, J.Y., Zhou, T., Efros, A.A.: Image-to-image translation with conditional adversarial networks. In: Proceedings of the IEEE Conference on Computer Vision and Pattern Recognition, pp. 1125–1134 (2017)
3. Zhang, H., Sakurai, K.: Conditional generative adversarial network-based image denoising for defending against adversarial attack. IEEE Access **9**, 169031–169043 (2021)
4. Zhang, H., Yao, Z., Sakurai, K.: Eliminating adversarial perturbations using image-to-image translation method. In: International Conference on Applied Cryptography and Network Security, pp. 601–620. Springer, Heidelberg (2023). https://doi.org/10.1007/978-3-031-41181-6_32
5. Zhang, S., Gao, H., Rao, Q.: Defense against adversarial attacks by reconstructing images. IEEE Trans. Image Process. **30**, 6117–6129 (2021)
6. Goodfellow, I.J., Shlens, J., Szegedy, C.: Explaining and harnessing adversarial examples. arXiv preprint arXiv:1412.6572 (2014)
7. Szegedy, C., et al.: Intriguing properties of neural networks. arXiv preprint arXiv:1312.6199 (2013)
8. Papernot, N., McDaniel, P., Wu, X., Jha, S., Swami, A.: Distillation as a defense to adversarial perturbations against deep neural networks. In: 2016 IEEE Symposium on Security and Privacy (SP), pp. 582–597. IEEE (2016)
9. Kurakin, A., Goodfellow, I., Bengio, S.: Adversarial machine learning at scale. arXiv preprint arXiv:1611.01236 (2016)
10. Madry, A., Makelov, A., Schmidt, L., Tsipras, D., Vladu, A.: Towards deep learning models resistant to adversarial attacks. arXiv preprint arXiv:1706.06083 (2017)
11. Tripathi, A., Singh, R., Chakraborty, A., Shenoy, P.: Edges to shapes to concepts: adversarial augmentation for robust vision. In: Proceedings of the IEEE/CVF Conference on Computer Vision and Pattern Recognition, pp. 24470–24479 (2023)
12. LeCun, Y.: The mnist database of handwritten digits (1998). http://yann.lecun.com/exdb/mnist/
13. Xiao, H., Rasul, K., Vollgraf, R.: Fashion-mnist: a novel image dataset for benchmarking machine learning algorithms. arXiv preprint arXiv:1708.07747 (2017)
14. Krizhevsky, A., Hinton, G., et al.: Learning multiple layers of features from tiny images (2009)
15. Russakovsky, O., et al.: ImageNet large scale visual recognition challenge. Int. J. Comput. Vision (IJCV) **115**(3), 211–252 (2015). https://doi.org/10.1007/s11263-015-0816-y
16. LeCun, Y., Bottou, L., Bengio, Y., Haffner, P.: Gradient-based learning applied to document recognition. Proc. IEEE **86**(11), 2278–2324 (1998)

17. He, K., Zhang, X., Ren, S., Sun, J.: Deep residual learning for image recognition. In: Proceedings of the IEEE Conference on Computer Vision and Pattern Recognition, pp. 770–778 (2016)
18. Szegedy, C., Vanhoucke, V., Ioffe, S., Shlens, J., Wojna, Z.: Rethinking the inception architecture for computer vision. In: Proceedings of the IEEE Conference on Computer Vision and Pattern Recognition, pp. 2818–2826 (2016)

Framework for Cognitive Self-Healing of Real Broadband Networks

Enock Cabral Almeida Vieira[1]([✉]), Paulo Carvalho[2],
and Flávio de Oliveira Silva[1,2]

[1] Federal University of Uberlândia, Uberlândia, MG, Brazil
{enock.vieira,flavio}@ufu.br
[2] Centro Algoritmi, University of Minho, Braga, Portugal
{pmc,flavio}@di.uminho.pt

Abstract. With the growing volume of data traffic demanded by corporate, business, and retail consumers, telecommunications operators are becoming an increasingly important player in the world economy. However, the operators must prepare themselves with solutions that allow dealing with incidents more quickly or even avoid them, always focusing on maintaining an acceptable customer service level. In this context, Self Healing (SH) solutions, supported by Machine Learning (ML) mechanisms, emerge as possibilities to address this challenge. This work presents a cognitive self-healing framework for telecommunications operators. This framework encompasses self-diagnosis, analysis, and automatic actuation for failure mitigation in fiber broadband telecommunications based on Gigabit Passive Optical Network (GPON). In addition, we did an experimental evaluation using a dataset from the operators' modems extracted from its Network Management System (NMS), bringing more reliability to our results. This work shows that using ML in telecommunication broadband networks is viable and can change how telecom operators manage and improve customer experience. We show that an intelligent model could do machine learning in telecom networks and make decisions without human intervention. Three automatic cognitive models were tested as experimental proof of concept with an average accuracy above 96%.

1 Introduction

The concept of network Self Healing (SH), or self-correcting networks, is becoming a major objective of telecommunications operators to achieve the expected customer's level of service. According to [1], a system with a self-healing property is expected to be able to monitor and recognize anomalies, locate the source of a failure, respond to changing conditions, and execute mechanisms to bring the system back to the *normal* operational state.

With the increasing deployment of new technologies of Automation, Artificial Intelligence, Analytics and related Application Programming Interfaces (API), possibilities are created for SH applications in various types of networks, such as fiber in the loop access networks (Fiber to the x (FTTx), where x generalizes for several configurations of fiber deployment), metropolitan networks (metro rings), intercity backbone networks, mobile networks (Fourth Generation Cellular Mobile Network (4G) and Fifth Generation Cellular Mobile Network (5G)) and data center enterprise networks. These types of networks have one need in common: assuring high service levels. Despite the particularity of each one, all of them require specific actions to identify failures/anomalies, analyze these failures, and propose their correction. Most of these activities are carried out manually by telecommunications operators' analysts. This may lead to more time to resolve an issue, increasing the Mean Time To Repair (MTTR), impacting the Service Level Agreement (SLA), and increasing service user complaints.

In this work, we present a framework that includes self-diagnosis, analysis, and application of intelligence, and, based on automation, it can deal with network failures to reduce the downtime of telecommunications networks. The work relies on information generated by network elements. It dynamically treats these volumes of data, resorting to artificial intelligence to learn and understand the behavior of data flows progressively and take self-correcting actions if required.

The work contributes to leveraging customer experience in fixed broadband networks using Machine Learning (ML) to support SH using data available in systems that support telecom industry operations and are present in every operator's infrastructure. In addition, our work contributes to the literature by providing a study that uses a real dataset, publicly available [2], and provides three different ML models that provide high accuracy and contribute to the research in this area of study.

The remainder of this paper is organized as follows: the related works are discussed and compared in Sect. 2; the proposed framework for cognitive self-healing in telco networks and the experimental methodology is described in Sect. 3; and the obtained results and insights are included in Sect. 4. Finally, the study's concluding remarks and suggestions for future work are presented in Sect. 5.

2   Related Work

In this section, we will discuss related work and observe research opportunities for evolution.

The work of [3], uses ML with supervised learning techniques, like Decision Tree (DT) and Support Vector Machines (SVM) in a wireless sensors network, using a 30 sensors data set. They generate Link quality estimation to improve fault prediction. However, this work did not apply the concepts of SH nor real data from telecommunication operators.

The paper [4] works with unsupervised learning in a broadband internet service provider to improve the fault detection process using historical logs from a network operator. Meanwhile, they do not propose a near real-time solution for this process.

In [5], actors used DT algorithms to identify faults in enterprise networks. They used a generic data set and classified situations as normal or anomalous, associating this with the correlated hardware and software components. Despite this, they did not apply SH concepts nor classify the situations in a near real-time way.

As regards network management, the authors of [6], adopting the concept of cognitive computing, propose an intelligent architecture of autonomous network based on Network Functions Virtualization (NFV), capable of achieving Quality of Service (QoS) objectives and operational efficiency. This architecture, called CogNet [6], uses ML algorithms to dynamically adapt resources to the immediate requirements of NFVs, minimizing application performance degradations. As in CogNet, our work will use ML for data analysis and subsequent suggestions of the best action to be worked on. However, CogNet only applies self-management to generic NFV structures, not real broadband network elements.

According to [7], an important point to be considered in carrying out a SH process is to work on the data collection part. The work identifies that telecommunication networks have four striking characteristics that can be difficult to handle with automation: volume (large quantities), speed (high frequency of collection), variety, and value (information hidden in the data). In [7], a group of databases is used in a single database for querying, which helps in intelligent agent learning. Despite studying the characteristics of telecommunications network data, [7] does not use real data from operators to validate its model.

In [8], the authors tested seven SH methods for 4G. They realized that unsupervised classification methods such as ADABOOST, Fuzzy logic-based with *big data* had behaved well, bringing good diagnoses to the root causes of failures in a mobile network. Conversely, in the present work, supervised learning methods are adopted, attending to the characteristics of data available for the training phase (see details in Sect. 3).

The work [9] uses real data from a telecom operator to do traffic prediction analysis of a broadband concentrator - Broadband Network Gateway (BNG). This work classifies and detects inconsistent data from a field failure report of a Gigabit Passive Optical Network (GPON) broadband access network. Our approach also uses real data from a telecom operator but applies SH associated with ML to the operator's network instead.

In [10], Convolutional Neural Networks (CNN)s are applied to fault diagnosis in telecommunications networks. The study compares solutions using the Naive Bayes algorithm, CNNs, and the Random Forest (RF) algorithm. Their results show that CNNs perform better than the other methods studied. However, despite working with network incidents in a dataset [2] adapted between real and interpolated data, this work did not apply the concept of SH.

Table 1 compares the related works discussed above, identifying each approach's characteristics and positioning the present work.

The Self Healing column classifies if the work addresses the concept of SH. The Machine Learning column classifies if the work applies the concept of ML. The next column shows if the work considers SH applied to broadband networks. The following column shows if the work uses real network telecommunication operator's data in their experiment. The last three columns identify if the papers use a real-time process to collect data or make an action in the network and also identify the network type and the SH domain studied.

Table 1. Related Work Comparison

Work	Self-Healing	Machine Learning	SH in Broadband Networks	Network Operator Real data	Real Time	Network Type	SH Domain
Wang [3]	No	Yes	No	No	No	Wireless	Fault Prediction
Hashmi [4]	Yes	Yes	Yes	Yes	No	Broadband	Fault Detection
Kiciman [5]	No	Yes	No	No	No	Enterprise	Fault Localization
Xu [6]	Yes	Yes	No	No	Yes	Cellular	Resource Mgmt
Omar [7]	Yes	Yes	No	No	No	Cellular	Fault Det., Aut. SH
Rahmani [8]	Yes	Yes	No	No	No	Cellular	Fault Diagnosis
Silva [9]	No	Yes	No	Yes	No	Broadband	Traf. Pred., Anom. Detec.
Bothe [10]	Yes	Yes	No	Yes	Yes	Cellular	Fault Diagnosis
Present Work	Yes	Yes	Yes	Yes	Yes	Broadband	Fault Detect., Aut. SH

To improve the analysis, Table 2 shows the ML techniques and some features used in each approach.

Table 2. Related Work Features and ML Techniques

Work	ML Technique	Features used
Wang [3]	Sup: DT, SVM	Signal strength, Channel load
Hashmi [4]	Unsup:: k-Means, FCM·SOM	Fault time and cause, Region, MTTR
Kiciman [5]	Sup: DT	Paths classified as Normal or Anomalous
Xu [6]	Unsup: ANN	QoS, Energy cons., VNF Load
Omar [7]	Sup: DT	Max. data rate, #users per BS and SNR
Rahmani [8]	Unsup: SVM,Adaboost, Fuzzy	Cell Power, Avg speed, Signal Quality,
Silva [9]	Unsup: SOM Sup: RNN	OLT RX Power, ONT RX Power, ONU dist.
Bothe [10]	Sup: RF, CNN, Neur.	Cell Outage, TxPower, Antenna Tilt
Present Work	Sup: ANN, DT, GradBoost	Jitter, Latency, PL, WiFi Quality

3   A Framework for Cognitive Self-Healing in Real Broadband Networks

The present work specifies a framework for implementing Cognitive Self-Healing in GPON networks (see Fig. 1). This framework is intended to be general enough for any telecommunications operator to implement it, bringing the benefits already mentioned in this paper's introduction, such as improved service and user satisfaction.

The first step was studying the problems and needs of a telecommunications operator to identify the parameters that must be monitored to ensure the best possible experience for the operator's users in using the broadband service. To this end, several interviews were conducted with the operator's Network Operations Center (NOC) and marketing area employees.

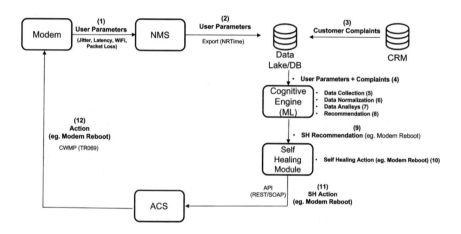

Fig. 1. Framework for Cognitive Self-Healing in Broadband Networks

One of the frequent fixed broadband internet users' pains is the low internet speed experience. The user usually does not know if problems are in the operator network, whether it is a Wireless Fidelity (Wi-Fi) router, or a specific modem issue. According to the visited operator, Modem and Wi-Fi represent together 42% of the total user complaints.

After analyzing information of the operator's Customer Relationship Management (CRM), it was noticed, by cross-referencing customer complaint information with the parameters of their modem (see Table 3), that there was a pattern indicating that a customer complaint occurred when a combination of parameters was outside the acceptable limits for a good user experience.

In this way, CRM data is collected and filtered to determine whether a customer has complained and combined with the real network parameters from the Network Management System (NMS). Now, we could see that when the user

Table 3. Parameters Definition

Parameter	Definition
Latency	Amount of time it takes for data to get from one network point to another
Jitter	Variation in latency
Packet Loss	The percentage of packets lost during transmission over a network
2.4 GHz Channel Quality	Quality of a 2.4 GHz channel, rated on a scale of 1 to 5
5GHz Channel Quality	Quality of a 5 GHz channel, rated on a scale of 1 to 5
N distant devices	Number of devices far from the modem (more than 10 m)

complained, its modem parameters were out of the acceptable limits for a good Quality of Experience (QoE).

The results are stored in a new database Data Lake (DTL) that will guide the intelligent software in the learning and automatic decision-making processes without human intervention. For this framework, we assume the NMS can export data to the DTL in a Near Real Time (NRT) model.

The framework includes implementing an ML model trained with normalized data from the DTL using supervised learning for data analysis. Three ML methods were used to evaluate which would be more accurate to identify whether the customer would complain. Once the model is trained, the model can be used to analyze the user parameters and predict whether the user will complain. Once it is identified that the client would complain, SH actions should take place. In practice, an SH module is proposed, capable of calling the API of an Auto-Configuration Server (ACS) platform that supports the CPE WAN Management Protocol (CWMP) in its Technical Report 069 (TR069) specification, defined in [11]. The ACS platform is very common in operators. It is used today by operators' customer care teams to reactively handle customer complaints with commands such as modem reboot and Wi-Fi channel change. Also, according [11], the TR069 specification is supported by the majority of the broadband modems in the world.

In the framework diagram (Fig. 1), the Modem module represents the user broadband equipment. The modem sends the user parameters (*Message #1*) to the NMS module, which collects the data from the modem. It is important to know that the network vendor implements the Modem and NMS interface in a proprietary IP protocol. However, this is not a problem for our framework because managing the modems with an NMS is mandatory to operate the network. To this framework, we assume that the NMS can export the user parameters (*Message #2*) to the Data Lake/DB module at a near real-time frequency.

The Data Lake/DB module receives data from the NMS and CRM modules. The CRM module exports the user complaints (*Message #3*) to the Data Lake/BD module being this data is important to train a core module, the Cognitive Engine (ML). The Cognitive Engine (ML) module is an intelligent component that analyses the data received from the Data Lake/BD module (*Message #4*), trains and tests a Machine Learning algorithm to understand where a specific user parameters combination could transform into a user complaint. The Cognitive Engine (ML) module, which will be further detailed in the next subsection, is expected to output SH recommendations (*Message #9*) to the Self-Healing module.

The Self Healing module is responsible for receiving the recommended action from the Cognitive Engine, analyzing the messages, identifying which type of action needs to be taken (*Message #10*), and calling the specific platform that will apply the SH action. Upon receiving an API call (normally in the HyperText Transfer Protocol (HTTP)/Representational State Transfer (REST)/Simple Object Access Protocol (SOAP)) from the SH module (*Message #11*), the ACS module triggers an action to the Modem module (*Message #12*), closing the loop of SH in the broadband network.

3.1 Cognitive Engine

This work focused on implementing the Cognitive Engine Module of the SH framework. Figure 2 illustrates the implemented components of the framework, highlighting the Cognitive Engine and the three types of algorithms we implemented.

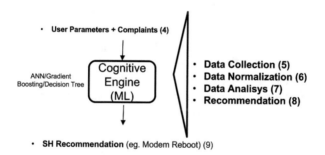

Fig. 2. Implemented Cognitive Engine Module

We implemented a ML model that analyzes the parameters of the user's broadband modem and predicts that the user would complain. From there, it would be enough to make an API call to complete the SH cycle by doing, for instance, a preventive reboot of the user's modem.

The Cognitive Engine module must take some important actions. First, the module must receive data from the DTL (*Message #5*) and once all the data are received, we pass to the data normalization step (*Message #6*), where it is important to adapt the data set replacing null values by zero, adapt some columns with fewer lines with zero and concatenates the values.

Three types of models were implemented (in Python), and their accuracy was evaluated, namely: DT, Artificial Neural Network (ANN), type (Multi-Layer Perceptron (MLP)), and Gradient Boosting. These models were selected because they represent supervised learning models, as we previously have the expected output (customer complaints or not). We choose one algorithm of higher complexity (ANN), one of low complexity (DT) and one of intermediate complexity (Gradient Boosting), to check if all of them could be applied to our framework.

For these models, the following modem user parameters were used as input for training each model: Latency, Jitter, Packet Loss, 2.4 GHz Channel Quality, 5 GHz Channel Quality, and number of distant devices. These parameters were described in Table 3.

The user parameters defined in Table 3 were crossed with the CRM database to find possible patterns. If some specific parameter configuration happens, the user normally complains to the contact center. With this rule in mind, training the ML models is the next step. If the models exhibit good accuracy in predicting a user complaint (*Message #7*), our model works, and Self-Healing recommendation (*Message #8*) can take place.

Now, we can describe each implemented algorithm.

3.1.1   Decision Tree

A DT is a non-parametric supervised learning method for classification and regression. The goal is to create a model that predicts the value of a target variable by learning simple decision rules inferred from the features of the data. A tree can be seen as a constant approximation by parts. The deeper the tree, the more complex the decision rules and the finer the model. Our decision tree model analyzes the parameters of the DTL and identifies a potential problem.

Figure 3 illustrates the DT algorithm implemented inside our framework. The algorithm was created based on the class *DecisionTreeClassifier* from *sklearn* library [12], using a random state parameter to start this method. In this case, we use 70% of the data for training and 30% to test.

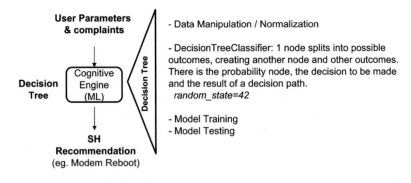

Fig. 3. Implemented Decision Tree

As output of the DT method, a data frame is generated as an Excel file. This data frame contains the user parameters, the expected output (user complains or not), and the real output. After this, the expected and real output can be compared, and the method's accuracy can be measured.

3.1.2   Fully Connected Neural Network (Multi-Layer Perceptron)

A MLP is an ANN like the perceptron type (classifier), but with more than one layer of neurons in direct feed. This type of network comprises layers of neurons connected by weighted synapses. Learning in this type of network is usually done through the backpropagation algorithm.

In our work, an ANN algorithm was created using the *Sequential* and *Dense* classes from Keras API [13], with the following parameters: 1 input layer (100 neurons, activation function = relu); 2 dense layers (100 neurons, activation function = relu); 1 output layer (1 neuron, activation function = sigmoid); Loss function='binary cross entropy'; optimizer='adam'; metrics=['accuracy']. Figure 4 illustrates the ANN algorithm implemented inside the framework. In this case we use 70% of the data to training and 30% to test.

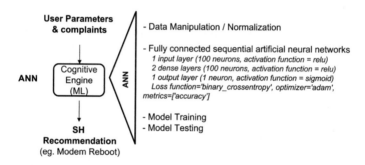

Fig. 4. Implemented ANN

A data frame is generated as an Excel file as an output of the ANN method. This data frame contains the user parameters, the expected output (user complains or not), and the real output. After this, the expected and real outputs are compared, and the method's accuracy is measured.

3.1.3   Gradient Boosting

Gradient boosting is a machine learning technique for regression and classification problems, which produces a prediction model in the form of a set of weak prediction models, usually decision trees. It constructs the model in steps and generalizes them, allowing the optimization of an arbitrary differentiable loss function. In short, the previous learning data analysis errors are gradually reduced.

In our work, a Gradient Boosting algorithm was created using the class *GradientBoostingClassifier* from the *sklearn* library [12], using a random state parameter to start this method. Figure 5 illustrates the Gradient Boosting algorithm implemented inside the framework. In this case, we use 70% of the data for training and 30% to test.

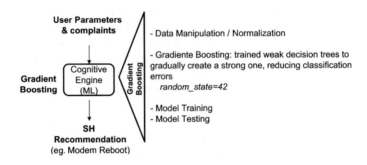

Fig. 5. Implemented Gradient Boosting

As before, the Gradient Boosting method generates a data frame as output (Excel file). This data frame contains the user parameters, the expected output (user complains or not), and the real output. After this, the expected and real outputs are compared, and the method's accuracy is measured.

4   Results and Discussion

In the tests carried out, real data of users from two cities served by the operator was collected: one dataset focusing on the retail market segment (residential broadband) and the other on the business market segment (micro and small business). These datasets were unified, anonymized and provided information about the modem parameters (Jitter, Latency, Packet Loss, and Quality of the Wi-Fi Channel used). The expected output regarding customer complaints in the period (customer complain = 1 or don't complain = 0) was added to this data, completing the final dataset [2]. In this way, it was expected that the models would predict, by analyzing the modem data, whether the user would complain.

The results of the algorithms are presented in Table 4, where "Precision" is the most important parameter for comparing the models.

Table 4. Results

Algorithm	#Layers	#Epochs	#Neurons	#Coded Lines	#Training Time (s)	Precision (%)
Decision Tree	NA	NA	NA	162	120	97,87
Artificial Neural Network	4 layers	1000	100	183	600	97,55
Gradient Boosting	NA	NA	NA	161	300	96,90

As shown in the table, all the algorithms led to a precision higher than 96%, which is acceptable in this context. This precision validates our assumption that a cognitive model could be important in an operator's customer care process.

The experiments also showed that the ANN is a flexible and high-potential tool to solve classification problems. In this model, the training time (in this experiment, the slower) is a point to observe and will depend on the problem to solve. As we used 2043 lines in our experiment, it was not a concern. However, it can be challenging when the problem needs larger data volumes to train the model.

As with any ML model, some post-training precautions are important to maintain good performance when the algorithm is in production once data sources on the operators can change (new data, dynamic environment, etc.). The model revaluation is very important to ensure process performance. The periodic model retraining with new data is important in this phase to ensure that the accuracy will continue.

The proposed framework can effectively change how operators work, allowing them to take preemptive actions based on ML. In this way, the system can automatically decide based on user parameters without human intervention. Once this framework is implemented in telecommunications operators, user satisfaction can be improved because SH actions can be carried out timely (e.g., modem reboot), avoiding user complaints.

It is important to note that this framework can be implemented in telecommunications operators with proprietary, open solutions or in hybrid scenarios. This flexibility strengthens the present framework proposal.

5   Conclusions

This paper presented a new framework for cognitive self-healing in real broadband networks to reduce customer complaints by anticipating corrective measures based on ML. The study relies on real broadband user parameters collected from a network operator in Brazil, discussed with its engineering and operating teams to learn how to extract value from CRM and Modem data. Once data was available in a Data Lake/DB, a Cognitive Engine (ML) was implemented to predict whether a user would complain with optimal accuracy. The results proved that the three cognitive data analysis methods yielded an optimal accuracy above 96%. With this proven, a Self-Healing script was applied, when applicable, using the CWMP platform to interact with the customer device to avoid a likely complaint.

In future work, new ML algorithms, such as CNN and Generative Artificial Intelligence (GenAI), may be considered an alternative to the implemented models, focusing on increasing the accuracy achieved. Another suggestion is to enhance the final SH script to evaluate the performance of the complete loop (total time from diagnosis to corrective action). An additional point to explore is the application of this framework to other types of networks, such as metro network switches, backbone routers (edge and core), access aggregators (Optical Line Terminal (OLT)s), and 5G voice and data switching centers.

Acknowledgements. The authors thank Algar Telecom S/A (Algar Telecom), a Brazilian telecommunications operator, who provided the real anonymous broadband network database. The authors also thanks the National Council for Scientific and Technological Development (CNPq) under grant number 421944/2021-8 (call CNPq/MCTI/FNDCT 18/2021) and Centro Algoritmi, funded by Fundação para a Ciência e Tecnologia (FCT) within the RD Units Project Scope 2020-2023 (UIDB/00319/2020) for partially support this work.

References

1. Souza Neto, N., Oliveira, D., Gonçalves, M., Silva, F., Frosi, P.: Self-healing in the scope of software-based computer and mobile networks, pp. 325–344 (2021)
2. Vieira, E., De Oliveira Silva, F.: Real dataset from broadband customers of a Brazilian telecom operator (2024). https://zenodo.org/records/10482897
3. Wang, Y., Martonosi, M., Peh, L.-S.: Predicting link quality using supervised learning in wireless sensor networks. Mobile Comput. Commun. Rev. **11**, 71–83 (2007)
4. Hashmi, U.S., Darbandi, A., Imran, A.: Enabling proactive self-healing by data mining network failure logs. In: 2017 International Conference on Computing, Networking and Communications (ICNC), pp. 511–517 (2017)
5. Kiciman, E., Fox, A.: Detecting application-level failures in component-based internet services. IEEE Trans. Neural Netw. **16**(5), 1027–1041 (2005)
6. Xu, L., et al.: Cognet: a network management architecture featuring cognitive capabilities. In: 2016 European Conference on Networks and Communications (EuCNC), pp. 325–329 (2016)
7. Omar, T., Ketseoglou, T., Naffaa, I.: A novel self-healing model using precoding & big-data based approach for 5g networks. Perv. Mob. Comput. **73**, 101365 (2021)
8. Rahmani, J., Sadeqi, A., Ayeh Mensah, D.N.: MPRA - self-healing in LTE networks with unsupervised learning techniques (2020)
9. Silva, W.S., Silva de Morais, A., Silva, W.D.O.: Proposal for the use of neural networks for data clustering in the context of qualitative analysis of complaints information in telecommunications services. New Trends Qual. Res. **4**, 499–506 (2020). https://publi.ludomedia.org/index.php/ntqr/article/view/66
10. Bothe, S., Masood, U., Farooq, H., Imran, A.: Neuromorphic AI empowered root cause analysis of faults in emerging networks. CoRR arxiv:2005.01472 (2020)
11. BroadBandForum. Technical report - tr-069 cpe wan management protocol (2020). https://www.broadband-forum.org/pdfs/tr-069-1-6-1.pdf
12. Pedregosa, F., et al.: Scikit-learn: machine learning in python. J. Mach. Learn. Res. **12**, 2825–2830 (2011)
13. Chollet, F., et al.: Keras (2015). https://github.com/fchollet/keras

Honey Bee Inspired Routing Algorithm for Sparse Unstructured P2P Networks

Aman Verma[⊠], Sanat Thakur, Ankush Kumar,
and Dharmendra Prasad Mahato

Department of Computer Science and Engineering, National Institute of Technology
Hamirpur, Hamirpur 177 005, Himachal Pradesh, India
{195522,20dcs007,20dcs025,dpm}@nith.ac.in

Abstract. Sparse Unstructured Peer-to-Peer (P2P) networks pose unique challenges for efficient and scalable routing. The nodes have limited and partial information about the network and the location of the desired data, which makes it hard to find the optimal or shortest path to the data. This article presents Honey Bee Optimization in P2P Networks (HBO_P2P), a unique routing algorithm inspired by the foraging behavior of honey bees. The proposed algorithm aims to address the inherent limitations of routing in unstructured P2P networks, focusing on improving packet delivery, minimizing hop count, reducing message overhead, and optimizing overall throughput. To evaluate the performance of our proposed algorithm, we conducted comprehensive experiments comparing it with existing algorithms commonly used in P2P networks, namely Particle Swarm Optimization (PSO), Genetic Algorithm (GA), and Ant Colony Optimization (ACO). Message overhead, packet delay, hop count, and throughput are among the important parameters that form the basis of the comparison. Our findings show that our suggested routing algorithm HBO_P2P is effective at resolving issues unique to unstructured P2P networks. The algorithm showcases notable improvements across multiple performance metrics when compared to established optimization techniques.

Keywords: Bee Routing Algorithm · Comparative Analysis · Intelligent Routing · Swarm-Based Algorithms Optimization

1 Introduction

P2P networks have obtained unmistakable excellent as a robust method for dispersing content, sharing belongings, and empowering decentralized correspondence [1,2]. Inside the great scene of P2P corporations, unstructured P2P networks stand aside for their adaptability, versatility, and simplicity of execution.

Unstructured allotted (P2P) networks address a decentralized and circulated model of correspondence and asset sharing at the net, where pals, or individual computers, interface straightforwardly with every different without relying on focal servers or predefined hierarchical modern structures [3]. Unstructured P2P networks act as a chief design for a heap of utilization consisting of document sharing, informing, and content dissemination.

The sign of unstructured P2P networks is their intrinsic effortlessness and flexibility. In those companies, participants can be part of and depart the organization voluntarily, and there are not any intense imperatives on peer affiliation. Peers interface with each other given their very own revelation components, frequently labored with using the trading of association facts or a focal tracker.

Whilst unstructured P2P networks provide some blessings, for instance, their capability to deliver content without the requirement for a concentrated basis, they likewise gift problems. These problems contain versatility problems, wasteful steering structures, and the ability for severe enterprise visitors, which could activate poor execution, particularly in larger agencies. In rundown, unstructured P2P networks play an essential part in empowering decentralized, peer-driven correspondence and asset sharing on the net. Their straightforwardness, flexibility, and freedom from unified manipulation make them a well-known choice for lots of applications, regardless of the innate problems associated with their dynamic and variable employer structure [4,5]. Routing in a sparse P2P unstructured network is an NP-hard problem.

In this paper, a sparse unstructured P2P network routing method inspired by honey bees is presented.

2   Related Works

Various methodologies were proposed for compelling steering in decentralized unstructured P2P organizations [6].

At any rate, it needs to be underscored that the numbered swarm-additives are applied in sure or explicit instances: a Physarum polycephalum ooze molds device is applied for steering in faraway sensor networks [7], a firefly-propelled calculation is applied for steering in VLSI [8], honey bee encouraged calculations are applied for steerage in transportable impromptu networks [9] and in telecom networks general [10], molecule swarm streamlining (PSO) is utilized for directing for faraway sensor groups (WSN) [11].

[8] Check out the utilization of the firefly calculation in VLSI steering. Rubio-Largo et al. in [12] makes the experience of directing low-pace traffic needs onto excessive speed lightpaths in telecom agencies. Wedde et al. [13] gift a flexible and strong route enlivened via trying to find the behavior of honey bees. Sim and Sun [14] additionally audit the ACO procedures for steering in stressed-out networks, even as Ren and Meng (2006) [11] examine the current ACO calculations, PSO calculations, and different bio-encouraged structures proposed for MANETs, wired networks, and observe their usage in WSNs for guidance, grouping, and security.

3 Algorithm for Routing in Sparse and Unstructured Peer-to-Peer Networks Inspired by Swarm Behavior

The suggested honey-bee-stimulated routing algorithms for sparse unstructured P2P networks go through several stages of shaping. Initially, a chosen model is suggested to fulfill the requirements for the sparse unstructured P2P routing system. Furthermore, we see the mapping between the research use case and natural swarm mechanisms using the description of the modified honey-bee set of rules.

3.1 Sparse P2P Network Definition

Sparse unstructured P2P networks are a type of peer-to-peer (P2P) networks that have a low average degree of connectivity among the nodes. This means that each node has only a few neighbors in the overlay network, and the network topology is random and dynamic. Sparse unstructured P2P networks operate well for high turnover applications, such as social media platforms, where users join and exit the network regularly. However, they also have some drawbacks, such as: 1) Inefficient search: In a sparse unstructured P2P network, a node must conduct a flooding-based query or a random walk to locate a resource or another node. These operations can result in high network traffic and bandwidth consumption. Furthermore, if the resource is uncommon or the network is too thin, the search success rate could be low. 2) Lack of guarantees: There are no guarantees regarding the consistency, availability, or dependability of the resources or the nodes in sparse unstructured P2P networks. For example, a node may not be able to find a resource that exists in the network, or a node may lose its data or connections due to network failures or malicious attacks. 3) Heterogeneity issues: Sparse unstructured P2P networks do not take into account the heterogeneity of the nodes in terms of their capacity, performance, or preferences. This can lead to unfairness, inefficiency, or instability in the network. For example, some nodes may be overloaded or underutilized, or some nodes may have incompatible or conflicting interests.

A sparse unstructured P2P community be represented by an undirected graph $GP2P = (VP2P, EP2P)$, wherein the vertices $v \in VP2P$ of the graph constitute the $P2P$ networks nodes and the rims $e \in EP2P$ represent connections among those nodes. Nodes $w, k \in VP2P$ are friends if and handiest if $\exists(w, k) \in EP2P$. Everyone has a logical particular identifier x_i and a physical deal with y_i. For every node $v \in VP2P$, a logical unique identification of node v_i is considered. Nonetheless, the most proficient associates can instantly switch packets by mapping the logical identifier x to the physical host address y. Consequently, the address resolution characteristic for community node c and unique identifier x_e evaluates whether node c can ascertain the physical address of node e and, consequently, whether it can engage with it immediately.

$$m(c, x_e) = \begin{cases} y_e, & \text{if } e \in neighbours(c) \\ x_e, & \text{otherwise} \end{cases} \tag{1}$$

If a routing algorithm makes use of smart agents, these vendors might also be aware of their source node's physical address y_e in addition to its logical address x_e. For this reason, they will return immediately to their source

The environment where a lot of hives and flowers are present is represented by the unstructured P2P network. More specifically, each node in the P2P network represents precisely one flower that bears nectar and precisely one hive with its bee population. A bee is sent out by each hive on a customizable beeSpawnInterval basis to search for food. Bees use hives to send out a single bee at a time, but they also explore the network simultaneously. Because of this, beeSpawnInterval can be used to arbitrarily set the number of exploring bees in a hive. Recruitment and navigation entail stages, much like in nature. A colony of hive bees seeks to optimize its output and determines the greenest routes from the hive to every type of plant in the surrounding area. The fitness function 3 defines the first class of a given path:

$$f_{path} = \frac{1}{H_{path}} \cdot \frac{1}{D_{path}} \tag{2}$$

where D_{path} is the total delay encountered on this path and H_{path} is the number of hops a bee made on the path from source to destination. The first algorithm depicts the behavior of the foraging bee.

3.2   Initialization

There is an initiation period before a hive releases bees to look for food. The dance floor is a locally stored routing table (RT) at a hive. A hive also keeps track of the heuristic distance to nearby hives in a distance matrix. First, a software agent is utilized to simulate a bee; it has a single source colony and a single destination flower. The hive's host address serves as its source address. The bee is allowed to go directly back to its hive as a result. The bee also maintains a list of travel information, which shows the route the bee has taken. One element in this list contains the unique identification of the visited node as well as the delay experienced. A bee that is a follower will travel along its favored path. A hive sends test bees to its neighbors before releasing its first batch of forager bees. As soon as they reach the neighbors, they return and modify the heuristic distance in the original hive's matrix.

3.3   Observe Waggle Dance

Before departing the colony, a follower bee selects both the flower it intends to visit and its function. There are not any waggle dances conducted on the dance ground to start with. As a result, foraging is the natural role of all bees. The selected target flower is made at random. A bee, but, goes via a specific decision-making manner if paths are indicated for the intended destination with the aid of waggle-dancing. First, the bee selects a marketed path with possibility [15].

$$p(path) = \frac{f_{path}}{f_{colony}} \tag{3}$$

$$f_{colony} = \frac{1}{n} \sum_{i=1}^{n} f_i \tag{4}$$

where f colony is the average health characteristic value of all currently marketed paths for a certain vacation destination flower r, and n is the number of offered paths to r. After the path is determined, the bee assesses it against the f colony. The possibility to emerge as a follower bee and consequently follow the chosen preferred path is defined in [16]. A bee will immediately become a forager if there is no direction assigned to the location it has selected in the routing table. If not, the bee chooses a path and follows it.

Algorithm 1. Basic Honey Bee Optimization Algorithm

    Input: *hive s, flowers R*
    Output: Optimal paths from s to all flowers $r \in R$
1: Initialization
2: **while** not terminated **do**
3:     **if** *beeSpawnInterval* has passed **then**
4:        *observeWaggleDance*
5:        *constructSolution*
6:        *performWaggleDance*
7:     **end if**
8: **end while**

---

where d_{ij} is the heuristic distance between the modern nodes i and j, as represented in the heuristic distance matrix M, and ρ_{ij} is the arc feature from node i to node j at time t. These values are given weight in the transition rule by the constants α and β. The arc feature ρ_{ij} is calculated in the following manner [15, 17].

3.4   Construct Solution

On its way from the hive to the goal flower, a bee hops from one node to the next. A bee saves the awesome identification x and the perceived delay at every node it visits in its reminiscence. As a result, the journey and any delays encountered can be completely recreated. A bee that reviews a loop clears all loop-related statistics from its reminiscence. The transition rule $P_{ij}(t)$ is used to decide the following hop j of a bee at node i and time t [17].

$$p_{ij}(t) = \frac{[\rho_{ij}(t)]^{\alpha} \cdot [\frac{1}{d_{ij}}]^{\beta}}{\sum_{j \in A_i(t)} [\rho_{ij}(t)]^{\alpha} \cdot [\frac{1}{d_{ij}}]^{\beta}} \tag{5}$$

where ρ_{ij} is the arc feature from node i to node j at time t, and d_{ij} is the heuristic distance between the modern nodes i and j, as included in the heuristic distance matrix M. These values are given weight in the transition rule by the constants α and β. The arc feature ρ_{ij} is calculated in the following manner [17].

1) if a bee is a forager: where $A_i(t)$ is the collection of nodes that are adjacent to the contemporary node i at time t, excluding the bee's previous hop node. A bee's last hop is only considered the next hop if it is the lone neighbor of i.

$$\rho_{ij} = \frac{1}{k}, k = |A_{i(t)}| \tag{6}$$

2) If a bee is a follower, a specific arc function (Eq. 7) is carried out. The probability of choosing a neighbor node j at the modern node i as the subsequent hop is λ if j is the following hop at the follower's preferred direction $Fi(t)$. The desired path is defined by the set of steering operations that specify the outing from the hive to the previously explored place with the help of one of its hive friends. However, a follower has the opportunity $(1-\lambda)$ to stray from the intended path and choose one of the friends $j \in A_i(t) : J is outside of F_i$. There is an equal chance of choosing one of the other acquaintances if the follower bee deviates from the favored path. The follower bee transitions with the arc function (Eq. 6) till it reaches its destination or crosses the favored course once more [17].

$$\rho_{ij}(t) = \begin{cases} \lambda, & \text{if } j \in F_i(t) \\ \frac{1-\lambda.|A_i(t) \cap F_i(t)|}{|A_i(t)| - |A_i(t) \cap F_i(t)|}, & \text{if } j \in F_i(t), \ \forall j \in A_i(t), \ 0 \le \lambda \le 1 \end{cases} \tag{7}$$

Since $A_i(t)$ and $F_i(t)$ may have zero or one element in their intersection, $|Ai(t) \cap Fi(t)|$ might be either 0 or 1. Once a bee has arrived at its destination, it immediately makes a P2P return to its hive.

3.5   Perform Waggle Dance

A bee analyzes the experienced direction and sub-paths when it returns to the hive by making use of Eq. 2 and 4.

Access from the node's routing desk is examined first. Once the colony fitness has been established, This expert course's intermediate nodes are reached via the tested experienced path and any contained sub-paths. If the fitness fee of an analyzed direction is excessive enough, it is introduced to the nodes routing desk [16]. If not, the path is deserted. Most effectively the maximum current encountered path is retained if the identical path is already present within the routing table.

3.6   Forwarding Data Packets

As a data packet travels from a supply node s to a specific destination r, the routing desk is queried to retrieve all pathways to r that are identified at s. The path on which the records packet is dispatched is based on a roulette wheel selection [15]. The suggested algorithm's flowchart is displayed in Fig. 1.

$$P_{forward}(path) = \frac{f_{path}}{f_{colony}} \tag{8}$$

3.7   Complexity of Our Proposed Algorithm

$O(m + k(m + l + p2 + s))$ is the total time complexity of the suggested routing algorithm, which depends on unique parameters. Here, m denotes the variety of foragers, l the variety of followers, k the number of iterations, p the total number of hops a bee makes in one direction, and s the number of sub-paths in the experienced route.

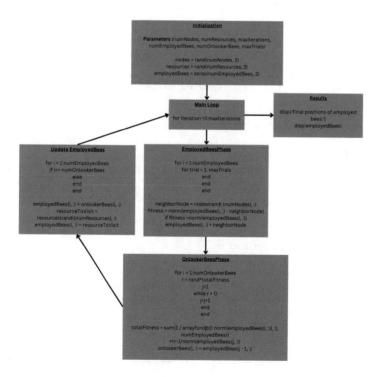

Fig. 1. Flowchart of our proposed algorithm

Algorithm 2. Our proposed Algorithm: HBO_P2P

Phase 1 – Initialization of nodes and resources (randomly generated)

1: $nodes = rand(numNodes, 2)$;   ▷ Node coordinates (2D for simplicity)
2: $resources = rand(numResources, 2)$;   ▷ Resource coordinates (2D for simplicity)

Phase 2 – Initialization of employed bees and onlooker bees

3: $employedBees = zeros(numEmployedBees, 2)$;   ▷ Current locations of employed bees
4: $onlookerBees = zeros(numOnlookerBees, 2)$;   ▷ Current locations of onlooker bees

Phase 3 – Main optimization loop

5: **for** $iteration = 1 : maxIterations$ **do**
6:     **for** $i = 1 : numEmployedBees$ **do**
7:         **for** $trial = 1 : maxTrials$ **do**   ▷ Generate a random neighbor node for the employed bee
8:             $neighborNode = nodes(rand(numNodes), :)$;   ▷ Calculate the fitness function (distance in this example)
9:             $fitness = norm(employedBees(i, :) - neighborNode)$;
10:            **if** $fitness < norm(employedBees(i, :))$ **then**   ▷ If better solution is found
11:                $employedBees(i, :) = neighborNode$;
12:            **end if**
13:        **end for**
14:    **end for**

Phase 4 – Onlooker bees phase

15:    $totalFitness = sum(1./arrayfun(@(i)norm(employedBees(i, :)), 1 : numEmployedBees))$;
16:    **for** $i = 1 : numOnlookerBees$ **do**
17:        $r = rand * totalFitness$;
18:        $j = 1$;
19:        **while** $r > 0$ **do**
20:            $r = r - 1/norm(employedBees(j, :))$;
21:            $j = j + 1$;
22:        **end while**
23:        $onlookerBees(i, :) = employedBees(j - 1, :)$;
24:    **end for**
                          ▷ Update employed bees with the best onlooker bees
25:    **for** $i = 1 : numEmployedBees$ **do**
26:        **if** $i <= numOnlookerBees$ **then**
27:            $employedBees(i, :) = onlookerBees(i, :)$;
28:        **else**   ▷ Randomly select a resource to visit (foraging)
29:            $resourceToVisit = resources(randi(numResources), :)$;
30:            $employedBees(i, :) = resourceToVisit$;
31:        **end if**
32:    **end for**
33: **end for**

Phase 5 – Results

34: $disp('Final positions of employed bees :')$;
35: $disp(employedBees)$;

$$onlookerBee(j) = employedBee(j) - F \times (employedBee(j) - resourceToVisit)$$

where j is the index of the employed bee, F is a random number between -1 and 1, and $resourceToVisit$ is a randomly chosen resource. The formula moves the onlooker bee closer to or farther from the resource, depending on the sign of F.

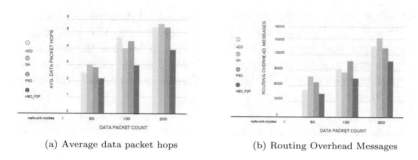

(a) Average data packet hops   (b) Routing Overhead Messages

Fig. 2. Average data packet hops and Routing Overhead Messages

4 Experimental Evaluations and Discussions

In this investigation, 8.00 GB of installed RAM (7.75 GB usable) and an Intel(R) Core(TM) i5-1035G1 CPU running at 1.00 GHz and 1.19 GHz were used for the studies. The system operated on a 64-bit architecture with a Windows 10, and 11 operating system. Graphics processing was facilitated by an NVIDIA GeForce MX230. MATLAB R2023b served as the primary simulation tool, and the experiments adhered to the software's system requirements, which include any Intel or AMD x86-64 processor with two or more cores, a minimum of 8 GB RAM, and 3.8 GB of storage dedicated solely to MATLAB.

4.1 Average Data Packet Hops

The result (as depicted in Fig. 2) shows the performance of four different algorithms (ACO, GA, PSO, HBO_P2P) in terms of average data packet hops as the data packet count increases. Data packet hops are the number of intermediate nodes that a packet passes through before reaching its destination in a network. Lower hops mean shorter and more efficient routes.

HBO_P2P has the highest number of average hops, indicating it may be less efficient in this context. This means that it takes more intermediate nodes for the packets to reach their destinations using HBO_P2P. One explanation for this could be that HBO_P2P might struggle to strike a balance between exploring and exploiting the search space and making use of the peer-to-peer network structure to improve agent cooperation and communication.

4.2   Routing Overhead Messages

The result (as depicted in Fig. 2) shows the comparison of four different algorithms (ACO, GA, PSO, and HBO_P2P) in terms of routing overhead messages as the data packet count increases. Routing overhead messages are the extra messages that are used to establish and maintain routes between nodes in a network1. Higher routing overhead means more network congestion and resource consumption.

HBO_P2P has the lowest routing overhead at all data packet counts. This means that it uses the least amount of messages to find and maintain routes among the four algorithms. HBO_P2P's ability to balance the search space's exploration and exploitation while also leveraging the peer-to-peer network structure to improve agent collaboration and communication could be one explanation for this.

4.3   Average Data Packet Delay

The result (as depicted in Fig. 3) shows the average data packet delay experienced by different algorithms (ACO, PSO, HBO_P2P) as the number of network nodes increases (50, 100, 200). The amount of time it takes a packet in a network to go from its source to its destination is known as the data packet delay. The lower delay means faster and more efficient data transmission.

HBO_P2P is the most efficient, with the lowest delay at each node count. This means that it takes the shortest time for the packets to reach their destinations using HBO_P2P. HBO_P2P's ability to balance the search space's exploration and exploitation while also leveraging the peer-to-peer network structure to improve agent collaboration and communication could be one explanation for this.

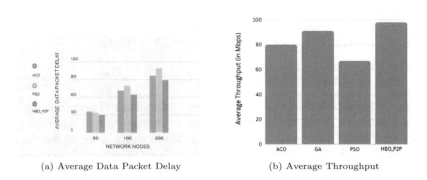

(a) Average Data Packet Delay   (b) Average Throughput

Fig. 3. Average data packet delay and Average Throughput

4.4   Average Throughput

The result (as depicted in Fig. 3) shows the average throughput (in *Mbps*) of four different algorithms: ACO, GA, PSO, and HBO_P2P. The amount of data that can be processed or sent in a particular amount of time is measured by throughput. Higher throughput means better performance.

5   Conclusion

This paper aimed to design an algorithm for routing in sparse unstructured P2P networks. By decreasing packet latency and boosting throughput while keeping hop count and routing overhead low, the technique enhances routing effectiveness and efficiency. The algorithm also adapts to the dynamic changes in the network topology and resource availability by adjusting the routing tables and the query scope. The algorithm adds to the corpus of information already available in the fields of swarm-inspired routing algorithms and unstructured P2P networks. Some directions for future research or possible extensions of HBO_P2P are: To test the algorithm on larger and more realistic unstructured P2P networks with different characteristics and parameters, such as the number of nodes, resources, queries, and churn rates. To compare the algorithm with other swarm-inspired routing algorithms, such as particle swarm optimization (PSO) or artificial bee colony (ABC), or other types of routing algorithms, such as DHT-based or query routing algorithms. Modify or enhance the algorithm by incorporating other factors or criteria into the fitness function, such as the popularity, relevance, or quality of the resources, or the preferences, feedback, or reputation of the nodes. To apply an adaptive or self-tuning mechanism to dynamically modify the algorithm's parameters—such as the number of trials, TTL, onlooker bees, and employed bees—in response to network circumstances and user behavior.

References

1. Barabási, A.-L., Albert, R.: Emergence of scaling in random networks. Science **286**(5439), 509–512 (1999) https://doi.org/10.1126/science.286.5439.509. https://www.science.org/doi/pdf/10.1126/science.286.5439.509
2. Buford, J.F., Yu, H.: Peer-to-peer networking and applications: synopsis and research directions. In: Shen, X., Yu, H., Buford, J., Akon, M. (eds.) Handbook of Peer-to-Peer Networking, pp. 3–45. Springer, Boston (2010). https://doi.org/10.1007/978-0-387-09751-0_1
3. Androutsellis-Theotokis, S., Spinellis, D.: A survey of peer-to-peer content distribution technologies. ACM Comput. Surv. **36**(4), 335–371 (2004). https://doi.org/10.1145/1041680.1041681
4. Farooq, M., Farooq, M.: A comprehensive survey of nature-inspired routing protocols. In: Bee-Inspired Protocol Engineering: From Nature to Networks, pp. 19–52 (2009). https://doi.org/10.1007/978-3-540-85954-3_2
5. Farooq, M., Di Caro, G.A.: Routing protocols for next-generation networks inspired by collective behaviors of insect societies: an overview. In: Swarm Intelligence: Introduction and Applications, pp. 101–160 (2008). https://doi.org/10.1007/978-3-540-74089-6_4

6. Bonabeau, E., Dorigo, M., Theraulaz, G.: Swarm Intelligence: From Natural to Artificial Systems. Oxford University Press (1999). https://doi.org/10.1093/oso/9780195131581.001.0001

7. Li, K., Torres, C.E., Thomas, K., Rossi, L.F., Shen, C.-C.: Slime mold inspired routing protocols for wireless sensor networks. Swarm Intell. **5**, 183–223 (2011)

8. Ayob, A., Majid, R.A., Hussain, A., Mustaffa, M.M.: Creativity enhancement through experiential learning. Adv. Nat. Appl. Sci. **6**(2), 94–99 (2012)

9. Saleem, M., Di Caro, G.A., Farooq, M.: Swarm intelligence based routing protocol for wireless sensor networks: survey and future directions. Inf. Sci. **181**(20), 4597–4624 (2011)

10. Wedde, H.F., Farooq, M., Lischka, M.: An evolutionary meta hierarchical scheduler for the Linux operating system. In: Deb, K. (ed.) GECCO 2004. LNCS, vol. 3103, pp. 1334–1335. Springer, Heidelberg (2004). https://doi.org/10.1007/978-3-540-24855-2_153

11. Ren, J., Meng, X.-H.: Cosmological model with viscosity media (dark fluid) described by an effective equation of state. Phys. Lett. B **633**(1), 1–8 (2006)

12. Rubio-Largo, A., Vega-Rodríguez, M.A., Gómez-Pulido, J.A., Sánchez-Pérez, J.M.: A multiobjective approach based on artificial bee colony for the static routing and wavelength assignment problem. Soft. Comput. **17**(2), 199–211 (2013)

13. Wedde, H.F., Farooq, M., Zhang, Y.: BeeHive: an efficient fault-tolerant routing algorithm inspired by honey bee behavior. In: Dorigo, M., Birattari, M., Blum, C., Gambardella, L.M., Mondada, F., Stützle, T. (eds.) ANTS 2004. LNCS, vol. 3172, pp. 83–94. Springer, Heidelberg (2004). https://doi.org/10.1007/978-3-540-28646-2_8

14. Sim, K.M., Sun, W.H.: Ant colony optimization for routing and load-balancing: survey and new directions. IEEE Trans. Syst. Man Cybern. Part A Syst. Hum. **33**(5), 560–572 (2003)

15. Šešum-Čavić, V., Kühn, E.: Chapter 8 self-organized load balancing through swarm intelligence. In: Bessis, N., Xhafa, F. (eds.) Next Generation Data Technologies for Collective Computational Intelligence, pp. 195–224. Springer, Heidelberg (2011). https://doi.org/10.1007/978-3-642-20344-2_8

16. Nakrani, S., Tovey, C.: On honey bees and dynamic server allocation in internet hosting centers. Adapt. Behav. **12**(3–4), 223–240 (2004)

17. Wong, L.-P., Low, M.Y.H., Chong, C.S.: A bee colony optimization algorithm for traveling salesman problem. In: 2008 Second Asia International Conference on Modelling & Simulation (AMS), pp. 818–823. IEEE (2008)

Anomaly Detection in WBANs Using CNN-Autoencoders and LSTMs

Kartikeya Dubey[✉] and Chittaranjan Hota

Birla Institute of Technology and Science, Pilani, Hyderabad Campus, Hyderabad 500078, India
{f20200031,hota}@hyderabad.bits-pilani.ac.in

Abstract. Wireless Body Area Networks (WBANs) are wireless networks that consist of microscopic sensors that are either placed on a subject's body or attached to their clothes. These low-power devices track the physiological readings from a subject and relay them to a server over the Internet. Since WBANs find various uses, from remote medical monitoring to early disease detection in the medical field, their readings must be accurate. Anomalies can occur in the data being transmitted for various reasons, such as sensor faults or malicious attacks.

In some instances, however, what is considered an anomaly may not be an anomaly, and the readings are correct. To mitigate such cases, we must detect anomalies promptly. The anomaly detection process developed here consists of point and contextual anomaly detection. Point anomaly detection deals with checking whether, given a set of readings, some of the readings are anomalous in an isolated sense. Contextual anomaly detection then checks whether readings from other sensors can corroborate the reading of our "faulty" sensor. This paper has employed a CNN Autoencoder and an LSTM-based classifier to do this two-step process. Our model shows an accuracy of 94% and a loss of 14%.

1 Introduction

With the advent of microscopic sensors, getting readings of various physiological parameters unobtrusively became possible. A natural progression of this technology is a network of microscopic sensors spread across a person's body. This is the concept behind Wireless Body Area Networks (hereafter referred to as WBANs).

The various sensors placed across the body collect the data. Since these sensors have extremely low power consumption, they cannot transmit the data to a remote location. So, to overcome this, the sensors first send their data to a central unit, which is closer in proximity, which then transmits the data to a remote data centre from where doctors worldwide can access it [1]. It is immediately apparent how WBANs will significantly improve patient care and quality of life. Using data from WBANs, medical professionals can be alerted on any development in their patients without the patients having to be physically present to get readings taken. The data is also valuable because medical professionals can get real-time readings of patients afflicted with various diseases, providing valuable data on the day-to-day dynamics of a disease.

The primary drawbacks of WBANs currently are security concerns and anomalous readings. Due to the sensitive nature of the data, a breach of data integrity would severely violate patients' privacies. Anomalous readings may lead to misdiagnosis or

L. Barolli (Ed.): AINA 2024, LNDECT 201, pp. 187–197, 2024.
https://doi.org/10.1007/978-3-031-57870-0_17

delivery of incorrect medical dosages, which is equally harmful to a patient's health. Our work in this research focuses on this second aspect of WBANs.

There are existing deep-learning algorithms for this task. Autoencoders are a popular technique employed to capture spatially sensitive data. The type of autoencoder, whether CNN-based or LSTM-based, matters as different features ofa the data are prioritized by each. With this in mind, we have proposed a novel CNN-based autoencoder. The output of this autoencoder is given as input to the LSTM, which finally classifies the anomalies.

The rest of this paper follows the following structure: Sect. 2 briefly explains the related works in anomaly detection in WBANs. Section 3 describes the design of the proposed model and its algorithms for point-anomaly and contextual-anomaly detection. Section 4 presents the experimental results and analysis. Section 5 concludes the paper and provides future research directions.

2   Related Work

To approach this problem, M. U. Harun Al Rasyid et al. [2] employ a sliding window. The $(n+1)^{th}$ measurement is predicted using the data from this sliding window. This process is repeated for each physiological reading, and the error of the predicted value is checked against a dynamic threshold. The reading is considered anomalous if most of the physiological readings' errors exceed their respective thresholds. The paper proposes using errors from regressed data to predict erroneous readings. However, it simply considers that if most of the readings are erroneous, the sensors must be erroneous rather than the readings supporting each other being accurate.

Mikel Canizo [3] proposes using multi-head CNN-RNNs to solve the problem. They use CNNs with 1D convolution to convolute each physiological sensor time series. This convoluted data is then fed to the RNN-based classifier that classifies all the sensors simultaneously. The paper then experimentally analyses the results of their model against pre-existing CNN-based models. The model performs well against traditional CNN models. However, the convolution used by this approach does not convolute all the sensor's data together, which could help increase accuracy by introducing an element of cohesion.

The approach employed by F. A. Khan et al. [4] uses a Markov model. First, the features of the ECG dataset are extracted using a Discrete Wavelet Transform (DWT). The features are divided into feature sequences. The probability for each feature sequence is then calculated, and the system decides if an anomaly has occurred. This model performs well in cases where anomalies make up 10% or even 5% of the dataset. The paper, however, focuses on ECG data. We aim to develop a more generic model for anomaly detection. Once again, the model implemented in this paper is for a single time series data while We aim to find anomalies across multiple time series datasets.

O. I. Provotar [5] uses an LSTM-based autoencoder to detect anomalies. First, they determine the dominant frequency using autocorrelation or Welch's power spectral density estimation. Then, the signal is decomposed into seasonal, trend, and residual. Then, the residual component is used to decide whether an anomaly occurs. This paper, once again, aims to find anomalies in only a single time series of data. We aim to find anomalies across multiple time series datasets.

3   Proposed Model

The proposed anomaly detection process consists of 2 steps: first, the CNN autoencoder reconstructs the input data; if, for any point, the error is greater than a set threshold, then the point is labelled as point anomaly. Then, the encoded data is passed to the LSTM classifier, classifying it as being contextually anomalous or not. Thus, we have proposed a combination of CNN and LSTM models to identify contextual anomalies present in the sensor values.

3.1   Data Preparation

The dataset used for the model's training is the MIMIC-I dataset [7]. The dataset contains detailed medical records for 72 patients. The data we are concerned with are the physiological vitals: the pulse, blood oxygen saturation (SpO_2), heart rate (HR), diastolic arterial blood pressure (ABPdia), systolic arterial blood pressure (ABPsys), mean arterial blood pressure (ABPmean), and the respiration rate (resp). Using the NumPy Python library [8], these 7 metrics were first converted into one 2D array. The values were then normalised to the range [0–1], both inclusive. The correlation matrix was then found for these metrics using the scikit-learn library [13]. Next, this correlated data is fed to the CNN AutoEncoder. For training, any row in the array that contains a 0 value is dropped as a sensor has been removed or is faulty.

3.2   Point Anomaly Detection

Point anomalies are points that are anomalous without considering the contextual readings. A CNN autoencoder is employed to perform this analysis. The training input provided to the autoencoder is the 2D array that contains all the vital readings. The array is split into 7×7 sections since providing square dimensions to the CNN autoencoder allows for easier output formatting.

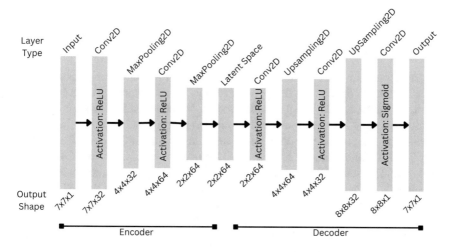

Fig. 1. An Overview of the autoencoder model

The encoder consists of five layers, an input layer and two pairs of Convolutional and MaxPooling Layers. The following five layers make up the decoder section of the model. This architecture can be seen in Fig. 1. The Convolutional layers detect patterns and features in the input, and the MaxPooling layers reduce the spatial dimensions of the input. For the threshold, the standard deviation of the error is considered. i.e., any point where Eq. 1 is satisfied is considered anomalous.

$$value_{predicted} - value_{actual} > value_{standard deviation} \tag{1}$$

These anomalies are labelled as such and passed on to the next step. The coded form of the input data obtained from the encoder is also passed along as input for the classifier model. The shape of the encoded data for each 7×7 segment is $2 \times 2 \times 64$. Algorithm 1 is an overview of the point anomaly detection process of a list of segments.

Algorithm 1. Point Anomaly Detection

1: **procedure** FINDPOINTANOMALIES(readings)
2:     segments ← SplitIntoSegments(readings)
3:     predicted_segments ← PredictWithCNN(segments)
4:     errors ← CalculateErrors(segments, predicted_segments)
5:     mean_error ← CalculateMean(errors)
6:     labels ← []
7:     **for** $i \leftarrow 1$ to N **do**
8:         **if** errors[i] > mean_error **then**
9:             labels.append(True) // True means a point anomaly has occurred
10:        **else**
11:            labels.append(False) // False means no point anomaly
12:        **end if**
13:    **end for**
14: **end procedure**

3.3 Contextual Anomaly Detection

Contextual anomaly detection entails checking whether a point anomaly is an anomaly or a valid value in the context of the other sensor values. To perform this analysis, we have employed an LSTM-based classifier model. We chose an LSTM as our Recurrent Neural Network (RNN) because it can handle long-term dependencies. This is advantageous to us since we are dealing with time series. A simple LSTM cell can be seen in Fig. 2. These cells' internal mechanisms, the gates, control the flow of information, allowing LSTM-based RNNs to capture and "remember" long-term dependencies in sequential data. The memory is stored in C_t, and the hidden state of the cell is stored in H_t where t denotes a time step. Several of these cells are chained together to form our LSTM RNN. The architecture of the LSTM classifier can be seen in Fig. 3.

Since this model takes a labelled dataset as an input for training, we must first label our unlabelled input. Contextual anomaly analysis must be done for each physiological

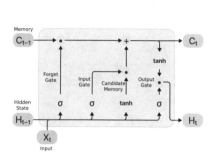

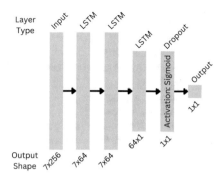

Fig. 2. A single LSTM cell.   **Fig. 3.** An Overview of the classifier model

metric separately, as opposed to the point anomaly analysis. To check whether a point is a contextual anomaly, we have checked whether the transformed value of the correlated value of that point is greater than the standard deviation of the same across the entire series. This is done in the following steps:

1. For each point, the correlated value is calculated using the correlation coefficient from the correlation matrix. This is done for each metric, using Eq. 2 as given below.

$$correlated\_value_i = \sum_{n=1}^{7} CC_{n,i} \cdot R_n \qquad (2)$$

where $CC_{n,i}$ is the Correlation coefficient of metric n and metric i.

2. These correlated values are then transformed using Eq. 3

$$anomaly\_score = \frac{correlated\_value_i - correlated\_value_{mean}}{correlated\_value_{standard\,deviation}} \qquad (3)$$

3. Thus, any point with an anomaly score greater than the standard deviation of anomaly scores is considered a contextual anomaly.

This is done for all the points first. Note that even if the anomaly_score exceeds the threshold, it is not a contextual anomaly if a point is not already considered a point anomaly. Next, since the points have been compressed from 7×7 to 2×2, if any of the 7 rows is considered a contextual anomaly, the entire 2×2 compression is also labelled a contextual anomaly.

With these as the labels, our classifier is trained, and the LSTM classifier is obtained. Algorithm 2 describes how to train the LSTM model, and algorithm 3 describes how to use it to classify a list of encoded segments.

4   Results

To implement our model, the Tensorflow [10] and keras [11] Python libraries were utilized. These were chosen since they allow for easy modifications and troubleshooting

Algorithm 2. LSTM Model

```
 1:  procedure TRAINCONTEXTUALANOMALIES(readings)
 2:      segment_length ← 7
 3:      number_of_segments ← ⌊length(readings)/segment_length⌋
 4:      segments ← []
 5:      encoded_segments ← []
 6:      anomaly_scores ← []
 7:      for i ← 0 to number_of_segments − 1 do
 8:          start_index ← i × segment_length
 9:          end_index ← start_index + segment_length
10:          segment ← readings.get(start_index to end_index)
11:          encoded_segment ← Encoder(segment)
12:          anomaly_score ← Calculate_Anomaly_Score(segment)
13:          segments.append(segment)
14:          encoded_segments.append(encoded_segment)
15:          anomaly_scores.append(anomaly_score)
16:      end for
17:      std_dev ← Standard_Deviation(anomaly_scores)
18:      labels ← Label_Anomalies(scores, std_dev)
19:      LSTM_classifier ← Train_LSTM_Classifier(encoded_segments, labels)
20:  end procedure
```

Algorithm 3. Contextual Anomaly Detection

```
 1:  procedure FINDCONTEXTUALANOMALIES(encoded_segments)
 2:      labels ← []
 3:      for i ← 1 to N do // Loop through all 2 × 2 encoded segments
 4:          if IS_CONTEXTUAL_ANOMALY(encoded_segments[i]) then
 5:              labels.append(True) // True means a contextual anomaly has occurred
 6:          else
 7:              labels.append(False) // False means no contextual anomaly
 8:          end if
 9:      end for
10:  end procedure
```

model creation and training. The scikit-learn library [13] was used to aid in dataset manipulation, such as creating train-test-validate splits and the correlation matrix. The Pandas [12] and Matplotlib [13] libraries were used to create diagrams and graphs.

First, the results of training the autoencoder are studied. To analyse the optimal specifications for training, the model is passed through varying epochs, and certain metrics for that model are graphed alongside it. The metrics chosen are the number of anomalies - while this does not provide much information regarding the model itself, it allows us to pinpoint diminishing returns, MAE - a good representative for an autoencoder output error, and MSE - chosen for the same reason as MAE.

As seen from Fig. 4, 5 and 6, after an initial steep decrease in the MAE and MSE, the models become more or less plateaued. After this point, overfitting begins; to avoid this, the number of epochs in the final model is set to 20.

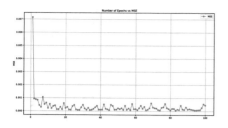

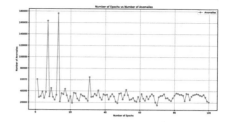

Fig. 4. MSE vs Epochs of training   **Fig. 5.** Anomalies vs Epochs of training

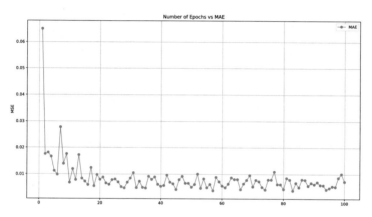

Fig. 6. MAE vs Epochs of training

With these as our parameters - 20 training epochs and a batch size of 128, we train the model. After reconstruction, at any point where the difference between the reconstructed value and the actual value is greater than the standard deviation, we mark it as a point anomaly. Figure 7, 8, 9, 10, 11 and 12 shows the results of the point anomaly detection process. Note that the figures show 500 time steps.

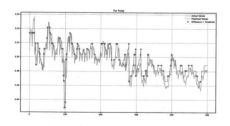

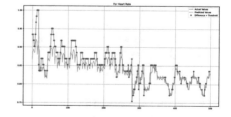

Fig. 7. Reconstructed Pulse   **Fig. 8.** Reconstructed Heart Rate

The results from Fig. 7, 8, 9, 10, 11 and 12 show that the reconstructions faithfully conform to the original data to a certain extent, and point anomalies can be discovered

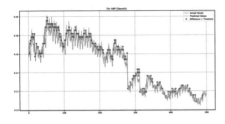

Fig. 9. Reconstructed ABPDia

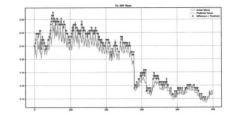

Fig. 10. Reconstructed ABPmean

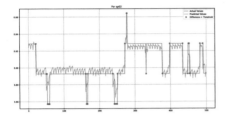

Fig. 11. Reconstructed SpO_2

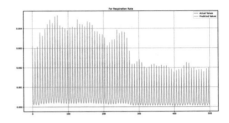

Fig. 12. Reconstructed Resp

using the CNN-based autoencoder. The encoded values from the autoencoder are then fed to the LSTM model as input.

Next, for training the LSTM model, yet again, we first aim to find an optimum number of training epochs. To do this, we use conventional evaluation metrics for classifiers: Accuracy, Precision, Recall, and F1-Score. We trained the model for varying numbers of epochs; the results of this can be seen in Fig. 13, 14, 15 and 16.

As can be seen from the graphs, the metrics all begin plateauing about Epoch 50. So, we use 50 as the number of epochs to train the LSTM model to avoid overfitting.

With our model trained, we can now classify contextual anomalies in our dataset. We must divide the input data into 7×7 sections. Then, these must be encoded into a 2×2 array using the encoder derived from our autoencoder. This is done to take advantage of the inherent feature extraction in autoencoders. These 2×2 arrays can now be classified using the LSTM model. The results of this classification for SpO_2 readings can be found in Fig. 17. Only SpO_2 readings are pictured for the sake of brevity, with other readings showing similar results.

As visible, the overlap between calculated and actual contextual anomalies is quite significant. The accuracy of the final model is 0.9400, and the loss of the model is 0.1457.

Furthermore, the classification report and the confusion matrix of performing testing on our classifier model can be seen in Table 1a and Table 1b. These results were obtained using the scikit-learn library [13].

Table 1. Performance metrics of the LSTM classifier model

(a) Classification Report of the LSTM classifier

Classes	Precision	Recall	F1-Score
Non-Anomalous	0.96	0.97	0.97
Anomalous	0.77	0.69	0.73

(b) Confusion Matrix of the LSTM classifier

		Label		
		Positive	Negative	Total
Prediction	Positive	471	140	611
	Negative	214	5231	5445
	Total	685	5371	6056

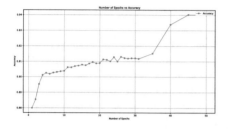

Fig. 13. Accuracy vs Epochs of training

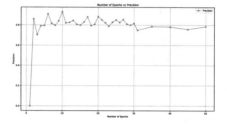

Fig. 14. Precision vs Epochs of training

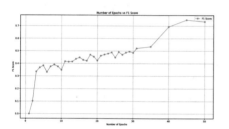

Fig. 15. F1 Score vs Epochs of training

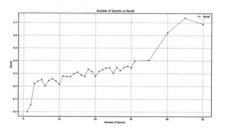

Fig. 16. Recall vs Epochs of training

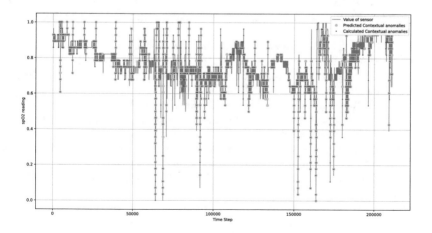

Fig. 17. A plot of classified contextual anomalies overlaid on SpO$_2$ readings.

5   Conclusions and Future Work

WBANs are a fast-developing field, and widespread usage of these will prove to be invaluable to the medical industry. As such, the detection of anomalous readings of these networks is an important task. This paper demonstrates that the usage of deep learning-based models can be helpful in this task. With our CNN-based autoencoder being able to detect point anomalies using an unlabeled dataset, we can detect and even replace anomalies. The LSTM classifier then further refines these anomalies to be contextually valid or not. Since accurate readings which appear anomalous must be allowed to pass through, the LSTM's task is vital. The model for detecting anomalies in SpO_2 shows an accuracy of 94% and a loss of 14%.

In the future, these models can be further refined with different paradigms to detect contextual anomalies, not just the anomaly score as implemented in this paper. With the use of feature extraction, this can be achieved in several ways. With newer and more complete medical records available as well, other metrics could also be considered beyond the seven used in this paper. Overall, deep-learning-based anomaly detection has enormous potential in the implementation of WBANs.

Acknowledgements. The authors would like to express their enormous gratitude to the BITS BioCyTiH Foundation and to DST, Govt. of India, for their support and funding provided for this research.

Data Availability Statement. The source code for the models described in this research is accessible on GitHub [9]. It has been implemented using TensorFlow [10], Keras [11], scikit-learn [13], and NumPy [8]. The Pandas Python library was used to help in data graphing [12].

References

1. IEEE Standard for Local and metropolitan area networks - Part 15.6: Wireless Body Area Networks. In: IEEE Std 802.15.6-2012, pp. 1–271 (2012). https://doi.org/10.1109/IEEESTD.2012.6161600
2. Harun Al Rasyid, M.U., Setiawan, F., Nadhori, I.U., Sudarsonc, A., Tamami, N.: Anomalous data detection in WBAN measurements. In: 2018 International Electronics Symposium on Knowledge Creation and Intelligent Computing (IES-KCIC), Bali, Indonesia, pp. 303–309 (2018). https://doi.org/10.1109/KCIC.2018.8628522
3. Canizo, M., Triguero, I., Conde, A., Onieva, E.: Multi-head CNN-RNN for multi-time series anomaly detection: an industrial case study. Neurocomputing **363**, 246–260 (2019). https://doi.org/10.1016/j.neucom.2019.07.034
4. Khan, F.A., Haldar, N.A.H., Ali, A., Iftikhar, M., Zia, T.A., Zomaya, A.Y.: A continuous change detection mechanism to identify anomalies in ECG signals for WBAN-based healthcare environments. IEEE Access **5**, 13531–13544 (2017). https://doi.org/10.1109/ACCESS.2017.2714258
5. Provotar, O.I., Linder, Y.M., Veres, M.M.: Unsupervised anomaly detection in time series using LSTM-based autoencoders. In: 2019 IEEE International Conference on Advanced Trends in Information Theory (ATIT), Kyiv, Ukraine, pp. 513–517 (2019). https://doi.org/10.1109/ATIT49449.2019.9030505

6. Zhai, J., Zhang, S., Chen, J., He, Q.: Autoencoder and its various variants. In: 2018 IEEE International Conference on Systems, Man, and Cybernetics (SMC), Miyazaki, Japan, pp. 415–419 (2018). https://doi.org/10.1109/SMC.2018.00080
7. Goldberger, A., et al.: PhysioBank, PhysioToolkit, and PhysioNet: components of a new research resource for complex physiologic signals. Circulation **101**(23), e215–e220 (2000). https://www.physionet.org/static/published-projects/circulation. Accessed 10 Sept 2023
8. Harris, C.R., et al.: Array programming with NumPy. Nature **585**(7825), 357–362 (2020). https://doi.org/10.1038/s41586-020-2649-2
9. Kartikeya, D.: Anomaly detection in WBANs. GitHub. https://github.com/kakuking/AnomalyDetectionInWBANs. Accessed 25 Oct 2023
10. TensorFlow Developers, TensorFlow. Zenodo (2023). https://doi.org/10.5281/ZENODO.4724125
11. Chollet, F., et al.: Keras. GitHub (2015). https://github.com/fchollet/keras
12. McKinney, W., et al.: Data structures for statistical computing in Python. In: Proceedings of the 9th Python in Science Conference, vol. 445, pp. 51–56 (2010)
13. Pedregosa, F., et al.: Scikit-learn: machine learning in python. J. Mach. Learn. Res. **12**, 2825–2830 (2011)
14. Hunter, J.D.: Matplotlib: a 2D graphics environment. Comput. Sci. Eng. **9**(3), 90–95 (2007)

LLM-Based Agents Utilized in a Trustworthy Artificial Conscience Model for Controlling AI in Medical Applications

Davinder Kaur[1], Suleyman Uslu[1], Mimoza Durresi[2], and Arjan Durresi[1(✉)]

[1] Indiana University-Purdue University Indianapolis, Indianapolis, IN, USA
{davikaur,suslu}@iu.edu, adurresi@iupui.edu
[2] European University of Tirana, Tirana, Albania
mimoza.durresi@uet.edu.al

Abstract. As Large Language Models (LLMs) become more widely used in decision-making contexts, they can enhance trustworthiness, reliability, and user satisfaction in AI applications. This paper proposes implementing an artificial conscience model as a control mechanism for AI-based systems in medical applications that utilize LLM-based AI agents. These models provide structured oversight to ensure that AI applications adhere to user-defined requirements, such as achieving a balance between accuracy and explainability. To demonstrate the practical application of this control mechanism, we present a case study using the Wisconsin Breast Cancer dataset. Our research highlights the effectiveness of LLM-based AI agents in tasks requiring monitoring and control. Furthermore, we emphasize the potential of the artificial conscience control model to improve transparency, adaptability, and user satisfaction in AI-driven decision-making contexts. By employing this mechanism, we aim to pave the way for more reliable, trustworthy, and user-centric AI systems.

1 Introduction

In recent years, the proliferation of Large Language Models (LLMs) has significantly transformed the landscape of artificial intelligence (AI), with these sophisticated systems being deployed across various decision-making domains [5]. As LLM-based AI agents become increasingly prevalent, they can be useful in ensuring the trustworthiness, reliability, and user satisfaction of AI applications. While LLMs demonstrate remarkable capabilities in tasks such as natural language processing and data analysis, their complex algorithms and black-box nature raise significant challenges regarding transparency and accountability [11].

This paper advocates using an artificial conscience model with LLM-based AI agents to control AI applications. This type of model provides structured oversight, enabling users to define and enforce specific requirements for AI decision-making processes. By incorporating principles of ethics and user-defined criteria, artificial conscience models aim to strike a balance between different trustworthy AI requirements [21].

Through a comprehensive analysis and case study utilizing the Wisconsin Breast Cancer dataset, we demonstrate the practical application of the artificial conscience model in regulating LLM-based AI agents. By evaluating the performance of these agents in real-world scenarios, we highlight their effectiveness in tasks that require monitoring, control, and user centric decision-making.

Furthermore, we explore the potential of the artificial conscience model to enhance transparency, adaptability, and user satisfaction in AI-driven decision-making contexts. By providing users with greater insight into AI decision-making processes and empowering them to define their preferences and constraints, these models pave the way for more accountable and user-centric AI systems.

In the subsequent sections of this paper, we delve deeper into the concept of the artificial conscience model, its implementation in controlling LLM-based AI agents, and its implications for enhancing trust, reliability, and user satisfaction in AI-driven decision-making. Through our research findings and analysis, we aim to contribute to the ongoing discourse on responsible AI development and foster the creation of more ethical and transparent AI systems. In Sect. 2, we provide background information and review prior work on controlling AI, artificial conscience, and LLMs. Section 3 outlines our framework, the "Trustworthy Artificial Conscience Control Module". Sect. 4 offers a practical demonstration of our framework using the Wisconsin breast cancer dataset. Finally, in Sect. 5, we present our concluding remarks.

2 Background and Related Work

This section provides an overview of the historical background and previous studies concerning artificial conscience and LLMs.

2.1 Need for AI Control

Controlling and overseeing AI agents is vital due to their significant impact on various aspects of our lives. While AI offers benefits, it also presents risks such as unfairness and privacy breaches if not properly managed. Rules and systems are necessary to ensure AI adheres to our values and operates correctly, averting potential issues. As AI and algorithmic decision-making increasingly pervade daily activities, their complexity and widespread adoption underscore the importance of robust control and oversight mechanisms.

Governmental and private organizations have proposed guidelines to ensure the safety and reliability of AI systems, emphasizing the essential role of humans in decision-making processes. These frameworks highlight the collaborative relationship between humans and AI technology, aiming to ensure ethical outcomes and mitigate potential harm to users or society. One critical aspect highlighted by these agencies is the necessity for human involvement or oversight in AI decision-making processes [12]. This is crucial because AI systems are designed to complement human abilities rather than replace them entirely. Various researchers have proposed diverse methods to incorporate human participation throughout the AI lifecycle. For instance, the European

Union (EU) has suggested human involvement across three phases: design, development, and oversight [9]. Alternatively, some researchers advocate for human involvement based on the level of risk associated with AI applications [8]. In high-stakes scenarios, humans are typically engaged in nearly all phases of the AI lifecycle, including the authentication of decisions before implementation. Conversely, human involvement may be limited to the design and development phases for low-stakes applications. Additionally, the International Organization for Standardization (ISO) has proposed integrating control points throughout the AI lifecycle to enhance trust and adoption of AI systems. These frameworks underscore the significance of human engagement in AI decision-making processes, as endorsed by major organizations, emphasizing the collaborative relationship between humans and AI technology to ensure responsible and ethical outcomes [14, 17].

2.2 Artificial Conscience

Artificial conscience, also called machine conscience, aims to incorporate certain elements of human cognition inspired by consciousness [16]. However, it does not seek to replicate the entirety of human consciousness, which arises from complex societal, cultural, and biological factors. Instead, it focuses on replicating specific feasible aspects for machines to emulate [6]. Researchers have outlined various objectives achievable through artificial conscience, including autonomy, resilience, self-motivation, and information integration. Achieving these objectives requires the design of conscious machines capable of replicating certain features of conscious experience.

Understanding consciousness is crucial for replicating its features. Cognitive neuroscientist [1] proposed the Global Workspace Theory (GWT), likening consciousness to a theater stage where working memory corresponds to the stage, focal consciousness to the bright spot, and unconscious processes to the audience. Another perspective influenced by theoretical computer science is the Conscious Turing Machine (CTM) theory, which integrates Turing's computational model and GWT. In CTM, short-term memory represents the stage, and processors represent the audience, competing to broadcast information [2]. Some researchers argue that traditional rationality is unattainable for real agents and propose bounded optimality, focusing on optimizing the selection algorithm rather than actions taken [37]. Collectively, these efforts aim to understand the human conscience and explore how certain aspects of it can be translated into AI systems.

2.3 Large Language Model-Based Agents

Over the years, humanity has strived to develop artificial intelligence (AI) capable of rivaling or surpassing human-level capabilities, with AI agents serving as a promising avenue for this endeavor. These agents, which possess the ability to perceive their environment, make decisions, and take action [38], have been the subject of extensive research since the mid-20th century. While previous efforts largely focused on refining algorithms and training methodologies for specific tasks, the field lacked a sufficiently versatile model for designing AI agents adaptable to diverse scenarios. Enter large language models (LLMs), hailed for their multifaceted and impressive capabilities [47], which are considered potential sparks for achieving Artificial General Intelligence (AGI) and creating general-purpose AI agents [4]. Leveraging LLMs as foundational frameworks, numerous research endeavors have significantly progressed in developing AI agents with adaptable and versatile functionalities.

Large language model agents are sophisticated artificial intelligence systems engineered to comprehend and produce human-like text on a vast scale. These models, exemplified by GPT (Generative Pre-trained Transformer) from OpenAI, undergo training on extensive datasets comprising diverse textual sources [27]. This training equips them to generate coherent and contextually appropriate responses across a wide array of prompts. Large language model agents find utility in tasks such as natural language understanding, text generation, translation, and summarization. Their operation relies on deep learning methodologies, particularly transformer architectures, which enable them to process and generate text with remarkable fluency and coherence [25]. While these models hold promise for revolutionizing various domains, including content creation, customer service, education, and research, they also raise ethical and societal concerns regarding potential issues like bias, misinformation, and privacy infringement.

3 Trustworthy Artificial Conscience Control Module

This section introduces the Trustworthy Artificial Conscience Control Module, designed to regulate AI agents according to user preferences. The module acknowledges that users may have varying requirements from the AI agents and allows them to assign weights to the evaluation metrics accordingly. These weights guide the negotiation process among the agents to determine the final "artificial feeling" solution. The negotiation process is governed by two key parameters: trust and trust sensitivity of the metrics. Subsection 3.1 discusses the trust management framework and the concept of trust sensitivity. Subsection 3.2 outlines the framework of the control module.

3.1 Trust Framework

Trust is a multifaceted concept that varies depending on the context and is defined differently across various disciplines. Generally, it refers to the confidence one entity has in another entity to behave as expected [12]. Social interactions heavily rely on trust, with individuals tending to trust each other more after experiencing positive interactions. Trust plays a crucial role in decision-making processes involving multiple entities and is essential for negotiations and trade-offs [22]. Managing trust information in such scenarios can be challenging, prompting researchers to propose various methods for its calculation and management. Numerous studies [28–30,49] have delved into and investigated trust management. Ruan et al. [36] introduced a trust management framework based on measurement theory, which has been applied across diverse decision-making domains such as healthcare [15,17], social networks [13,19,20,34,48], crime detection [13,18,26], cloud computing [31,32,40], the Internet of Things [33,35], and the food-energy sector [41–46]. This framework has also been utilized to quantify user acceptance of AI systems based on requirements, highlighting the versatility and potential of trust in facilitating negotiations across different aspects of AI system metrics.

A trust framework gauges trust between entities based on their past interactions. Ratings exchanged between agents are crucial factors in trust calculation in decision-making scenarios involving multiple agents.

Agents iteratively propose and rate each other's solutions, with ratings ranging between 0 and 1, where higher ratings denote stronger impressions. The mean of these ratings denoted as the impression is calculated to represent an agent's overall perception of another agent. Equation 1 shows the formula to calculate impression.

$$m^{A:B} = \frac{\sum_{i=1}^{N} r_i^{A:B}}{N} \tag{1}$$

Additionally, confidence, indicating the consistency of the impression, is inversely related to the standard error of the mean, as shown in Eq. 2. Both impression and confidence values are utilized to calculate trust between agents, facilitating trust propagation and aggregation even without direct connections.

$$c = 1 - 2*e \text{ where } e = \sqrt{\frac{\sum_{i=1}^{N}(m^{A:B} - r_i^{A:B})^2}{N*(N-1)}} \tag{2}$$

Moreover, the framework incorporates concepts from social psychology, such as trust pressure and trust sensitivity, to capture agents' adaptive behavior. Trust pressure, determined by the disparity between an agent's target and current trust levels, influences their behavior, with the magnitude of this effect governed by their trust sensitivity. This sensitivity is introduced through user-assigned weights reflecting the significance of metrics in the application context, emphasizing the importance of human input in shaping AI systems. Equation 3 and Eq. 4 show the calculation for trust pressure and sensitivity.

$$P = T_{target} - T_{current} \tag{3}$$

$$P_e = P \times S_T \tag{4}$$

3.2 Artificial Conscience Control Module

The Artificial Conscience Control Module is crucial in steering AI agents according to user-defined criteria. It acknowledges users' diverse expectations and requirements, which impart significance to AI processes beyond mere algorithms. This framework operates within a decision-making context for AI systems, where "artificial AI agents" engage in negotiations based on user-defined criteria and trust garnered from their peers to determine an "Artificial Feeling" (AF) through a weighted average among agents, as illustrated in Fig. 1.

The negotiation process within this framework consists of several key steps:

- Agent Goal Initialization: At the outset of the negotiation, each artificial AI agent is tasked with achieving the best possible solution. Agents may prioritize different parameters based on their goals, leading to negotiation dynamics.
- Round-Based Negotiation: Artificial agents engage in negotiation rounds, with the number of rounds, denoted as n, being application or user-specific.

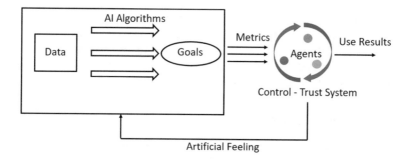

Fig. 1. A model to control AI agents

- Initial Solution Proposal: In the first round, each artificial agent proposes its best solution based on its objectives.
- Trust-Driven Solution Adaptation: Artificial agents assess the level of trust they receive from the user/community, reflecting user approval. If trust is insufficient, agents adjust their solutions in subsequent rounds to enhance trust, potentially at the expense of benefits.
- Solution Comparison and Rating: Solutions are compared based on their benefits using a distance metric calculated through the Euclidean distance formula. In each round, artificial agents rate other agents' solutions based on their distance, influencing trust calculation.
- Trust Sensitivity and Sacrifice: Artificial agents aim to minimize the distance between their proposed solution and their goal while considering received trust. Trust sensitivity governs the level of sacrifice, with higher sensitivity prompting greater sacrifice to improve ratings.
- Artificial Feeling (AF) Calculation: After n rounds, the artificial feeling (AF) is computed as the weighted average among artificial agents, facilitating comparison and selection of solutions from different AI algorithms.
  This framework facilitates the control of the AI agents according to the user's specific requirements, ensuring that the system's operation aligns closely with user preferences and objectives.

4   Implementation

We assessed various AI agents for feature selection in breast cancer prediction to demonstrate the proposed framework. AI agent explanations served as an interface for domain experts. This section offers insights into the data, experimental configuration, and evaluation outcomes.

4.1   Data

To validate our framework, we utilized The Wisconsin Breast Cancer Data Set [10] to differentiate between malignant and benign samples. This dataset comprises features computed from images of fine needle aspirates of breast masses. Ten features, including radius, texture, perimeter, area, smoothness, compactness, concavity, concave points, symmetry, and fractal dimension, characterize the shape, size, and texture of each cell nucleus [39]. For each feature, calculations are made for the largest value, mean value, and standard error across the cell range.

4.2 Experimentation and Results

In our investigation, AI agents are tasked with the decision-making process of classifying breast cancer cases into malignant and benign categories. Multiple AI agents are deployed and assessed based on accuracy and explanation metrics to accomplish this task. The objective is to fine-tune and regulate AI agents' behavior according to end users' preferences. For instance, while accuracy may be paramount for one user, explainability holds greater significance for others. Thus, our framework facilitates the customization of AI systems based on user or stakeholder requirements. We assume that certain AI agents have been employed to select the most relevant features from the Wisconsin Breast Cancer dataset, meeting predefined performance thresholds. These agents must undergo evaluation based on their rationale for identifying cancerous and non-cancerous cases. Moreover, users of these agents require a comprehensive understanding of the explanations provided by the AI system to establish trust. Our experiment aims to gauge the degree of alignment between the AI agent's explanations for its predictions and users' thought processes and comprehension in the domain. In our study, we employed two distinct types of metrics. Firstly, we utilized performance metrics, specifically accuracy, to measure the effectiveness of the agents in predicting true labels. Additionally, we incorporated an explainability metric to assess the agents' reasoning capabilities and the user's comprehension of the selected attributes.

For experimentation, we utilized Language Model models (LLMs), specifically emphasizing the Generative Pre-trained Transformer (GPT), for feature selection in breast cancer prediction. GPT was selected due to its robust zero-shot performance, allowing it to achieve excellent results without extensive fine-tuning efforts [23]. Our findings demonstrate the effectiveness of GPT-based models in feature selection using extensive breast cancer research data obtained from the web. We employed the GPT-3.5 model, an advanced language model developed by OpenAI [3], which is based on the generative pre-trained transformer architecture and boasts 175 billion parameters, making it one of the largest language models available. GPT-3.5 excels in various natural language processing tasks. We applied zero-shot learning techniques for feature selection, where the model is trained to recognize or classify objects, categories, or concepts it has never encountered during training [7]. This approach allows the model to generalize to previously unseen classes by leveraging semantic embeddings or auxiliary information. The prompts used in our experiment were carefully crafted to suit the task, context, specificity, and open-ended nature.

To illustrate the functionality of the artificial conscience control module, we constructed two AI agents. Agent 1 was designed with the prompt "Select the best 5 features based on the research on this dataset that will give me the best cancer prediction accuracy", while Agent 2 operated with the prompt "Select the minimum number of features based on the research on this dataset that will give me the best cancer prediction accuracy". Following these prompts, Agent 1 identified the features (concave points_worst, perimeter_worst, concavity_worst, area_worst, texture_worst) out of a total of 30 available features, while Agent 2 selected the three features (concave points_worst, perimeter_worst, concavity_worst). These selected features were then compared with those outlined in a research paper, where the authors indicated three parameters (texture_mean, area_worst, smoothness_worst) that yielded maximum accuracy [24].

Once the features are selected by the AI agents, they are integrated into the training process of the machine learning model using logistic regression. Subsequently, the model's performance is evaluated in terms of accuracy and explainability. This comprehensive assessment ensures a thorough understanding of how effectively the selected features contribute to the model's predictive capabilities while also considering the ease of interpretation for end-users. Features selected according to the research paper yield approximately 97.37% accuracy, whereas those selected by AI Agents 1 and 2 achieve 96.23% and 95.6% accuracy, respectively. These findings indicate that AI agents based

Selected Features	AI agent generated Explaination
• concave points_worst • perimeter_worst • concavity_worst	The selected features are known to be highly informative and relevant in distinguishing between benign and malignant tumors, as evidenced by extensive research on the Wisconsin Diagnostic Breast Cancer (WDBC) dataset. Here is the rationale behind choosing these specific features: • **concave points_worst**: The presence of a higher number of concave points in the worst-case scenario is often associated with malignant tumors. The prominence of this feature suggests the irregularity and complexity of the tumor boundary, which can be a crucial indicator of malignancy. • **perimeter_worst**: The perimeter of the tumor, particularly in its worst-case scenario, is a key characteristic that helps to assess the irregularities and complexities in the tumor's shape. A larger perimeter is often linked to irregular and potentially malignant tumor boundaries. • **concavity_worst**: The level of concavity in the worst-case scenario provides valuable information about the severity of irregularities in the tumor boundary. A higher value for this feature indicates more pronounced irregularities, which are often associated with malignant tumors. These features have consistently demonstrated their significance in accurately predicting cancer in the context of the WDBC dataset, making them reliable indicators for the diagnosis of breast cancer. By focusing on these specific features, you can build a simplified yet effective model that prioritizes the most critical characteristics for accurate cancer prediction.

Fig. 2. Features Chosen by the AI agent and their explanation

on LLM models can effectively perform feature selection, yielding comparable accuracy levels based on extensive data.

Subsequently, the AI agent is prompted to generate an explanation for its selection of the specific subset of features. This explanation serves as a tool for end users or medical professionals to comprehend the reasoning behind the AI agent's output and assess its validity and relevance to their needs. Figure 2 illustrates the features chosen by the AI agent alongside their corresponding explanations. These agents were then evaluated for their explainability. An AI agent achieving higher accuracy by utilizing more features, such as Agent 1, maybe less explainable as it relies on a larger number of features for decision-making, which can pose challenges for users to comprehend. Conversely, an agent employing fewer features, like Agent 2, may exhibit slightly lower accuracy but greater explainability. Alternatively, another agent that utilizes all features may attain maximum accuracy but minimal explainability, as understanding all features can be challenging for users. Therefore, our artificial conscience model aids in controlling these AI agents by considering the trade-off between accuracy and explainability based on user requirements.

5   Conclusion

In conclusion, our study illustrates the significance of employing an artificial conscience control model to use LLM-generated AI agents effectively according to user requirements. By leveraging this framework, we can navigate the intricate balance between different user requirements, crucial factors in ensuring the trustworthiness and usability of AI systems. Our experiment with the Wisconsin Breast Cancer dataset serves as a practical demonstration of how this framework can be applied in real-world scenarios. Hence, integrating artificial conscience models into AI development processes has vast potential to improve transparency, adaptability, and user satisfaction in contexts where AI guides decision-making.

References

1. Baars, B.J.: In the theatre of consciousness. Global workspace theory, a rigorous scientific theory of consciousness. J. Conscious. Stud. **4**(4), 292–309 (1997)
2. Blum, L., Blum, M.: A theory of consciousness from a theoretical computer science perspective: insights from the conscious turing machine. Proc. Natl. Acad. Sci. **119**(21), e2115934,119 (2022)
3. Brown, T., et al.: Language models are few-shot learners. Adv. Neural. Inf. Process. Syst. **33**, 1877–1901 (2020)
4. Bubeck, S., et al.: Sparks of artificial general intelligence: early experiments with GPT-4. arXiv preprint arXiv:2303.12712 (2023)
5. Chang, Y., et al.: A survey on evaluation of large language models. ACM Trans. Intell. Syst. Technol. (2023)
6. Chella, A., Manzotti, R.: Artificial Consciousness. Andrews UK Limited (2013)
7. Dang, H., Mecke, L., Lehmann, F., Goller, S., Buschek, D.: How to prompt? Opportunities and challenges of zero-and few-shot learning for human-AI interaction in creative applications of generative models. arXiv preprint arXiv:2209.01390 (2022)

8. EC: Ethics guidelines for trustworthy AI (2018). https://ec.europa.eu/digital-single-market/en/news/ethics-guidelines-trustworthy-ai
9. Floridi, L.: Establishing the rules for building trustworthy AI. Nat. Mach. Intell. **1**(6), 261–262 (2019)
10. Frank, A., Asuncion, A.: UCI machine learning repository, vol. 213, no. 2. University of California. School of information and computer science, Irvine (2010). http://archive.ics.uci.edu/ml
11. Hadi, M.U., et al.: A survey on large language models: applications, challenges, limitations, and practical usage. Authorea Preprints (2023)
12. Information Technology – Artificial Intelligence – Overview of trustworthiness in artificial intelligence . Standard, International Organization for Standardization (2020)
13. Kaur, D., Uslu, S., Durresi, A.: Trust-based security mechanism for detecting clusters of fake users in social networks. In: Barolli, L., Takizawa, M., Xhafa, F., Enokido, T. (eds.) WAINA 2019. AISC, vol. 927, pp. 641–650. Springer, Cham (2019). https://doi.org/10.1007/978-3-030-15035-8_62
14. Kaur, D., Uslu, S., Durresi, A.: Requirements for trustworthy artificial intelligence – a review. In: Barolli, L., Li, K.F., Enokido, T., Takizawa, M. (eds.) NBiS 2020. AISC, vol. 1264, pp. 105–115. Springer, Cham (2021). https://doi.org/10.1007/978-3-030-57811-4_11
15. Kaur, D., Uslu, S., Durresi, A.: Trustworthy AI explanations as an interface in medical diagnostic systems. In: Barolli, L., Miwa, H., Enokido, T. (eds.) NBiS 2022. LNNS, vol. 526, pp. 119–130. Springer, Cham (2022). https://doi.org/10.1007/978-3-031-14314-4_12
16. Kaur, D., Uslu, S., Durresi, A.: A model for artificial conscience to control artificial intelligence. In: Barolli, L. (ed.) AINA 2023. LNNS, vol. 654, pp. 159–170. Springer, Cham (2023). https://doi.org/10.1007/978-3-031-28451-9_14
17. Kaur, D., Uslu, S., Durresi, A., Badve, S., Dundar, M.: Trustworthy explainability acceptance: a new metric to measure the trustworthiness of interpretable AI medical diagnostic systems. In: Barolli, L., Yim, K., Enokido, T. (eds.) CISIS 2021. LNNS, vol. 278, pp. 35–46. Springer, Cham (2021). https://doi.org/10.1007/978-3-030-79725-6_4
18. Kaur, D., Uslu, S., Durresi, A., Mohler, G., Carter, J.G.: Trust-based human-machine collaboration mechanism for predicting crimes. In: Barolli, L., Amato, F., Moscato, F., Enokido, T., Takizawa, M. (eds.) AINA 2020. AISC, vol. 1151, pp. 603–616. Springer, Cham (2020). https://doi.org/10.1007/978-3-030-44041-1_54
19. Kaur, D., Uslu, S., Durresi, M., Durresi, A.: A geo-location and trust-based framework with community detection algorithms to filter attackers in 5G social networks. Wirel. Netw. 1–9 (2022)
20. Kaur, D., Uslu, S., Durresi, M., Durresi, A.: Artificial intelligence control for trust-based detection of attackers in 5G social networks. Comput. Intell. (2023)
21. Kaur, D., Uslu, S., Rittichier, K.J., Durresi, A.: Trustworthy artificial intelligence: a review. ACM Comput. Surv. (CSUR) **55**(2), 1–38 (2022)
22. Kong, D.T., Yao, J.: Advancing the scientific understanding of trust and culture in negotiations. Negot. Confl. Manage. Res. **12**(2), 117–130 (2019)
23. Liu, Y., et al.: Summary of ChatGPT/GPT-4 research and perspective towards the future of large language models. arXiv preprint arXiv:2304.01852 (2023)
24. Mangasarian, O.L., Street, W.N., Wolberg, W.H.: Breast cancer diagnosis and prognosis via linear programming. Oper. Res. **43**(4), 570–577 (1995)
25. Naveed, H., et al.: A comprehensive overview of large language models. arXiv preprint arXiv:2307.06435 (2023)
26. Rittichier, K.J., Kaur, D., Uslu, S., Durresi, A.: A trust-based tool for detecting potentially damaging users in social networks. In: Barolli, L., Chen, H.-C., Enokido, T. (eds.) NBiS 2021. LNNS, vol. 313, pp. 94–104. Springer, Cham (2022). https://doi.org/10.1007/978-3-030-84913-9_9

27. Roumeliotis, K.I., Tselikas, N.D.: ChatGPT and open-AI models: a preliminary review. Future Internet **15**(6), 192 (2023)
28. Ruan, Y., Alfantoukh, L., Durresi, A.: Exploring stock market using twitter trust network. In: 2015 IEEE 29th International Conference on Advanced Information Networking and Applications (AINA), pp. 428–433. IEEE (2015)
29. Ruan, Y., Alfantoukh, L., Fang, A., Durresi, A.: Exploring trust propagation behaviors in online communities. In: 2014 17th International Conference on Network-Based Information Systems (NBiS), pp. 361–367. IEEE (2014)
30. Ruan, Y., Durresi, A.: A survey of trust management systems for online social communities-trust modeling, trust inference and attacks. Knowl.-Based Syst. **106**, 150–163 (2016)
31. Ruan, Y., Durresi, A.: A trust management framework for cloud computing platforms. In: 2017 IEEE 31st International Conference on Advanced Information Networking and Applications (AINA), pp. 1146–1153. IEEE (2017)
32. Ruan, Y., Durresi, A.: A trust management framework for clouds. Comput. Commun. **144**, 124–131 (2019)
33. Ruan, Y., Durresi, A., Alfantoukh, L.: Trust management framework for internet of things. In: 2016 IEEE 30th International Conference on Advanced Information Networking and Applications (AINA), pp. 1013–1019. IEEE (2016)
34. Ruan, Y., Durresi, A., Alfantoukh, L.: Using twitter trust network for stock market analysis. Knowl.-Based Syst. **145**, 207–218 (2018)
35. Ruan, Y., Durresi, A., Uslu, S.: Trust assessment for internet of things in multi-access edge computing. In: 2018 IEEE 32nd International Conference on Advanced Information Networking and Applications (AINA), pp. 1155–1161. IEEE (2018)
36. Ruan, Y., Zhang, P., Alfantoukh, L., Durresi, A.: Measurement theory-based trust management framework for online social communities. ACM Trans. Internet Technol. (TOIT) **17**(2), 1–24 (2017)
37. Russell, S.J.: Rationality and intelligence. Artif. Intell. **94**(1–2), 57–77 (1997)
38. Schwartz, S., Yaeli, A., Shlomov, S.: Enhancing trust in LLM-based AI automation agents: new considerations and future challenges. arXiv preprint arXiv:2308.05391 (2023)
39. Street, W.N., Wolberg, W.H., Mangasarian, O.L.: Nuclear feature extraction for breast tumor diagnosis. In: Biomedical Image Processing and Biomedical Visualization, vol. 1905, pp. 861–870. SPIE (1993)
40. Uslu, S., Kaur, D., Durresi, M., Durresi, A.: Trustability for resilient internet of things services on 5G multiple access edge cloud computing. Sensors **22**(24), 9905 (2022)
41. Uslu, S., Kaur, D., Rivera, S.J., Durresi, A., Babbar-Sebens, M.: Trust-based game-theoretical decision making for food-energy-water management. In: Barolli, L., Hellinckx, P., Enokido, T. (eds.) BWCCA 2019. LNNS, vol. 97, pp. 125–136. Springer, Cham (2020). https://doi.org/10.1007/978-3-030-33506-9_12
42. Uslu, S., Kaur, D., Rivera, S.J., Durresi, A., Babbar-Sebens, M.: Trust-based decision making for food-energy-water actors. In: Barolli, L., Amato, F., Moscato, F., Enokido, T., Takizawa, M. (eds.) AINA 2020. AISC, vol. 1151, pp. 591–602. Springer, Cham (2020). https://doi.org/10.1007/978-3-030-44041-1_53
43. Uslu, S., Kaur, D., Rivera, S.J., Durresi, A., Babbar-Sebens, M., Tilt, J.H.: Control theoretical modeling of trust-based decision making in food-energy-water management. In: Barolli, L., Poniszewska-Maranda, A., Enokido, T. (eds.) CISIS 2020. AISC, vol. 1194, pp. 97–107. Springer, Cham (2021). https://doi.org/10.1007/978-3-030-50454-0_10
44. Uslu, S., Kaur, D., Rivera, S.J., Durresi, A., Babbar-Sebens, M., Tilt, J.H.: A trustworthy human–machine framework for collective decision making in food–energy–water management: the role of trust sensitivity. Knowl.-Based Syst. **213**, 106,683 (2021)

45. Uslu, S., Kaur, D., Rivera, S.J., Durresi, A., Durresi, M., Babbar-Sebens, M.: Trustworthy acceptance: a new metric for trustworthy artificial intelligence used in decision making in food–energy–water sectors. In: Barolli, L., Woungang, I., Enokido, T. (eds.) AINA 2021. LNNS, vol. 225, pp. 208–219. Springer, Cham (2021). https://doi.org/10.1007/978-3-030-75100-5_19

46. Uslu, S., Kaur, D., Rivera, S.J., Durresi, A., Durresi, M., Babbar-Sebens, M.: Trustworthy fairness metric applied to AI-based decisions in food-energy-water. In: Barolli, L., Hussain, F., Enokido, T. (eds.) AINA 2022. LNNS, vol. 450, pp. 433–445. Springer, Cham (2022). https://doi.org/10.1007/978-3-030-99587-4_37

47. Xi, Z., et al.: The rise and potential of large language model based agents: a survey. arXiv preprint arXiv:2309.07864 (2023)

48. Zhang, P., Durresi, A.: Trust management framework for social networks. In: 2012 IEEE International Conference on Communications (ICC), pp. 1042–1047. IEEE (2012)

49. Zhang, P., Durresi, A., Barolli, L.: Survey of trust management on various networks. In: 2011 International Conference on Complex, Intelligent and Software Intensive Systems (CISIS), pp. 219–226. IEEE (2011)

Task Allocation Based on Simulated Annealing for Edge Industrial Internet

Vitor Gabriel Reis Lux Barboza and Janine Kniess[✉]

Postgraduate in Applied Computing, Santa Catarina State University, Florianópolis, Brazil
vitor.barboza@edu.udesc.br, janine.kniess@udesc.br

Abstract. Industrial Internet of Things (IIoT) applications are critical in terms of response time and accuracy. Cloud computing is often associated with IIoT as a technology to provide significant resources, such as long-term storage and processing power. When industrial devices send data to cloud computing, latency appears as an important aspect. In this context, Edge computing is a potential alternative, as it offers resources for processing at the edge of the network. This paper provides a Simulated Annealing-based approach to the allocation of tasks in industry. The edge node receives multiple tasks with different priorities to process from vehicles and must find the best order of task completion to meet the application deadline. The results obtained in the iFogSim simulator prove that the Task Allocation Approach was able to select the best order that obeys the application deadline in different configuration scenarios.

1 Introduction

The Industrial Internet of Things (IIoT) is a section of devices connected to the internet used in smart factories, remote maintenance and production lines, surveillance, as well as Automated Guided Vehicles performing tasks [7]. Typically, IIoT devices have few computing resources, such as memory, processing, and power. Cloud computing is a technology widely used by the industry to provide resources, such as long-term storage and processing power [11]. However, the response time of IIoT applications is an essential requirement in such critical environments, and sending data to the cloud can affect the time limit required by the application. As a solution to this problem, a recent technology has been employed: Edge Computing. This is because it brings computational resources to the edge of the network, near the IIoT devices, and, therefore, significantly reduces the latency for the applications [10].

To illustrate the use of edges, consider the IIoT application scenario from Fig. 1. The Device-layer is composed of IIoT devices, such as industrial vehicles, connected to the network. Vehicles can perform different tasks and these tasks have different priorities, high or low, depending on the industrial application categorization. For instance, if a machine is already out of supply, the load is urgent, so it takes high priority. However, if the machine is still supplied, the task takes lower priority. In this context, suppose a vehicle is equipped with a camera that can take pictures of the production machine load compartment. It sends pictures to the edge node to process the images and provide back the results obtained. The edge node can receive multiple task requests from different

L. Barolli (Ed.): AINA 2024, LNDECT 201, pp. 210–221, 2024.
https://doi.org/10.1007/978-3-031-57870-0_19

vehicles and needs to allocate them according to the time limit and the priority set for each request.

In Fig. 1, the Edge-layer is composed of edge nodes. In the scenario, one node in each manufacturing plant. The Cloud-layer is composed of cloud computing resources to provide long-term storage for the data already processed from the edge nodes. To address the challenge of task allocation in the proposed scenario, this paper presents as a contribution the TASɛC, a Task Allocation algorithm for mobile IIoT devices in Edge Computing based on the metaheuristic Simulated Annealing [6], that selects the optimized combination to meet the application tasks deadline requirements and reduce the network latency.

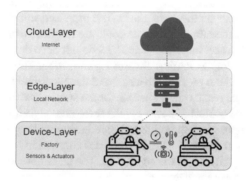

Fig. 1. Application IIoT Scenario

This paper is organized as follows: Sect. 2 shows the related works. Section 3 describes the proposed Task Allocation approach for IIoT, which details the problem formulation and the algorithm modeling with Simulated Annealing. Then, Sect. 4 shows the results obtained through the simulations. Finally, Sect. 5 brings the conclusions and future works.

2   Related Work

Computing resources deployed at the edge node can process data from IIoT devices, analyze the results, and make real-time decisions [1]. Task allocation in edge comput-ing can be challenging because different tasks require varying amounts of computing power, storage, and network bandwidth. To address this challenge, researchers have proposed task allocation methods and optimization techniques [9]. This work examines the current methods for task allocation over edge computing as well as heuristics to search for optimal solutions.

The work presented by [4] proposes an approach using edge computing to provide task allocation for industrial devices. Each task has deadlines to finish the processing, delay sensitive tasks are prioritized by the mechanisms to get previous processing. The heuristic selected by the author to make the decision process in task allocation is the

Genetic Algorithm, simulated with the CloudSim. The paper proposes a scenario that varies between 15 and 40 tasks in each benchmark. The results show satisfactory task attendance within the limited time and total response time minimized.

In [11], the authors point to edge computing as a solution to bring the required resources near the applications. In this work, the scenario covers mobile devices, and each task has defined its limit for conclusion. The solution described in the question is a dynamic task scheduling algorithm that prioritizes tasks based on their urgency and uses Dynamic Programming for the scheduling strategy. The authors simulated the task scheduling algorithm on the EdgeCloudSim simulator and evaluated metrics such as response time, energy consumption, and processing cost. The results showed success in reducing response time and meeting the limits of the tasks.

In the work [12], the authors are using edge computing in industrial scenarios to process the device's tasks on the edge nodes. The proposed solution considers an optimization strategy for scheduling tasks. The goals are response time minimization and a reduction in the total processing time. The tasks do not have deadlines. The algorithm to find allocation decisions is Particle Swarm Optimization. This approach considered more than one edge node to process the tasks, and the experiments simulated 50 to 250 simultaneous tasks. The results show success for the strategy, with PSO minimizing the total response time and achieving better results than other algorithms compared.

The paper [8] proposes a solution based on edge computing that prioritizes urgent tasks over other tasks in a queue. The IIoT devices considered in this work are mobile and send their tasks to a queue in the edge node. The tasks have deadlines defined, and they wait for the optimization decision made by the scheduler to start being processed at the edge. The queue time is registered and counted in the total response time. The goal is to reduce the processing overload and response time, as well as meet the task limit requirements. The optimization algorithm to determine the allocation decision is Animal Migration Optimization. In the experiments, the iFogSim simulator was used, simulating up to 50 simultaneous tasks. Results prove reductions and satisfactory attendance within the task limits.

Comparing the related work to the proposed TASεC approach (described in Sect. 3), it includes the criteria: (i) attends tasks with a defined time limit; (ii) considers the tasks waiting time at the edge queue; and (iii) considers the priority of completing tasks. The scheduler strategy for optimization is developed based on the Simulated Annealing algorithm. The SA performs a search for the optimized decision for the edge node to process the tasks, meeting the requirements imposed by the application.

3   System Model and Problem Statement

The task allocation problem formulation considers the response time for each request sent by the vehicles. The scenario includes one edge node, which has an input queue for the incoming tasks. Industrial vehicles, $V = \{1, 2, \ldots, v\}$, where the vehicle index is $v \in V$. Each vehicle sends multiple requests to be processed on the edge node. The tasks are represented by $M = \{1, 2, \ldots, i\}$, where $i \in M$.

Each task corresponds to a tuple with the attributes $T_i = (T_i^{limit}, D_i, C_i, P_i, v)$, in which, T_i^{limit} is the deadline to complete the task, in seconds; D_i is the total size of the packet, in bits; C_i is the amount of CPU cycles to process one bit of data; P_i is the task priority level, of which 1 is high priority and 2 is low priority; and finally v is the vehicle index that originated the task. The problem formulation is treated as a permutation of integers, where each combination order for the task execution has a cost associated with it, the response time, calculated by the scheduler on the edge and compared with the time limit of each task.

The steps for the task allocation are: (i) Message request transmission to edge. The transmission time is calculated by dividing the total packet size by the network transmission rate [12]. The Eq. (1) represents this calculation; (ii) Queue delay in edge. The edge node must handle multiple tasks, and there will be a time for a task to wait its turn to be processed, that is, the time the request is queued at the edge. This waiting time is known as queue delay, RT_{queue}; (iii) Edge processing time. It's calculated by finding the total CPU cycles to process the task by multiplying the task size, D_i, and the amount of cycles to process one bit, C_i, then divided by the edge CPU frequency, Eq. (2); (iv) Edge response. The message result found on the edge after the task processing is returned to the respective vehicle that sent it.

The total response time to complete the task Eq. (3) is the sum of the Eqs. (1) and (2), and the RT_{queue}. Each task has its own time limit for execution; the combinatorial optimization must consider them while searching for the solution. Finishing the tasks within the limit is a constraint on the objective function, as presented in Eq. (5). The response time, RT_i must be less or equal to the task limit, T_i^{limit}. Therefore, to address the constraint, a penalty function is added to $f(x)$. When the scheduler is calculating the solution and the current solution comes across an infraction of the time limit, the penalty will add more time to the total response time, driving the scheduler to avoid them and select solutions that meet the constraint. We defined a penalty coefficient, $g \in [0,1]$, where 0 is no penalty, and 1 when the worst case is reached. To normalize the penalty function, it's defined that the worst case corresponds to $RT_i \geq 2 \times T_i^{limit}$, that means any response time, two times above the limit time, shall receive the maximum penalty, $g = 1$. For any other between $T_i^{limit} < RT_i \leq 2 \times T_i^{limit}$, the coefficient will be $0 < g < 1$, where g will be a proportional value varying between 0 and 1. The objective function for the response time minimization in task allocation is represented by Eq. (4) and the $f(x)$ in Eq. (6).

$$RT_{trans} = \frac{D_i}{R_j} \tag{1}$$

$$RT_{exec} = \frac{D_i \times C_i}{f_j} \tag{2}$$

$$RT_i = RT_{trans} + RT_{exec} + RT_{queue} \tag{3}$$

$$min \sum_{v=1}^{V} \sum_{i=1}^{M} RT_{v,i}^{j} \tag{4}$$

$$s.t. : RT_i \leq T_i^{limit} \tag{5}$$

$$f(x) = \sum_{v=1}^{V} \sum_{i=1}^{M} RT_{i,v}^{j} + g \times \sum_{v=1}^{V} \sum_{i=1}^{M} (RT_{v,i}^{j} - T_i^{limit}) \tag{6}$$

Algorithm 1. Task Allocation with SA

1: **procedure** TASK_ALLOCATION $(i, D_i, C_i, T_i^{limit}, P_i, v, N, \alpha)$
2:      $S \leftarrow$ GENERATE_INITIAL_SOLUTION$(i, D_i, C_i, T_i^{limit}, P_i, v;)$
3:      $RT\_total \leftarrow$ CALCULATE_RT_TOTAL$(S); S^* \leftarrow S;$ ▷ Set initial solution S as best solution
4:      **for** $(Iter = 1; Iter < N; Iter++)$ **do**
5:          $S' \leftarrow$ GENERATE_NEIGHBOR_SOLUTION$(S);$ ▷ Generate neighbor solution S'
6:          $RT\_neighbor \leftarrow$ CALCULATE_RT_TOTAL$(S');$
7:          **if** $(RT\_neighbor < RT\_total)$ **then**
8:              $S \leftarrow S'; RT\_total \leftarrow T\_neighbor;$
9:              **if** $(RT\_total < RT\_best)$ **then**
10:                 $S^* \leftarrow S;$ ▷ Set current solution S' as best solution
11:                 $\Delta = RT\_neighbor - RT\_total;$ ▷ Calculate the delta over the solutions
12:                 $r \leftarrow$ generate $r \in [0, 1];$ ▷ Select randomly a float between 0 and 1
13:                 **if** $(r < e^{-\Delta/Temp})$ **then**
14:                     $S \leftarrow S';$ ▷ Set solution S' as the current solution
15:                     $RT\_total \leftarrow RT\_neighbor;$
16:                 **end if**
17:             **end if**
18:         **end if**
19:         $Temp = (1 - (Iter/N))^\alpha;$ ▷ Update temperature
20:     **end for**
21:     $S^* \leftarrow$ PRIORITIZATION$(S^*);$
22:     **return** $(S^*, RT\_best);$
23: **end procedure**
24: **procedure** GENERATE_INITIAL_SOLUTION
25:     new Task$(Task[0] = i, Task[1] = D_i, Task[2] = C_i, Task[3] = T_i^{limit}, Task[4] = P_i, Task[5] = v);$
26:     $RT_{trans} = D_i/R_j;$ ▷ Equation (1)
27:     $RT_{exec} = (D_i \times C_i)/f_j;$ ▷ Equation (2)
28:     $RT_i = RT_{trans} + RT_{exec} + RT_{queue};$ ▷ Equation (3)
29:     $S \leftarrow$ new ArrayList $<$ Tasks $>;$
30:     **for** $(j = 1; j \leq get\_size(ArrayList < Tasks >); j++)$ **do**
31:         $task \leftarrow get\_int\_random(get\_size(ArrayList < Tasks >));$
32:         **if** $(S[j] \not\supset get\_value(task, ArrayList < Tasks >))$ **then**
33:             $S[j] \leftarrow get\_value(task, ArrayList < Tasks >);$
34:         **end if**
35:     **end for**
36:     **return** $S;$
37: **end procedure**

The Algorithm 1 stands for the *Simulated Annealing (SA)* functioning in the task allocation optimization. The SA [6] is an algorithm inspired by physical phenomena inherent to metal tempering, in which the temperature applied to the metal is a determining factor, shaking the molecules when the metal is heated or stabilizing the molecules when it is cooled. Following natural inspiration, the SA algorithm starts with an elevated temperature, offering an exploration of the search field with stochastic searches. As the temperature reduces, the algorithm is directed to the search for local refinement. The metal cooling time is perceived in the form of iterations. The SA's native parameters must be seen carefully, as they influence whether it approaches and reaches the global optimum [6].

In the Algorithm 1, the inputs are the tuples from each task and two native SA parameters: the total number of iterations is N and the cooling coefficient is α. The initial solution, S (line 2), is generated by reading all the tasks in a queue and putting them in a list structure as a sequence. The initial solution total response time (line 3) is obtained by the procedure that runs the Fitness Function in the Algorithm 2, lines 1 to 22, where RT_{queue} initially is equal to zero for each task. On each iteration of the

---

Algorithm 2. Total Response Time

```
 1: procedure CALCULATE_RT_TOTAL(S)
 2:     RT_queue ← 0;
 3:     for (j = 0; j ≤ get_size(S); j++) do                                    ▷ Equation (4)
 4:         fromTask = getTask(j);
 5:         if (j + 1 < get_size(S)) then
 6:             destTask = getTask(j + 1);
 7:         else
 8:             destTask = getTask(j);
 9:         end if
10:         if ((fromTask.RT_i(destTask) + RT_queue) ≥ T_i^{limit}) then         ▷ Equation (5)
11:             if ((fromTask.RT_i(destTask) + RT_queue) ≥ 2 × T_i^{limit} then
12:                 g = 1;                                                       ▷ Set penalty coefficient
13:             else
14:                 g = T_i^{limit} − fromTask.RT_i(destTask);
15:             end if
16:         end if
17:         penalty+ = g × (T_i^{limit} − fromTask.RT_i(destTask));             ▷ Calculate the penalty
18:         RT_queue = fromTask.RT_i(destTask) + RT_queue;
19:     end for
20:     RT_total = RT_queue + penalty; return RT_total;                         ▷ Set the RT_total
21: end procedure
22: procedure GENERATE_NEIGHBOR_SOLUTION
23:     S' ← new Array[ ]; x1 ← int_random(size(S));
24:     x2 ← int_random(size(S));
25:     if (x1 == x2) then
26:         GENERATE_NEIGHBOR_SOLUTION(S);
27:     else
28:         if (x1 > x2) then
29:             if (destTask.P_i(fromTask) < fromTask.P_i(destTask)) then
30:                 S'[x1] ← S[x2]; S'[x2] ← S[x1];
31:             else
32:                 if (destTask.P_i(fromTask) == fromTask.P_i(destTask)) then
33:                     if (destTask.T_i^{limit} < fromTask.T_i^{limit}) then
34:                         S'[x1] ← S[x2]; S'[x2] ← S[x1];
35:                     end if
36:                 end if
37:             end if
38:         end if
39:     end if
40:     return S';
41: end procedure
```

---

loop, it will increase the RT_{queue} summing the response times of the previous tasks and considering those times for the next RT_{queue} task in the sequence.

In line 5, the SA's iteration loop is started. It runs until the total number of iterations, N, is achieved. In line 6, the neighbor solutions are generated by randomly swapping the tasks in the original sequence and finding when the swap generates a better solution (lines 7 to 18). If a better solution is found, the algorithm defines it as the best; otherwise, it uses the neighbor in the stochastic search for other solutions. After that, the temperature changes (line 21) according to a predefined equation.

Figure 2 shows a representation of the scenario components. In the picture it's seen the Factory where the edge node and the vehicles are, Fig. 2(a), (b) and (c) show the same factory. The four red circles are the vehicles, and the central bar is the edge node queue. The (a) shows the initial moment when the queue is being populated by tasks which were sent simultaneously by the vehicles. The task sent by vehicle 4 (V4) was the first to arrive, then it received the index 1 (T1), its deadline is 9 s. After that, V1 sent the task 2 (T2), the V3 sent the T3, the V2 sent the T4, then another task from V1 arrives, the T5, and sent by V3, the T6. The table (a) shows the distribution of tasks by vehicles and their T_i^{limit} attribute.

The transition from (a) to (b), stands for the moment the Algorithm 1 is running and calculating the tasks RT_i, while executing the first queue swap. In (b) the vehicles are only waiting for the tasks response. The table (b) reveals the RT_i, obtained from the Eq. 3. The algorithm identified the tasks T4 and T5 are out of the limit, T4 will be processed after 3.09 s, but its deadline is 1.5 s, and T5 will be processed after 3.10 s but its deadline is 3 s. The algorithm searches for better positions for T4 and T5 by the swap. After, it decides that T1 will be moved to the T5 original's position, T4 will be moved to the T1 original's position and T5 will be moved to the T4 original's positions.

In the transition from (b) to (c) the allocation result is committed to the queue, as the table (c) shows, this allocation solution attends the T4 and T5 limits and other tasks limits. All the response time are within the limit. In (c) the edge starts processing each one of the tasks in the given order. As concludes the processing, the responses are returned to the vehicles, finishing the execution.

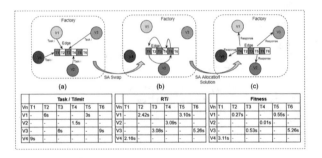

(a)   (b)   (c)

	Task / Tlimit							RTi							Fitness					
Vn	T1	T2	T3	T4	T5	T6	Vn	T1	T2	T3	T4	T5	T6	Vn	T1	T2	T3	T4	T5	T6
V1	-	6s	-	-	3s	-	V1	-	2.42s	-	-	3.10s	-	V1	-	0.27s	-	-	0.55s	-
V2	-	-	1.5s	-	-	-	V2	-	-	3.09s	-	-	-	V2	-	-	0.01s	-	-	-
V3	-	-	6s	-	-	9s	V3	-	-	3.08s	-	-	5.26s	V3	-	-	0.53s	-	-	5.26s
V4	9s	-	-	-	-	-	V4	2.16s	-	-	-	-	-	V4	3.11s	-	-	-	-	-

Fig. 2. Optimized Solution Representation.

4   Performance Evaluation

In this section, we present the results of experiments conducted on top of the iFogSim Simulator [3] version 2, which allows to simulate IoT devices, Edge Computing and Cloud Computing. The computer used in the experiments had an Intel Core i7 1.8 GHz processor, 16 GB of RAM, and the operating system Windows 10.

The Table 1 presents the description of tasks that are performed by industrial vehicles, as well as their respective parameters: the task limit, T_i^{limit}; the task size, D_i; the number of CPU cycles, C_i; the priority level, P_i; and the calculated response time of the task type, RT_i. The reason for a task taking different priorities, high or low, depends on the industrial application categorization. Four different groups of tasks were defined (see Table 1). For instance, in Tasks Group 1, the tasks are related to industrial machine supply. In the event that the machine is already out of supply, the load is urgent, so the task takes high priority. However, if the machine is still supplied, the task takes lower priority. The works from [8, 12], and [2] were inspirations and references for adapting such task parameters.

Table 1. Vehicles Tasks Description and Parameters

Task Group	Description	T_i^{limit}	D_i	C_i	P_i	RT_i
1	Take picture of an industrial machine to supply	12	66.355.200	180	1 or 2	2.1565
2	Detect movements around	6	16.588.800	44	1 or 2	0.2571
3	Detect obstacle forward	3	7.372.800	20	1 or 2	0.0921
4	Send the current location	1,5	400.000	100	1 or 2	0.0090

This work considers the use of only one edge node. Three distinct benchmark scenarios (Table 2) were evaluated, looking for task attendance within the time limit and the prioritization of tasks with a high urgency. The scenarios in the simulator receive the following parameters: the number of vehicles and the number of tasks. Vehicles are randomly chosen for sending the message to the Edge, as are the tasks associated with each vehicle. The probability of choice for Task Group 1 is 14.5%, and for Tasks Group 2, 3, and 4, it is 28.5%. The reason for the difference is that vehicles spend more time moving through the factory than acting in the supply.

Table 2. Experimental Scenarios

Scenario	Task Group	Priority	Vehicle 1	Vehicle 2	Vehicle 3	Vehicle 4	Vehicle 5	Total
S1	Group 1	High	0	0	1	0	0	1
		Low	0	1	1	0	4	6
	Group 2	High	1	1	1	0	0	3
		Low	3	0	1	3	1	8
	Group 3	High	2	1	0	2	2	7
		Low	1	1	2	3	1	8
	Group 4	High	2	3	4	1	3	13
		Low	0	0	0	1	1	2
	Total	–	9	7	10	10	12	**48**
S2	Group 1	High	0	1	0	0	1	2
		Low	2	0	5	3	2	12
	Group 2	High	0	0	0	0	0	0
		Low	1	7	5	1	7	21
	Group 3	High	0	2	1	5	2	10
		Low	6	4	1	6	3	20
	Group 4	High	7	3	0	2	3	15
		Low	2	3	5	2	4	16
	Total	–	18	20	17	19	22	**96**
S3	Group 1	High	1	0	0	0	0	1
		Low	1	3	11	5	2	22
	Group 2	High	2	0	3	0	4	9
		Low	7	4	5	7	7	30
	Group 3	High	1	0	1	2	1	5
		Low	8	7	8	7	5	35
	Group 4	High	8	7	9	6	7	37
		Low	0	1	2	2	0	5
	Total	–	28	22	39	29	26	**144**

The difference between the scenarios is the number of tasks M sent by the vehicles to the edge. In scenario S1, $M = 48$, in scenario S2, $M = 96$, and in scenario S3, $M = 144$. The benchmarks are based on [2] and [8]. The network transmission rate, R_j, was 100 Mbps. The edge processor, f_j, was set to 8 GHz. For each scenario, the Algorithm 1 was executed 30 times, and the same parameters were used. Regarding the SA in all scenarios, the total interactions N, were set to 2500 and the cooling coefficient α was set to 2. It acts as an exponential, which influences the temperature cooling. The vehicles have low mobility, 1.5 m/s, and it is considered that messages are not lost. TASeC was compared with the First-in-First-out (FIFO) and the Quicksort algorithms [5]. Two different task attributes were used in in task scheduling: priority and time limit (T_i^{limit}).

Figure 3, presents the response time median for scenario S1 (from Table 2), and Fig. 4 presents the success attendance percentage for the same scenario. In scenario S1, the groups of tasks have high and low priority.

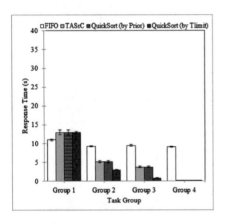

Fig. 3. Response Time (S1)

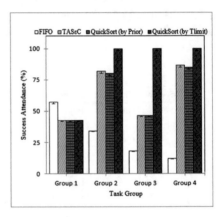

Fig. 4. Attendance (S1)

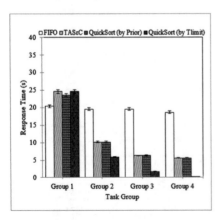

Fig. 5. Response Time (S2)

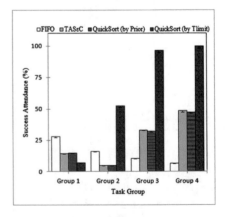

Fig. 6. Attendance (S2)

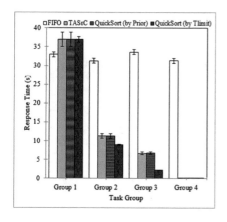

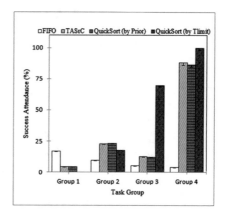

Fig. 7. Response Time (S3)   **Fig. 8.** Attendance (S3)

In scenario S1, considering the response time (in the median), all evaluated algorithms executed the requests received within Group 1, $T_i^{limit} = 12$ s. The FIFO algorithm, on the other hand, exceeded the time limit for the other groups when processing the queue of requests at the edge: Group 2 = 6 s, Group 3 = 3 s, and Group 4 = 1.5 s. As for the success in fulfilling requests, in scenario S1, the TASεC algorithm presented related behavior to Quicksort by priority. It is interesting to note that TASεC obtained a better result in Group 4, which has the lowest T_i^{limit} (1.5 s). This is because TASεC considers the request timeout (T_i^{limit}) and priority (whether high or low) when evaluating it. In Quicksort by priority, only the priority item is considered in the evaluation. In Quicksort, due to the time limit (T_i^{limit}), task priority was not considered in the evaluation.

In Table 3, we separate the success percentage according to the priority characteristic, high or low. The TASεC achieved 100% of attendance in all groups of high-priority tasks. The Quicksort (prior) showed similar results, except for Group 4, in which the percentage was 97%. In relation to low-priority tasks, in scenario S1, the TASεC achieved 33% of attendance in Group 1 and 75% in Group 2. No low-priority tasks were able to be handled in groups 3 and 4.

TASεC presented superior results with high-priority tasks because, during the generation of neighboring solutions, TASεC first attends to urgent tasks through the P_i attribute, according to Algorithm 2 (lines 31, 32, and 33). If the tasks have equal priorities, it directs the tasks with smaller T_i^{limit} to the front of the queue.

Table 3. Tasks Success Atendance

Success Attendance (%)		S1		S2		S3	
		High	Low	High	Low	High	Low
TASεC	Group 1	100	33	100	0	100	1
	Group 2	100	75	100	0	100	1
	Group 3	100	0	100	0	100	1
	Group 4	100	0	100	0	100	1
QuickSort (prior)	Group 1	100	33	100	0	100	0
	Group 2	100	70	100	0	100	0
	Group 3	100	0	96	0	99	0
	Group 4	97	0	97	0	100	0

Aiming to investigate TASεC behavior in more detail, two other scenarios were evaluated, S2 and S3. Figures 5, 6, 7, 8, show the results. Quicksort (prior), when scheduling tasks, can benefit tasks with a higher T_i^{limit} if these have high priority, causing them to be ahead of tasks with a smaller T_i^{limit}. Consequently, some tasks exceed the limit, as in the Quicksort (prior) results in Figs. 4, 6, and 8.

In Figs. 6 and 8, tasks attendance of groups 1, 2, and 3 showed a low percentage. TASεC presented this behavior because some tasks with a higher T_i^{limit} and high priority were ahead of tasks with a lower T_i^{limit}, but not urgent. For example, in Table 3, it is shown that all high-priority tasks were met by TASεC in scenarios S2 and S3. Quicksort was unable to handle all urgent tasks in scenarios S2 and S3. It can be concluded from the experiments that TASεC performed better on the most urgent tasks than Quicksort (prior) as the number of requests increased. In scenario S2, the number of requests is 96, and in scenario S3, the number is 144. In these cases, Quicksort answered on average 98% of the requests in scenario S2 and 99% in scenario S3. It should be noted that in industrial environments, failure to complete urgent tasks can result significant financial losses, and risks to employees' lives.

Also, it was registered the average time for all TASεC executions, the 30 repetitions were around 100 millisecond, varying accordingly the number of tasks, one single execution took around 3 millisecond, which in face to the critical environments delay sensitive, this algorithm based in SA is adherent. Other heuristics that take larger times to find solution are not indicated for industrial scenarios.

5   Conclusions

In this work, we presented the TASεC approach for task allocation in an industrial environment. Tasks are sent by industrial vehicles to an edge node in a manufacturing plant. The goal is to carry out tasks in terms of priority within a time limit.

The metaheuristic Simulated Annealing was used as a basis for finding the best order to complete tasks within a time limit. Experiments with three different scenarios were conducted in the iFogSim simulator. From the results, we concluded that the

TASεC performs better than the other algorithms evaluated in relation to meeting the most urgent tasks (high priority), because it was able to consider the composition of attributes, time limit, and priority of tasks when scheduling.

In future work, we will propose improvements in TASεC in relation to vehicle mobility to deal with failures in the exchange of messages with the edge node.

References

1. Bayar, A., Şener, U., Kayabay, K., Eren, P.E.: Edge computing applications in industrial IoT: a literature review. In: Bañares, J.Á., Altmann, J., Agmon Ben-Yehuda, O., Djemame, K., Stankovski, V., Tuffin, B. (eds.) GECON 2022. LNCS, vol. 13430, pp. 124–131. Springer, Cham (2023). https://doi.org/10.1007/978-3-031-29315-3_11
2. de Figueiredo Marques, V., Kniess, J.: Mobility aware RPL (MARPL): mobility to RPL on neighbor variability. In: Miani, R., Camargos, L., Zarpelão, B., Rosas, E., Pasquini, R. (eds.) GPC 2019. LNCS, vol. 11484, pp. 59–73. Springer, Cham (2019). https://doi.org/10.1007/978-3-030-19223-5_5
3. Gupta, H., Vahid Dastjerdi, A., Ghosh, S.K., Buyya, R.: iFogSim: a toolkit for modeling and simulation of resource management techniques in the internet of things, edge and fog computing environments. Softw.: Pract. Exp. **47**, 1275–1296 (2017)
4. He, J.: Optimization of edge delay sensitive task scheduling based on genetic algorithm. In: International Conference on Algorithms, Data Mining, Information Technology (2022)
5. Hoare, C.A.: Quicksort. Comput. J. **5**(1), 10–16 (1962)
6. Kirkpatrick, S., Gelatt, C.D., Vecchi, M.P.: Optimization by simulated annealing. Science **220**(4598), 671–680 (1983). https://doi.org/10.1126/science.220.4598.671
7. Masuduzzaman, M., Nugraha, R., Shin, S.Y.: Industrial intelligence of things (IIoT 2.0) based automated smart factory management system using blockchain. In: 13th International Conference on Information and Communication Technology Convergence (ICTC), pp. 59–64 (2022)
8. Matrouk, K.: Mobility aware-task scheduling and virtual fog for offloading in IoT-fog-cloud environment. Wirel. Pers. Commun. **130**, 801–836 (2023)
9. Patsias, V., Amanatidis, P., Karampatzakis, D., Lagkas, T., Michalakopoulou, K., Nikitas, A.: Task allocation methods and optimization techniques in edge computing: a systematic review of the literature. Future Internet **15**(8) (2023). https://doi.org/10.3390/fi15080254
10. Shi, W., Cao, J., Zhang, Q., Li, Y., Xu, L.: Edge computing: vision and challenges. IEEE Internet Things J. **3**(5), 637–646 (2016). https://doi.org/10.1109/JIOT.2016.2579198
11. Xue, Y., Wu, X., Yue, J.: An offloading algorithm of dense-tasks for mobile edge computing. In: icWCSN 2020, pp. 35–40. Association for Computing Machinery, New York (2020)
12. You, Q., Tang, B.: Efficient task offloading using particle swarm optimization algorithm in edge computing for industrial internet of things. J. Cloud Comput. **10**, 1–11 (2021)

Realtime BGP Anomaly Detection Using Graph Centrality Features

Janel Huang[1], Murugaraj Odiathevar[2], Alvin Valera[1], Jyoti Sahni[1], Marcus Frean[1], and Winston K. G. Seah[1(✉)]

[1] School of Engineering and Computer Science, Victoria University of Wellington, Wellington, New Zealand
{janel.huang,alvin.valera,jyoti.sahni,marcus.frean, winston.seah}@ecs.vuw.ac.nz
[2] Research and Business Foundation, Sungkyunkwan University, Seoul, South Korea
muru.raj@g.skku.edu

Abstract. Border Gateway Protocol (BGP) anomalies, such as hijacking, is currently growing in trend due to limited detection capabilities. BGP hijacking maliciously reroutes Internet traffic, causing Denial of Service (DoS) to major Internet Service Providers (ISPs) or redirection attacks to Internet users. While it has been shown that BGP anomalies can be detected using machine learning (ML) methods, the features used to train these ML models are not comprehensive. This is because node level features, such as the number of BGP announcements, average Autonomous System (AS) path length and average edit distance do not consider the structure or relationships present in the network graph. In this paper, an approach to extract information from BGP updates to build a network graph is proposed. Then, centrality information is used as features to model the graphical structure of the network to build an early detection tool for BGP anomalies using ML. The proposed method has been validated on real world data from the CenturyLink outage and shows promising results for anomaly detection (as early as one hour before the event was reported) in both individual and a defined group of networks. Furthermore, the anomaly source can be determined using the proposed method.

Keywords: Anomaly detection · Border Gateway Protocol · Graph Centrality · Autoencoders · Gaussian Mixture Model

1 Introduction

Operating as the backbone of the Internet, Border Gateway Protocol (BGP) is the routing protocol used for transferring information across different networks or Autonomous Systems (ASes) on the Internet. The presence of BGP routers are known through BGP

L. Barolli (Ed.): AINA 2024, LNDECT 201, pp. 222–233, 2024.
https://doi.org/10.1007/978-3-031-57870-0_20

update messages transmitted from router to router across the Internet. Over the years, there have been many incidents caused by anomalous BGP updates. BGP anomalies are caused by events such as hijacking or misconfigurations which have been shown to cause severe outages and redirection attacks [8], proving to be an important consideration for securing Internet traffic. The disconnectivity of panix.com in United States of America (USA) is an example showcasing the intentional redirection of BGP routes [22]. Another BGP anomaly example is the global BGP CenturyLink Outage that was a consequence of the misconfiguration of BGP routes [18]. BGP anomalies have caused severe outages for Internet users and revenue loss for many businesses across the globe, making the detection of BGP anomalies very crucial.

All BGP anomalies historically have shown changes in the network structure and thus provides the ability to capture the BGP anomalies. This has motivated the use of graphical network features that are fed to graph neural networks (GNN) [10] and shown to be highly effective [11]. However, the process requires the collection of datasets, which can be tedious and resource intensive [9], and more importantly, only suitable for offline detection like most ML-based anomaly detection methods. Methods that are designed for realtime detection (e.g. [7, 16, 19]) have been shown to detect the anomalies as they occur which leaves little to no time for network operators to react.

The main contribution of this paper is a technique to identify (even predict) network anomalies in realtime (as they develop, even before these anomalies become full-blown events) and determine the source of the anomaly by extracting information from BGP update messages; hereafter, also referred to simply as "BGP updates". Instead of using node features, a network structure is built using BGP updates. From the graph generated, centrality features are extracted to model the structure of the network. This information is then passed into two machine learning (ML) algorithms to detect anomalies, viz., an autoencoder to detect anomalies in the entire network, and a Gaussian Mixture Model (GMM) to detect anomalies in individual networks.

The rest of this paper is structured as follows. Section 2 describes the current methods used to detect BGP anomalies and Sect. 3 outlines the design of the proposed anomaly detection method. Section 4 then discusses the validation and experimental results. Finally, Sect. 5 ends with concluding remarks and research directions recommended to be examined in the future.

2   Related Work

Current BGP anomaly detection methods include time series, statistical pattern recognition, using historical data, reachability check and ML [2]. In this section, we review representative methods for BGP anomaly detection.

The earliest method used to detect BGP anomalies is *time series* [1, 17] which gained popularity initially as it could find characteristics of abnormal behaviour in a set of BGP updates collected within a time period. However, only a limited number of incidents can be detected in data collected over two years using statistical features such as the number of announcements and message volume. Such features have a distinct behaviour for a specific type of anomaly, thus limiting the ability of using statistical features to detect a wide range and new types of BGP anomalies.

Building upon the time series method, *statistical pattern recognition* has been shown to be successful in determining existing BGP anomalies as it can find relationships amongst BGP updates, e.g. [25], using features such as AS-path and edit distance [4] to determine the behaviour of the network topology. On the contrary, new types of BGP anomalies are still undetected because the features do not consider the entire network topology as they are examined independently in an instance. Furthermore, without constructing of a topology, relationships amongst AS-path and edit distance features cannot be accurately described. Thus, this motivates the use of network topology built using BGP updates to accurately describe the inherent relationships.

To overcome the limitation of identifying new BGP anomalies, a whitelisting approach that uses *historical BGP data* to determine the abnormality of new BGP updates has been proposed [8, 21]. This method utilises a BGP attribute of prefix origin change but only specific types of anomalies, such as prefix hijacking, can be detected. Other anomalies such as link failures and sub-prefix hijacks remain undetected as the feature used is not comprehensive to reflect all changes in the network topology. Thus, reinforcing those features which model the network topology must be utilised.

In contrast to methods discussed above, the *reachability check* method gained popularity as it is less computationally expensive as only a single hop count calculation is required for each BGP update [24, 28]. But such a method only considers reachability changes and does not identify relationships amongst multiple reachability checks. Thus, identification of all and new BGP anomalies is not possible [24, 28], thereby corroborating that a network topology must be built to extract a more comprehensive set of features.

All methods presented above have not proven to allow the capability for a method to automatically learn from experience and improve its performance over time. To counter this limitation, *machine learning (ML)* methods can be used where an ML model is trained using existing BGP updates to detect anomalies within a network [3, 14, 26]. This method is sought-after as the objective function for modelling abnormal and normal behaviour can be found and optimized automatically. However, ML methods typically rely on statistical features such as the number of announcements, withdrawals or the average AS path length which are not sufficiently comprehensive to model the entire network [16]. Only direct and indirect anomalies are detectable while other or new (not previously known) types of anomalies cannot be detected [3, 26]. This reinforces the notion that graph-level features must be extracted to capture all and new BGP anomalies [10, 23] as well as the potential of graph neural networks [11, 12].

While training ML models with graphical network features and the use of GNN have been shown to be effective, they require the collection and analysis of large datasets, which are time and resource intensive, making them primarily suitable for offline detection and forensic analysis. Hence, this research proposes a method to use BGP updates to build the network structure and extract graph-level features for determining anomalies in realtime. ML will be used as a detection tool to enable automatic identification of normal and abnormal features. As the network topology is complex and high dimensional, a method like ML that can automatically detect correlations in such data is suitable in time and accuracy.

3   Design

The design of the anomaly detection method focuses on graphical feature extraction from BGP updates. This process includes the construction of the network graph and the selection of graph-level features, as shown in Fig. 1. Subsequently, this information is used to train an ML model to detect anomalies.

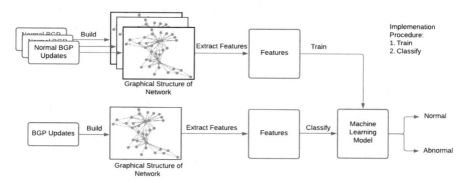

Fig. 1. Workflow of Proposed Anomaly Detection Method

3.1   BGP Updates

To formulate the graphical view of the network, appropriate BGP update attributes must be selected. BGP updates are used to inform BGP routers of paths to ASes in the Internet. This enables BGP routers to select the best path when forwarding traffic to destinations. An example of a BGP update is shown in Fig. 2. According to RFC4271 [20], the

```
TIME: 08/30/20 14:19:38.526402
TYPE: BGP4MP_ET/MESSAGE/Update
FROM: 210.7.33.5 AS38022
TO: 210.7.46.211 AS65002
ORIGIN: IGP
ASPATH: 38022 3356 1299 35908
NEXT_HOP: 210.7.33.5
COMMUNITY: 3356:3 3356:22 3356:86 3356:575
3356:666 3356:2012 38022:10800
ANNOUNCE
    74.119.238.0/24
```

Fig. 2. BGP Update Message Example

AS_PATH attribute can be used where each node present in the path represents a direct connection to the next node in the path. The source (FROM) and destination (TO) nodes within each update can also be included as a node in the graph. Each node can have connections added or removed within each BGP update. Therefore, the ANNOUNCE-MENT and WITHDRAWAL attributes should be used to ensure that connections are added or removed appropriately from the corresponding nodes. Therefore, these five attributes are used to construct the network graph.

3.2   Data Structures

After constructing the network, the graph must then be represented in a form suitable for training an ML model. The representation must model the graph structure and the node relationships to capture a graph-like form of the network. Hence, adjacency lists are used to store a list of nodes in the network and their connected immediate neighbours.

During announcements and withdrawals of BGP routes, a BGP router will add or remove routes respectively. Therefore, the IP addresses of an AS must be present in each node to find withdrawn or announced routes. Each node stores a dictionary of IP addresses as an AS may compose of multiple IP addresses. Our system uses

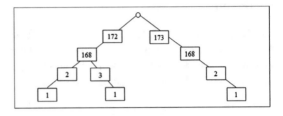

Fig. 3. Example of a Computed Trie Structure for 172.168.2.1, 173.168.2.1 and 172.168.3.1

a trie structure to store all IP addresses with their associated nodes. E.g., when storing an IP address of 172.168.2.1, 173.168.2.1 and 172.168.3.1, the trie structure will compute the tree as shown in Fig. 3. Using a trie structure yields better efficiency, $O(1)$. The pseudo code for building the graph using BGP updates is shown in Algorithm 1.

Algorithm 1 BGP Update Graph Creation

1: f = Read BGP Update
2: l = Read line in f
3: a ▷ Announcing Node

4: **while** l is not *null* **do**
5:     **if** l == "ASPATH" **then**
6:         Add AS Path using l
7:     **else if** l starts with "TO"
8:         **or** l starts with "FROM" **then**
9:         n = add ip to node using l
10:         **if** l starts with "TO" **then**
11:             $a = n$
12:         **end if**
13:     **else if** l starts with "ANNOUNCE" **then**
14:         Add path, l to a
15:     **else if** l starts with "WITHDRAW" **then**
16:         Remove path, l from a
17:     **end if**
18:     l = Read next line in f
19: **end while**

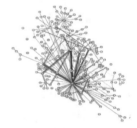

Fig. 4. Graph on 2020-08-30 at 9:04 (before anomaly event)

Fig. 5. Graph on 2020-08-30 at 10:05 (during anomaly event)

3.3   Features

Centrality features are extracted from the constructed network graph and used in the ML algorithm to determine whether the BGP updates are abnormal. In this paper, we propose using the features of node centrality as it is computable and can reflect a node's presence in relation to the entire network. Other features such as storing the connectivity of nodes are not feasible due to over 60,000 nodes in the network. The feature of clustering coefficients can also be used, but individual network information is lost as aggregations of neighbourhoods is required.

Node centrality measures how central a node's position is within a network. BGPlay visualisations [6] of the CenturyLink outage in Figs. 4 and 5 show that the node centralities changed significantly. This is because a large portion of the traffic was rerouted, leading to several nodes having more or fewer paths routing through them, thereby changing its centrality. Therefore, node centrality can be used as a feature in the ML algorithm to detect anomalies because a large difference in the centrality of a node can indicate abnormal behaviour [5]. We used centrality metrics that are feasible to compute, namely:

(i) Degree Centrality (DC) - Number of immediate neighbours of a node and is calculated, as follows [5]:

$$\mathbf{DC}(u) = \frac{n}{N-1} \tag{1}$$

(ii) Closeness Centrality (CC) - Inverse distance to all the reachable neighbours of a node; the CC formula proposed by Wasserman $et\ al.$ [27], as shown in Eq. (2), is used as it scales each node's CC separately by the size of the corresponding node and its neighbours,

$$\mathbf{CC}(u) = \left\{ n - 1 / \sum_{v=1}^{n-1} distance(u,v) \right\} \frac{n-1}{N-1} \tag{2}$$

where n denotes the number of reachable neighbours of a node u and N denotes the total number of nodes in the graph [27].

The pseudo code for extracting centrality features is shown in Algorithm 2. DC is inexpensive to compute as it only requires enumerating the number of immediate neighbours of each node. CC is also computationally inexpensive as enumerating the distance to all reachable neighbours of a node is only required. The distance to all reachable neighbours is calculated using Dijkstra's algorithm to allow efficient computation of distances.

Other centralities such as betweenness and eigenvector [5] are not used as they are computationally infeasible for realtime detection. Betweenness centrality requires all possible paths to be computed, which is infeasible due to the presence of over 60,000 nodes in the network. Calculating the eigenvector centrality requires the computation of the adjacency matrix which is infeasible due to the computational resource constraints, viz., extensive memory needed to create a large matrix size for over 60,000 nodes present in the BGP updates.

3.4 Machine Learning Models

The ML model must identify the main patterns in normal data (extracted from the normal non-anomalous BGP updates) as such patterns can reveal outliers or anomalies. ML models as suggested by literature [2] such as autoencoder and GMM are selected. To detect anomalies in the entire network, an autoencoder is used as it can find the complex relationships amongst normal centralities and thereby determine the anomalous centralities. An antoencoder is a NN with an encoder, hidden and decoder layers.

Each input in the encoder layer is reduced in dimensionality within the hidden layer to $1 + \sqrt{n}$, where n represents the number of nodes in the network graph, using the Relu activation function [15]. This removes the noise in the data and preserves the important variations. Subsequently, the features are reconstructed in the decoder layer which enables the model to identify complex relationships amongst the inputs and reconstruct the input. By training the weights within each layer using centralities during normal operation, this enables the model to learn the definition of normal centralities. Hence, anomalous centralities would return a large reconstruction error as the model cannot reconstruct the centrality using the learnt normal relationships.

The main disadvantage of Autoencoders is that the detection of anomalies for a specific AS is not addressed. Identification of problematic ASes is useful for network administrators to avoid routing to such ASes in an abnormal event. To detect individual network anomalies, a white-box method such as GMM is used [15]. GMM uses normal centralities to compute a Gaussian distribution for each AS. Each new centrality has a probability fitting into the trained distribution. Hence, it is assumed that a low probability of a centrality value fitting into the trained normal distribution would indicate anomalies for the AS. The training parameter indicating the number of components within each GMM is 1, as there was one cluster present in the normal distribution of every AS, essentially a univariate Gaussian (UG).

To train the ML models, BGP updates that are gathered during known normal network operations, such as two weeks before the anomaly incident (excluding two days before the anomaly incident.) Using less than 2 weeks of data is inadequate, as there is insufficient training data provided for the ML models to find correlations or define a correct distribution for detecting anomalies.

Algorithm 2. Centrality Feature Extraction

1: f = Read BGP Update
2: g = Get graph from f ▷ Contains each node with its immediate neighbours
3: N = Enumerate number of nodes in g

4: **for** $n = 1, \ldots, N$ **do**
5:     $node$ = get n in $graph$
6:     Calculate $paths$ as the shortest path length from each $node$ to its reachable neighbours
7:     Calculate **CC** as in Eq. (2) using number of $paths$, N and total length of $paths$
8:     Calculate **DC** as in Eq. (1) using number of immediate neighbours of $node$ in g and N
9: **end for**

---

4   Evaluation

Our proposed method is evaluated against the detection of abnormal behaviour during the BGP CenturyLink outage from 08-30-2020 10:04 (UTC). This helps to determine whether the learnt model can correctly classify anomaly incidents. The BGP CenturyLink outage was detected by ISPs through Twitter feeds by CenturyLink [18]. This reinforces the fact that an earlier and more reliable detection method is required to alert BGP anomalies. Anomaly detection for the entire network and specific ASes from the New Zealand (NZ), Japan (WIDE) and Serbia (SOXRS) core routers on the day of the BGP CenturyLink outage will be evaluated using autoencoders and GMM respectively. Experiments included an evaluation of the CC and DC features used for determining the anomaly score. The detection of an anomaly incident is based on the autoencoder reconstruction error (anomaly score) breaching a "threshold" that is determined (viz., maximum reconstruction error) using a validation set which consists of a small subset of normal data, two days before the anomaly incident.

4.1   Entire Network Detection

Determination of whether an entire network is anomalous allows ISPs to have a generalised view of a network's stability. This allows faster determination of anomalous behaviour in comparison to monitoring multiple individual networks. Hence, an experiment using the degree and closeness centralities from the BGP updates of the NZ and WIDE core routers are used to determine the anomaly score in the BGP CenturyLink Outage. For the NZ core router, both the closeness and degree centrality (Figs. 6 and 7) show an increase in the anomaly score before the estimated time breach.

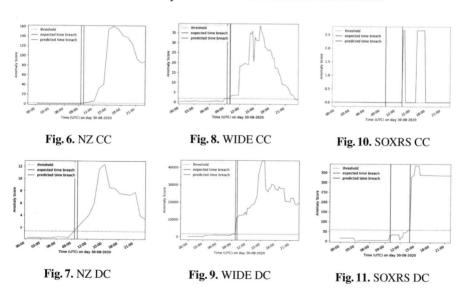

Fig. 6. NZ CC   **Fig. 8.** WIDE CC   **Fig. 10.** SOXRS CC

Fig. 7. NZ DC   **Fig. 9.** WIDE DC   **Fig. 11.** SOXRS DC

We observe a similar increase in anomaly score for the WIDE core router, as shown in Figs. 8 and 9. This indicates that the proposed method can detect the anomaly incident earlier, thus increasing the time for remediating the incident. A rise in the anomaly score is observed as the BGP CenturyLink Outage caused network instability where many nodes in the network needed to redirect traffic to compensate the loss of a Tier 1 ISP. This caused the number of connected neighbours for each node to change significantly. E.g., AS174 (Cogent Communications) had an increase of 200 immediate neighbours [18], further validating that our method is able to detect the anomalous activity in the network during the BGP CenturyLink Outage.

However, the detection time of anomalies can be significantly impacted by the geographic position of the observing router. E.g., a 3-h delay of detection for the BGP CenturyLink Outage is observed for the SOXRS router as shown in Figs. 10 and 11. BGP updates are transferred router to router and routers like SOXRS which are located further away from the source of the incident will receive the anomalous updates later. This suggests that multiple core routers located at different positions around the world should be used to monitor anomalies.

4.2   Individual Network Detection

If the entire network is deemed anomalous, specific ASes should be checked to determine whether they are affected or is the source of the BGP incident. This helps to ensure that ISPs can update their routing tables to prevent traffic from being routed to such ASes, thus minimizing the chance of Denial of Service (DoS). The evaluations of individual networks will focus on AS3561 and AS38022 that are involved in the BGP CenturyLink Outage. AS3561 is the BGP CenturyLink AS which caused the outage through a misconfiguration and AS38022 (Research and Education Advanced Network New Zealand (REANNZ)) is a peer of AS3561 that had to redirect its traffic to compensate for the disconnectivity of AS3561.

Using UG to compute (cf: Sect. 3.4) the anomaly score of an AS (e.g. AS3561) from the BGP updates received by the other (NZ, WIDE and SOXRS) ASes' core routers can be used to determine the anomaly source. Depending on the source of the anomaly, different core routers will detect anomalies at different times. E.g., NZ detected abnormal behaviour at least 1 h before the anomaly event because AS3561 is an immediate neighbour of NZ, hence it can detect abnormal behaviour before WIDE (further away) and SOXRS (furthest away.) As SOXRS is geographically further away from AS3561 than WIDE, the detection of the anomalous activity is 2 h after the anomaly event.

Figures 12 and 13 show the CC and DC anomaly scores of AS38022, respectively, from the NZ core router (itself) while Figs. 14 and 15 show the CC and DC anomaly scores of AS38022, respectively from the WIDE core router. Figures 13, 14 and 15 show a rise in the anomaly score before the estimated time breach as the network was unstable. A later time breach is predicted for NZ (by itself) as shown in Fig. 12 as the anomaly event did not originate from AS38022. The distance from AS38022 to its reachable neighbours did not change until the error from the source of the anomaly, AS3561, propagated through the Internet. As AS3561 is a trusted network peer of AS38022, the error that is propagated by AS3561 is deemed normal when transferred to AS38022. However, an earlier time breach is predicted from the WIDE core router as it can view the anomalous activity between AS38022 and AS3561 from an outsider's point of view. The earliest detection times of the anomalies occurring in the entire network as well as individual networks are summarized in Table 1. The NZ router detected an anomaly in

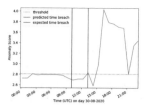

Fig. 12. NZ AS38022 CC

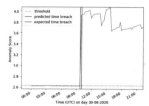

Fig. 14. WIDE AS38022 CC

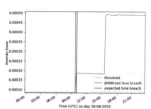

Fig. 15. WIDE AS38022 DC

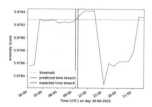

Fig. 13. NZ AS38022 DC

Table 1. Earliest Anomaly Detection Time during BGP CenturyLink Outage on 08-30-2020 with an estimated time breach at **10:04 UTC** based on Twitter feeds.

Network	Earliest Detection Time (UTC)
Entire Network	09:30
AS3561 Network	08:30 (detected by NZ)
AS38022 Network	10:00 (detected by WIDE)

AS3561 about 1 h 30 min before it was reported while the WIDE router detected an anomaly in AS38022 (NZ) a few minutes before the event was report.

No anomaly score is generated from SOXRS for AS38022 as it does not have AS38022 within its routing table during the detection period. This is because SOXRS is geographically further away from AS38022 and did not have any traffic that travelled to AS38022 within the detection period. This suggests that multiple core routers should be used for detection to allow anomalies to be discovered throughout the Internet.

4.3   Results Analysis and Findings

The key objective of our approach is early detection of anomalous events using the BGP update messages. This is achieved by comparing the anomaly scores (reconstruction errors) during anomalous events with those during normal operations. Our aim is not the analysis of ML models through typical performance metrics of accuracy, precision, recall, F1, etc. [13, 16].

Both the closeness and degree centralities show similar results, with a similar time of anomaly detection. However, DC is cheaper to compute than CC as it only requires calculation of nodes' immediate neighbours instead of reachable neighbours. DC yields a complexity of $O(n)$ and CC yields a complexity of $O(ne)$ where n represents the number of nodes and e represents the number of edges in the network. This suggests that DC should be used to alert abnormalities in the entire network. CC, however, performs better than DC in determining the severity of an incident where a large number of ASes can be indicated as anomalous as demonstrated in the BGP CenturyLink Outage. This is because CC takes into account the reachable neighbours of each node, meaning a major ISP outage will be reflected in more ASes in comparison to DC which only considers the immediate neighbours of nodes.

5 Conclusion and Future Work

In this paper, we have proposed a network anomaly detection method that extracts graph-like features from BGP updates on-the-fly for realtime anomaly detection. Construction of the network graph using BGP update attributes is conducted, and degree and closeness centrality features are extracted for anomaly detection using ML models, viz., autoencoders and GMM. The detection method is successful in detecting anomalous BGP incidents such as misconfiguration events. An immediate future work item would be the study of different ML models and their effectiveness, as well as, other network centrality measures. Assessing the ML models using typical performance metrics (e.g. accuracy, precision, recall, F1, etc.) could provide insights into the potential for overfitting and performance on unseen data, which is crucial for our approach. Other future work items include distributed processing to enhance the computational speed of feature extraction and network neighbourhood aggregation to enable Internet-wide anomaly detection. Using traffic link analysis to increase the confidence of the proposed method also needs further study.

Acknowledgements. The work of Winston K.G. Seah was partially supported by a grant from the APNIC Foundation, via the Information Society Innovation Fund (ISIF Asia). The authors acknowledge the support of REANNZ for providing the BGP Update datasets.

References

1. Al-Musawi, B., Branch, P., Armitage, G.: Detecting BGP instability using recurrence quantification analysis (RQA). In: Proceedings of the IEEE 34th International Performance Computing and Communications Conference (IPCCC), Nanjing, China (2015)
2. Al-Musawi, B., Branch, P., Armitage, G.: BGP anomaly detection techniques: a survey. IEEE Commun. Surv. Tutor. **19**(1), 377–396 (2017)
3. Al-Rousan, N.M., Trajković, L.: Machine learning models for classification of BGP anomalies. In: Proceedings of the IEEE 13th International Conference on High Performance Switching and Routing, Belgrade, Serbia, pp. 103–108 (2012)
4. Blazakis, D., Karir, M., Baras, J.S.: Analyzing BGP ASPATH behavior in the Internet. In: Proceedings of the 9th IEEE Global Internet Symposium (2006)
5. Chen, H., Yin, H., Chen, T., Nguyen, Q.V.H., Peng, W.C., Li, X.: Exploiting centrality information with graph convolutions for network representation learning. In: Proceedings of the IEEE 35th International Conference on Data Engineering (ICDE), Macao, China, pp. 590–601 (2019)
6. Di Battista, G., Mariani, F., Patrignani, M., Pizzonia, M.: BGPlay: a system for visualizing the interdomain routing evolution. In: Liotta, G. (ed.) GD 2003. LNCS, vol. 2912, pp. 295–306. Springer, Heidelberg (2004). https://doi.org/10.1007/978-3-540-24595-7_27, http://bgplay.routeviews.org/
7. Fezeu, R.A.K., Zhang, Z.L.: Anomalous model-driven-telemetry network-stream BGP detection. In: Proceedings of the IEEE 28th International Conference on Network Protocols (ICNP), pp. 1–6 (2020)
8. Haeberlen, A., Avramopoulos, I., Rexford, J., Druschel, P.: NetReview: detecting when interdomain routing goes wrong. In: Proceedings of the 6th USENIX Symposium on Networked Systems Design and Implementation, Boston, MA, USA, pp. 437–452 (2009)
9. Hoarau, K., Tournoux, P.U., Razafindralambo, T.: BML: an efficient and versatile tool for BGP dataset collection. In: Proceedings of the IEEE International Conference on Communications Workshops (ICC Workshops), pp. 1–6 (2021)

10. Hoarau, K., Tournoux, P.U., Razafindralambo, T.: Suitability of graph representation for BGP anomaly detection. In: Proceedings of the IEEE 46th Conference on Local Computer Networks (LCN), Edmonton, AB, Canada, pp. 305–310 (2021)
11. Hoarau, K., Tournoux, P.U., Razafindralambo, T.: BGNN: detection of BGP anomalies using graph neural networks. In: Proceedings of the IEEE Symposium on Computers and Communications (ISCC), Rhodes Island, Greece (2022)
12. Latif, H., Paillissé, J., Yang, J., Cabellos-Aparicio, A., Barlet-Ros, P.: Unveiling the potential of graph neural networks for BGP anomaly detection. In: Proceedings of the 1st International Workshop on Graph Neural Networking (GNNet), Rome, Italy, pp. 7–12 (2022)
13. Li, Z., Rios, A.L.G., Trajković, L.: Machine learning for detecting anomalies and intrusions in communication networks. IEEE J. Sel. Areas Commun. **39**(7), 2254–2264 (2021)
14. Lutu, A., Bagnulo, M., Cid-Sueiro, J., Maennel, O.: Separating wheat from chaff: winnowing unintended prefixes using machine learning. In: Proceedings of the IEEE Conference on Computer Communications (INFOCOM), Toronto, ON, Canada, pp. 943–951 (2014)
15. Odiathevar, M., Cameron, D., Seah, W.K.G., Frean, M., Valera, A.: Humans learning from machines: data science meets network management. In: Proceedings of the International Conference on COMmunication Systems NETworkS (COMSNETS), Bengaluru, India, pp. 421–428 (2021)
16. Peng, S., et al.: A multi-view framework for BGP anomaly detection via graph attention network. Comput. Netw. **214**, 109129 (2022)
17. Prakash, B., Valler, N., Andersen, D., Faloutsos, M., Faloutsos, C.: BGP-lens: patterns and anomalies in internet routing updates. In: Proceedings of the ACM SIGKDD International Conference on Knowledge Discovery and Data Mining, Paris, France, pp. 1315–1324 (2009)
18. Prince, M.: Aug 30th 2020: Analysis of century-link/level(3) outage. https://blog.cloudflare.com/analysis-of-todays-centurylink-level-3-outage/
19. Putina, A., et al.: Unsupervised real-time detection of BGP anomalies leveraging high-rate and fine-grained telemetry data. In: Proceedings of the IEEE Conference on Computer Communications Workshops (INFOCOM WKSHPS), pp. 1–2 (2018)
20. Rekhter, Y., Hares, S., Li, T.: A border gateway protocol 4 (BGP-4). RFC 4271 (2006). https://rfc-editor.org/rfc/rfc4271.txt
21. Shi, X., et al.: Detecting prefix hijackings in the internet with Argus. In: Proceedings of the Internet Measurement Conference (IMC), Boston, MA, USA, pp. 15–28 (2012)
22. Simon, T.L.: oof. panix sidelined by incompetence... again. https://www.mail-archive.com/nanog@merit.edu/msg40003.html
23. Sun, J., et al.: An efficient BGP anomaly detection scheme with hybrid graph features. In: Quan, W. (ed.) ICENAT 2022. CCIS, vol. 1696, pp. 494–506. Springer, Singapore (2022). https://doi.org/10.1007/978-981-19-9697-9_40
24. Tahara, M., Tateishi, N., Oimatsu, T., Majima, S.: A method to detect prefix hijacking by using ping tests. In: Ma, Y., Choi, D., Ata, S. (eds.) APNOMS 2008. LNCS, vol. 5297, pp. 390–398. Springer, Heidelberg (2008). https://doi.org/10.1007/978-3-540-88623-5_40
25. Theodoridis, G., Tsigkas, O., Tzovaras, D.: A novel unsupervised method for securing BGP against routing hijacks. In: Gelenbe, E., Lent, R. (eds.) Computer and Information Sciences III, pp. 21–29. Springer, London (2013). https://doi.org/10.1007/978-1-4471-4594-3_3
26. de Urbina Cazenave, I.O., Köşlük, E., Ganiz, M.C.: An anomaly detection framework for BGP. In: Proceedings of the International Symposium on Innovations in Intelligent Systems and Applications, Istanbul, Turkey, pp. 107–111 (2011)
27. Wasserman, S., Faust, K.: Social Network Analysis: Methods and Applications. Structural Analysis in the Social Sciences. Cambridge University Press, Cambridge (1994)
28. Zhang, Z., Zhang, Y., Hu, Y.C., Mao, Z.M., Bush, R.: iSPY: detecting IP prefix hijacking on my own. IEEE/ACM Trans. Netw. **18**(6), 1815–1828 (2010)

An Empirical Analysis on Leveraging User Reviews with NLP-Enhanced Word Embeddings for App Rating Prediction

Pratyush Mishra[1], Vikram Singh[1,2(✉)], Aneesh Krishna[2], and Lov Kumar[1]

[1] National Institute of Technology, Kurukshetra, Kurukshetra, India
{12112232,viks,lovkumar}@nitkkr.ac.in
[2] Curtin University Perth, Bentley, Australia
A.Krishna@curtin.edu.au

Abstract. The growth in mobile applications (apps) is rapid and immense. What was once a technology to satisfy the basic needs of a mobile user has now become a huge source of entertainment, education, games, etc. App development is iterative and incremental; therefore, they are released in improved versions to achieve higher app success in both functional and non-functional aspects. In this generic setting, the increasing prevalence of user reviews and generated ratings offers a wealth of insights, especially for post-development analysis. However, user reviews can be highly variable and inconsistent, as the informal nature of the language allows for misinterpretation, leading to a daunting task. In this paper, an empirical analysis is conducted to assert the effectiveness of NLP techniques in modeling the semantic contexts of user reviews in embedding vectors for predicting mobile app ratings. Here, each contextual *macro-category* is preprocessed with SMOTE to handle skewed distribution and then subjected to 4 feature selection techniques, e.g. *PCA, ANOVA,* etc. In the last step, 13 different prediction models (classifiers) are fitted to predict the mobile app rating. The study confirms significantly improved predictive performance in *Glove-T* and *GloVe-W* equipped word embedding models. Among prediction models, logistic regression (LOGR) outperforms the equivalent with an overall predictive ability of a greater extent, with an average accuracy score of 67, an average AUC of 0.70, and an F-measure of 0.72, indicative of its high functioning with numerical data and word vectors.

Keywords: Android APP · Deep Learning · Word Embedding

1 Introduction

Mobile applications (Apps) have ingrained themselves as an essential and omnipresent aspect of everyday life, ranging from social networking and online shopping to enhancing work efficiency. Yet, in a landscape with an overwhelming multitude of apps accessible through various stores, the challenge emerges to a naive user: *How can individuals identify the truly valuable ones?* In the modern landscape, consumers place a notably higher level of trust in peer reviews,

L. Barolli (Ed.): AINA 2024, LNDECT 201, pp. 234–244, 2024.
https://doi.org/10.1007/978-3-031-57870-0_21

ratings, and recommendations as compared to traditional advertising strategies. A significant majority of consumers engage with at least one rating or review before deciding to download a free app [9]. Interestingly, the feedback conveyed through app ratings plays a pivotal role in shaping user expectations regarding the app's functionalities, features, and the in-app experience they can anticipate upon installation. On the other hand, app developers constantly seek input on functional and non-functional aspects as app versions are released incrementally with frequent updates and patches, allowing for improvements and optimizations [7]. The increasing prevalence of user-generated reviews provides a rich collection of user opinions and significantly enables developers to assess *'How functionally well their apps are performing'*. Similarly, analysis of generated app 'ratings' provides a quantitative, objective, and aggregate measure of user satisfaction and helps identify areas for improvement or comparison [5,8]. This study uses word-embedding techniques to capture the semantic understanding of user reviews for mobile rating predictions. The study reaffirms the tightly coupled relationship between the review's semantic context and the rating's score. The following research questions (RQs) were set and to be addressed in our work:

- **RQ1:** *How can data balancing with word embeddings and classification models be used effectively to handle imbalanced datasets where mobile app ratings in specific macro-categories (07 macro-categories) may be skewed towards a specific class (e.g., a large number of 5-star ratings)?*
- **RQ2:** *How do different types of mobile app macro-categories (e.g., Art, Health, Shopping, social media, etc.) influence the effective feature selections from user reviews that strongly correlate with app rating prediction?*
- **RQ3:** *What types of NLP techniques with word embedding (Word2Vec, GloVe, etc.) are most suitable for improving the accuracy of mobile app rating predictions, and how do they compare in terms of performance?*

2   Related Work

In recent years, the field of sentiment analysis and predictive analytics, especially in the context of mobile app reviews and ratings, has received a surge in research interest. Abulhaija, Sabreen et al. [1] proposed and categorically validated a taxonomy based on relevant research published between 2018 and 2022. This work served as a comprehensive reference point that provided insights into current trends and methods for evaluating mobile app predictions based on user reviews.

Feature Prioritization-Based App Rating Prediction: The predictive models are extended beyond sentiment analysis to incorporate various app-related attributes and characteristics of mobile app rating prediction. An interesting work by Maryam Navaei [6] conducted a performance analysis on the ML-based mobile app rating predictions model based on UI features of apps, extracted from the RICO repository. The traditional metrics, e.g. Accuracy,

Recall and Precision confirm the improved predictive performance. In other work, Kayalvily, et al. [4] focused on understanding the current Google Play Store landscape, considering factors like *the number of downloads* and *ratings* in today's market trends. Their predictive models are based on Decision trees, KNN and RF regression techniques.

User Reviews Content-Based App Rating Prediction: The mobile app rating prediction using User review content (text) has gained significant interest in recent years. Xhang Fan, et al. [2] recently, introduced a collective model for the app rating predictions, which simultaneously learns multi-criteria sub-scores, and enhances prediction quality by transferring criterion knowledge from the reviews. They have experimented on three real-world datasets and demonstrated reduced prediction errors by up to 13.14% compared to baseline approaches. Md Shamim Hossain et al. [3] evaluated over 224,281 Google play Stote reviews using ML algorithms to examine User's sentiments for rating prediction in blended learning (supervised and unsupervised).

3   Study Design

3.1   Research Methodology

Figure 1 illustrates the proposed research methodology of the study and begins with the inherent division of 'user reviews' into seven mutually exclusive semantic reviews, 'macro'-categories', e.g. *Art, General, Health,* etc. These categories are defined to cover accurate review classes and achieve higher predictive accuracy.

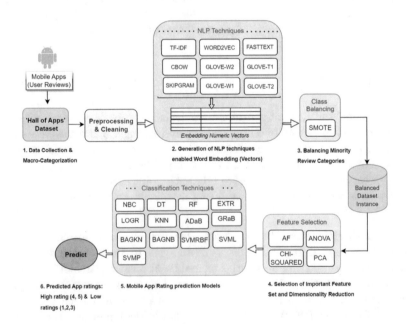

Fig. 1. Proposed Research Framework of Mobile App Rating Prediction Models

The study adapts nine different NLP techniques to effectively model the inherent semantics of user reviews written informally using natural languages. In addition, basic data cleaning is performed to improve the overall quality of the dataset before generating contextually equivalent word embedding (vectors) for each user review object. Further, data balancing is performed based on SMOTE techniques to ensure an even distribution of data samples for each micro-category in the user review dataset. The four different feature selection techniques, e.g., ANOVA, AF, CHI-SQARED, and PCA, are adapted to identify the potential features in the balanced dataset. Finally, a total of 13 different classification approach-based app rating prediction models were developed and evaluated for the prediction on two app rating classes, i.e., high-rating apps (with a score of 4 or 5) and low-rating apps (with scores of 3, 2, 1).

3.2 Dataset Details

The dataset utilized in this paper to evaluate the performance of various prediction models is the 'Hall Of Apps' dataset, MSR 2020[1]. The dataset containing metadata about top-charting apps was extracted weekly from GP for 4 different countries and for 30 weeks, with nearly 19 million user reviews. The study is designed to focus on the user reviews of each app. Therefore, priority and importance are given to the 'Review' subsection of the dataset. In the dataset, the apps have a different 'purpose' and are therefore classified into about 34 different individual categories.

4   Comparative Analysis

4.1   Effect of Oversampling Techniques

In this study, we have used SMOTE to handle the class imbalance nature of data. The box-plot performance indicators illustrate the statistics of performance parameters to find the impact of data sampling techniques in predictive analysis of mobile app 'ratings'.

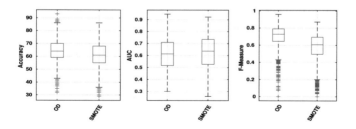

Fig. 2. Data-balancing: OD vs SMOTE

[1] https://2020.msrconf.org/details/msr-2020-Data-showcase/14/Hall-of-Apps-The-Top-Android-Apps-Metadata-Archive.

Performance Analysis of Data Balancing Based on AUC, Accuracy, and F-Measure on Box Plots and Descriptive Analysis: The predictive capability of models trained on an original data (OD) and SMOTE-generated balanced dataset is shown in Fig. 2.

The contents of the figure confirmed that the models trained through class balancing data have reduced interquartile range. The overall range of the values is less extreme; the descriptive statistics, as listed in Table 1, in terms of Accuracy and F-measure. Table 1 shows that class balancing has regressed the performance of the models with the OD, resulting in the highest accuracy, with an average Accuracy of 0.64 for the OD and 0.61 for the SMOTE-based data. The F-measure is 0.73 for the OD compared to 0.61 for the balanced data. In the case of AUC, it can be inferred that class balancing does produce a higher overall result. However, it can be seen through the values of 0.61 average for the OD and an average of 0.64 for the SMOTE data that the difference in performance is marginal between the balanced and imbalanced datasets in such cases. This implies that the overall correct classification rate, precision and recall balance, and ability to distinguish have decreased in the balanced data. Thus, overall, the oversampling techniques only regress the performance of the models, and the best performance is seen using the original dataset with imbalanced classes.

Table 1. Descriptive statistics of AUC and Friedman Test: OD vs SMOTE

Technique	Descriptive statistics						Friedman Test		
	Min	Max	Mean	Median	Q1	Q3	Accuracy	AUC	F-Measure
OD	0.30	0.95	0.61	0.61	0.52	0.71	1.39	1.56	1.22
SMOTE	0.26	0.92	0.63	0.64	0.53	0.74	1.61	1.44	1.78

4.2   Performance of Feature Selection Techniques

In this study, we exploited performance indicators to find the best feature selection technique that identifies the best feature sets for rating prediction.

Performance Analysis of Feature Selection Techniques on AUC, Accuracy, and F-Measure on Box Plots and Descriptive Analysis: Figure 3 shows the performances of the adapted 04 feature selection techniques. From Fig. 3, it is clear that the models trained using the CHI-SQUARED test-based feature sets are better at predicting rating scores than other models.

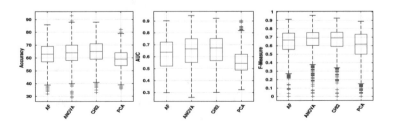

Fig. 3. Performance of Feature Selection Techniques

A prediction model trained with the features from CHI-SQUARED achieved average Accuracy, AUC, and F-measure values of 65%, 0.67, and 0.69, respectively, outperforming other techniques. The statistical description and Friedman test listed in Table 2 confirm the superior performance of CHI-SQUARED. The functioning of CHI-SQUARED with a slightly higher score might suggest that it is well suited for word occurrences in textual data that require discriminative power in their analysis and that its sensitivity to the categorical nature of the data.

Table 2. Descriptive statistics and Friedman Test: Feature Selection Techniques

Technique	Descriptive statistics						Friedman Test		
	Min	Max	Mean	Median	Q1	Q3	Accuracy	AUC	F-Measure
AF	0.30	0.90	0.62	0.64	0.52	0.72	2.51	2.57	2.48
ANOVA	0.26	0.95	0.65	0.67	0.55	0.75	2.26	2.12	2.22
CHI2	0.30	0.92	0.66	0.67	0.57	0.75	2.16	2.05	2.32
PCA	0.32	0.90	0.56	0.54	0.49	0.62	3.07	3.25	2.99

4.3   Performance of NLP-Based Word-Embedding Techniques

In this work, nine word-embedding strategies based on NLP techniques have been used to generate numerical vectors (embeddings) of user reviews from different macro-categories.

Performance Analysis of NLP Enabled Word-Embedding Schemes on AUC, Accuracy, and F-Measure on Box Plots and Descriptive Analysis: Figure 4 shows the ability of the model to predict ratings using different word embedding techniques. The models trained with primarily trained word vectors computed from the GloVe techniques, GloVe-Twitter and GloVe-Wikipedia, are the most accurate. All four models based on these techniques have the same average AUC value of 0.64 as worn in Table 3.

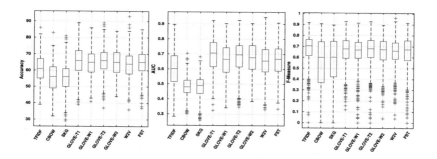

Fig. 4. NLP-based Word-Embedding methods

Their accuracy values are also the same, averaging 63, except for GloVe-Twitter 200, which has an average accuracy of 64. It is interesting to note that both the overall and interquartile range (IQR) of the GloVe Wikipedia models is larger than that of the Twitter models, as is the number of outliers among the models. The second significant score provided is by FastText, which also lies in the range of the results provided by the GloVe models. However marginally lower, with an average accuracy of 61, AUC of 0.61, and F-measure of 0.68. Word2Vec, as well as Consistent bag of words (CBOW), perform the worst.

Table 3. Descriptive statistics of AUC and Friedman Test: Word-Embedding techniques

Technique	Descriptive statistics						Friedman Test		
	Min	Max	Mean	Median	Q1	Q3	Accuracy	AUC	F-Measure
TFIDF	0.28	0.90	0.60	0.60	0.51	0.69	5.47	5.65	4.10
CBOW	0.30	0.70	0.48	0.48	0.44	0.52	7.02	8.01	6.01
SKG	0.26	0.68	0.48	0.49	0.44	0.53	6.79	7.93	5.81
GLOVE-T1	0.32	0.92	0.70	0.71	0.62	0.78	3.96	3.25	4.77
GLOVE-W1	0.34	0.90	0.66	0.67	0.58	0.74	4.39	4.29	5.02
GLOVE-T2	0.35	0.92	0.69	0.70	0.62	0.76	3.96	3.51	4.53
GLOVE-W2	0.39	0.90	0.67	0.68	0.60	0.76	4.39	3.86	4.83
W2V	0.33	0.95	0.65	0.66	0.58	0.73	4.67	4.38	4.97
FST	0.34	0.91	0.66	0.67	0.59	0.74	4.35	4.12	4.96

4.4   Performance of Rating Prediction Models

The mobile app rating prediction model in the study adapts to a classification-based prediction model, using a total of 13 classifiers to develop the app rating prediction ecosystem with data preprocessing techniques previously discussed and evaluated.

Performance Analysis of Rating Prediction Models Based on AUC, Accuracy, and F-Measure on Box Plots: Figure 5 highlights the performance measures on box plots using AUC, F-Measure, and accuracy of different classifiers for the App rating predictions. In contrast, Table 4 asserts the descriptive statistics Friedman test.

It can clearly be observed that the models trained using Logistic Regression perform better in predictive analysis, and their overall predictive ability is of a greater extent, with an average accuracy score of 67, average AUC of 0.70, and F-measure of 0.72. Similar scores are given by SVM, specifically linear and RBF kernels, however marginally less. However, It is also observed that SVM has a larger number of outliers present below the bottom range. The rest of the techniques have lowered values across all three measures as compared to both of

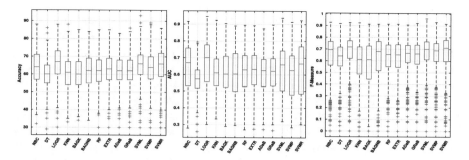

Fig. 5. Performance values of Prediction (Classification) Models

the top techniques and internally differ by a small margin. The models developed using Decision trees have low predictive ability as compared to other techniques, with values of 60 in accuracy, 0.57 in AUC, and 0.64 in F-measure, respectively.

Table 4. Descriptive statistics and Friedman Test of Classification Techniques

Technique	Descriptive statistics						Friedman Test		
	Min	Max	Mean	Median	Q1	Q3	Accuracy	AUC	F-Measure
NBC	0.28	0.92	0.64	0.67	0.53	0.76	5.91	5.36	6.00
DT	0.26	0.79	0.57	0.57	0.52	0.63	8.96	9.60	8.69
LOGR	0.30	0.95	0.67	0.70	0.56	0.78	4.38	3.86	4.38
KNN	0.33	0.92	0.62	0.61	0.53	0.69	8.23	7.25	8.68
BAGK	0.29	0.90	0.61	0.60	0.52	0.71	8.66	7.40	8.79
BAGNB	0.30	0.89	0.61	0.60	0.50	0.73	7.74	7.73	6.85
RF	0.27	0.90	0.62	0.63	0.53	0.71	7.54	7.03	7.94
EXTR	0.27	0.89	0.63	0.63	0.54	0.72	7.29	6.77	8.01
ADaB	0.27	0.87	0.61	0.63	0.53	0.69	7.68	7.66	7.44
Grab	0.27	0.90	0.62	0.62	0.53	0.70	7.80	7.55	7.60
SVML	0.32	0.94	0.63	0.66	0.50	0.74	5.11	6.65	5.40
SVMP	0.28	0.91	0.61	0.63	0.48	0.71	6.20	7.79	6.12
SVMR	0.30	0.92	0.63	0.67	0.48	0.77	5.50	6.34	5.08

4.5   Interpretation of Experimental Analysis

This section briefly presents the interpretations of the entire research study conducted under the formalized Research Questions (RQs) to discuss what the findings imply and how they relate to the original research objectives.

Interpretations to the RQ1: *How can data balancing, word embeddings, and classification models be used effectively to handle imbalanced datasets*

where mobile app reviews in specific 'Genre' or 'macro-categories' (07 macro-categories) may be skewed towards a specific rating class (e.g., a large number of 5-star ratings)?: The study primarily adapts to the SMOTE, a data balancing technique to handle the skewness inherent to each 'macro-categories', with synthetic data instances in inferior review categories. The performance evaluations assert the effectiveness of SMOTE-based data instances on overall predictive performance, with word embeddings and classification model techniques. The results shows the AUC values of models trained on OD and SMOTE- and SMOTE-based data, across different reviews Macro-categories. Section 4.1 reiterated the dominant predictive ability of the model trained with SMOTE-based data.

Interpretations to the RQ2: *How do different types of mobile app macro-categories (e.g., Art, Health, Shopping, social media, etc.) influence the effective feature selections from user reviews that strongly correlate with app rating prediction?*: Understanding the unique characteristics and expectations of each app macro-category, 'genre', is a crucial aspect of app rating prediction. Each app 'macro-category' has distinct inherent and influential features for user review in the dataset. In this study, four feature selection techniques, e.g., AF, ANOVA, CHI-Sqared, and PCA, were applied to investigate the macro-category of review and understand the preferred features that express functional and non-functional ratings of a mobile app and are highly correlated with the macro-category app rating. Through employed feature selection techniques, the CHI-squared technique demonstrates a higher score and appears well suited for word occurrences in the review of textual data that require discriminative power in their analysis and its sensitivity to the categorical nature of the data might provide a slight advantage of capturing the relationships between words or phrases and sentiment labels. Here, each genre-specific category aims to highlight the segregation in customer reviews and allow for granular analysis, providing more feature-level information to reveal the inherent patterns and correlations.

Interpretations to the RQ3: *What types of NLP techniques with word embedding (Word2Vec, GloVe, etc.) are most suitable for improving the accuracy of mobile app rating predictions, and how do they compare in terms of performance?*: The experimental analysis outlines that both GloVe embedding generators, based on pre-trained global vectors from Twitter data as well as those trained on Wikipedia data, with a dimensionality of 100 are most suitable for improving the accuracy of mobile app rating predictions. This is clearly implicated through box plot analysis as well as the Descriptive statistics of AUC. As the visual difference in the box plots is indiscernible between the two GloVe models, further, the Friedman test affirms that GloVe-Twitter 100 is the best, as seen in results. It is determined that increasing the dimensionality of the GloVe models leads to a decrease in performance as well as an increase in complexity and expense. Techniques such as FastText and Word2Vec provide average stability in comparison and CBOW and SkipGram are the least suitable for improving accuracy, providing the worst performance.

5   Conclusion

Word embeddings offer several advantages for classification models using NLP-based techniques, including the ability to capture semantic information and leverage contextual details. These benefits make them a crucial component for User review processing of many state-of-the-art NLP classification models. In this paper, an effort for the mobile app rating prediction is comprehensively analyzed over the User review dataset categorized into seven semantic macro-categories. The inherent contextual properties of user reviews are utilized with the NLP-enhanced word embedding scheme for the prediction of mobile app ratings.

The robust prediction model, through meticulous data balancing, feature selection, and prediction model development, showcased the feasibility and effectiveness of employing ML techniques to forecast app ratings with notable Accuracy, AUC, and F-measure levels. Among a range of NLP techniques considered to enhance the embeddings (Vector) in the experimental work, the Global Vector (GLOVE-T and GLOVE-W) embedding generators trained with Twitter and Wikipedia data with the dimensionality-100 perform the best and most suitable for improving the accuracy of mobile app rating predictions. The Glove pre-trained embedding schemes assert superior predictive capabilities due to their training on Twitter and Wikipedia like large and semantically rich text corpus. The study asserts the essential future work includes the need for NLP-enhanced multilingual word embedding using user review-based app rating prediction, Domain-Specific word embeddings, and Demographic word embedding for the rating prediction.

References

1. Abulhaija, S., Hattab, S., Abdeen, A., Etaiwi, W.: Mobile applications rating performance: a survey. Int. J. Interact. Mob. Technol. **16**(19) (2022)
2. Fan, G., Zhang, C., Chen, J., Li, P., Li, Y., Leung, V.C.: Improving rating prediction in multi-criteria recommender systems via a collective factor model. IEEE Trans. Netw. Sci. Eng. (2023)
3. Hossain, M.S., Uddin, M.K., Hossain, M.K., Rahman, M.F.: User sentiment analysis and review rating prediction for the blended learning platform app. In: Applying Data Science and Learning Analytics Throughout a Learner's Lifespan, pp. 113–132 (2022)
4. Kayalvily, T., Denis, A., Mohd Norshahriel, A.R., Sarasvathi, N.: Data analysis and rating prediction on google play store using data-mining techniques. J. Data Sci. **2022**(01) (2022)
5. Li, X., Zhang, Z., Stefanidis, K.: Mobile app evolution analysis based on user reviews. In: New Trends in Intelligent Software Methodologies, Tools and Techniques, pp. 773–786. IOS Press (2018)
6. Navaei, M.: A performance analysis on machine learning algorithms to predict mobile application's star rating by its user interface features. East Carolina University (2022)

7. Nayebi, M., Adams, B., Ruhe, G.: Release practices for mobile apps–what do users and developers think? In: 2016 IEEE 23rd International Conference on Software Analysis, Evolution, and Reengineering (SANER), vol. 1, pp. 552–562. IEEE (2016)
8. Qiao, Z., Wang, G.A., Zhou, M., Fan, W.: The impact of customer reviews on product innovation: empirical evidence in mobile apps. Anal. Data Sci.: Adv. Res. Pedag. 95–110 (2018)
9. Vasa, R., Hoon, L., Mouzakis, K., Noguchi, A.: A preliminary analysis of mobile app user reviews. In: Proceedings of the 24th Australian Computer-Human Interaction Conference, pp. 241–244 (2012)

Hybrid Evolutionary Algorithm for the Overlap Constrained Resource Allocation Problem in Wireless Networks

Yiting Wang, Yawen Li, Zequn Wei[(⊠)], and Junqi Li

Beijing University of Posts and Telecommunications, Beijing, China
zequn.wei@gmail.com

Abstract. In wireless networks, efficiently allocating limited network resources holds significant practical importance. This work focuses on the NP-hard overlap constrained resource allocation problem (OCRAP) in wireless networks. As one of the practical decision-making problems, OCRAP aims to find a subset of candidate wireless resources, each capable of servicing multiple areas, to maximize the profit function while satisfying the given budget and overlap constraints. We have formulated the OCRAP model based on the budgeted maximum coverage problem and propose an effective hybrid evolutionary algorithm (HEA) for solving it. The proposed HEA algorithm combines a tabu search procedure for local optimization with an effective crossover operator to generate promising offspring solutions. We show computational results on 60 benchmark instances and present comparative studies with several heuristic algorithms as well as the general CPLEX solver. We also provide a convergence analysis to further demonstrate the robust performance of the proposed algorithm.

1 Introduction

The wireless resource allocation problem [18] has garnered considerable interest over the last few years which aim to select wireless resources in order to cover as many users as possible within a given budget. In this work, we investigate for the first time the wireless resource allocation problem with the overlap constraints (OCRAP), aimed at reducing the frequency of different resources overlapping in the same area, thereby improving the efficiency of resource utilization.

OCRAP involves a set of available wireless resources with given weights, each resource is associated with a set of service areas. Each area, encompassing multiple users, is characterized by profits, which reflect the importance of serving particular users. Note that an area may be covered multiple times by different resources, thus we limit the number of times resources can overlap. The objective of OCRAP is to choose a subset of resources that optimizes the total area coverage while not exceeding resource budget as well as limiting overlap to a given threshold. The OCRAP is related to the budgeted maximum coverage problem (BMCP) [11, 17] where a wireless resource corresponds to an item with its weight and a service area corresponds is an element with its weight. However, the BMCP does not take into account overlap constraints, which are a crucial component of the OCRAP. Specifically, the OCRAP can be described as follows.

L. Barolli (Ed.): AINA 2024, LNDECT 201, pp. 245–256, 2024.
https://doi.org/10.1007/978-3-031-57870-0_22

Given a wireless resource set $I = \{1, 2,..., m\}$ where each resource $i \in I$ has a non-negative weight w_i, and a service area set $E = \{1, 2,..., n\}$ where each area j has a non-negative profit p_j. The aim of OCRAP is to identify a subset S of resources, that maximizes the total profit from serviced areas while satisfying the specified budget and overlap constraints. Specifically, each service area can be covered at most k times. To formulate the problem mathematically, define the binary variable y_i as 1 when resource i is chosen, and 0 otherwise. Given a binary relationship matrix M, where $M_{ij} = 1$ if resource i covers service area j, H_j represents the number of times that area j is covered by the selected resource i. Then, we introduce a binary variable x_j for each area j, where x_j is set to 1 if $H_j > 0$, and to 0 otherwise. Formally, the OCRAP can be written as follows.

$$Maximize\ P(S) = \sum_{j=1}^{n} x_j p_j \tag{1}$$

$$s.t. \quad W(S) = \sum_{i=1}^{m} y_i w_i \leq C \tag{2}$$

$$H_j = \sum_{i=1}^{m} y_i M_{ij}, j = 1, 2, ..., m \tag{3}$$

$$H_j \leq k \tag{4}$$

$$x_j = \begin{cases} 1, & if\ H_j > 0; \\ 0, & if\ otherwise. \end{cases} \tag{5}$$

$$y_i \in \{0, 1\}, i = 1, 2, ..., m \tag{6}$$

Function (1) represents the objective value corresponding to the selected resource set S. Constraint (2) requires that the total weight of the resource set S do not exceed the given budget and Eq. (3) give the definition of H_j. Constraint (4) is the overlap constraint that restricts the serving of an area by the resource set S to a maximum of k times. Constraint (5) and (6) show the domains of variable x_j and y_i.

Solving the OCRAP problem presents computational difficulties as it is a generalization of the NP-hard BMCP. In fact, the OCRAP reduces to the BMCP when we ignore the overlap constraint. In combinatorial optimization, modifying the constraints often requires adapting the solution approaches. This paper introduces an efficient hybrid evolutionary algorithm (HEA) tailored to address the OCRAP. The principal contributions of this work are outlined below.

First, we introduce a novel resource allocation model with overlap constraints, specifically designed for wireless network scenarios. Second, in addressing the proposed problem model, we devised an effective hybrid evolutionary algorithm (HEA) that integrates a tabu search for local optimization with an effective crossover operator to obtain promising offspring solutions. Our HEA algorithm is able to achieve 56 out of the 60 best-known results reported by all the reference algorithms. Third, we present the application of the CPLEX solver to solve benchmark instances designed for the OCRAP, providing the upper and lower bounds for future research.

The remainder of the paper is organized as follows. In Sect. 2, we present a brief literature review of the OCRAP. The proposed HEA algorithm is introduced in Sect. 3. Section 4 shows experimental results of HEA with several heuristic algorithms and the CPLEX solver. The convergence analysis of HEA is also presented in this section. The last section concludes the work.

2 Related Work

We discuss the resource allocation problems in networks and review literature of the budgeted maximum covering problem (BMCP), which is closely related to the OCRAP.

Several models have been formulated to address the challenges of resource allocation problem (RAP) in networks [2, 5, 21]. Konnov et al. [13] presented an optimal allocation of homogeneous resources in spatially distributed systems considering both utilities of users and network expenses. Letchford et al. [15] introduced the RAP problem in mobile wireless communications to allocate the available channels and power, while ensuring compliance with quality of service (QoS) constraints. Falsafain et al. [6] introduced an NP-hard cognitive radio resource allocation problems incorporating maximum aggregation range (MAR) constraints. Existing studies are mostly dominated by exact algorithms [4, 12, 13, 15], while heuristic algorithms are applied to discover local optima [1]. Besides, Bouras et al. [3] proposed two machine learning methods for improving real time user allocation on the network to provide effective utilization of the network resources. Note that some resource allocation problems in wireless networks incorporate variations of the knapsack problem, such as formulating the RAP problem by the 0–1 multichoice multidimensional knapsack problem [20]. This study incorporates task and resource requirements to achieve optimal allocation while prioritizing high-value tasks. These problem formulations above consider various constraints related to data transmission and quality of service measurement and management. However, research on the issue of network coverage overlap remains limited. When coverage constraints, different networks utilize the same spectrum area, resulting in inefficient spectrum utilization, ultimately impacting overall communication capacity and capability. This paper fills the gap by proposing a new model to address these issues and provide practical insights.

As an NP-hard problem, various solution methods have been developed for the BMCP [16, 17, 19], including the variable depth local search [24] and the probability learning-based tabu search [17]. In particular, Wei et al. [23] proposed an iterated hyperplane search approach for the BMCP and reported 18 new lower bounds for 30 benchmark instances with short running times. BMCP exhibits strong practical utility and can address deployment issues in software-defined networks (SDN) [10], software package installation and financial decision-making [17]. It is worth mentioning that the set union knapsack problem (SUKP) [8] and the BMCP possess a dual relationship [17], and numerous approaches have been designed for solving SUKP [9, 22]. However, to the best of our knowledge, existing studies on the BMCP or SUKP have not taken into account overlap constraint, which is relevant in the resource allocation of wireless network.

3   A Hybrid Evolutionary Algorithm for the Overlap Constrained Resource Allocation Problem

3.1   General Procedure

The main framework of our HEA algorithm is shown in Algorithm 1. HEA starts with the initialization process of the best solution S^* (line 3) found and the population POP (line 4). After updating S^* (line 5), the algorithm goes into a *"while"* loop to improve S^* iteratively (lines 6–14). During each generation of the loop, two parent solutions S^i and S^j are selected randomly (line 7) and applied to obtain the offspring solution S^o by a backbone-based crossover operator (line 8, Sect. 3.3). Then a tabu search procedure (line 9, Sect. 3.4) is employed to examine the two neighborhoods N_1 and N_2 and discover high-quality neighboring solution S_b. The best solution S^* is updated by S_b immediately according the objective value (lines 10–12). Then the population will be updated accordingly (line 13, Sect. 3.5). Finally, the algorithm ends by returning the best solution S^* upon reaching the time limit t_{max}.

Algorithm 1：Hybrid evolutionary algorithm for the OCRAP

1: **Input**: Instance I, the maximum run time t_{max}, population POP, the maximum number of iterations I_{max}, neighborhood N_1, N_2.

2: **Output**: The overall best solution S^* found.

3: $S^* \leftarrow \emptyset$ /* Initialize S^* */

4: $POP=\{S^1,...,S^{|POP|}\} \leftarrow Initialize\_POP(I)$ /*Section 3.2*/

5: $S^* \leftarrow argmax\{f(S^k)|k=1,...,|POP|\}$ /* Record the best solution S^* */

6: **while** *time* $< t_{max}$ **do**

7:     $(S^i, S^j) \leftarrow Random\_Pick\_S(POP)$

8:     $S^o \leftarrow Crossover\_Operator(S^i, S^j)$ /*Section 3.3*/

9:     $S_b \leftarrow Tabu\_Search(S^o, N_1, N_2, I_{max})$ /*Section 3.4*/

10:    **if** $f(S) \geq f(S^*)$ **then**

11:        $S^* \leftarrow S_b$ /* Update the best solution S^* */

12:    **end if**

13: $POP \leftarrow Update\_POP(S_b, POP)$ /*Section 3.5*/

14: **end while**

15: **return** S^*

3.2   Population Initialization

Algorithms 2 shows HEA constructing the initial solution of the population POP in three main steps. First, the algorithm employs a *Random_Initial(I)* strategy (line 5) which randomly selects resources that meets the budget constraint and overlap constraint until no feasible resource can be selected for a $m/20$ consecutive iterations, where m is the number of all candidate resources. Second, we employ a short tabu search procedure (see

Sect. 3.4) to improve the solution S^i (line 6). Third, we add the S^i into the population POP (line 7). The above three steps iteratives until all the population solutions are generated.

Algorithm 2: Population initialization procedure (*Initialize_POP*)

1: **Input**: Instance I, the maximum number of iterations I_{max}, Neighborhood N_1, N_2.

2: **Output**: Initial population POP.

3: $0 \leftarrow i$ /* Initialize S^* */

4: **while** $i<|POP|$ **do**

5:    $S^i \leftarrow$ *Random_Initial(I)*

6:    $S^i \leftarrow$ *Tabu_Search(S^i,N_1,N_2,I_{max})*   /*Section 3.4*/

7:    $POP \leftarrow$ *Add_Solution(S^i,POP)*

8:    $i \leftarrow i+1$ /* Update the best solution S^* */

9: **end while**

10: **return** POP

3.3 The Backbone-Based Crossover Operator

As a population-based optimization algorithm, the HEA algorithm typically employs a crossover operator to produce offspring solutions. We employ the backbone-based crossover operator that is dedicated to transferring as much valuable information as possible from the parent solutions to the offspring solutions.

Given two solutions, S^i and S^j, the resources set can be divided into three subsets: public set $X_1 = S^i \cap S^j$, unique set $X_2 = (S^i \cup S^j)\backslash(S^i \cap S^j)$ and unrelated set $X_3 = V \backslash (S^i \cup S^j)$. The backbone-based crossover operator producing an offspring solution S^o by accepting all resources in X_1 and some resources that are picked randomly from X_2 while excluding the terms in X_3.

Here we give an illustrative example of backbone-based crossover operator in Fig. 1. Given a benchmark instance with 6 resources (i.e., $m = 6$), the selected resources from parent solutions S^i and S^j are $\{1, 2, 3, 4\}$ and $\{1, 2, 6\}$, respectively. Then the public resource set $X_1 = \{1, 2\}$ and the unique resource set $X_2 = \{3, 4, 6\}$. As a result, all the resources belonging to the set X_1 (i.e., 1 and 2) are added into the offspring solution S^o and the resources belonging to X_2 will be randomly selected (i.e., 4 and 6). Note that during the crossover process, the budget constraint and overlap constraint are always satisfied.

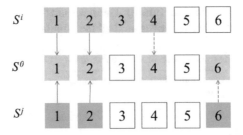

Fig. 1. An illustrative example of the backbone-based crossover operator.

3.4 Tabu Search Procedure

Tabu search [7] is recognized as a highly favored local search technique to solve 0–1 decision-making problems. The proposed HEA algorithm employs an efficient tabu procedure to achieve local optimization.

As shown in Algorithm 3, we first initialize several components of the search (lines 3–5). For each iteration of the search (lines 6–14), all the neighboring solutions belonging to the neighborhood $N_1 \cup N_2$ are examined and the best admissible neighboring solution S'_c is identified (lines 7–9). Then the best solution S_b and the tabu list will be updated by S'_c accordingly (lines 11–17). The tabu search procedure terminates and returns the best solution S if there is no improvement for I_{max} consecutive iterations, where I_{max} is a parameter (see Sect. 4.2).

The proposed tabu search uses the common *flip* or *swap* move operator to obtain neighboring solutions. The *flip(p)* operator functions by either adding a non-selected resource p to the current solution S_c, or removing a selected resource p from solution S_c. The *swap(p, q)* operator involves exchanging a selected resource p with a non-selected resource q to create a neighboring solution, while satisfying the constraints. The neighborhoods N_1 and N_2 corresponding to the two move operators are defined as follows, respectively.

$$N_1(S_c) = \left\{ S' : S'_c = \oplus flip(p), p \in S_c \right\} \tag{7}$$

$$N_2(S) = \left\{ S' : S'_c = \oplus swap(p, q), q \in V, p \in \overline{V} \right\} \tag{8}$$

where V and $\overline{V}$ correspond to the sets of selected and non-selected resources, respectively.

The tabu list record the resources involves in the move operation *move(i, j)* to avoid revisiting the solutions found before. In our case, we adopt a two-dimensional tabu matrix as the tabu list to record the two resources i and j engaged in the move operation, where j could be a dummy resource for the *flip* operation. Then for all the remaining tabu iterations, only the operation *move(i, j)* and *move(j, i)* are prohibited, whereas move operations of i and j with other resources are allowed independently.

Algorithm 3: Tabu search procedure (*Tabu_search*)

1: **Input**: Input offspring solution S^o, the maximum number of iterations I_{max}, Neighborhood N_1, N_2.

2: **Output**: The best solution S_b found during tabu search.

3: $S_b, S_c \leftarrow S^o$ /* S_c is the current solution */

4: $iter \leftarrow 0$

5: Initial *tabu_list*

6: **while** $iter < I_{max}$ **do**

7:    Find all admissible neighboring solutions $N'(S_c)$ in $N_1(S_c) \cup N_2(S_c)$

8:    **if** $N'(S) \neq \emptyset$ **then**

9:       $S_c \leftarrow argmax\{f(S'): S'_c \in N'(S_c)$

10:      **if** $f(S'_c) > f(S_b)$ **then**

11:         $S_b \leftarrow S'_c$

12:         $iter \leftarrow 0$

13:      **else**

14:         $iter \leftarrow iter + 1$

15:      **end if**

16:      $S_c \leftarrow S'_c$

17:      Update *tabu_list*

18:    **end if**

19: **end while**

20: **return** S_b

3.5 Population Updating

In order to ensure the population diversity, we employ the population updating rule introduced in [14], which is suitable for 0–1 decision-making investigated in this work. This population updating rule is based on the following three steps.

(1) Calculate the distance $D(S^i)$ of each candidate solution S^i, where $D(S^i)$ indicates the hamming distance between S^i and the population *POP*.

$$D\left(S^i\right) = Min\left\{dis\,tan\,ce\left(S^i, S^j\right) : S^j \in POP, S^i \neq S^j\right\} \qquad (9)$$

(2) Calculate the score of each candidate solution S^i. $Score(S^i)$ is determined through a scoring strategy that considering both the objective function value and the distance $D(S^i)$, as defined below.

$$Score(S_i) = \beta \times \frac{f\left(S^i\right) - f_{min}}{f_{min_{max}} + (1 - \beta) \times \frac{D(S^i) - D_{min}}{D_{min_{max}}}} \qquad (10)$$

In the population, the objective values range from f_{min} to f_{max}, and the distances between solutions vary from D_{min} to D_{max}. The parameter β is empirically determined, and we have simply set it to 0.7.

(3) Remove the worst candidate solution from the *POP* according to $Score(S^i)$ obtained in step (2).

As a result, we can obtain a population that not only exhibits high-quality solutions but also maintains a diverse range of solutions, enhancing the overall effectiveness of the optimization process.

4   Experimental Results and Analysis

4.1   Benchmark Instances

Drawing on the relationship between OCRAP and BMCP introduced in Sect. 1, the benchmark instances for the OCRAP used in this work are modified from the 30 BMCP instances, where the number of items (corresponding the resources for the OCRAP) ranges from 585 to 1000. All existing information in these instances was retained, and an additional parameter k, representing the overlap constraint, was introduced to each instance. With k set to 2 and 3 in this experiment, we obtained 60 OCRAP instances in total.

4.2   Experimental Settings

To demonstrate the performance of our HEA algorithm, we designed three popular heuristic algorithms as the reference algorithms, i.e., the genetic algorithm (GA), the greedy algorithm (Greedy) and the iterated local search algorithm (ILS). All the reference algorithm as well as our algorithm was code in C++ and compiled using the g++ compiler with the -O3 option. All experiments were tested on an Intel Xeon 6148 processor (2.40 GHz CPU) running the Linux operating system. For each test, every reference algorithm independently solved each instance 30 times, subject to a cut-off time of 600 s. Moreover, we also adopted the CPLEX solver to solve the 60 OCRAP benchmark instances, setting the cut-off time to 7200 s per instance.

We have adopted a simple adaptive mechanism for setting the parameters in the proposed HEA algorithm. Specifically, for a given OCRAP benchmark instance with m resources, the population size is set to $m/50$, while the maximum iterations I_{max} for the tabu search is set to $5m$.

4.3   Computational Results

Table 1 displays the summarized comparisons of the HEA algorithm with each of the reference algorithms. We report the detailed computational results on GitHub[1]. In Table 1, the first column lists the names of the algorithms used for comparison. Column 2 shows the indicator of the best (f_{best}) and average (f_{avg}) objective values achieved over 30 independent runs. The following three columns provide the counts of instances where our HEA algorithm outperformed, matched, or underperformed compared to each reference algorithm. The last column gives the p-values from the Wilcoxon signed-rank test.

---

[1] Available at: https://github.com/YitingQueen/HEA_results_summary.

Table 1. Summarized comparisons between the HEA algorithm and each reference algorithm.

Algorithm pair	Indicator	#Wins	#Ties	#Losses	p-value
HEA vs. GA	f_{best}	60	0	0	$1.63e-11$
	f_{avg}	60	0	0	$1.63e-11$
HEA vs. Greedy	f_{best}	51	6	3	$3.85e-08$
	f_{avg}	58	0	2	$1.63e-11$
HEA vs. ILS	f_{best}	51	8	1	$2.19e-09$
	f_{avg}	56	0	4	$3.40e-10$

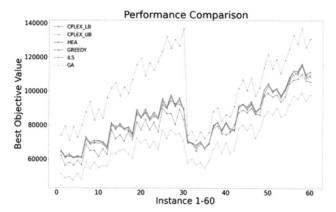

(a)   The best objective value (f_{best}) on the 60 benchmark instances.

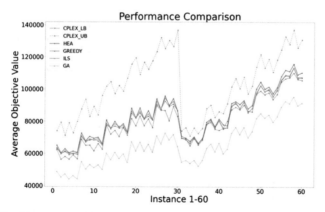

(b)   The average objective value (f_{avg}) on the 60 benchmark instances.

Fig. 2. The best objective value and average objective value of the compared algorithms.

From Table 1, we can observe that our HEA significantly outperforms the reference algorithms GA, Greedy and ILS. Specifically, HEA dominates GA on all the 60 instances with no exception. When comparing with Greedy and ILS, our algorithm is able to achieve better f_{best} and f_{avg} results on most cases. The p-values below 0.05 confirm the statistical significance of the differences in results between the HEA and each compared algorithm.

To complete the comparison, we show in Fig. 2 the f_{best} value and the f_{avg} value of the four compared algorithms, as well as the lower bound (LB) and the upper bound (UB) obtained by the CPLEX solver. Figure 2 clearly illustrates that our HEA algorithm outperforms the reference algorithms with respect to both f_{best} and f_{avg} indicators.

4.4 Convergence Analysis

To investigate the behavior of our HEA algorithm with the reference algorithms, we performed an additional convergence analysis. We have excluded the results of GA algorithm due to its poor performance.

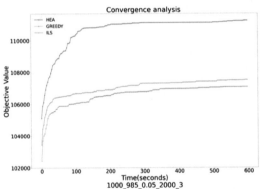

(a)   Convergence analysis on instance 1000_985_0.05_2000_3

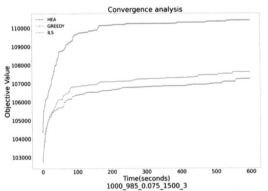

(b)   Convergence analysis on instance 1000_985_0.075_1500_3

Fig. 3. Convergence analysis of HEA and the reference algorithms.

The experiment is performed on two challenging instances 1000_985_0.05_2000 and 1000_985_0.075_1500 with large standard deviation values. Each algorithm was executed 30 times, adopting the same experimental settings as introduced in Sect. 4.2. We generate the convergence graphs using the function: $t \to F_{best}$, where t represents the running time, and F_{best} is the average of the f_{best} values obtained by each algorithm over 30 runs at time t.

Figure 3 confirms the computational efficiency of our HEA algorithm. The curves of HEA run strictly above the Greedy curves and ILS curves, which means that HEA not only exhibits faster convergence than Greedy and ILS but also achieves better results. This implies that our HEA algorithm can achieve the best objective value in less time, which is meaningful in practical applications. These outcomes validate the competitiveness of HEA when compared to the reference algorithms Greedy and ILS.

5 Conclusion

In this study, we investigate for the first time the overlap constrained resource allocation problem (OCRAP) with application to wireless networks. We proposed the mathematical model for the OCRAP and developed an efficient hybrid evolutionary algorithm (HEA) to address it. The HEA algorithm combines a powerful tabu search procedure for local optimization with an efficient crossover operator to produce promising offspring solutions. The computational results obtained from 60 OCRAP benchmark instances reveal that our algorithm surpasses the performance of the genetic algorithm, the greedy algorithm and the iterated local search algorithm, as well as the CPLEX solver. The computational efficiency of our HEA algorithm is further confirmed through the convergence analysis.

This work has led to the emergence of several new research directions. First, given its NP-hard nature, finding the optimal solution of OCRAP poses a significant challenge. Developing exact algorithms to attain optimal solutions or to attain better upper and lower bounds holds considerable merit. Second, due to its practical significance, it is interesting to develop more powerful heuristic algorithms to solve the OCRAP problem.

Acknowledgement. This work is partially supported by the National Natural Science Foundation Program of China (Grant No. 72301036, 62172056). Support from the Research Innovation Fund for College Students and the High-performance Computing Platform of Beijing University of Posts and Telecommunications is also acknowledged.

References

1. Abrardo, A., Alessio, A., Detti, P., et al.: Radio resource allocation problems for OFDMA cellular systems. Comput. Oper. Res. **36**, 1572–1581 (2009)
2. Ahmed, I.Z., Sadjadpour, H., Yousefi, S.: Constrained resource allocation problems in communications: an information-assisted approach. In: Proceedings, MILCOM 2021–2021 IEEE Military Communications Conference (MILCOM) (2021)
3. Bouras, C., Kalogeropoulos, R.: User allocation in 5G networks using machine learning methods for clustering. In: Proceedings, International Conference on Advanced Information Networking and Applications (AINA) (2021)

4. Capone, A., Carello, G., Filippini, I., et al.: Solving a resource allocation problem in wireless mesh networks: a comparison between a CP-based and a classical column generation. Networks **55**, 221–233 (2010)
5. Dutta, R.N., Ghosh, S.C.: Resource allocation for millimeter wave D2D communications in presence of static obstacles. In: International Conference on Advanced Information Networking and Applications (AINA) (2021)
6. Falsafain, H., Heidarpour, M.R., Vahidi, S.: A branch-and-price approach to a variant of the cognitive radio resource allocation problem. Ad Hoc Netw. **132**, 102871 (2022)
7. Glover, F., Laguna, M.: Tabu Search. Springer, Heidelberg (1998)
8. Goldschmidt, O., Nehme, D., Yu, G.: Note: On the set-union knapsack problem. Naval Res. Logist. (NRL) **41**, 833–842 (1994)
9. He, Y., Xie, H., Wong, T.-L., et al.: A novel binary artificial bee colony algorithm for the set-union knapsack problem. Futur. Gener. Comput. Syst. **78**, 77–86 (2018)
10. Kar, B., Wu, E.H.-K., Lin, Y.-D.: The budgeted maximum coverage problem in partially deployed software defined networks. IEEE Trans. Netw. Serv. Manag. **13**, 394–406 (2016)
11. Khuller, S., Moss, A., Naor, J.S.: The budgeted maximum coverage problem. Inf. Process. Lett. **70**, 39–45 (1999)
12. Kia, S.S.: A distributed dynamical solver for an optimal resource allocation problem over networked systems. In: Proceedings, IEEE Conference on Decision and Control (CDC) (2015)
13. Konnov, I., Kashina, O., Laitinen, E.: Vector resource allocation problems in communication networks. In: Proceedings, Symposium and Workshops on Modeling and Optimization in Mobile, Ad Hoc and Wireless Networks (WiOpt) (2013)
14. Lai, X., Hao, J.-K., Glover, F., et al.: A two-phase tabu-evolutionary algorithm for the 0–1 multidimensional knapsack problem. Inf. Sci. **436**, 282–301 (2018)
15. Letchford, A.N., Ni, Q., Zhong, Z.: An exact algorithm for a resource allocation problem in mobile wireless communications. Comput. Optim. Appl. **68**, 193–208 (2017)
16. Li, L., Wang, D., Li, T., et al.: Scene: a scalable two-stage personalized news recommendation system. In: Proceedings, International ACM SIGIR Conference on Research and Development in Information Retrieval (2011)
17. Li, L., Wei, Z., Hao, J.-K., et al.: Probability learning based tabu search for the budgeted maximum coverage problem. Expert Syst. Appl. **183**, 115310 (2021)
18. Stanczak, S., Wiczanowski, M., Boche, H,: Fundamentals of Resource Allocation in Wireless Networks: Theory and Algorithms. Springer, Heidelberg (2009)
19. Suh, K., Guo, Y., Kurose, J., et al.: Locating network monitors: complexity, heuristics, and coverage. Comput. Commun. **29**, 1564–1577 (2006)
20. Vanderster, D.C., Dimopoulos, N.J., Parra-Hernandez, R., et al.: Resource allocation on computational grids using a utility model and the knapsack problem. Futur. Gener. Comput. Syst. **25**, 35–50 (2009)
21. Wang, P., Peng, W., Zhang, W., et al.: Joint channel and power allocation algorithm for flying ad hoc networks based on bayesian optimization. In: Proceedings, International Conference on Advanced Information Networking and Applications (2021)
22. Wei, Z., Hao, J.-K.: Multistart solution-based tabu search for the Set-Union Knapsack Problem. Appl. Soft Comput. **105**, 107260 (2021)
23. Wei, Z., Hao, J.-K.: Iterated hyperplane search for the budgeted maximum coverage problem. Expert Syst. Appl. **214**, 119078 (2023)
24. Zhou, J., Zheng, J., He, K.: Effective variable depth local search for the budgeted maximum coverage problem. Int. J. Comput. Intell. Syst. **15**, 43 (2022)

An Efficient Method for Underwater Fish Detection Using a Transfer Learning Techniques

Mahdi Hamzaoui[1(✉)], Mohamed Ould-Elhassen Aoueileyine[1], Lamia Romdhani[2], and Ridha Bouallegue[1]

[1] Innov'COM Laboratory, Higher School of Communication of Tunis, Technopark Elghazala, 2083 Raoued, Ariana, Tunisia
{mahdi.hamzaoui,mohamed.ouldelhassen,ridha.bouallegue}@supcom.tn
[2] University of Qatar, Doha, Qatar
lamia.romdhani@qu.edu.qa

Abstract. Detecting fish species is crucial in aquaculture, playing a vital role in safeguarding populations and monitoring their health and nutritional systems. However, traditional machine learning methods struggle to identify objects in images with complex backgrounds, especially in low-light conditions. This paper aims to enhance the performance of a YOLO NAS model for fish recognition. Employing transfer learning, our model leverages a pre-trained model on the COCO(Common Objects in COntext) dataset and is subsequently tested in various scenes. The experimental results demonstrate the model's effectiveness. Using Recall and mAP50 evaluation metrics, our novel model achieves precision rates of 0.973 and 0.922, respectively.

1 Introduction

Underwater video is used to study marine and riverine ecosystems, especially to record animal abundance and behavior as part of environmental monitoring. Given humans' innate tendency to evaluate their surroundings visually, optical video gives detailed information in a visually interpretable format. However, human video analysis takes time, therefore some level of automation is required to process huge video collections quickly and efficiently. This is critical for providing the information required for permitting and adaptive management decisions. Computer vision and machine learning have demonstrated their efficacy in automating comparable monitoring tasks, such as video surveillance. Underwater visuals pose distinct challenges, including abrupt shifts in illumination, uneven spectral propagation, limited contrast, the presence of clutter in the form of floating vegetation, and alterations in visibility caused by turbidity [1].

Previously, multiple approaches for automatic fish recognition using various image and video processing algorithms were used. Hsiao et al. suggested a method for detecting fish in films that uses temporal properties of the fish. To model background pixels in video frames, they used adaptive background subtraction with Gaussian Mixture Models (GMM). It is believed that the video frames utilized for training GMM are entirely background, with no fish instances. GMM detects mobility in video frames (most likely by fish) when a specific section of the frame fails to match the distribution of the backdrop model. On a dataset consisting of many underwater films captured

L. Barolli (Ed.): AINA 2024, LNDECT 201, pp. 257–267, 2024.
https://doi.org/10.1007/978-3-031-57870-0_23

within the Southern Taiwan region, they produced good results on the fish detection job, with an average success rate of 83.99% [2]. Huang et al. demonstrated an efficient method for reducing the effect of environmental variability by using decision trees with Support Vector Machine (SVM) trained on fish images. On a dataset of 24,000 photos comprising 15 distinct fish species, they raised the accuracy of fish species classification to 74.8% [3]. Rova et al. introduced unconstrained underwater fish classification algorithms based on image processing and pattern recognition by capturing the scale texture pattern and fish shape data [4].

Deep Neural Networks (DNNs) have recently been employed for fish detection and species classification. Simegnew Yihunie Alaba et al. proposed a CNN-based methodology that begins with object detection. This is a good strategy to utilize when there are multiple fish in the image. MobileNetv3 is used to extract the image's most relevant elements. These characteristics allow for accurate fish classification. The findings showed that this model's accuracy was 80.61% [5]. Salman et al. (2016) conducted a comparison of multiple traditional machine learning approaches and a deep network known as Convolution Neural Network (CNN) on Fish4Knowledge LifeCLEF 2014 and LifeCLEF 2015 fish images [6]. Frank Storbeck et al. collected length and height data for each fish type from various places. The model for recognizing fish types based on their height and length is developed once the neural network has been trained. The network architecture was subjected to specific interventions such as the deletion of non-useful connections from nodes. The model was 95 percent accurate. In this investigation, other fish-specific metrics detected by the camera can be taken. Fish type recognition can be improved by training neural networks with the fish's diagonal length, width, and surface area [7]. Praba Hridayami et al. suggested a deep convolutional neural network-based fish recognition algorithm. For transfer learning, the VGG16 model, which has been pre-trained on ImageNet, is employed. This study made use of a dataset of 50 species. In this dataset, the CNN model is trained on four different types of images: RGB color space images, blending images, canny filter images, and RGB images blended with blending images. The RGB image blended with the blending image produces the best results, with an accuracy of 96.4% [8]. Similarly, a fish identification and classification utilizing YOLO (You Only Look Once), a CNN variation, was proposed by Sung et al.. They trained the YOLO network on 829 images and achieved 93% accuracy in fish species classification on 100 images [9].

Many additional research have employed computer vision techniques to detect fish. Jing Hu et al. introduced a multi-class support vector machine technique (MSVM) for fish classification based on skin, color, and texture parameters. Images are captured using a smartphone. Six feature vector groups are constructed. The best classifier in this study is Directed Acyclic Graph Multiple Support Vector Machines (DAGMSVM), with an accuracy ranging from 92.22 to 100% depending on the type of species [10]. Ling Yang et al. created a computer vision model for detecting fish. They have extensively examined the difficulties faced, such as fish color similarity, high-density scene, darkness, and low resolution. It is implemented a 3D imaging acquisition system for fish behavior analysis. Deep learning techniques are used to train the model. Other preprocessing strategies can be incorporated in this study to address the issues connected with scene complexity. Binarization of captured images, for example, can produce more accurate results than gray image conversion [11].

The motivation behind our work is to implement a model capable of detecting fish-like objects. YOLO NAS is used to implement this model while integrating the concept of transfer learning in order to improve the model's performance. The "Methodology" section describes the data acquisition phase of the study. This section critically compares various transfer learning approaches. Next, the "Results and discussion" section presents the results of the study and evaluates the performance of our approach, which links the YOLO NAS model to a model pre-trained on the COCO dataset. Finally, the paper concludes with the 'Conclusion' section, which offers a forward-looking perspective and outlines avenues for future research.

2   Methodology

In this section, we commence by providing a comprehensive description of the dataset and detailing various preprocessing techniques employed. Subsequently, we proceed to train our YOLO NAS model utilizing several transfer learning technologies. Towards the conclusion of this section, we derive the optimal transfer learning method for our model. The methodology can be succinctly summarized in the following steps:

2.1   Data Gathering and Preprocessing

The Deep Fish dataset is a robust dataset offering high-resolution fish images. This dataset is originally sourced from 'Papers with Code', a website that provides a platform for the collaborative development of machine learning and artificial intelligence research.

Fig. 1. Sample images taken from the dataset

The dataset includes thousands of high-quality images showing a wide range of fish. An annotation file (.txt file) corresponds to each image file. The coordinates of the bounding boxes that encompass the fish seen in each image, together with the corresponding classifications of fish, are contained in these annotation files. It's crucial to remember that bounding box annotations fall between 0 and 1 and are normalized based on the image size.

This dataset's main goal is to support the creation of object detection models that can accurately identify and categorize various fish species in underwater photos and videos. These models can be used in a variety of disciplines, from biological and ecological study to the development of automation in the fishing sector. Figure 1 shows sample images taken from the dataset. Figure 2 presents images with bounding boxes.

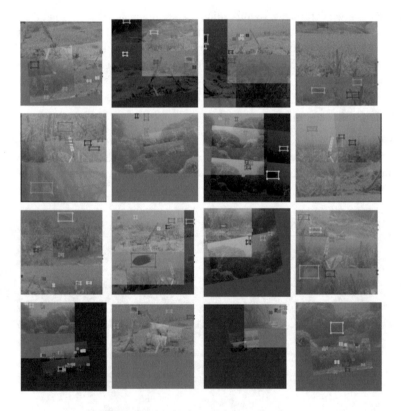

Fig. 2. Sample images with bounding box

2.2 The Extensive COCO Dataset

Microsoft Inc. created the massive object detection, segmentation, and captioning dataset COCO. Common Objects in Context is referred to as COCO. The photographs in the collection are taken in natural settings with ordinary things in scenes seen every day. Eighty identified object categories with exact object localization labels are present in the collection. The datasets were made available in two years, starting in 2014. 164K images total, 118K for train, 5K for validation, and 20K for test are included in the 2017 version [12].

2.3 The Efficiency of the YOLO NAS Algorithm

Deci, a business that creates tools and production-grade models for building, optimizing, and deploying deep learning models, released YOLO-NAS in May 2023. YOLO-NAS is appropriate for real-time edge-device applications because of its ability to identify small objects, increase localization accuracy, and improve the performance-per-compute ratio. Its open-source architecture is also accessible for usage in research [13]. Figure 3 illustrates that pretraining YOLO NAS with the COCO dataset, measured on an NVIDIA T4 card, results in higher accuracy rates than the earlier versions of YOLO, ranging from version 3 to 8.

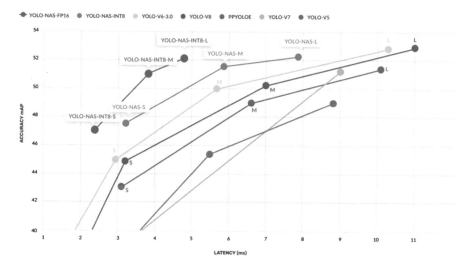

Fig. 3. Accuracy comparison of different versions of YOLO tested on the extensive COCO dataset

2.4 Transfer Learning

A model created for one task is utilized as the foundation for a new model for another task, a process known as transfer learning in machine learning. It uses the information obtained from solving one problem to another that is related but distinct. Transfer learning is a popular technique in computer vision where a pre-trained model from a large database is used as a starting point to finish a specific assignment. Compared to starting from scratch and training a new model, using a pre-trained model that has already learned generic features from a similar assignment saves time. In this work, we used the pre-trained models YOLO-NAS-S, YOLO-NAS-L and Resnet. YOLO-NAS-S and YOLO-NAS-L are two variants of YOLO that are pre-trained on the large COCO dataset. Resnet, which stands for "Residual Networks," is a deep neural network architecture developed to solve the vanishing gradient problem in deep neural networks. Resnet is initially based on the ImageNet dataset, which contains millions of images spread over thousands of categories [14].

2.4.1   The General Workflow of Our Work

As depicted in Fig. 4, the dataset comprises high-resolution images accompanied by bounding box coordinates. During the training phase, three pre-trained models, namely Resnet, YOLO-NAS-S, and YOLO-NAS-L, were individually incorporated into the system to enhance its performance through transfer learning. A comparative analysis was conducted among these three models to determine the most suitable one for integration into our final YOLO-NAS model. Once the optimal algorithm was identified, the model underwent training, validation, and subsequent testing phases.

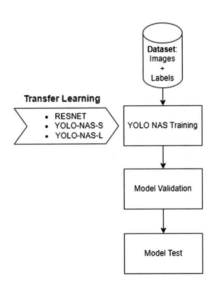

Fig. 4. The Workflow of our Model for Fish Detection

3   Result and Discussion

3.1   Results of Combining the YOLO NAS Model with Pre-trained Models

Our fish detection model was trained by combining it with various pre-trained models. Initially, we fused it with the Resnet model, which had been pre-trained using the renowned ImageNet dataset. At epoch 20, we achieved values of 1.8697, 1.8564,

0.8434, and 0.6274 for the Loss-train, Loss-validation, Recall, and mAP50 metrics, respectively as shown in Table 1 and Fig. 5. By passing the second pre-trained model YOLO-NAS-S which is trained on the coco dataset to our YOLO NAS model, we obtained respectively the results 1.6812, 1.6448, 0.9569 and 0.8722. Finally, the performance of the YOLO NAS model is improved with the pre-trained YOLO-NAS-L model, which is also pre-trained on the COCO dataset. The results obtained are respectively 1.5914, 1.5118, 0.9733 and 0.9220 for the Loss-train, Loss-validation, Recall, and mAP50 metrics.

Table 1. Results obtained using different approaches.

Approach	Loss in train phase	Loss in validation phase	Recall	mAP50
Resnet-YOLO NAS	1.8697	1.8564	0.8434	0.6274
YOLO-NAS-S-YOLO NAS	1.6812	1.6448	0.9569	0.8722
YOLO-NAS-L-YOLO-NAS	1.5914	1.5118	0.9733	0.9220

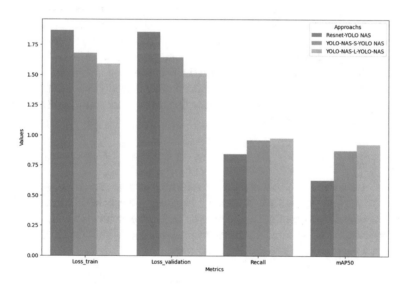

Fig. 5. The results obtained with the different evaluation metrics

3.2 Evolution of YOLO NAS Model Training Associated with YOLO-NAS-L

Our YOLO NAS model for the detection of fish-like objects is improved by associating it with a pre-trained YOLO-NAS-L model in the context of transfer learning. It was trained over 20 epochs. Figure 6 shows the evolution of its training using the two metrics Recall and mAP50.

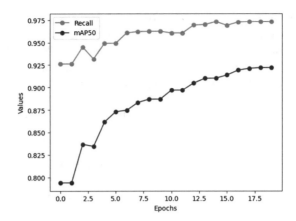

Fig. 6. The evolution of our model training

3.3 Detection Results

During the testing phase, we captured underwater images under various conditions and assessed them using our enhanced model. Figure 7 illustrates the scenes depicting fish detection, showcasing the model's remarkable performance with minimal errors. Our model succeeded in detecting fish with high accuracy rates and the boxes encompassed the objects optimally. It was able to perform the fish detection task even in scenes with unfavourable conditions. Figure 7 shows that accurate detection were achieved by the model in complex cases such as fish overcrowding, poor water quality and convergence between the fish skin and the background.

3.4 Overfitting Test

It is crucial to consider the phenomenon of overfitting when the complexity of the model increases. Overfitting occurs when a model becomes too specialized in capturing the training data, leading to poor generalization on new, unseen data [15]. The Fig. 8 illustrates the evolution of our model's performance throughout the training cycle. Validation is of utmost importance to prevent overfitting.

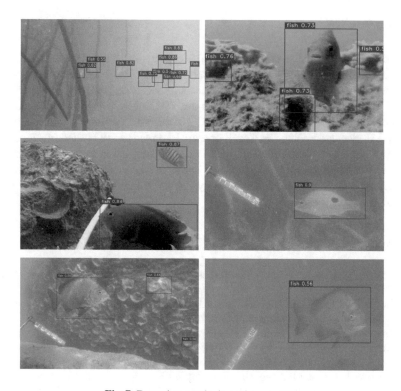

Fig. 7. Detection results in various scenes

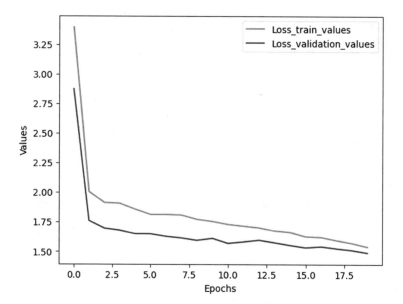

Fig. 8. Overfitting test

4 Conclusion

The effectiveness of the fish detection model is crucial due to the intricacy of the undersea environment. In this research, we offer a transfer learning-based fish detection solution based on a YOLO NAS model linked to another model that has been pre-trained on the COCO dataset. Beyond its link with the YOLO-NAS-S and Resnet models, the results of coupling YOLO NAS with YOLO-NAS-L are noteworthy. The model's performance is evaluated using the Recall and mAP50 metrics. We tested our model on a number of delicate scenarios and got good results. Our goal is to use computer vision in future research to instantly diagnose fish illnesses. The fish farmer and his intelligent farm will be able to communicate instantly thanks to the model's integration into an on-board system.

References

1. Srivastava, V.: Classification of Fish Species Using Deep Learning Models. Master's thesis, NTNU (2023)
2. Hsiao, Y.H., Chen, C.C., Lin, S.I., Lin, F.P.: Real-world underwater fish recognition and identification, using sparse representation. Ecol. Inform. **23**, 13–21 (2014)
3. Huang, P.X., Boom, B.J., Fisher, R.B.: Hierarchical classification with reject option for live fish recognition. Mach. Vis. Appl. **26**(1), 89–102 (2015)
4. Rova, A., Mori, G., Dill, L.M.: One fish, two fish, butterfish, trumpeter: Recognizing fish in underwater video. In: MVA, pp. 404–407, May 2007
5. Alaba, S.Y., et al.: Class-aware fish species recognition using deep learning for an imbalanced dataset. Sensors **22**(21), 8268 (2022)
6. Jalal, A., Salman, A., Mian, A., Shortis, M., Shafait, F.: Fish detection and species classification in underwater environments using deep learning with temporal information. Ecol. Inform. **57**, 101088 (2020)
7. Storbeck, F., Daan, B.: Fish species recognition using computer vision and a neural network. Fish. Res. **51**(1), 11–15 (2001)
8. Hridayami, P., Putra, I.K.G.D., Wibawa, K.S.: Fish species recognition using VGG16 deep convolutional neural network. J. Comput. Sci. Eng. **13**(3), 124–130 (2019)
9. Sung, M., Yu, S.C., Girdhar, Y.: Vision based real-time fish detection using convolutional neural network. In: OCEANS 2017-Aberdeen, pp. 1–6. IEEE, June 2017
10. Hu, J., Li, D., Duan, Q., Han, Y., Chen, G., Si, X.: Fish species classification by color, texture and multi-class support vector machine using computer vision. Comput. Electron. Agric. **88**, 133–140 (2012)
11. Yang, L., et al.: Computer vision models in intelligent aquaculture with emphasis on fish detection and behavior analysis: a review. Arch. Comput. Methods Eng. **28**, 2785–2816 (2021)
12. Lin, T.-Y., et al.: Microsoft COCO: common objects in context. In: Fleet, D., Pajdla, T., Schiele, B., Tuytelaars, T. (eds.) ECCV 2014. LNCS, vol. 8693, pp. 740–755. Springer, Cham (2014). https://doi.org/10.1007/978-3-319-10602-1_48
13. Terven, J., Córdova-Esparza, D.M., Romero-González, J.A.: A comprehensive review of YOLO architectures in computer vision: from YOLOv1 to YOLOv8 and YOLO-NAS. Mach. Learn. Knowl. Extr. **5**(4), 1680–1716 (2023)

14. Wolf, T., Sanh, V., Chaumond, J., Delangue, C.: TransferTransfo: a transfer learning approach for neural network based conversational agents. arXiv preprint arXiv:1901.08149 (2019)
15. Li, G., Song, Z., Fu, Q.: A new method of image detection for small datasets under the framework of YOLO network. In: 2018 IEEE 3rd Advanced Information Technology, Electronic and Automation Control Conference (IAEAC), pp. 1031–1035. IEEE, October 2018

Comparative Analysis of Data Preprocessing Methods in Machine Learning for Breast Cancer Classification

Timothy Stockton, Brandon Peddle, Angelica Gaulin, Emma Wiechert, and Wei Lu[✉]

Department of Computer Science, Keene State College, The University System of New Hampshire, Keene, NH, USA
wlu@usnh.edu

Abstract. Breast cancer, the most diagnosed cancer according to the American Cancer Society, impacts about 13% of women and has a rising incidence rate. With a 10-year survival rate of 84%, early detection is essential, especially in cases of dense breast tissue where 41% go undetected. This paper explores various feature selection and extraction techniques in machine learning for breast cancer classification, including correlation-based selection, recursive elimination, linear discriminant analysis, principal component analysis, and their combinations. The experimental evaluation with the well-known Wisconsin Diagnostic Breast Cancer dataset shows that the linear discriminant analysis feature transformation provides the best overall performance for machine learning algorithmic classifications regarding Accuracy, Precision, Recall, and F1-Score.

1 Introduction

Breast cancer is a common disease in the present day. Women are more likely to be diagnosed, and it affects millions every day. It is the most common type of cancer diagnosis, making up 30% (or approximately 1 out of 3) of cancer diagnoses among women. It has also caused more deaths than almost any other kind of cancer; it is second only to lung cancer. About 13% of women (1 out of every 8) will be diagnosed with breast cancer within their lifetimes. Its incidence is rising, with yearly diagnosis rates increasing by 0.5% [1].

There are many potential risk factors for breast cancer. Those that an individual could change include excess weight, exercise, alcohol consumption, smoking, breastfeeding, and use of menopausal hormones. Other risk factors include personal and family history of cancer, tall height, high bone mineral density, naturally high levels of certain endogenous sex hormones, use of oral contraceptives, and diagnosis of ductal carcinoma in situ (DCIS), lobular carcinoma in situ (LCIS), or benign breast disease [2]. Early diagnosis and treatment of breast cancer go along with a better prognosis. The longer a cancer is present, the more time it must grow and spread in the body. There are also more treatment options available for smaller cancers. As a result, early diagnosis and treatment are critical.

L. Barolli (Ed.): AINA 2024, LNDECT 201, pp. 268–279, 2024.
https://doi.org/10.1007/978-3-031-57870-0_24

Breast cancer can be challenging to diagnose, even for trained specialists. Although a variety of diagnosis methods exist, breast cancer is typically screened for via mammogram. It is recommended that all women from age 45 to 54 receive annual mammograms for breast cancer screening, then biennially afterward for as long as they are in good health [2]. However, mammograms have their limitations. Nearly half of women aged 40 or older have dense breast tissue, which makes mammogram results more difficult to interpret [5]. According to one study, 41% of breast cancers were not found when mammograms were performed on very dense breast tissue [3]. In addition, mammograms are weak at detecting small lesions [4]. This can cause breast cancers to go unnoticed, allowing them to become more likely to be fatal.

Machine Learning (ML), a subset of artificial intelligence, centers on identifying statistical algorithms capable of making accurate predictions based on specific datasets [6–10]. If forecasts are precise enough, the ML model can perform tasks without human instruction [11]. Some ML models are commonly used [12], such as linear regression (LR), decision tree (DT), random forest (RF), support vector machine (SVM), artificial neural network (ANN), and XGBoost tree (XGB). Their performance varies depending on their application and sometimes even the specific datasets used with them. Information within a dataset is split up for training and testing ML models. The standard train/test split is 30% for training and 70% for testing, although some researchers prefer alternative percentages.

The medical industry recently began using ML to diagnose many diseases and conditions. There are now computer systems designed to help health professionals better detect signs of disease. This is called computer-aided diagnosis (CAD). ML algorithms can inspect information thoroughly and much more quickly than a human. This allows diagnoses to be made more quickly and accurately. Breast cancer research and diagnosis are benefitting from CAD systems. Investigation of test results by both ML and a human specialist can cut down on undiagnosed breast cancer cases. This clears the way to better prognoses and fewer deaths.

This study aims to expand on previous research on the application of machine learning in breast cancer diagnosis. The research done in this study explores the idea of alternative approaches to data preprocessing and how they affect various models. This will help advance our ability to diagnose breast cancer effectively and quickly. In this paper, eight different approaches to data preprocessing were made, including a control set. The experimental evaluation with the well-known Wisconsin Breast Cancer Dataset demonstrates that the accuracy of the model's results can be further optimized with a preprocessing method that tailors to the model.

The remainder of this paper is structured as follows. Section 2 discusses some of the most recent studies in the domain. Section 3 describes our methodology, including creating the eight modified training sets that involve applying distinct preprocessing techniques to each, using the original Wisconsin Breast Cancer Dataset as a base. Additionally, there is a control set, which remains unmodified mainly except for scaling, to serve as a benchmark for comparison against the modified settings. In Sect. 4, we conduct a comprehensive analysis of how different preprocessing techniques impact the performance of machine learning models, including six Machine Learning algorithms trained on each of the nine possible training sets, and the resulting models were then evaluated using four popular statistical methods: Accuracy, Precision, Recall, and the F1-Score. Finally, in Sect. 5, we offer concluding remarks and discuss future work.

2 Related Work

Data preprocessing is a common practice in ML research. Preprocessing helps control for missing, redundant, or lousy information within a dataset, allowing the ML models to learn more efficiently. This can be done in various ways, including feature selection, feature extraction, sampling, scaling, and manual data manipulation. This section illustrates some typical examples of recent studies in this domain.

In [13], Elsadig et al. calculated Chi-square, ReliefF, ANOVA, Gini Index, and Gain ratio for feature selection on their dataset, resulting in 17 integral features. Meanwhile, Hassan et al. utilized the Least Absolute Shrinkage and Selection Operator (LASSO) to isolate the 14 most important features. Hassan et al. also preprocessed their data with Standard Scaler for normalization [14]. Fulorunso et al. preprocessed their dataset via sampling. They experimented with the use of Edited Nearest Neighbor (ENN), Synthetic Minority Oversampling Technique (SMOTE), and a combination of both approaches (SMOTEENN), finding that SMOTEENN sampling performed best [15].

Some researchers also manually correct erroneous data before feeding it to ML algorithms. This is what Poornajaf and Yousefi did with their dataset before giving it to their ML models [16]. Abhishek S. Powar et al. [17] then handled missing or inconsistent/erroneous values and removed duplicates to reduce redundancy and potential bias and improve the dataset quality. They selected features based on relevancy and prediction impact using correlation analysis. The features were scaled and normalized so that extensive value features did not dominate the learning process. They used a 60–40 train-test split.

In [18], Sahar Arooj et al. believe that the complexity of ML procedures, such as preprocessing and feature extraction, factors in a decrease in efficiency and accuracy. They used a train-test split of 80–20. Reza Rabiei et al. [19] used the Synthetic Minority Oversampling Technique (SMOTE) to "balance the training data due to the difference in the number of study class records." David A. Omondiagbe et al. [20] centered, scaled, and cleaned their data. They used feature selection techniques (CFS and RFE) and feature extraction methods (PCA and LDA). Their results show that SVM with RBF kernel and LDA preprocessing can produce promising results and that an ANN model built with LDA preprocessed data performs better than the other ANN models. They have a table of the model and technique combinations they used.

In their research, Ahmed Elazab et al. [14] discuss various feature selection techniques for machine learning, including PCA and LDA. They also explore filtering methods like the chi-square test. These techniques are crucial in selecting the most discriminative features to prevent overfitting and reduce redundancy in the feature space, thereby enhancing the effectiveness of machine learning models. Dahri et al. utilized the Standard Scaler module to transform their dataset, ensuring all features were close in numerical scale [15]. The researchers then proceeded to implement feature selection to retain only the features that exerted the most significant variance necessary to predict records correctly.

By contrast, Amethiya et al. cataloged a pipeline in which the dataset was loaded, and then feature extraction techniques were executed directly upon the raw data [23]. They include a figure that notes preprocessing as an interstitial process but does not elaborate on the exact methods used in the associated description. Khater et al. explain in detail their methodology of cleaning the data for erroneous and misleading data points, which taint the fidelity of the dataset and its ability to train models later [24]. The researchers also utilize what they describe as a gamut of feature selection techniques before moving on to model training.

In [25], Silva-Aravena et al. utilized label encoding to prepare the label in their unique dataset for each record, transforming the malignant and benign designation into a numerical format digestible by machine learning algorithms. Harinishree et al. cataloged the findings of numerous similar studies in a comparative study. The authors also describe numerous datasets utilized in previous studies to train models to detect breast cancer. These datasets include the MIAS dataset and the ubiquitous Wisconsin Breast Cancer Dataset [26].

3 Methodology

The data set we used is the Wisconsin Diagnostic Breast Cancer data set. It contains 32 features, a benign/malignant classification, and information about clump radius, texture, perimeter, area, smoothness, compactness, concavity, symmetry, and fractal dimensions. These features are derived from mammograms, allowing us to visualize the internal structures of the breast tissue using X-rays [22]. More recently, data sets derived from thermographic imaging have been used because thermography is non-invasive, low-cost, and radiation-free [21].

The data set must be "cleaned" for missing, incomplete, erroneous values and duplicates. Data could be polluted due to network anomalies during original data collection [27]. We checked for null values within the data set, which we did not find, using isna().sum(). Next, we used LabelEncoder to convert the diagnosis label strings, M for malignant and B for benign, into integers of 1 and 0, respectively, so the machine learning models could utilize this feature. Then, we used RobustScaler to scale the data, ensuring that extreme values do not skew the results. It puts all features on the same scale between 0 and 1 so that although a number may be large or small, it is processed relative to its position within that feature only. This cleaning is performed on the original data set before making duplicates so that each data set receives the same cleaning.

We analyzed the number of malignant and benign tumors in the Wisconsin Breast Cancer dataset. The number of benign tumors in this dataset is 350. The number of malignant tumors in this dataset is roughly 200. The histogram analysis shows that the spread of feature concave_points_worst in the Wisconsin Breast Cancer dataset. The highest value of this feature is 0.29, and the value with the highest number of occurrences is 0.8. Similarly, the highest value of the concave_points_mean feature is 0.195, and the lowest is 0.005. This feature's highest number of occurrences is 0.03, with roughly 118 occurrences.

The boxplot analysis compares the features of Perimeter Worst with Label; this shows the correlation between the data collected on Perimeter Worst and the diagnosis assigned to the mammogram sample. The mean of the malignant tumors' perimeter worse value is around 132. The mean of the benign tumors is about 87.5. The minimum perimeter worst value for malignant tumors is 87.5; for benign tumors, it is around 60. The maximum perimeter worst value for malignant tumors is 210, and the maximum perimeter for benign tumors is 110. The outliers for malignant tumors are 215, 230, and 250. The outliers for benign tumors are 65 and 130. This box-plot analysis shows that malignant tumors score higher in the perimeter worst feature. Still, there is some indication that other features are needed to determine if a tumor is malignant or not successfully. This indicates that perimeter worse, in conjunction with different features, is an excellent feature to use in the feature selection for the models tested. Similarly, we also conducted a boxplot analysis to compare the label feature with the Radius Worst feature, illustrating the correlation between Radius Worst and the tumor diagnosis label. The mean of the malignant tumors is 22. The mean of the benign tumors is 13. The minimum value of Radius Worst for malignant tumors is 13, and the minimum value of Radius Worst for benign tumors is 8. The maximum value of Radius Worst for malignant tumors is 33, and the maximum value of Radius Worst for benign tumors is 18. The outliers for malignant tumors are 33.5 and 36. The outliers for benign tumors are 7 and 21. This shows that Radius Worst is a feature that has higher values in malignant tumors and lower values in benign tumors. In conjunction with other features, this feature is excellent for testing the models in this study. The same analysis is conducted to analyze the label feature with the Area Worst feature, and the result shows that the feature Area Worst is an excellent feature to apply to machine learning models to diagnose breast cancer.

In feature selection, only features with valuable information for classification are kept, while the others are dropped. This is determined by the technique used. Our research utilizes two techniques: correlation-based feature selection (CFS) and recursive feature elimination (RFE). We created nine duplicates of the original cleaned data set. Three duplicates have CFS applied, while another three have RFE applied. Specifically, CFS looks at the feature's intrinsic properties and removes features that correlate highly with other features because they provide redundant information [14]. CFS does not require using an ML model and derives its weights. We retained six features, including perimeter worst, radius worst, area worst, concave points worst, concave points mean, and concavity mean. RFE builds an ML model with all the features and then ranks them based on how important they are to reducing the error in the model [14]. We used a random forest classifier model to retain the six highest-ranked features.

Feature extraction involves the combination of features and reduction of dimension. We use principal component analysis (PCA) and linear discriminant analysis (LDA). First, we applied PCA to one of the data set duplicates that did not have a feature selection technique. Then, we do the same for LDA, leaving duplicate a single control data set with no feature selection or extraction techniques. Next, we apply PCA to a duplicate with CFS applied and then do the same for LDA on another CFS duplicate. Finally, we use PCA on a duplicate with RFE applied and then LDA on another RFE duplicate.

Specifically, PCA transforms the data set by combining features in a way that does not correlate [14]. These new features are called principal components. We specified an output of six components with an explained variance of roughly 89%. The LDA then conducts feature transformations to maximize the scatter between classes and minimize the scatter within each class [14].

We have nine data sets derived from the cleaned original at this stage. Specifically, we have one with only CFS applied, one with only RFE applied, one with only LDA applied, and one with only PCA applied. We also have combination data sets: one with CFS and LDA applied, one with CFS and PCA applied, one with RFE and LDA applied, and one with RFE and PCA applied. Finally, we have a dataset with no feature selection or extraction techniques used, which acts as a control data set.

Each of our 9 data sets is now split into an X variable, which contains all features other than the labels, and a Y variable, which contains only the labels. These variables are used to build the training and testing sets. We used a 70–30 split, meaning the training set includes 70% of the records, and the testing set contains 30%.

4 Comparative Analysis with Machine Learning Models

Our first ML model used is linear regression, which applies a non-linear transformation to the input variables (features) such that the output is restricted to either 0 or 1. In our case, the model's output is a determination of either being benign or malignant. This is achieved with a cutoff value that can be adjusted and acts as a decision boundary. The logistic function (sigmoid function) at the heart of logistic regression will model the probability of a record belonging to a particular class. This probability is then compared to the cutoff value and, if it is higher than the cutoff, will be classified as one or, if it is lower than the cutoff, will be classified as 0. Our study uses the default cutoff of 0.5 because we are not fine-tuning our models. This study compares the outcome of different pre-processing techniques, which can be seen regardless of the model tuning provided all data sets use the same model settings. We created nine separate linear regression models, each using a different one of our data sets to train and test it.

Our second ML model is a decision tree, a series of decision rules like if-else statements. It is a tree because each decision branch, starting with the first, is followed by another decision branch or an endpoint (leaf). These endpoints state a specific class, and any inputs that reach that endpoint via the decisions are determined to be a part of that class. A decision tree can be incredibly complex or very simple. When not using combination features (meaning each feature is explainable, which is not the case with PCA components), the tree itself can be easily explained, and it is clear why the classification is made. Our study used the default max_depth of none and the default min_samples_leaf of 1 because the study focuses on pre-processing techniques. We created nine separate decision-tree models, each using a different one of our data sets to train and test it. Figure 1 shows a decision tree similar in accuracy to the models we used, but it is smaller. The bottom right leaf has the most effective decision rule, whether the unexplainable PCA feature 1 is more significant than 0.7. It has a Gini index value of zero, meaning it is pure, and has 117 labels for only a single class.

Our third ML model is a random forest comprised of multiple decision trees as an ensemble [19]. Each decision tree within the forest has unique splitting criteria and uses a different subset of the data. They will each produce an individual classification based on the input. These classifications are combined as a majority vote, giving us a probabilistic

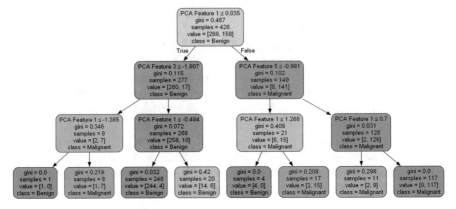

Fig. 1. Decision tree classification results

classification. In this study, we used the default model settings and created nine separate random forest models, each using a different one of our data sets to train and test it.

Our fourth ML model is a support vector machine (SVM), which creates a hyperplane in an optimal position that linearly separates the data points into two classes based on their features [17]. The points closest to the hyperplane are the support vectors, and the distance between the support vectors and the hyperplane is called the margin. This optimization is possible because the optimal hyperplane will maximize the margin. Suppose the data is not linearly separable in its current dimension. In that case, it will be mapped to a higher-dimensional space by a kernel function such as a polynomial kernel or radial basis function (RBF). There is a cost parameter called c, which defines how hard the margin should be. Smaller c values have a higher tolerance for misclassification and a wider margin, while lower c values have a lower tolerance for misclassification and a smaller margin. We used a linear kernel and a c of 10 in this study. We created nine separate SVM models, each using a different one of our data sets to train and test it.

Our fifth ML model is an artificial neural network (ANN), a set of inter-connected nodes, also called neurons, organized into different layers. We specifically used a multi-layer perceptron (MLP) NN where data flows in one direction, starting with an input layer, then the computational hidden layers, and finally an output layer [19]. NN's are based on how a brain works; each node will be active or inactive. The state of each node will impact the nodes in the next layer. This state of being either active or inactive is controlled using an activation function such as hyperbolic tangent (tanh), rectified linear (relu), and logistic function (sigmoid). Our study used an ANN with two 5 x 5 hidden layers, a logistic activation function, a stochastic gradient descent (SGD) solver, a batch size 25, and a learning rate 0.1. We created nine separate neural network models, each using different data sets for training and testing.

Our sixth and final ML model is an extreme gradient boosting (XGBoost) tree. It has improved regularization compared to a standard gradient boosting tree (GBT), which consists of many small decision trees called weak learners. Each tree will additively correct the errors from previous trees such that the prediction accuracy will increase with each tree [19]. The final prediction will be a summation of each tree's prediction. Small trees are used to avoid overfitting. There is a learning rate parameter that scales a tree's output. A high learning rate makes the model less robust but requires fewer trees, while a lower learning rate makes the model more robust and requires more trees. This study used the default learning rate of 0.1, the default max_depth of 3, and an objective of "multi: softmax." We created nine separate XGBoost tree models, each using a different one of our data sets to train and test it.

In summary, we obtained the Wisconsin Breast Diagnostic Cancer data set in this study. Then, we performed some general cleaning on the data set, including handling missing entries, scaling, and encoding. We then used different combinations of pre-processing techniques to make nine unique data sets. These data sets had a 70–30 train-test split applied. Next, we created two lists, one with all the training data sets and one with all the testing data sets. For each model, we instantiated it and then made a loop in which the model fit with each data set, and then its accuracy was calculated. This resulted in 9 variations of each model with 54 different models. Figure 2 is the flowchart for our methodology. We used each model's default settings because this study focuses on pre-processing techniques.

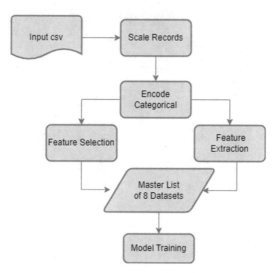

Fig. 2. Flowchart of our methodology

We created eight modified training sets, each transformed using unique preprocessing techniques and an unmodified control (excepting scaling) from the original Wisconsin Breast Cancer Dataset. Necessarily, eight corresponding preprocessed testing sets and a control were also created to test the models. 6 Machine Learning algorithms were then trained on each of the nine possible training sets, and the resulting models were evaluated using four popular statistical methods: Accuracy, Precision, Recall, and the F1-Score, as summarized in Fig. 3.

Algorithm	Preprocessing Combination	Accuracy	Precision	Recall	F1-Score
LR	Control	0.96479	0.98	0.92453	0.95146
	CFS	0.92254	1	0.79245	0.88421
	RFE	0.92254	1	0.79245	0.88421
	LDA	0.97183	0.96226	0.96226	0.96226
	PCA	0.97183	1	0.92453	0.96078
	CFS -> LDA	0.14789	0.01429	0.01887	0.01626
	CFS -> PCA	0.92254	1	0.79245	0.88421
	RFE -> LDA	0.92254	0.9375	0.84906	0.89109
	RFE -> PCA	0.92254	1	0.79245	0.88421
DT	Control	0.89437	0.83929	0.88679	0.86239
	CFS	0.91549	0.90196	0.86792	0.88462
	RFE	0.91549	0.90196	0.86792	0.88462
	LDA	0.94366	0.92453	0.92453	0.92453
	PCA	0.9507	0.92593	0.9434	0.93458
	CFS -> LDA	0.1338	0.03947	0.0566	0.04651
	CFS -> PCA	0.88028	0.84615	0.83019	0.8381
	RFE -> LDA	0.90845	0.87037	0.88679	0.8785
	RFE -> PCA	0.88028	0.84615	0.83019	0.8381
RF	Control	0.92958	0.95745	0.84906	0.9
	CFS	0.91549	0.93617	0.83019	0.88
	RFE	0.90141	0.89796	0.83019	0.86275
	LDA	0.94366	0.92453	0.92453	0.92453
	PCA	0.92254	0.90385	0.88679	0.89524
	CFS -> LDA	0.1338	0.03947	0.0566	0.04651
	CFS -> PCA	0.92254	0.97727	0.81132	0.8866
	RFE -> LDA	0.90845	0.87037	0.88679	0.8785
	RFE -> PCA	0.92254	0.97727	0.81132	0.8866
SVM	Control	0.95775	0.96078	0.92453	0.94231
	CFS	0.92958	0.97778	0.83019	0.89796
	RFE	0.92958	0.97778	0.83019	0.89796
	LDA	0.97183	0.96226	0.96226	0.96226
	PCA	0.96479	0.98	0.92453	0.95146
	CFS -> LDA	0.14789	0.01429	0.01887	0.01626
	CFS -> PCA	0.92958	0.97778	0.83019	0.89796
	RFE -> LDA	0.92254	0.9375	0.84906	0.89109
	RFE -> PCA	0.92958	0.97778	0.83019	0.89796
NN	Control	0.96479	0.98	0.92453	0.95146
	CFS	0.90845	0.95455	0.79245	0.86598
	RFE	0.92254	1	0.79245	0.88421
	LDA	0.97183	0.96226	0.96226	0.96226
	PCA	0.96479	0.98	0.92453	0.95146
	CFS -> LDA	0.1338	0.01389	0.01887	0.016
	CFS -> PCA	0.91549	0.97674	0.79245	0.875
	RFE -> LDA	0.92958	0.92157	0.88679	0.90385
	RFE -> PCA	0.91549	0.97674	0.79245	0.875
XGBoost	Control	0.93662	0.95833	0.86792	0.91089
	CFS	0.90845	0.9	0.84906	0.87379
	RFE	0.90141	0.89796	0.83019	0.86275
	LDA	0.94366	0.92453	0.92453	0.92453
	PCA	0.94366	0.92453	0.92453	0.92453
	CFS -> LDA	0.1338	0.03947	0.0566	0.04651
	CFS -> PCA	0.90845	0.95455	0.79245	0.86598
	RFE -> LDA	0.91549	0.88679	0.88679	0.88679
	RFE -> PCA	0.90845	0.95455	0.79245	0.86598

Fig. 3. Comprehensive results for pre-processing methods with machine learning models.

5   Conclusions and Future Work

In general, the results of this study suggest that, depending on the model, the PCA and LDA preprocessing transformations provide the best overall preparation of the dataset for algorithmic classification, as illustrated in Table 1. Despite the simplicity of use and prevalence in the previous literature on PCA, we found the LDA method to be a highly defensible and visibly effective preprocessing and feature reduction strategy for the stated goal. The LDA preprocessing technique provided exemplary scores when used with the Logistic Regression, Support Vector Classifier, and Multi-Layer Perceptron ML algorithms. Incomparable performance between two training sets from the same ML algorithm, the F1-Score was generally considered the tiebreaker. It is worth noting that the PCA preprocessing technique did exceptionally well in many areas and could quickly overtake LDA as the preferred choice, depending on the model choice and use-case of an implementation.

Table 1. Summary of best performing preprocessing algorithms by ML model.

Algorithm	Preprocessing Combination	Accuracy	Precision	Recall	F1-Score
LR	LDA	0.97183	0.96226	0.96226	0.96226
DT	PCA	0.9507	0.92593	0.9434	0.93458
RFC	LDA	0.94366	0.92453	0.92453	0.92453
SVM	LDA	0.97183	0.96226	0.96226	0.96226
NN	LDA	0.97183	0.96226	0.96226	0.96226
XGBoost	LDA	0.97183	0.92453	0.92453	0.92453
	PCA	0.94366	0.92453	0.92453	0.92453

For this study, the Logistic Regression algorithm provided the most impressive testing scores when exposed to new data while remaining the most straightforward algorithm; LR was therefore declared the "best" performing model. Nevertheless, it is worth restating that the models in this study intentionally received little to no hyperparameter tuning. This decision to use default model training settings was made to demonstrate the unaltered effects of preprocessing on the training and testing process results. Therefore, while the Logistic Regression Model operated most impressively here, as LR has relatively few hyperparameters to tune, other models would likely surpass that performance with further optimization. We are particularly interested in the Support Vector Classifier, whose linear functions synergize well with the underlying linear mathematics of the LDA and PCA modules and whose processing time easily outpaced the more complex Multi-Layer Perceptron. Finally, even without tuning, all the best-performing test models performed better than those trained on the unaltered control sample. This result, while providing an interesting final finding, also underscores the potential for preprocessing to enhance the accuracy of breast cancer detection to almost 100% with the proper combination of preprocessing algorithms and model hyperparameter tuning.

In the future, we will integrate selected with other unsupervised machine learning technologies, such as the E-means clustering algorithm [28–30], to help interpret the model outcomes.

Acknowledgments. This research is supported by New Hampshire - INBRE through an Institutional Development Award (IDeA), P20GM103506, from the National Institute of General Medical Sciences of the NIH.

References

1. American Cancer Society. About Breast Cancer (2023). https://www.cancer.org/content/dam/CRC/PDF/Public/8577.00.pdf
2. American Cancer Society. Breast Cancer Facts & Figures 2022–2024. https://www.cancer.org/content/dam/cancer-org/research/cancer-facts-and-statistics/breast-cancer-facts-and-figures/2022-2024-breast-cancer-fact-figures-acs.pdf
3. Gordon, P.B.: The impact of dense breasts on the stage of breast cancer at diagnosis: a review and options for supplemental screening. Curr. Oncol. **29**(5), 3595–3636 (2022). https://doi.org/10.3390/curroncol29050291
4. Chen, H.L., Zhou, J.Q., Chen, Q., Deng, Y.C.: Comparison of the sensitivity of mammography, ultrasound, magnetic resonance imaging and combinations of these imaging modalities for the detection of small (≤ 2 cm) breast cancer. Medicine **100**(26), e26531 (2021)
5. National Cancer Institute. Dense Breasts: Answers to Commonly Asked Questions - National Cancer Institute. www.cancer.gov, https://www.cancer.gov/types/breast/breast-changes/dense-breasts. Accessed 11 Dec 2023
6. Lu, W.: A novel framework for network intrusion detection using learning techniques. IEEE Pacific Rim Conference on Communications, Computers and signal Processing (PACRIM 2005), pp. 458–461 (2005). https://doi.org/10.1109/PACRIM.2005.1517325
7. Lu, W., Traore, I.: An unsupervised approach for detecting DDoS attacks based on traffic based metrics. In: Proceedings of IEEE Pacific Rim Conference on Communications, Computers and Signal Processing (PACRIM 2005), pp. 462–465
8. Lu, W., Xue, L.: A heuristic-based co-clustering algorithm for the internet traffic classification. In: 2014 28th International Conference on Advanced Information Networking and Applications Workshops, pp. 49–54 (2014). https://doi.org/10.1109/WAINA.2014.16
9. Lu, W., Tavallaee, M., Ghorbani, A.A.: Hybrid traffic classification approach based on decision tree. In: 2009 IEEE Global Telecommunications Conference (GLOBECOM 2009), Honolulu, pp. 1–6 (2009). https://doi.org/10.1109/GLOCOM.2009.5425624
10. Mercaldo, N., Lu, W.: Classification of web applications using AiFlow features. In: Barolli, L., Amato, F., Moscato, F., Enokido, T., Takizawa, M. (eds.) Web, Artificial Intelligence and Network Applications. WAINA 2020. Advances in Intelligent Systems and Computing, vol. 1150, pp. 389–399. Springer, Cham (2020). https://doi.org/10.1007/978-3-030-44038-1_35
11. Lu, W., Burnett, B., Phipps, R.: Feature Selections for Detecting Intrusions on the Internet of Medical Things. Lecture Notes on Data Engineering and Communications Technologies, vol. 174. Springer, Cham (2023). https://doi.org/10.1007/978-3-031-33242-5_7
12. Lu, W.: Applied machine learning for securing the Internet of Medical Things in healthcare. In: Barolli, L. (ed.) Advanced Information Networking and Applications. AINA 2023. LNNS, vol. 654. Springer, Cham (2023)
13. Elsadig, M.A., Altigani, A., Elshoush, H.T.: Breast cancer detection using machine learning approaches: a comparative study. Int. J. Power Electron. Drive Syst. **13**(1), 736–745 (2023). https://doi.org/10.11591/ijece.v13i1

14. Hassan, M.D.: A comparative assessment of machine learning algorithms with the Least Absolute Shrinkage and Selection Operator for breast cancer detection and prediction. Decis. Analyt. J. **7**, 100245 (2023)

15. Fulorunso, S.O., Awotunde, J.B., Adigun, A.A.: A hybrid model for post-treatment mortality rate classification of patients with breast cancer. Healthc. Analyt. **4**, 100254 (2023). https://doi.org/10.1016/j.health.2023.100254

16. Poornajaf, M., Yousefi, S.: Improvement of the performance of machine learning algorithms in predicting breast cancer. Front. Health Inf. **12**, 132 (2023). https://doi.org/10.30699/fhi.v12i0.400

17. Powar, AS., Batwal, A.P.: Research paper on enhancing breast cancer prediction through machine learning and cross-validation techniques: a comparative analysis. Int. Res. J. Moderniz. Eng. Technol. Sci. **05**(05) (2023)

18. Arooj, S., Atta-Ur-Rahman, Z.M., Khan, M.F., Alissa, K., Khan, M.A., Mosavi, A.: Breast cancer detection and classification empowered with transfer learning. Front Publ. Health. **10**, 924432 (2022). https://doi.org/10.3389/fpubh.2022.924432

19. Rabiei, R., Sohrabei, S.: Prediction of breast cancer using machine learning approaches. J. Biomed. Phys. Eng. **12**(3), 297–308 (2002)

20. Omondiagbe, D.A.: Machine learning classification techniques for breast cancer diagnosis. IOP Conf. Ser.: Mater. Sci. Eng. **495**, 012033 (2019)

21. Elazab, A., Lei, B.Y.: Breast cancer detection and diagnosis using mammographic data: systematic review. J. Med. Internet Res. **21**(7), e14464 (2019)

22. Dhahri, H., Al Maghayreh, E., Mahmood, A.: Automated breast cancer diagnosis based on machine learning algorithms. J. Healthc. Eng. **2019**, 1–11 (2019). https://doi.org/10.1155/2019/4253641. https://www.hindawi.com/journals/jhe/2019/4253641/

23. Amethiya, Y., Pipariya, P., Patel, S.: Comparative analysis of breast cancer detection using machine learning and biosensors. Intell. Med. **2**(2) (2021). https://doi.org/10.1016/j.imed.2021.08.004

24. Khater, T., Hussain, A., Bendardaf, R.: An explainable artificial intelligence model for the classification of breast cancer. IEEE Access **4**, 1 (2023). https://doi.org/10.1109/access.2023.3308446

25. Silva-Aravena, F., Núñez Delafuente, H., Gutiérrez-Bahamondes, J.H.: A hybrid algorithm of ML and XAI to prevent breast cancer: a strategy to support decision making. Cancers **15**(9), 2443 (2023). https://doi.org/10.3390/cancers15092443

26. Harinishree, M.S., Aditya, C.R., Sachin, D.N.: Detection of breast cancer using machine learning algorithms – a survey. In: Proceedings of the 5th International Conference on Computing Methodologies and Communication (ICCMC), Erode, pp. 1598–1601 (2021). https://doi.org/10.1109/ICCMC51019.2021.9418488

27. Lu, W.: Unsupervised Anomaly Detection Framework for Multiple-Connection-Based Network Intrusions. Ottawa Library and Archives Canada (2007). ISBN: 9780494147795

28. Lu, W., Tong, H., Traore, I.: E-means: an evolutionary clustering algorithm. In: Kang, L., Cai, Z., Yan, X., Liu, Y. (eds.) Advances in Computation and Intelligence. ISICA 2008. LNCS, vol. 5370, pp. 537–545. Springer, Heidelberg (2008). https://doi.org/10.1007/978-3-540-92137-0_59

29. Lu, W., Traore, I.: Determining the optimal number of clusters using a new evolutionary algorithm. In: Proceedings of IEEE International Conference on Tools with Artificial Intelligence (ICTAI 2005), Hong Kong, pp. 712–713 (2005)

30. Lu, W., Traore, I.: A new evolutionary algorithm for determining the optimal number of clusters. In: Proceedings of IEEE International Conference on Computational Intelligence for Modeling, Control and Automation (CIMCA 2005), vol. 1, pp. 648–653 (2005)

Hybrid Technical-Visual Features for Stock Prediction

Chih-Fong Tsai[1]([✉]), Ya-Han Hu[1], Ming-Chang Wang[2], and Kang Ernest Liu[3]

[1] Department of Information Management, National Central University, Taoyuan, Taiwan
cftsai@mgt.ncu.edu.tw, yahan.hu@g.ncu.edu.tw
[2] Department of Business Administration, National Chung Cheng University, Chiayi, Taiwan
mcwang@ccu.edu.tw
[3] Department of Agricultural Economics, National Taiwan University, Taipei, Taiwan
kangernestliu@ntu.edu.tw

Abstract. Stock prediction is an interesting and important problem for analysts and investors. However, the factors affecting the movement of stock prices are very complicated, making it very difficult to effectively predict them. In the literature, technical analysis has been widely used for stock prediction, the analysis of candlestick charts being one of the main methods. In this paper, we apply an artificial intelligence technique based on the wavelet transform to automatically extract the visual features of candlestick charts. The prediction performance obtained using general technical indicators, the visual features of candlestick charts, and a combination of the two is examined. The experiments, based on data for the Taiwan stock market including ten electronics companies, show that combining both technical and visual features can allow different classifiers to achieve better prediction performance than the other methods.

1 Introduction

Stock prediction has generally been approached by either fundamental analysis or technical analysis [1, 2]. In fundamental analysis, the emphasis is placed on price construction and the factors that affect a company's value. Basically, this method is based on analyzing financial statements and industry competitiveness to assess reasonable stock prices [3]. Technical analysis [4], on the other hand, is based on extracting the tangible and intangible messages conveyed by past investment information (such as trading price and volume). Different graphical or numerical analysis methods can be used to determine stock price trends and the trading decision.

In the literature, there have been many related studies using technical indicators to construct a variety of different classifiers for stock prediction, such as k-nearest neighbor [5], support vector machine [6], decision tree [7], and neural networks [8] methods.

In Tsai and Quan [9], they treated stock prediction as an image retrieval problem, using candlestick charts as images and basing stock prediction is on a search for the most similar historical image(s) to a query image. The assumption is that the stock price indicated by the most similar historical candlestick chart reflects the future stock price of the query candlestick chart. In their study, the analysis of the visual (texture) features of the candlestick charts was demonstrated to outperform other technical indicators for the Dow Jones Industrial Average Index (INDU). Other recent works employed deep learning techniques to analyze candlestick chart images for stock prediction, such as Chen and Tsai [10] and Jin and Kwon [11].

From the data mining and machine learning viewpoints, it is not yet known whether combining technical and visual features can make the classifiers perform better than techniques using technical indicators or visual features alone. Therefore, the aim of this study is to examine the prediction performance of different classifiers, using technical indicators, visual features and hybrid technical-visual features, respectively. Specifically, data for ten electronics companies in Taiwan and the price index of the Taiwan Stock Exchange (TAIEX) are examined.

The remainder of this paper is organized as follows. In Sect. 2, we briefly introduce the concept of stock price prediction. Section 3 describes the experimental procedure and presents the experimental results. Finally, some conclusions are offered in Sect. 4.

2 Stock Prediction

Stock prediction (or stock analysis) has long been regarded as an important research problem. One major goal of this process is to determine the future value of a company's stock or other financial instrument traded on the financial exchange market. There are two main approaches to stock prediction discussed in the literature, fundamental analysis [12] and technical analysis [4]. In the first approach, it is assumed that the stock has an intrinsic value depending on the level of the company's profitability. The stock price thus follows changes in profitability. Fundamental analysis is a type of cross-sectional analysis which must take into account the political and economic situation and the changing structure of the industry. It is also a signal for scrutinizing the company's performance, profitability and the soundness of its financial structure, and information about all these aspects is needed in order to estimate a reasonable stock price range.

Technical analysis, on the other hand, emphasizes previous trading information, such as trading price and volume, and is based on charts, from which information is used to predict the future movement of stock prices. In other words, stock price movement follows rules and regulations, and it is assumed that historical stock price patterns and graphics will be repeated. In particular, the accessory appliances of technical analysis are technical indicators, which can be used to construct a prediction model. It has been pointed out in related studies, such as those by Tsai and Hsiao [13] and Dash and Dash [7], that most researchers have focused on the use of technical indicators to construct prediction models.

A technical analysis, a technical indicator is a mathematical calculation based on indications of price and/or volume, the results of which are used to decide when to make trades. Many popular and obscure indicators have been explained in past studies, including their construction, their mathematics, and their basic use in trading, as well as the application of the values obtained to forecast probable price changes. Significantly, technical indicators are usually composed of price indicators (i.e., a moving average (MA) and relative strength indicator (RSI)), volume indicators (i.e., the Chande momentum oscillator (CMO)), price-volume indicators, and so on.

3 Experiments

3.1 Experimental Setup

Figure 1 shows the experimental procedure followed in this work. First of all, the technical indicators and candlestick charts are collected for a ten-year period, from 2013 to 2022, then divided into training and testing sets based on time, 2013 to 2017 and 2018 to 2022, respectively. In addition, data for ten electronics companies in Taiwan obtained from the Forbes Global 2000 and the price index of the Taiwan Stock Exchange (TAIEX) from 2013 to 2022 are examined.

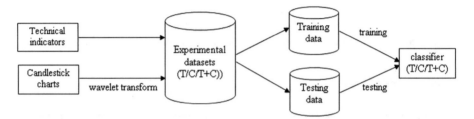

Fig. 1. The experimental procedure.

About the technical indicators, seven indicators, including the moving average (MA), stochastic oscillator (KD), difference indicator (DIF), moving average convergence/divergence (MACD), relative strength indicator (RSI), bias indicator (BIAS), and Williams overbought/oversold indicator (WMS%R), are used (c.f. Table 1). Specifically, 13 technical features, which are MA_5, MA_10, MA_15, K_9, D_9, DIF, MACD_9, RSI_6, RSI_12, BIAS_6, BIAS_12, WMS%R_6, and WMS%R_12, are collected.

On the other hand, for analysis of the visual features of candlestick charts, the discrete wavelet transform (DWT) method is used, where each image is usually represented by the m-th low sub-band (LL_m) (Sundararajan, 2015). Here, each candlestick chart is transferred into a 3-scale Haar wavelet transform (HWT) and then related features are extracted from the LL_3 sub-band. As a result, we get k main pieces of information from each chart corresponding to the k consecutive trading days, each one comprised of an upper shadow, a lower shadow and a candle body.

Table 1. The seven technical indicators.

Abbreviation	Description
MA_n	nSUM = Sum of the closing prices of the n trading period MA_n = nSUM/n
KD	1. UP = the closing price on the t-th day – the lowest price of the n trading period 2. DN = the highest price of the n trading period – the lowest price of the n trading period 3. RSV = (UP/DN) × 100 4. K_n(t) = [2 × K_n(t – 1) + RSV(t)]/3 5. D_n(t) = [2 × D_n(t – 1) + K_n(t)]/3 K(0) and D(0) are all 50
DIF	1. DI = (2 × the closing price on the t-th day + the highest prices on the t-th day + the lowest prices on the t-th day)/4 2. EMA_12(t) = [10 × EMA_12(t – 1) + 2 × DI(t)]/12 3. EMA_26(t) = [24 × EMA_26(t – 1) + 2 × DI(t)]/26 4. DIF = EMA_12 – EMA_26 EMA_12(0) = the average of DI for 12 days; EMA_26(0) = the average of DI for 26 days
MACD	1. MACD(t) = [8 × MACD(t – 1) + 2 × DIF(t)]/10 MACD(0) = DIF(0)
RSI_n	1. UP = sum of upward changes in the n trading period 2. DN = sum of downward changes in the n trading period 3. RS = UP/DN RSI_n = 100 – [100/(1 + RS)]
BIAS_n	1. UP = the price on the t-th day – the average price for the n trading period 2. DN = the average price for the n trading period BIAS_n = UP/DN
W%R_n	1. $H(n)$ = the highest prices for the n trading period 2. $L(n)$ = the lowest prices for the n trading period 3. C = the closing price on the t-th day W%R_n = 100 – [($H(n)$ – C)/($H(n)$ – $L(n)$)] × 100

According to Tsai and Quan (2014), the extracted features consist of texture-based and location-based features. For simplicity, we set i_x as the location of the x-th main-information ($x = 1, 2, ...,$ and k), while $j_{x,1}$ to $j_{x,y}$ indicate the length of the shadow line, and $k_{x,1}$ to $k_{x,y}$ indicate the length of the candle body. In addition, we extract 4 pixel values from the texture-based features, which are the average pixel values of the whole shadow line (AS_x) as follows:

$$AS_x = \sum_{r=1}^{y} LL_3(i_x, j_{x,y})/y \tag{1}$$

the pixel values of the highest location of the upper shadow (US_x)

$$US_x = LL_3(i_x, j_{x,y}) \tag{2}$$

the lowest location of the lower shadow (LS_x)

$$LS_x = LL_3(i_x, j_{x,y}) \tag{3}$$

and the mid-point of the candle body (CBS_x)

$$CBS_x = LL_3\left(i_x, \frac{|j_{x,y} - j_{x,1}|}{2}\right) \tag{4}$$

In addition, we extract the positions of the upper shadow (ULP_x) and the lower shadow (LLP_x) to represent the location-based features. Consequently, in a candlestick chart, there are 6 different values used to describe the information for each trading day. In other words, each candlestick chart is represented by $6 \times k$ features. Since Tsai and Quan (2014) found the best visual representation to be obtained with a chart containing 30 trading days, so k is set to 30 in this paper, which results in a 180-dimensional feature vector for each candlestick chart.

Consequently, for each of the ten electronics companies and the TAIEX, there are three different types of datasets, containing technical (T), visual (V), and technical + visual (T + V) features, respectively. Moreover, each of these features is associated with a class label, indicating that the stock price is either up or down. In this paper, although there are short-, mid-, and long-term predictions, for simplicity, we only consider the prediction of the next day's stock price.

In terms of classifier design, four different classifiers are constructed, which are the C4.5 decision tree, k-nearest neighbor (KNN), multilayer perceptron (MLP) neural network, and support vector machine (SVM) classifiers.

3.2 Results

Tables 2, 3, 4, and 5 list the classification results of the four classifiers using technical (T), visual (V), and combined technical and visual (T + V) features, respectively. As we can see, on average, the classifiers performed better when a combination of technical and visual features was used than for technical or visual features alone. In addition, the results are consistent with those of Tsai and Quan (2014), indicating that the visual feature method has more discriminative power than do the technical indicators for stock prediction. Specifically, the best classifier is MLP, with SVM performing second best.

Figure 2 shows the performance of the four classifiers by the T, V, and T + V features for TAIEX. Again, these results also show that the T + V feature classifier outperforms the other two features. For the prediction of TAIEX, the SVM classifier performs the best, whereas MLP is the second best.

Table 2. Classification performance of the C4.5 classifiers.

	T	V	T + V
adi	73.29%	86.56%	86.81%
altr	84.32%	85.44%	86.89%
amat	89.22%	87.85%	89.46%
amd	81.58%	85.68%	87.45%
glw	74.58%	84.15%	85.44%
hrs	80.53%	83.11%	83.11%
intc	82.30%	82.46%	87.53%
msi	76.27%	86.48%	88.50%
mu	81.58%	84.23%	89.46%
txn	84.15%	84.55%	84.88%
Avg	80.78% (3)	85.05% (2)	86.95% (1)

Table 3. Classification performance of the KNN classifiers.

	T	V	T + V
adi	63.07%	68.30%	71.84%
altr	73.87%	73.93%	76.27%
amat	73.05%	76.91%	79.65%
amd	65.16%	71.68%	75.38%
glw	67.10%	69.59%	72.33%
hrs	74.58%	74.98%	76.19%
intc	67.34%	70.23%	73.85%
msi	61.54%	72.89%	75.70%
mu	65.81%	72.49%	71.60%
txn	65.65%	77.55%	77.80%
Avg	67.72% (3)	72.86% (2)	75.06% (1)

Table 4. Classification performance of the MLP classifiers.

	T	V	T + V
adi	85.20%	90.59%	91.15%
altr	85.29%	90.51%	91.71%
amat	81.74%	91.23%	90.83%
amd	77.55%	89.54%	89.46%
glw	79.73%	88.42%	87.61%
hrs	83.02%	91.55%	91.47%
intc	83.99%	89.94%	90.43%
msi	70.80%	90.67%	91.15%
mu	81.66%	91.47%	91.15%
txn	82.06%	89.30%	89.22%
Avg	81.10% (3)	90.32% (2)	90.42% (1)

Table 5. Classification performance of the C4.5 classifiers.

	T	V	T + V
adi	87.85%	87.77%	87.45%
altr	87.78%	86.56%	87.45%
amat	88.74%	86.32%	90.59%
amd	86.32%	86.89%	87.45%
glw	87.05%	87.45%	88.01%
hrs	87.61%	88.50%	90.10%
intc	83.19%	86.48%	88.09%
msi	81.01%	83.75%	87.45%
mu	83.19%	88.33%	89.22%
txn	86.40%	88.66%	89.30%
Avg	85.91% (3)	87.07% (2)	88.51% (1)

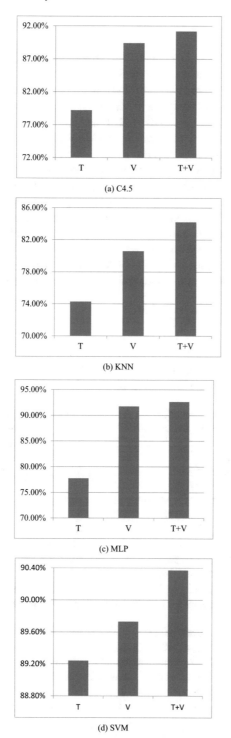

(a) C4.5

(b) KNN

(c) MLP

(d) SVM

Fig. 2. Classification performance of C4.5, KNN, MLP, and SVM for TAIEX.

4 Conclusion

In this paper, we demonstrate that combining different types of features, including technical indicators and the visual wavelet features of candlestick charts can better 'describe' the content information and predict the up and down movement of the stock price. Specifically, the combination of technical and visual features allows better performance for four different classifiers (i.e., C4.5, KNN, MLP, and SVM) than that obtained using technical indicators or visual features alone for predicting the stock prices of ten electronics companies in Taiwan and the price index of the Taiwan Stock Exchange (TAIEX).

Although the results are promising, there are several issues that could be investigated in the future. First, since we only consider the prediction of the next day's stock price, other prediction tasks including mid- and long-term predictions, such as predicting the next 7 days', one month's, and quarter's stock prices should be examined. Second, as the dimensions of the combined features are certainly large, it would be better to perform feature selection to reduce the dimensionality and examine the prediction performance of the reduced features. Third, using other stock markets and longer periods of training and testing sets could allow us to better understand the prediction performance of these three types of features.

References

1. Ji, G., Yu, J., Hu, K., Xie, J., Ji, X.: An adaptive feature selection schema using improved technical indicators for predicting stock price movements. Exp. Syst. Appl. **200**, 116941 (2022)
2. Nti, I.K., Adekoya, A.F., Weyori, B.A.: A systematic review of fundamental and technical analysis of stock market predictions. Artif. Intell. Rev. **53**, 3007–3057 (2020)
3. Jordan, B., Miller, T., Dolvin, S., Yuce, A.: Fundamentals of Investments, 3rd edn. McGraw-Hill, New York (2012)
4. Edwards, R.D., Bassetti, W.H.C., Magee, J.: Technical Analysis of Stock Trends, 10th edn. CRC Press, Boca Raton (2012)
5. Sarala, V., Phani Bhushan, G.N.V.: Stock market trend prediction using k-nearest neighbor (KNN) algorithm. J. Eng. Sci. **13**(8), 249–256 (2022)
6. Nurrimah, N., Rustam, Z.: Stock price trend prediction method based on support vector machines with Fisher score. Int. Symp. Curr. Prog. Math. Sci. 030014 (2019)
7. Dash, R., Dash, P.K.: A hybrid stock trading framework integrating technical analysis with machine learning techniques. J. Financ. Data Sci. **2**(1), 42–57 (2016)
8. Sezer, O.B., Ozbayoglu, M., Dogdu, E.: A deep neural-network based stock trading system based on evolutionary optimized technical analysis parameters. Procedia Comput. Sci. **114**, 473–480 (2017)
9. Tsai, C.-F., Quan, Z.-Y.: Stock prediction by searching for similarities in candlestick charts. ACM Trans. Manag. Inf. Syst. **5**(2), 9 (2014)
10. Chen, J.-H., Tsai, Y.-C.: Encoding candlesticks as images for pattern classification using convolutional neural networks. Financ. Innov. **6**, 26 (2020)
11. Jin, G., Kwon, O.: Impact of chart image characteristics on stock price prediction with a convolutional neural network. PLoS ONE **16**(6), e0253121 (2021)

12. Abarbanell, J.S., Bushee, B.J.: Fundamental analysis, future earnings, and stock prices. J. Account. Res. **35**(1), 1–24 (1997)
13. Tsai, C.-F., Hsiao, Y.-C.: Combining multiple feature selection methods for stock prediction: union, intersection, and multi-intersection approaches. Decis. Supp. Syst. **50**(1), 258–269 (2010)

Intelligent Health Promotion: Machine Learning in the Prevention of Stress-Related Diseases

Gabriel Fernandes Silva[1], Victor Ströele[1]([⊠]) [iD], Regina Braga[1] [iD], Mário Dantas[1] [iD], and Michael Bauer[2] [iD]

[1] Computer Science Postgraduate Program, Federal University of Juiz de Fora, Juiz de Fora, Brazil
gabrielfernandes.silva@estudante.ufjf.br,
{victor.stroele,regina.braga,mario.dantas}@ufjf.br
[2] Department of Computer Science, The University of Western Ontario, London, ON, Canada
bauer@uwo.ca

Abstract. Diseases caused by slow and progressive damage are the leading cause of mortality. The stress experienced throughout the day can cause many illnesses, as it is responsible for diminishing the body's defenses. In such cases, prevention is a fundamental component achieved through monitoring individuals, usually through a process that heavily depends on human intervention. Therefore, developing solutions capable of automated monitoring becomes necessary to assist individuals in their daily lives. This work aims to promote individuals' health by monitoring them through smart wearable devices and providing notifications that enable them to learn more about themselves. The work focuses on developing a computational environment composed of wearable devices and an application integrated with a machine-learning model. This model predicts the user's heart rate data and generates notifications accordingly. The results show that real-time user monitoring is possible, and moments of stress can be identified using machine learning, leading to generating notifications.

Keywords: Health Promotion · Machine Learning · User Monitoring · Stress

1 Introduction

Progress in medicine has redirected public health priorities, with infectious diseases being overtaken by progressive illnesses such as cancer, diabetes, and cardiovascular disease. As highlighted by Bauer [1], daily habits, especially stress, play a crucial role in developing these conditions that are typically monitored by a healthcare provider, which can require frequent visits and ongoing costs. Contemporary technology provides a less invasive approach to prevention, allowing continuous health monitoring without requiring substantial daily effort [4].

Supported by UFJF.

Prevention is fundamental to health promotion and is facilitated by early diagnosis carried out by the primary health sector. Regular monitoring, such as visits by health workers, allows gathering information for more assertive preventive measures and is beneficial for people with comorbidities [4,6].

This study proposes an innovative real-time monitoring system, using smart wearable devices to collect vital signs and related data, such as heart rate and the user's location. Our system uses a machine learning model to identify instabilities in vital signs through the data collected, generating relevant notifications for users. The approach aims to promote self-knowledge and encourage healthy habits.

The challenges faced by our proposal include the consistent extraction of data from sensors, offline monitoring of the user (when the Internet connection is lost), the optimization of the machine learning model, and the implementation of an efficient architecture for processing large volumes of data. According to Pereira [8] and Creswell [2], this paper's research is classified as exploratory, bibliographical, and applied; it seeks to investigate whether real-time monitoring, combined with machine learning, can effectively generate useful notifications for personal health promotion.

The research question that guided the development of this work is: *Can real-time extraction of users' vital signs data using smart wearable devices, combined with a machine learning model, generate relevant notifications to help users promote their health?*

Considering the research question, the main objective of this work is to propose an architecture capable of processing the data collected from users' body sensors, enabling the offline first requirement. A computational environment is also proposed and implemented based on the architecture, consisting of an application for Android mobile devices connected to a smart bracelet, which provides data for training and processing a machine learning model to generate information to assist the user in health monitoring.

This paper is an evolution of Sergio's work [11], incorporating an application for automated and ubiquitous data collection and enriching the dimensionality of the data for a more robust identification of moments of stress using the user's location and the principle of *Offline First* in the application.

In addition to the introduction section, this work is organized as follows: Sect. 2 presents related works. Section 3 presents the materials and methods that detail the proposed architecture. Section 4 presents the results obtained, and Sect. 5 presents the conclusion based on the analysis of the results and future work.

2 Related Work

In order to retrieve papers related to this research, searches were carried out in the *Google Scholar* and *IEEE* databases using keywords such as "Machine Learning E-Health", "machine learning in health", "Machine Learning Heart Rate", "IoT in health", "real-time monitoring".

Batista and Filho [5] present the panorama of *Machine Learning* techniques in the health area. It was developed to use classification algorithms to identify the presence of cardiovascular diseases from the patient's clinical data, with continuous and categorical data such as age and whether the patient is a smoker. The synthetic data created by [3] was used to develop the predictive models. Several algorithms were tested, and *Gradient Boosted Trees* was the most efficient. In conclusion, the authors show that the algorithms made it possible to fit predictive models to a set of synthetic data on cardiovascular risk. A negative point is the lack of real data to evaluate the efficiency of the algorithms.

Sabic *et al.* [10] conducted a study to detect anomalies in heart rate data using machine learning models. Five algorithms were analyzed using two simulated databases, one with 0.5% anomalies and the other with 2.5% anomalies. According to the authors, the anomaly detection process requires an operational definition that assumes two truths: (i) anomalies are rare in occurrence, and (ii) anomalies are different from normal data. To improve model learning, it was stipulated that heart rate data outside the range of 60 to 100 bpm (beats per minute) should be considered an anomaly. The authors evaluated the *Local Outlier Factor (LOF)*, *K-Nearest Neighbors (KNN)*, *Support Vector Machines (SVM)*, *Isolation Forest (IF)* and *Random Forest (RF)* algorithms. After experimenting on the models, *KNN* and *SVM* performed best on the dataset with 0.5% anomalies, while *Random Forest* performed best on the 2.5% dataset. A negative aspect of the work was that the models were not evaluated on real data, even though the models worked well on the simulated data and in developing the models.

Site, Nurmi, and Lohan [7] gathered 67 studies on health monitoring using machine learning algorithms. Their study analyzed different types of sensors to obtain individuals' health data. These sensors were used to monitor the vital signs of people with chronic diseases, such as heart disease, diabetes, and blood disorders, among others. Most of the data was obtained through sensors such as accelerometers, EEG (Electroencephalogram) monitors, ECG (Electrocardiograms), and wearable devices/*smartwatches*. The authors show that using body devices to extract data related to users' vital signs is feasible.

Risch *et al.* [9] used machine learning models to predict whether an individual was infected with the COVID-19 virus or healthy, using a Recurrent Neural Network (RNN) with long short-term memory (LSTM) cells for binary classification. The results showed the efficiency of using machine learning in healthcare, predicting relevant information for more assertive decision-making. However, using RNN requires much computing power to train the model, which is a challenge for implementing monitoring systems.

Leão *et al.* [11] propose an architecture that allows data collected from body sensors to be processed by a machine learning model to generate notifications for the user when moments of stress are identified. The authors used the *Fog-Cloud* concept combined with the implementation of the *Lambda* architecture to overcome the challenges of developing a real-time monitoring system. This way, a mobile application collected data from a *smartwatch*. Collecting the user's heart rate data every 5 min sends it to a *Data Lake* service via an HTTP request implemented in a WEB API format. The storage service makes this data available via WEB API to another service where the data is processed to make predictions. Several machine learning models were used during the development of the work, with the *K-Nearest Neighbor Regressor (KNN)* generating the best results. A negative point cited by the authors is the limited number of users to whom the experiment was submitted. In addition, the authors suggested that future work could include using location data to increase the dimensionality of the data and enrich the model.

Table 1 shows the analysis of the related works detailed above, showing the characteristics considered relevant to the development of this work, namely, use of *Machine Learning*, identifying whether the study uses machine learning algorithms in its development, whether the study uses real-time monitoring as a premise, as prioritized in the work by [11], and whether the work compares different machine learning algorithms, analyzing the metrics and context applied, as seen in the works [7,10] and [5]. Finally, Table 1 shows whether the studies used real data or a simulated database. As previously stated, the research reported in this paper considers an extension of the work [11] by expanding the data to include user location data and improving data collection in an environment that does not provide an internet connection. The choice of machine learning model was also based on the work of [11] and [7].

Table 1. Comparison of related work

References	Year	Machine Learning	Real-Time	Comparison	Real data
Batista and Filho [5]	2019	✔	–	✔	–
Sabic *et al.* [10]	2020	✔	–	✔	–
Site, Nurmi and Lohan [7]	2021	–	–	✔	✔
Risch *et al.* [9]	2022	✔	–	–	✔
Sergio *et al.* [11]	2023	✔	✔	–	✔

3   Materials and Methods

Figure 1 shows a conceptual model of the architecture proposed in this work. This architecture is based on the *Edge-Fog-Cloud* paradigm, where wearable devices, such as *smartbands*, responsible for collecting heart rate values and sending them to the user's *smartphone* represent the edge layer or *edge*. On the other hand, the user's *smartphone* represents the fog layer, where the heart rate collected by the wearable devices persists until the environment allows it to be synchronized with the cloud. The same goes for location data, collected from the device's GPS and stored until a synchronization connection with the database server in the cloud takes place.

The database server, training, and storage of the trained models take place in the *cloud*. The *Data Lake* is responsible for storing all the data collected by the heart rate and GPS sensors. We chose a Data Lake because we do not yet have a standard model for storing sensor data. The data lake gives us greater flexibility, allowing the incorporation of other devices with different data models into the proposed solution.

From this data stored in the *Data Lake*, one model per user is trained weekly with a floating window of the last 30 days of data collection. In this way, the model training flow is responsible for fetching this data, pre-processing it, training it, and storing these models along with their respective performance metrics.

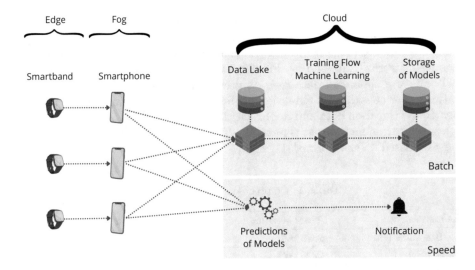

Fig. 1. Conceptual Architecture with its main components and the distribution of data flow considering the Edge-Fog-Cloud layers.

After this process, the models trained and stored in the cloud are available to make predictions of the user's heart rate. Location information was added to expand the dimensionality of the data so that the model fits the user's habits better. Thus, it is possible to group a user's data by location and time, obtaining the average, maximum, and minimum heart rate every minute of the day in all the locations where they were collected.

During the development of the model, a test *dataset* was assembled with 225,003[1] records of heart rate measurements, location, and time of data collection. The location data was obtained by exporting the data collected by *Google Maps* from the users' devices. The heart rate data was obtained by exporting the data from *Zepp Life*, which is the application provided by *Xiaomi* to connect with the smart bracelet *Mi Band 5*. This data was used to test the model and

---

[1] https://github.com/gabrielfsil/dataset-heartrate-location.

understand real data distribution. After exporting this data, it was unified by time to create a *dataset* with the desired dimensionality.

Figure 2 shows the distribution of the heart rate collected in the *dataset*, where you can see *outliers* defined as frequencies above 117 beats per minute.

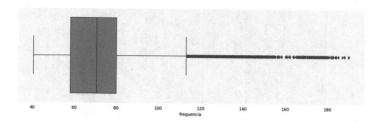

Fig. 2. Boxplot of the dataset selected for testing.

Figure 3 shows the behavior of the data about the latitude and longitude collected, represented on the X and Y axis, and the Z axis shows the heart rate value, so it is possible to see where some *outliers* are located, showing a high heart rate.

This *dataset* of tests was fundamental to understanding how localization, i.e., a context data type, can enrich the data and improve the model's functioning. During this experiment, the *KNeighborsRegressor (KNN)* was used, and a Coefficient of Determination (R^2) and Mean Square Error (MSE) of 0.977 and 2.129, respectively, were observed for the model.

The KNN model was chosen in this initial test because of the need to use machine learning models with low memory consumption and processing time in this initial experiment. However, the architecture can change the ML model according to the features presented in the data set. The ML model selection is based on metrics selected by the developer.

In this context, it is important to highlight how the model works to identify moments of user stress. Training the model with the user's data identifies the behavior pattern that is specific to that user. Once trained, the model is made available to make heart rate predictions. When this predicted value deviates significantly from the actual value collected by the sensor, a moment of possible stress is characterized, and a notification is sent to the user. Thus, when the model makes a heart rate prediction lower than the actual value collected, it indicates that the user's heart rate is higher than the behavior pattern identified by the model.

The *fog* was implemented by an application to collect user data in real-time. The data was collected from a smart bracelet, which makes up the *edge*, connected to the app via a BLE (Bluetooth Low Energy) wireless connection. The application was developed in *React Native* with the principle of *offline first*, which allows the user to stay *offline* and continue monitoring the data collected by the bracelet. This data is stored in the device's database, implemented using

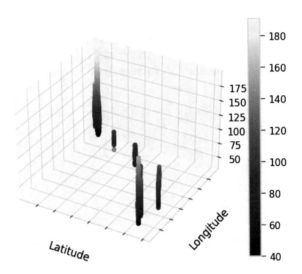

Fig. 3. Test dataset distribution concerning location. The X and Y axes represent latitude and longitude, and the Z axis represents the heart rate value.

Realm. This tool allows the data to be stored on the user's *smartphone* until an internet connection is established so that synchronization can start automatically. Another feature implemented in the application is a background task, which searches for the configured bracelet to re-establish the connection if lost approximately every 10 min. This task is also responsible for updating the *access token*, which expires in 30 min, from a *refresh token*, which expires in 60 days, to allow synchronization to continue without the user having to authenticate again in that 60-day interval.

The data is synchronized flexibly with a *MongoDB* database. *Realm* starts a session to synchronize the new data collected and does so according to the connection specifications and the number of records to be inserted. This strategy allows the synchronization process not to be overloaded by a large number of records or network instability.

The model training flow occurs after all the data collection and synchronization stages. This service was implemented using *AirFlow*. Every week, a task is executed in *AirFlow*, which searches for the data collected over the last 30 days, pre-processes it, and trains the models. The model is trained with the help of *MLFlow*, which stores the model along with the observed performance metrics. *AirFlow* and *MLFlow* use a *PostgreSQL* database to record the results obtained in the tasks, the performance metrics, and the trained model.

The trained model is made available to the user. Thus, according to the output generated by the model, a comparison is made with the actual frequency. If the frequency is significantly lower than that collected by the sensors, the application generates a notification for the user. Figure 4 illustrates the computing environment proposed for this use case.

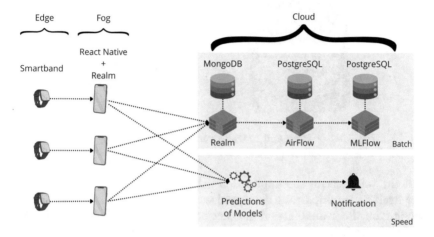

Fig. 4. Technologies used to implement the architecture.

4   Experimental Results

The application[2], named Olivia, was developed using *React Native* and is responsible for authenticating users and connecting to the smart bracelet, where heart rate data is collected and alerts the user if an unexpected pattern in heart rate is identified. The app requires some permissions from the user to work properly, such as access to location in the background, permission to run in the background, and access to *Bluetooth.*

A login screen was developed where the user provided their email and password to access the application. After this step, they are directed to the connection monitoring screen, where they connect the bracelet to take heart rate measurements (Fig. 5). The user clicks "Connect Device" on this screen, and the app will list the nearby devices. When the user selects the desired device, the app connects via *BLE* and monitors the connected device's heart rate sensor. Location monitoring is based on the GPS of the user's *smartphone,* which informs the app whenever the user's location changes.

---

[2] https://github.com/gabrielfsil/olivia-mobile.

Fig. 5. User login and bracelet connection monitoring screen.

All the information the application collects is stored in a database on the device itself. It is periodically synchronized with the main database, *MongoDB*, depending on the availability of internet access. Suppose the connection to the bracelet is terminated for some external reason. In that case, the application runs a task in the background to search for the bracelet registered by the user and re-establish the connection.

Considering the results obtained in previous work [11] and the need to use machine learning models with low memory consumption and processing time, we used *KNeighborsRegressor (KNN)* in this article. The data collected by the application and stored in the main database were preprocessed to train the model. Preprocessing in this initial experiment consists of searching for heart rate and location data collected over the last 30 days and grouping them. The data was grouped by time (one-minute interval) and location (latitude and longitude, i.e., context data). Consider that the sensor monitored one user during the week. Imagine that the user didn't move around that week; he was bedridden. In this case, the grouping is done by the minutes of the day since the location doesn't vary. But if during that week, some days the user goes to work and others he doesn't, data will be grouped by time (every minute of the day) and by location (latitude and longitude). This way, we have records of the maximum, minimum, and average frequency per minute of the day when the user was monitored in a given location.

After preprocessing, the data is used to train a model for each user, with the user's average heart rate as the target value. During the experiment, the models were trained with data collected from volunteer users, which corresponded to 34,126 heart rates collected. After grouping them, 1,440 records were obtained. In this study, we did not treat gaps in data collection. Therefore, there is no guarantee that we will have a heart rate every minute of the day. We made this choice because the system is offline first, ensuring that data is sent even without an internet connection. Gaps only occur if the bracelet is switched off.

Concerning the performance of the implemented model, the Coefficient of Determination (R^2) and Mean Square Error (MSE) metrics were used. The model showed 0.828 and 21.913 in the R^2 and MSE performance metrics, respectively. Figure 6(a) shows the distribution of the model's behavior relating predicted values to actual values. The graph of a perfect model would be a diagonal line, the current model shows a distribution on this line with some points more dispersed on this diagonal.

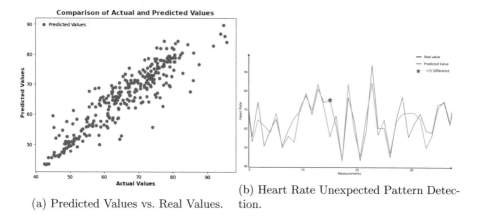

(a) Predicted Values vs. Real Values.

(b) Heart Rate Unexpected Pattern Detection.

Fig. 6. Machine Learning Model Results.

Interviews were conducted with medical professionals to identify a value for a heart rate to be significantly altered and perceived as an unusual pattern. During these interviews, the values for an altered rate ranged from 15 to 30 beats per minute above the individual's base rate. Therefore, following the suggestions of the medical partners in this project, we decided to consider 15 beats per minute above the predicted rate as a moment of stress for the user.

Figure 6(b) shows the behavior of the predicted values and the real values collected by the bracelet. When the difference between them is greater than 15 beats per minute (bpm), the point is marked with a star. Considering medical professionals' recommendations, this marked point represents an unusual pattern in the user's heart rate, generating a notification for the user.

After identifying moments of stress for the user, it is important to notify them. These notifications are essential for users to understand the events causing this stress in their daily lives. Thus, the notification of a stress event detected by the model would be as follows:

– *I noticed a significant variation in your heart rate on [date] at [time], while you were at [location]. This could indicate a moment of stress. We recommend that you take some time to relax and practice stress-reduction techniques.*

Therefore, the RQ: "Can real-time extraction of users' vital signs data using smart bracelets, combined with a machine learning model, generate relevant notifications to help users promote their health?" can be answered, supported by data extracted from user bracelets and the KNN processing. The user could monitor his/her heart rate through alerts provided by the Olivia APP. In this vein, specific actions can be taken by the user considering these alerts based on user and contextual data analysis and processing of an ML model. Therefore, based on evidence in this initial experiment, we can state that real-time processing of users' vital signs data, using a machine learning model, can generate relevant notifications to help users promote their health. This was made possible using personal data with contextual information processed by an intelligent algorithm. This combination can potentially improve the insights that can be gained and enable more data-driven concerns in application development.

4.1 Limitations

During the development of the work, limitations were identified in the implementation of the computing environment. One of these concerns the periodicity of background tasks on *Android* devices, whose execution depends on the availability of resources and other tasks in the operating system, according to the *Android* documentation, which indicates an interval between 10 and 15 min for such tasks.

Also, the experiment, conducted with a few users, makes it impossible to comprehensively evaluate the architecture's performance on a larger scale of data. In a scenario with a larger volume of data, potential bottlenecks could arise, especially in training the model in the cloud, which could become overloaded in the face of a significant increase in the amount of data collected by the devices, making it another essential limitation of the project.

5 Final Remarks

This study developed a computational environment responsible for monitoring users, using data collected through body sensors and location to notify the user's moments of stress detected by the machine learning model also developed in the study.

Through the experiments, the proposed architecture can monitor the user in real-time and identify moments of stress to generate a notification. The monitoring proved to be very efficient, withstanding moments of instability in the connections, such as remaining monitored even without an internet connection, and re-establishing the connection with the bracelet if some external factor, such as the distance between the devices, terminated it. Even with the limitations found in the functioning of the application, it presented a satisfactory result.

Based on the data collected by the sensors and synchronized in the architecture's main database, it was possible to validate the models' training *pipeline* for each user with satisfactory performance metrics. These results showed that the proposed architecture can identify moments of user stress based on heart rate and location.

Future work includes new approaches to increase the efficiency and impact of the solution. Among these is the implementation of the predictive model on the user's device to allow notifications even when they are *offline*, along with a periodic update of the model used on their device. Another approach that could contribute to the evolution of this work is monitoring other indicators relating to the user's health, such as temperature, number of steps, and oxygen saturation, among others. To improve the current work, it is also interesting to analyze the behavior of other machine learning models to make user predictions. In order to expand the universe of user profiles for the system, it would also be interesting to use federated learning techniques based on the models trained for each user to prevent the model from adjusting to stress behavior.

References

1. Bauer, M.E.: Estresse. Ciência hoje **30**(179), 20–25 (2002)
2. Creswell, J.W., Poth, C.N.: Qualitative Inquiry and Research Design: Choosing Among Five Approaches. Sage publications, Thousand Oaks (2016)
3. Laderas, T., Vasilevsky, N., Pederson, B., Haendel, M., McWeeney, S., Dorr, D.A.: Teaching data science fundamentals through realistic synthetic clinical cardiovascular data. bioRxiv, p. 232611 (2017)
4. Massola, S.C., Pinto, G.S.: O uso da internet das coisas (iot) a favor da saúde. Revista Interface Tecnológica **15**(2), 124–137 (2018)
5. de Moraes Batista, A.F., Chiavegatto Filho, A.D.P.: Machine learning aplicado à saúde. Sociedade Brasileira de Computação (2019)
6. Norman, A.H., Tesser, C.D.: Prevenção quaternária na atenção primária à saúde: uma necessidade do sistema único de saúde. Cad. Saude Publica **25**(9), 2012–2020 (2009)
7. Nurmi, J., Lohan, E.S.: Systematic review on machine-learning algorithms used in wearable-based ehealth data analysis. IEEE Access **9**, 112221–112235 (2021)
8. Pereira, A.S., Shitsuka, D.M., Parreira, F.J., Shitsuka, R.: Metodologia da pesquisa científica. UAB/NTE/UFSM (2018)
9. Risch, M., et al.: Investigation of the use of a sensor bracelet for the presymptomatic detection of changes in physiological parameters related to covid-19: an interim analysis of a prospective cohort study (covi-gapp). BMJ Open **12**(6), e058274 (2022)
10. Šabić, E., Keeley, D., Henderson, B., Nannemann, S.: Healthcare and anomaly detection: using machine learning to predict anomalies in heart rate data. AI Soc. **36**(1), 149–158 (2021)
11. Sergio, W.L., di Iorio Silva, G., Ströele, V., Dantas, M.A.: An architecture proposal to support e-healthcare notifications. In: Barolli, L. (ed.) AINA 2023. LNCS, vol. 661, pp. 157–170. Springer, Heidelberg (2023). https://doi.org/10.1007/978-3-031-29056-5_16

RsViT – A Scalable Vision Transformer Backbone for Diffusion Model

Thanawin Sakpuaram and Chantana Chantrapornchai[✉]

Department of Computer Engineering, Faculty of Engineering, Kasetsart University, Bangkok, Thailand
thanawin.sak@ku.th, fengcnc@ku.ac.th

Abstract. Denoising diffusion probabilistic models and vision transformers have shown tremendous progress in the image-generative field in recent years. This research proposes a more scalable vision transformer-based backbone for a diffusion model called RsViT. While RsViT has a similar architecture to U-Net, a hierarchical symmetrical encoder-decoder architecture. RsViT's primary mechanism is more similar to a transformer than CNNs. Instead of the ResNet block, RsViT uses Restormer's transformer block, which eliminates the weaknesses of vision transformer: the computational overhead of the attention map and the lack of localization, making the model more scalable and superior in generative tasks. From our experiments, on average, RsViT achieves more PSNR and SSIM than the UNet-based backbone by 0.12 and 0.0032 on 256×256 ImageNet-100 dataset in restoration tasks.

1 Introduction

Diffusion model has shown its excellent performance on image generative tasks such as image restoration [7,9] and image synthesis [6,13], beating GANs and achieving a state-of-the-art performance [4]. Diffusion model was initially introduced with a CNN-based backbone, U-Net [11]. However, recently, Vision Transformer has shown that a Transformer-based approach can rival or even outperform CNN-based approaches by achieving comparable or superior result in image classification [5]. This suggests that a Transformer-based approach could a viable backbone network.

Despite the success ViT has shown in image classification, a vanilla ViT has some drawbacks, notably the quadratic growth of computational overhead to the spatial resolution of the image due to the self-attention mechanism used in transformer. With higher-resolution images, finding hardware that can satisfy the memory and computation requirements will be a bigger problem.

In this paper, we propose a vision transformer-based architecture for a diffusion model called RsViT. RsViT alleviates the weakness of self-attention in ViT [5], which is the quadratic expansion of attention map to the spatial resolution of an image, by employing an improved transformer block from Restormer [17] allowing the model to perform on high-resolution images, making the model scalable.

The rest of the paper is organized as follows. In Sect. 2, we briefly explain how the diffusion model works, how the Vision Transformer works, the advantages and disadvantages of CNNs and Self-Attention, and the related works in using a Vision Transformer as a diffusion backbone. In Sect. 3, we describe the proposed method. In Sect. 4,

© The Author(s), under exclusive license to Springer Nature Switzerland AG 2024
L. Barolli (Ed.): AINA 2024, LNDECT 201, pp. 302–312, 2024.
https://doi.org/10.1007/978-3-031-57870-0_27

we explain the experimental setup and the method used to evaluate the result. Lastly, we conclude our work in Sect. 5.

2 Backgrounds

2.1 Denoising Diffusion Probabilistic Models (DDPM)

DDPM [13] is a class of generative models that learns the parameterized Markov Chain to gradually remove Gaussian noises from the image. The model consists of 2 processes, the forward and the reverse process. The forward process is a fixed Markov Chain that gradually corrupts the image by adding Gaussian noises [4, 13].

$$q(x_t|x_{t-1}) = \mathcal{N}(x_t; \sqrt{1-\beta_t}x_{t-1}, \beta_t \mathbf{I}) \tag{1}$$

$$q(x_{1:T}|x_0) = \prod_{t=1}^{T} q(x_t|x_{t-1}) \tag{2}$$

$$q(x_t|x_0) = \mathcal{N}(x_t; \sqrt{\bar{\alpha}_t}x_{t-1}, (1-\bar{\alpha}_t)\mathbf{I}) \tag{3}$$

In Eq. (2) x_0 is the initial input and $x_{1:T}$ are the more noisy images. $q(x_{1:T}|x_0)$ is a conditional Gaussian distribution used to add noise to the image. The mean of this conditional Gaussian distribution changes according to the output of the previous timestep i.e., x_{t-1}, and the fixed variance (Eq. (2)). The noisy image at specific timestep can be generated without going through all the previous step by using Eq.(3) where $\alpha_t = 1 - \beta_t$, $\bar{\alpha}_t = \Pi_{s=1}^{t}\alpha_s$ and β_t is a variance schedule used to specify the amount of noise to be added to the image.

$$p_\theta(x_T) = \mathcal{N}(x_t; 0, \mathbf{I}) \tag{4}$$

$$p_\theta(x_{t-1}|x_t) = \mathcal{N}(x_{t-1}; \mu_\theta(x_t, t), \Sigma_\theta(x_t, t)) \tag{5}$$

$$p_\theta(x_{0:T}) = p(x_T)\prod_{t=1}^{T} p_\theta(x_{t-1}|x_t) \tag{6}$$

The reverse process aims to denoise the image or generate a less noisy version of the image by learning to predict the noise when given the noisy image with a timestep. The model learns the transition of a parameterized Markov Chain. This reverse process is defined as $p_\theta(x_{0:T})$. Starting the process with $p_\theta(x_T)$ which is Gaussian noise with a zero mean and variance of $\mathbf{I}$. as shown in Eq. (4) then the model predicts the transition $p_\theta(x_{t-1}|x_t)$ sequentially until it reaches x_0.

$$p_\theta(x_{0:T}) = p_\theta(x_T)\prod_{i=1}^{S} p_\theta^{(\tau_i)}(x_{\tau_{i-1}}|x_{\tau_i}) \times \prod_{t\in\bar{\tau}} p_\theta^{(t)}(x_0|x_t) \tag{7}$$

The reverse process can be accelerated by the method of DDIM [13]. By using a deterministic sampling method, the model can skip some of the steps instead of iteratively going through every time step which greatly reduces the sampling time. With the DDIM method, we can sample image on a subset $x_{\tau_1}, ..., x_{\tau_s}$, where τ is a sub-sequence of the original timesteps [1,...,T]

2.2   Vision Transformer

Transformer [14] was first introduced for sequence transduction tasks such as natural language processing. By utilizing the concept of self-attention, the transformer was able to compete with other state-of-the-art language models, e.g., LSTM and RNN.

Vision Transformer (ViT) [5] is a pure transformer model that can challenge or outperform CNNs in various vision tasks. Since the transformer is used for sequence-like input, the image, which is two-dimensional ($3 \times H \times W$ in case of RGB image), must be transformed into a one-dimensional sequence of token ($N \times D$) to be able to be fed into Transformer.

ViT splits an input image $x \in \mathbb{R}^{H \times W \times C}$ into flattened patches $x_p \in \mathbb{R}^{N \times (P^2 C)}$ where (H, W) is the size of an input image, C is the number of the channels, P is the patch size and $N = \frac{HW}{P^2}$ is the number of patches. ViT then maps the patches to $\mathbf{z} \in \mathbb{R}^{N \times D}$ by using linear projection and adding position vector to the vector $\mathbf{z}$ resulting in a sequence vector $\mathbf{z}_0$ that can be applied directly to transformer. This $\mathbf{z}_0$ is then normalized before going through the multi-head attention and MLP layer. Lastly, the class token is fed into the MLP head to produce the classification output.

2.3   Convolution vs Self-attention

Convolution is the primary mechanism of CNNs. It has a kernel that goes through the whole image and converts all the pixels in its receptive field to a single value. CNNs gradually look at the image in a small region until it reaches the end of the image. This convolution operation makes it so that the model can learn local connection and translation equivariance. However, it also negatively affects the model, which is the lack of global relation from the limited receptive field and static weights at sampling time.

On the other hand, the transformer calculates the attention value between each position in the sequence despite the distance [5, 14]. In other words, the model is capable of looking at the image as a whole, not just in a small region. This helps the model learn global dependencies that CNNs lack. Despite the said advantage, conventional self-attention has a huge computational overhead where the relation between computational complexity and the image resolution is a quadratic equation $O(H^2 W^2)$. Furthermore, relying only on the self-attention mechanism makes the model lack localization, which is important for fine-grained details within the image.

2.4   Vision Transformer as a Diffusion Backbone

There have been several research studies on using Vision Transformer as a diffusion backbone. GenViT [16] uses vanilla ViT as a backbone for DDPM. Similar to U-Net used in DDPM, the time embedding is fed to each layer in ViT separately. The result shows that GenViT performed worse than the UNet-based DDPM when performing image synthesis on the CIFAR10 dataset. In the image generative task, it is essential that the model can learn and generate the fine-grained details of an image. A vanilla ViT that only incorporates the attention mechanism to model lacks localization, which makes it hard to generate complex and fine-grained details in the image.

U-ViT [1] improves the vanilla ViT by adding skip connections between shallow and deep layers of the transformer block. Furthermore, U-ViT adds a convolution block before the output to work as a refinement block. Instead of incorporating time embedding by feeding them to each layer like how UNet-based DDPM and GenViT do, U-ViT treats time embedding as a token, like how a patch of image is treated. U-ViT demonstrates that their ViT-based diffusion model can perform on par with UNet-based diffusion models in class-conditioned and unconditional image synthesis.

IU-ViT [2] improves U-ViT performance by adding a depth-wise convolution block to the feed-forward network, which helps alleviate the lack of localization, making the model perform better with small details of an image. Moreover, IU-ViT also proposes a method to reduce information loss when combining all the image patches back to the full image by removing the last linear projection layer that was used to reduce the dimension of the data and rearranging the data first instead.

ASCEND [2] suggests that for image generative tasks, the model with the fixed scale, like ViT, could perform worse than the model with hierarchical architecture due to multiple scales of information. Taking the issue of ViT having a fixed scale to the consideration, ASCEND proposes a ViT-based backbone with a hierarchical encoder-decoder architecture by using upsampling and downsampling and replacing the vanilla transformer block with Swin transformer blocks [8].

3   Methodology

3.1   Proposed Method

Our main goal is to develop a diffusion model with a transformer-based backbone that can perform as well as the conventional UNet-based diffusion model and overcome the 2 weaknesses of a vanilla ViT which are:

- **Scalabilty**. The conventional self-attention that was used in ViT grows quadratically to the resolution of the image making it hard to apply the vanilla ViT to high-resolution image.
- **The Lack of Local context**. Despite being able to find the relevance between long-distance data, in this case, the global interaction of the pixels, self-attention does not consider the local interaction, which makes generating fine-grained details in small regions of an image worse than CNNs.

Overall Architecture. Similar to U-Net [11], we employ a symmetrical encoder-decoder architecture for the backbone neural network. Each layer of encoder and decoder consists of multiple transformer blocks that are sequentially connected. The input image $x \in \mathbb{R}^{H \times W \times C}$ is first projected using convolution block to expand its dimension to feature $z \in \mathbb{R}^{H \times W \times D}$. In the encoder part, the image $x \in \mathbb{R}^{H \times W \times D}$ is fed through each transformer block in the layer. After that, it is down-sampled by half of its spatial resolution while the dimension is doubled $x \in \mathbb{R}^{H/2 \times W/2 \times 2D}$. This process of encoding is repeated until the input reaches the latent layer where the down-sampling is stopped, keeping the same shape $z_L \in \mathbb{R}^{H/2^L \times W/2^L \times 2^L D}$ where L is the number of layers. The decoder then takes in the latent feature, concatenates it with output from the same level of encoder output via a skip connection, reduces the dimension, and subsequently sends them through the transformer blocks in its layer and up-samples it by reducing the dimension and increasing the spatial resolution. The model repeats the decoding process until the output has the same shape as the input. Finally, the output of the last decoder is fed to the refinement block, which has the same components as the encoder and decoder, as shown in Fig. 1.

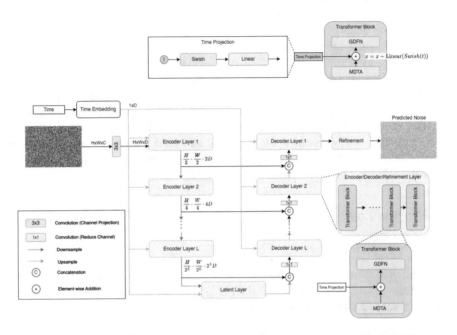

Fig. 1. Overall architecture of our model. Each layer consists of multiple Restormer transformer blocks. Initially, the input goes through the convolution block to expand its dimensions (channels) from $x \in \mathbb{R}^{H \times W \times C}$ to $x \in \mathbb{R}^{H \times W \times D}$. Then the input is passed through each encoder and then gets down-sampled after that it is up-sampled and concatenated with its respective encoder output. The time embedding is fed to each transformer block separately.

Transformer Block. To help alleviate the self-attention mechanism's computational overhead and improve the transformer's localization, we instead use the Restormer transformer block[17], which consists of 2 main modules, MDTA and GDFN. MDTA calculates the attention score across the channels instead of the spatial resolution, reducing computational overhead. GDFN introduces the usage of gating mechanism and depth-wise convolution into the feed-forward network, allowing the model to focus on the fine details of an image.

Time Embedding. The time embedding is first generated by using sinusoidal position embedding, which is the method the transformer uses to create unique position embedding. Incorporating the time embedding into the model is done by feeding the time projection to each transformer block directly. Similar to how time embedding is fed in U-Net DDPM, where the time embedding is fed after the normalization and convolution block in the ResNet block, the time embedding first goes through the time projection block and then fed into the Resotrmer block after the normalization and attention block as shown in Fig. 2.

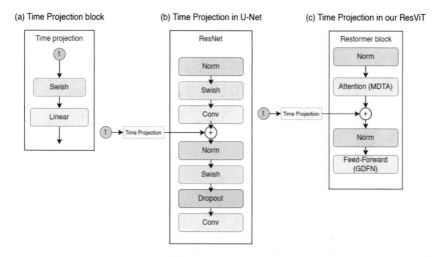

Fig. 2. (a) Time projection block. Time embedding goes through this block before going into the transformer block. (b) The method to feed time projection into U-Net-based backbone, where each layer contains ResNet block. (c) Our proposed method is to feed time projection.

4   Experiments

We evaluate the proposed RsViT by restoring the images in three tasks: super-resolution4x, deblurring, and inpainting, using DDRM [7] sampling algorithm. Some visualization of the results can be seen in Fig. 3. We compare the result to a UNet-based diffusion model (DDIM) [13] by using PSNR and SSIM as metrics for evaluating the model's performance. We also explore how some parameters affect the model's size and performance.

4.1   Experimental Setup

Datasets. We trained RsViT on 256×256 pixels ImageNet-100 [12], which is a subset of ImageNet [3], containing 100 classes with 130K training images (1300 images per class) 5K validation images and (50 images per class).

Configuration. We trained all of our models on NVIDIA Tesla V100-SXM2-32 GB. Each model is evaluated at 1M iterations. Multiple configurations of RsViT are shown in Table 1. The varied parameters are:

- **dim** - The dimension (channel) of an image after the first projection layer.
- **#heads** - The number of attention heads in one transformer block that can calculate the attention value in parallel.
- **n_#blocks** - The number of transformer blocks connected sequentially to each layer.
- **depth** - This indicates how many layers of encoder/decoder.

Table 1. Configurations of our initial three models

model	dim	head	n_block	depth	#params
RsViT-base	24	2	2	3	803,832
RsViT-med	48	4	3	4	1,404,0230
RsViT-large	64	8	4	5	116,107,096

4.2   Experimental Results

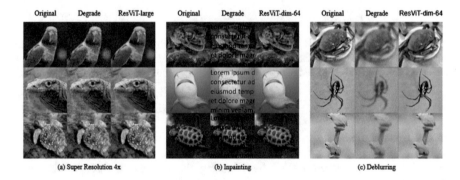

(a) Super Resolution 4x   (b) Inpainting   (c) Deblurring

Fig. 3. RsViT results on ImageNet-100 (a) Super-Resolution 4x (Noisy with $\sigma_y = 0.05$) (b) Inpainting (Noisy with σ_y) = 0.05 (c) Uniform Deblurring (Noisy with σ_y) = 0.05

Model Variants. The results in Table 2 show that overall our initial 3 configurations perform worse than DDIM. However, when comparing RsViT-med to DDIM, RsViT-med has 99M (81.61%) less number of parameters than DDIM while having 3.78% and 4.56% less PSNR and SSIM on average. Furthermore, when comparing RsViT-med to RsViT-large, The performance of RsViT-large is not far better than RsViT-med, while having 112M (87.93%) more parameters, on average of three tasks RsViT-large only has 0.34% and 0.12% increase in PSNR and SSIM. This implies that increasing some of the parameters might not result in a performance increase since the number of the parameters grows by a large margin while the performance barely improves. Thus, we need to explore how each parameter affects the model.

Exploring the Effect Each Parameter. We explore how each parameter affects the model's performance and size by finding **PIR (Performance Increase Ratio)**. First, We normalize PSNR and SSIM using min-max normalization to set them to the same scale by using results from RsViT-base as a baseline (min value). We then find PIR for each metric (PSNR and SSIM) $PIR^i_{metric} = \frac{(metric^i - metric^{base})}{\ln(params_i - params_{base})}$; $m \in \{PSNR, SSIM\}$, i is model variant, and average them. $PIR = (PIR^i_{PSNR} + PIR^i_{SSIM})/2$

Table 2. Comparison on Super-resolution4x, Deblurring, Inpainting on ImageNet-100 test set containing 5,000 images of 100 classes of our first three configurations and conventional U-Net-based backbone.

model	#params	time	Super Resolution		Deblurring		Inpainting	
			PSNR↑	SSIM↑	PSNR↑	SSIM↑	PSNR↑	SSIM↑
RsViT-base	803K	4	24.88	0.6637	22.90	0.8336	28.70	0.5853
RsViT-med	14M	9	25.29	0.6787	23.53	0.8438	29.15	0.6054
RsViT-large	116M	10	**25.32**	0.6786	23.68	0.8460	29.23	0.6061
DDIM (U-Net) [13]	113M	4	25.31	**0.6789**	**24.73**	**0.8952**	**30.95**	**0.6511**

Table 3. Comparison on Super-resolution4x, Deblurring, Inpainting on ImageNet-100 test set containing 5,000 images of 100 classes when the parameter **dim** is varied, and others are the same as the base model.

dim	#params	time	Super Resolution			Deblurring			Inpainting		
			PSNR↑	SSIM↑	PIR↑	PSNR↑	SSIM↑	PIR↑	PSNR↑	SSIM↑	PIR↑
48	3.1M	6	25.15	0.673	0.0492	24.48	0.0516	0.0563	30.78	0.8928	**0.0662**
64	5.4M	7	**25.27**	0.6737	0.0588	**24.99**	**0.0651**	**0.0648**	30.86	0.8935	0.0647
80	8.5M	8	25.26	**0.6761**	**0.0622**	24.8	0.0573	0.0594	**30.86**	**0.8942**	0.0631

From Table 3; 4, 5 and 6, It can be seen that compared to another parameter, increasing **dim** gives the most increase in performance while still maintaining a fair amount of PIR. In Table 4, we find that increasing **head** can help reduce the sampling time because of the parallelization when calculating the attention map. Still, it can be seen that the performance drops when **head** increases from 6 to 8. This could be because some of the heads are not relevant, i.e., some heads might generate irrelevant information [10, 15], leading to worse performance. From Table 5 and 6, increasing depth also increases the number of parameters by a large amount while the performance does not do so by much. On the other hand, n_block only affects the number of parameters by a little but also has little effect on the model's performance.

Table 4. Comparison on Super-resolution4x, Deblurring, Inpainting on ImageNet-100 test set containing 5,000 images of 100 classes when the parameter **head** is varied, and others are the same as the base model.

head	#params	time	Super Resolution			Deblurring			Inpainting		
			PSNR↑	SSIM↑	PIR↑	PSNR↑	SSIM↑	PIR↑	PSNR↑	SSIM↑	PIR↑
4	803,856	4	24.90	0.6652	0.0271	24.06	0.6319	0.1986	30.48	0.8872	**0.2688**
6	803,880	4	**25.05**	**0.6677**	**0.0980**	**24.61**	**0.6392**	**0.2113**	**30.58**	0.8882	0.2288
8	803,904	2	24.91	0.6658	0.0288	24.12	0.6333	0.1534	30.46	**0.8870**	0.1983

Table 5. Comparison on Super-resolution4x, Deblurring, Inpainting on ImageNet-100 test set containing 5,000 images of 100 classes when the parameter **n_block** is varied, and others are the same as the base model.

n_block	#params	time	Super Resolution			Deblurring			Inpainting		
			PSNR↑	SSIM↑	PIR↑	PSNR↑	SSIM↑	PIR↑	PSNR↑	SSIM↑	PIR↑
3	1M	5	25.08	0.6698	0.0405	24.56	0.6421	0.0667	30.63	0.8893	0.0730
4	1.3M	7	25.10	0.6682	0.0354	24.74	0.6433	0.0672	30.60	0.8892	0.0686
5	1.6M	7	**25.19**	**0.6723**	**0.0551**	**24.86**	**0.6484**	**0.0701**	**30.73**	**0.891**	**0.0698**

Table 6. Comparison on Super-resolution4x, Deblurring, Inpainting on ImageNet-100 test set containing 5,000 images of 100 classes when the parameter **depth** is varied, and others are the same as the base model.

depth	#params	time	Super Resolution			Deblurring			Inpainting		
			PSNR↑	SSIM↑	PIR↑	PSNR↑	SSIM↑	PIR↑	PSNR↑	SSIM↑	PIR↑
4	2.7M	5	25.11	0.6718	0.0429	24.63	0.6452	**0.0600**	30.61	0.8898	**0.0626**
5	10.2M	7	25.20	**0.6755**	**0.0552**	24.70	0.6487	0.0568	30.72	0.891	0.0586
6	39.8M	7	**25.22**	0.6744	0.0496	**24.93**	**0.6512**	0.0564	**30.78**	**0.8913**	0.0548

Optimizing the Model Parameters. We try to find the model configuration with the best possible performance that does not have too many parameters by selecting the value of parameters (dim, head, n_block, depth) with the most PIR. According to PIR, the configuration we get is *dim* = 64, *head* = 6, *n_block* = 5, *depth* = 5. However, since our model calculates attention across dimensions, the dimension of the input (dim) needs to be divisible by the number of heads to split the input to each head equally; We make a small change to *dim* = 66 resulting in *dim* = 66, *head* = 6, *n_block* = 5, *depth* = 5, we called this *RsViT-opt*.

Table 7 shows that in term of performance, RsViT-opt performs better than DDIM, having higher PSNR and SSIM in all 3 tasks however it has more parameters and is a lot slower than DDIM. Figure 4 depicts PSNR and SSIM for all of our experimented model variants.

Table 7. Comparison on Super-resolution4x, Deblurring, Inpainting on ImageNet-100 test set containing 5,000 images of 100 classes when the parameter **n_block** is varied, and others are the same as the base model.

model	#params	time	SR4x		Deblurring		Inpainting		avg.	avg.
			PSNR↑	SSIM↑	PSNR↑	SSIM↑	PSNR↑	SSIM↑	PSNR↑	SSIM↑
RsViT-large	116M	10	25.32	0.6786	23.68	0.6061	29.23	0.846	26.08	0.7102
dim-64	5.4M	7	25.27	0.6737	**24.99**	0.6505	30.86	0.8935	27.04	0.7392
head-6	803K	2	25.05	0.6677	24.61	0.6392	30.58	0.8882	26.75	0.7317
n_block-5	1.6M	7	25.19	0.6723	24.86	0.6484	30.73	0.891	26.93	0.7372
depth-5	10.2M	4	25.20	0.6755	24.7	0.6487	30.72	0.891	26.87	0.7384
DDIM	113M	4	25.31	0.6789	24.73	0.6511	30.95	0.8952	27.00	0.7417
RsViT-opt	147M	19	**25.39**	**0.6817**	24.89	**0.6553**	**31.07**	**0.8977**	**27.12**	**0.7449**

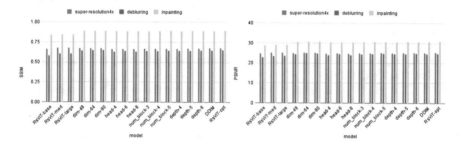

Fig. 4. PSNR and SSIM score of all RsViT configurations and DDIM in all three tasks.

5   Conclusion

In this paper, we propose a scalable vision transformer backbone for the denoising diffusion model called RsViT. RsViT has a hierarchical symmetric encoder-decoder architecture. With the use of Restormer block, the attention map is scaled linearly with dimension parameters, making attention map scales linearly with dimension parameters, and our main objective, building a scalable ViT-based backbone, is met. We add the time embedding to the model by changing the architecture of the Restormer block by adding time separately to each block. We experimented with three image restoration tasks: super-resolution 4x, Deblurring, and Inpainting. We introduce a way to find a balanced model in terms of performance and size (number of parameters) by using PIR. Our best model outperform a UNet-based diffusion model in all three experiments. However, RsViT's sampling time is much longer than that of the UNet-based approach. The sampling speed should be further studied and improved.

References

1. Bao, F., et al.: "All are worth words: a ViT backbone for diffusion models". In: 2023 IEEE/CVF Conference on Computer Vision and Pattern Recognition (CVPR). Los Alamitos, CA, USA: IEEE Computer Society, June 2023, pp. 22669–22679 (2023). https://doi.org/10.1109/CVPR52729.2023.02171.
2. Cao, H., et al.: "Exploring Vision Transformers as Diffusion Learners". arXiv:2212.13771 (2022). https://ui.adsabs.harvard.edu/abs/2022arXiv221213771C
3. Deng, J., et al.: "ImageNet: a large-scale hierarchical image database". In: 2009 IEEE Conference on Computer Vision and Pattern Recognition, pp. 248-255 (2009). https://doi.org/10.1109/CVPR.2009.5206848.
4. Dhariwal, P., Nichol, A.: Diffusion models beat GANs on image synthesis. Adv. Neural. Inf. Process. Syst. **34**, 8780–8794 (2021)
5. Dosovitskiy, A., et al.: "An image is worth 16x16 words: transformers for image recognition at scale". arXiv:2010.11929 (2020). https://ui.adsabs.harvard.edu/abs/2020arXiv201011929D
6. Ho, J., Jain, A., Abbeel, P.: Denoising diffusion probabilistic models. Adv. Neural. Inf. Process. Syst. **33**, 6840–6851 (2020)
7. Kawar, B., et al.: Denoising diffusion restoration models. Adv. Neural. Inf. Process. Syst. **35**, 23593–23606 (2022)
8. Liu, Z., et al.: "Swin transformer: hierarchical vision transformer using shifted windows". In: Proceedings of the IEEE/CVF International Conference on Computer Vision, pp. 10012-10022
9. Özdenizci, O., Legenstein, R.: Restoring vision in adverse weather conditions with patch-based denoising diffusion models. IEEE Trans. Pattern Anal. Mach. Intell. **45**(8), 10346–10357 (2023)
10. Qiu, D., Yang, B.: Text summarization based on multi-head self-attention mechanism and pointer network. Complex Intell. Syst. **8**(1), 555–567 (2022)
11. Ronneberger, O., Fischer, P., Brox, T.: U-net: convolutional networks for biomedical image segmentation. In: Navab, N., Hornegger, J., Wells, W., Frangi, A. (eds.) MICCAI 2015 Part III. LNCS, vol. 9351, pp. 234–241. Springer, Cham (2015). https://doi.org/10.1007/978-3-319-24574-4_28
12. AMBESH SHEKHAR. ImageNet100. Dataset (2021). https://www.kaggle.com/datasets/ambityga/imagenet100
13. Song, J., Meng, C., Ermon, S.: "Denoising diffusion implicit models". arXiv:2010.02502 (2020). https://ui.adsabs.harvard.edu/abs/2020arXiv201002502S
14. Vaswani, A., et al.: "Attention is all you need". In: Advances in Neural Information Processing Systems, vol. 30 (2017)
15. Voita, E et al.: "Analyzing multi-head self-attention: specialized heads do the heavy lifting, the rest can be pruned". In: Korhonen, A., Traum, D., M'arquez, L (eds.) Proceedings of the 57th Annual Meeting of the Association for Computational Linguistics. Ed. by . Florence, Italy: Association for Computational Linguistics, July 2019, pp. 5797-5808. https://doi.org/10.18653/v1/P19-1580, https://aclanthology.org/P19-1580
16. Yang, X., et al.: "Your VIT is secretly a hybrid discriminative-generative diffusion model". arXiv preprint arXiv:2208.07791 (2022)
17. Zamir, S., et al.: "Restormer: efficient transformer for high-resolution image restoration". In: Proceedings of the IEEE/CVF Conference on Computer Vision and Pattern Recognition, pp. 5728–5739 (2022)

A Deep Neural Network (DNN) Based Contract Policy on Hyperledger Fabric for Secure Internet of Things (IoTs)

Sabina Sapkota[1], Haojun Huang[1], Yining Hu[2(✉)], and Farookh Hussain[2]

[1] Huazhong University of Science and Technology, Hongshan, Wuhan, Hubei, China
hjhuang@hust.edu.cn
[2] University of Technology Sydney, 81 Broadway, Ultimo, NSW 2007, Australia
{yining.hu,farookh.hussain}@uts.edu.au

Abstract. IoT has revolutionized our interaction with the environment by enabling the connection of devices and sensors to the internet. However, the ubiquitous nature of IoT and the lack of a security framework result in various vulnerabilities particularly in data security. Blockchain of IoT (BIoT) has been studied recently to enhance security in IoT networks. However, this integration results in issues with network throughput, storage, and transaction security. To address these challenges, this paper proposes a deep neural network (DNN)-based contract policy to detect anomalous transactions upon their entry to the network and an encoding method to compress validated transactions. The proposed contract policy detects anomalies in IoT transactions, allows transactions to be stored in an encoded format, enhancing the security of an IoT network while minimizing storage space when storing IoT transactions. The proposed DNN-based algorithm rejected 2790 anomalous transactions out of 119,103 transactions and achieved a compression ratio of 47.81 for encoded transactions. Furthermore, we implemented the proposed solution on Hyperledger Fabric as an chaincode's contract policy.

1 Introduction

IoT has emerged as an essential technology that significantly enhances the quality of life and services for humankind. According to Cisco, by 2030, up to 500 billion IoT devices will be connected to the internet [1]. An IoT network contains a combination of sensors, hardware, and software working together to share data over networks [2]. These devices are limited in size and have low processing capability, often relying on third parties such as cloud and edge computing to perform advanced tasks [3]. The architecture of IoT networks is divided into three layers: device, network, and application. At the device layer, sensors collect data from the surroundings. The network layer transmits this collected data and ensures secure communication between devices. Finally, the application layer receives user requests, processes them, and delivers the relevant information. However, due to a lack of standards for compact and heterogeneous IoT devices, manufacturers often neglect security to reduce costs, resulting in security issues. In the application layer, issues include insecure interfaces, firmware vulnerabilities, and security risks to middleware affecting data privacy and access control [4]. The

L. Barolli (Ed.): AINA 2024, LNDECT 201, pp. 313–325, 2024.
https://doi.org/10.1007/978-3-031-57870-0_28

network layer faces issues such as data congestion, eavesdropping, and insecure node discovery [4]. The physical layer struggles with limited resources, low power, and the absence of high-level encryption [5]. Moreover, IoT devices rely on a centralized architecture where a single authority controls all network activities, including authentication, storage, and network connectivity [5]. Additionally, the trust between devices relies on centralized third-party services. For example, many IoT devices depend on cloud-based services for authentication and data storage. The failure of the central infrastructure will impact all connected devices, and the number of IoT devices allowed to connect is limited.

More recently, researchers have explored BIoT to address the security, trust, scalability, and centralization challenges of IoT [6,7]. Blockchain eliminates the need for third-party authentication. It provides IoT device identity services through public key cryptography, with all peer nodes in the network verifying identities before approving transactions. Establishing a peer-to-peer (P2P) decentralized network ensures secure data storage between devices in various peer nodes, hence the availability of IoT services, even in the event of node failures. Furthermore, smart contracts can automate conditional transactions [8]. To utilize the features of blockchain beyond cryptocurrency, a permissioned, tokenless blockchain platform has been designed for enterprise-based applications, known as Hyperledger Fabric [9]. Fabric features a modular design, enabling components to be independently developed, updated and replaced, thereby introducing flexibility and scalability to blockchain-based applications. Within Fabric, the tasks of transaction validation and block creation are divided into three phases: endorsing, ordering, and validating. Fabric also allows users to define endorsement policies and develop smart contracts in general-purpose programming languages. These features make Hyperledger Fabric particularly suitable for IoT applications [10].

Nonetheless, the BIoT architecture is also susceptible to various types of attacks. In blockchain, there is no mechanism to detect whether transactions from registered devices are secure [11]. Consequently, an IoT device under the control of an attacker can send malicious transactions, involving harmful actions such as unauthorized transfers, data manipulation, or disruptions to the network's normal operations. By controlling the device, the attacker gains significant power over the network, enabling the manipulation and acceptance of malicious transactions. Validating and storing transaction data by multiple nodes in its original form demands significant storage space and energy, and raise concerns about data security due to the data's visibility to anyone on the network. In Fabric, peers endorse a transaction (usually several kilobytes (kb) in size) and send it to the client. The client forwards the collected endorsement along with the whole transaction to the ordering node, and from there, a block of transactions is sent to the validating nodes. This creates significant communication overhead, impacting overall network performance [12].

This paper proposes a framework based on Fabric, where a DNN-based endorsement node is used for securing IoT transactions during communication and storage. The framework works by detecting anomalies in incoming transactions, eliminating malicious transactions, and encoding legitimate transactions before storing them in the shared ledger. We train the DNN model with the labeled dataset so that it can detect anomalous transactions. The proposed framework undergoes different consensus levels across the IoT layers, including identity authorization at the device layer, decentralized network management, network traffic detection at the network layer, and anomaly detection and transaction encoding based on smart contracts at the application layer. A long-short-term memory (LSTM) autoencoder model is employed to detect anomalies and encode validated transactions. The proposed approach aligns with the Confidentiality, Integrity, and Availability (CIA) security triad model, maintaining confidentiality through public key cryptography, ensuring integrity via cryptographic hashing and encoding, and preserving availability through blockchain decentralization. This paper makes the following main contributions:

- Developed an LSTM autoencoder-based DNN model for detecting anomalous transactions and encoding validated transactions. The algorithm rejected 2790 anomalous data out of 119,103 data based on the error threshold of 0.781 and achieved a compression ratio of 47.81 for encoded transactions with an signal-to-noise ratio (SNR) of 18.12 dB.
- Implemented the proposed DNN model as a chaincode's contract policy in an endorsing node on Fabric. After endorsement, the network processes a reduced number of encoded transactions, leading to secure transactions, efficient storage management, reduced latency, and improved overall network speed.

The remainder of this paper is organized as follows. Section 2 provides background information on Fabric and DNN. Section 3 presents the architecture of the proposed DNN-based blockchain solution. Section 4 implements an anomaly detection model based on an LSTM autoencoder and evaluates the performance of the proposed model. Section 5 presents the detailed system design and implementation in Fabric. Section 6 evaluate the performance of proposed Fabric network. Section 7 discusses the related work. Finally, Sect. 8 concludes the paper and discusses future directions.

2 Background

2.1 Blockchain - Hyperledger Fabric

Blockchain, introduced by Satoshi Nakamoto in 2008 [13], is a decentralized, P2P, and immutable ledger. There are two main types of blockchains: permissionless and permissioned [14]. Permissionless blockchains, such as Bitcoin, allow anyone to join without fully trusting participants but may face challenges in handling high transaction volumes. In contrast, permissioned blockchains, such as Hyperledger Fabric, restrict participation to specific user groups, often designed for confidentiality by isolating critical transactions from public access. These blockchains rely on various consensus protocols, such as proof of work (POW) and practical Byzantine fault tolerance (PBFT), to validate and agree upon transactions, ensuring a secure and consistent ledger. Hyperledger Fabric, developed by the Linux Foundation and IBM, supports smart contracts, known as chaincode, in Go, Java, and Node.js. It also allows for consensus algorithm selection, with PBFT being the default [9]. Transactions in Fabric undergo three phases: endorsement, ordering, and validating. Entities in the network receive organizational identity from the certificate authority (CA), who issues unique public and private key pairs to nodes. The private key is used for signing transactions, while the public key is used to verify signatures and serve as a permission ID. The membership service provider (MSP) maintains permission IDs. Endorsing nodes are specified by the endorsement policy and host chaincode. The endorsement policy defines who is allowed to endorse the transactions, access the network, and specifies the required number of endorsements to validate the transactions. The chaincode has its own contract policy, which includes the conditions to endorse the transaction. Once the endorsement policy is satisfied, the chaincode runs on endorsing nodes to check its contract policy and, if satisfied, endorses the transaction. These policies ensure that transactions are processed and validated by the appropriate nodes.

2.1.1   Trasation Validation

To initiate a transaction, the client sends a proposed message to the endorsement nodes in the format `propose<transaction, clientSig>`, where `clientSig` is the requesting node's signature. Endorsement nodes verify the signature, analyze the transaction, and follow the endorsement policy. If satisfied, the chaincode runs automatically to endorse the entire transaction, and sends it back to the client as `<transaction-endorsed,epSig>`, where `epSig` is the endorsement node's signature. Once the client has collected enough endorsements based on the endorsement policy, it sends the endorsed transaction to the ordering node. If the endorsement is refused, the endorsement node sends a message (`transaction-invalid, rejected`) to the client. Ordering nodes use PBFT consensus to organize endorsed transactions into blocks, which are hashed and broadcast to validating nodes via the gossip protocol [9]. Validating nodes independently verify transactions based on the endorsement policy and follow the consensus rule to add the block to the ledger. Eventually, the ledger state is updated across all validating and endorsing peers, and a transaction validation notification is sent to the client application, indicating the completion of the transaction.

2.2   Long Short-Term Memory (LSTM) Autoencoder

LSTM is a type of Recurrent Neural Networks (RNNs) that predicts values based on distant past data [15]. Another type of DNN, the autoencoder, replicates input as output while efficiently encoding by eliminating noise [16]. The IoT devices collect data from their surroundings, which is dependent on their past values and often contains noise in the form of random variations or interference. Therefore, the LSTM autoencoder is an ideal solution for our IoT-focused transaction encoding and anomaly detection. The LSTM autoencoder consists of two key components: 1) an encoder that transforms the input into a compressed encoded code, effectively reducing the size of the input data, and 2) a decoder that converts the encoded data back into a reconstructed version of the input. The encoder ($\phi : X \rightarrow F$) encoded input into a code using the Eq. 1:

$$h = \sigma(Wx + b) \qquad (1)$$

Here, h represents the code or latent representation, which is the encoded version of the input data, σ is an activation function (e.g., sigmoid or ReLU), W is a weight matrix, x is the input vector, and b is a bias vector. This encoding process removes any noise present in the data and makes it compact, working as the lossy compression of input data [19]. The encoded data can only be decoded by a decoding algorithm. The decoder ($\psi : F \rightarrow X$) reconstructs the input. During training, a cost function minimizes the error between the input and its reconstruction. Weights and biases are iteratively adjusted through back propagation. The model aims to reduce reconstruction errors, such as squared errors also known as threshold of error. If the incoming data exceeds a predefined threshold, it will be identified as an anomalous value and discarded by the model. This is because the data exceeding the threshold is different from the model's trained data, making the model unable to reconstruct it [17].

3   System Architecture

The proposed DNN-based Fabric solution architecture is illustrated in Fig. 1. To address the security threats present in each layer of the IoT network, Fig. 2 details the optimized interaction between Fabric and the IoT layers. The operations of each layer are described below.

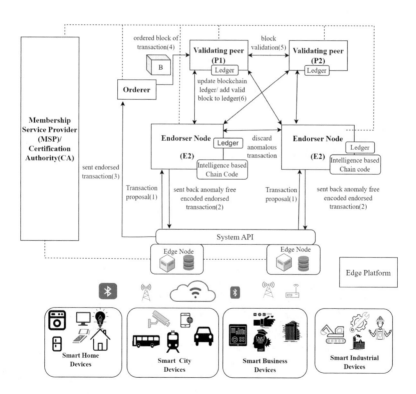

Fig. 1. Framework of proposed DNN-based Hyperledger Fabric

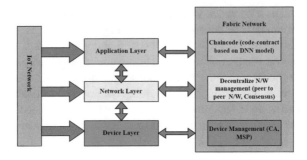

Fig. 2. Architecture of optimized Fabric integrated with IoT layers

3.1 Device Layer

This layer contains various smart devices and sensors used in areas such as smart homes and smart healthcare. The collected IoT data differs from environmental data (temperature, humidity, pressure) or medical data (blood pressure, heart rate). Each IoT device on the blockchain has a unique public key (P_A) (cf. Sect. 2.1) known to all nodes, with the private key (P_B) kept confidential. Before sending a transaction across the network, the IoT device signs it using its private key and sends it in the format: $\langle data, P_B \rangle$. Endorsement nodes can then approve transactions from known senders by verifying the signature using their respective public key P_A.

3.2 Network Layer

At the network layer, decentralized network management is achieved through P2P nodes and a DNNs-based anomaly detection algorithm detects network traffic anomalies. Multiple endorser nodes and validating nodes collaborate synchronously to validate incoming transactions and update the ledger.

3.3 Application Layer

The DNN-based chaincode operates at the application layer to provide the requested services. After verifying the endorsement policy for an incoming transaction, endorsement nodes execute a chaincode to validate its contract policy. The contract policy is based on LSTM autoencoder model which determines if the transaction is anomaly-free, the chaincode validates the encoded version of the transaction rather than the original. Anomalies are determined by evaluating whether a transaction exceeds the defined error threshold of the LSTM autoencoder. The encoding process converts the legitimate transaction into a compressed latent variable, also known as the encoded version, making it more manageable for network transfer and ledger storage. After the endorsement, the encoded transaction is sent to the ordering node, which distributes transaction blocks to the validating nodes. Validating nodes implemented the PBFT-based consensus; hence, they require approval from at least one-third of them to add the block to the ledger. The block includes timestamps, the current transaction hash, number of encoded transactions, and the hash of the preceding block. Once stored in the decentralized blockchain ledger, data blocks become immutable, secure, and tamper-proof from external attacks.

The DNN-based chaincode in the application layer offers advantages including anomalous traffic detection, transaction encoding, and automatic execution. In this way, security is guaranteed at each IoT layer by the proposed solution.

4   Anomaly Detection Algorithm

4.1   Dataset, Prepocessing and Training

To visualize our proposed solution, we utilized a pump sensor dataset from Kaggle [18] which consists of data from 52 different sensors implemented in a pump. The dataset includes 55 columns and 220,320 data values. By eliminating NAN values and making the timestamp an index, we focused on 119,103 data values. The 'machine_status' column, with fields 'Normal,' 'Recovering,' and 'Broken,' was used to distinguish normal from anomalous behavior. We encoded the 'machine_status' values into numeric form for model compatibility, as shown in Table 1, resulting in 116,133 values suggesting normal behaviors of the pump and 2,790 abnormal values (broken, recovering) suggesting when the pump starts working maliciously or sending values deviated from its normal range. The training dataset was partitioned into 70% for training (81,293) and 30% for testing (34,840). Finally, the sensor dataset was scaled to a range of (–1, 1) for efficient model training.

Table 1. Machine status column

Machine_Status	Data Value
1 (Normal)	116,133
2 (Recovering)	2,965
0 (Broken)	5

In the training phase, the LSTM autoencoder model was fine-tuned by adjusting its parameters using the Adam optimizer, a popular optimization algorithm known for its effectiveness in training neural networks [11]. The model was trained over 10 epochs with batches of 512 samples, where the Mean Absolute Error (MAE) loss function measured the difference between the predictions and the actual values. A validation split of 0.1 was employed to detect over-fitting. After training, the model's accuracy on abnormal data was evaluated with epoch = 10, batch size = 512, resulting in a final loss of 0.5773 and a validation loss of 0.5750.

4.2   Threshold Calculation

The threshold for anomalous data was determined by evaluating the MAE between predicted values (x_i), representing the output of our predictive model for a given set of data, and measured values (y_i), which denote the actual observed values of those data. The MAE is calculated as: $MAE = \frac{\sum_{i=1}^{n} |x_i - y_i|}{n}$, where n is the total number of observed values. The final threshold value for the proposed trained model, indicating the acceptable error for the system network, is calculated using Eq. 2:

$$Threshold = mean(trainMAE) + std(trainMAE) = 0.781 \tag{2}$$

where, the *mean* is the average of the MAE of the train values, and *std* is the standard deviation (SD). After implementation, the network effectively encodes data below the specified threshold, but struggles when noise exceeds it, resulting in the rejection of such data as anomalies. A comparison of MAE values between abnormal and normal data, shown in Fig. 3, illustrates distinct data patterns. The x-axis represents the data values of both normal and abnormal data, and the y-axis represents their respective MAE values. Analysis reveals that only a few data points in the normal dataset exceed the threshold, while in the abnormal dataset, nearly all values exceed it.

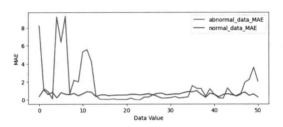

Fig. 3. Plotting normal dataset's MAE values against anomalous dataset's MAE values

4.3 Evaluation Metrics

Compression Ratio (CR)

The encoding algorithm in the proposed solution works as a compression algorithm [19]. The compression ratio (CR) is calculated as: $CR = \frac{Unencoded\ Data\ Size}{Data\ Size}$. In the implemented dataset (cf. Sect. 4.1), the training data's original size was 994,659,120 bytes, which was reduced to 20,803,328 bytes in its encoded form, resulting in a compression ratio of 47.81. This confirms the LSTM autoencoder can be effectively used to compress transactions.

Signal to Noise Ratio (SNR)

SNR compares the strength of the target signal to the background noise, calculated as: $SNR = 10 \times \log\left(\frac{SD_{OriginalData}}{RMS_{ReconstructionError}}\right)$. Where SD is the original data's standard deviation, and RMS is the root mean square of the reconstruction error. We obtained a SNR value of 18.12 dB, which suggests the captured signal is 18.12 dB stronger than the noise.

5   Detailed System Design and Implementation

5.1   System Design

Our system integrates DNNs into Fabric's endorsement nodes without altering Fabric's existing architecture. The transaction flow of the integrated system is detailed in Fig. 4. The endorsement policy is based on: validating the transaction certificate (MSP), spec-

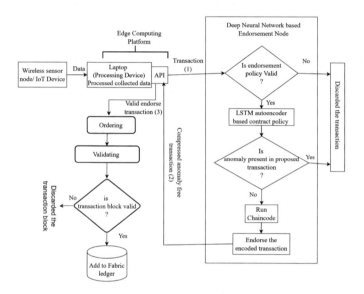

Fig. 4. Transaction flow in optimized Fabric network

ifying endorsing nodes, and endorsing the encoded transaction. When the client sends a `propose` message to endorsing nodes, the client's signature and endorsement policy are verified. The chaincode in the endorsement node then runs to satisfy the contract policy which is based on Algorithm 1:

The implemented LSTM autoencoder algorithm checks for anomalies based on the calculated threshold (cf. Sect. 4.2). If the data is anomaly-free, the chaincode endorses the encoded transaction, sends it back to the client as `<encode-transact ion-endorsed, epSig>`, where encode-transaction-endorsed represents an encoded version of the endorsed transaction. After obtaining enough endorsements as defined in the endorsement policy, the client forwards them to the ordering node and from their to the validating node as explained in Sect. 2.1.1. The framework then employs PBFT consensus for secure ledger updates which require approval from at least one-third of validating nodes to add the block to the ledger. If the transaction is anomalous, it is then rejected, and the endorser sends a `transaction-invalid` message to the client.

Algorithm 1: LSTM Autoencoder

Input: Training sequence $X = \{x_1, x_2, \ldots, x_N\}$
Output: Reconstructed sequence $O = \{o_1, o_2, \ldots, o_N\}$
Step 1: Model Architecture
- Initialize LSTM Autoencoder model:
    Encoder: Define the architecture with LSTM layers, Dropout layers, and other desired components.
    Decoder: Define the architecture with RepeatVector, LSTM, Dropout, and TimeDistributed layers.
    Compile the model using specified parameters (optimizer, loss function, metrics).
Step 2: Training
- Train the model on the training data ($trainX$) using the compiled model:
    Fit the model with specified parameters (epochs, batch size, validation split).
Step 3: Set Threshold
- Generate predictions on training data: trainPredict = autoencoder.predict(trainX)
- Calculate MAE: trainMAE = mean($|$trainPredict $-$ trainX$|$, axis $= 1$)
- Set threshold: threshold = mean(trainMAE) $+$ std(trainMAE)
Step 4: Anomaly Detection
- Identify anomalies: anomaly = transactionMAE $>$ threshold
- Count anomalous and non-anomalous transaction:
    count_anomaly $= 0$
    count_nonanomaly $= 0$
- For each i in anomaly:
    If any($i ==$ True):
        Print 'error: anomaly in transaction occur'
        count_anomaly = count_anomaly $+ 1$
    Else:
        Print 'no anomaly in transaction'
        count_nonanomaly = count_nonanomaly $+ 1$

5.2 Implementation

We implemented the proposed solution on Hyperledger Fabric v2.2 LTS blockchain and Docker containers. The implementation of the LSTM autoencoder was carried out in a Jupyter notebook using Python3 in the Anaconda distribution, and we employed the TensorFlow 2.3.0 development library.

6 Performance Evaluation

Instead of relying on distant cloud services, the proposed framework reduces data processing delays and improves network response times by processing data locally on the edge platform, which includes local servers and laptops. The DNN-based anomaly detection model is trained on a normal dataset. When both normal and abnormal data are sent as transactions (cf. Sect. 4.1), the model rejects 2970 anomalous transactions during the endorsement phase, resulting in fewer endorsements. This minimizes unnecessary network traffic, hence reducing the workload on ordering and validating nodes.

The proposed Fabric network not only verifies members' identity through cryptographic means but also ensures the validity of transactions from authorized members through the anomaly detection model. The model encodes normal transactions with a compression ratio of 47.81, which significantly reduces the transaction size. The endorsed encoded transaction passes through the ordering and validating phases instead of the original transaction, leading to reduced latency. Reducing transaction size also leads to efficient storage management in Fabric. The encoded data can only be decoded by using the decoding algorithm. When a registered user requests the data, the associated encoded data is sent to the user. The user application, designed with decoding features, enables users to validate and interpret responses from the Fabric network, ensuring accurate and reliable information retrieval. Overall, the proposed network architecture not only reduces delays but also enhances data security.

7 Related Work

Significant research is currently being conducted to explore the potential applications of blockchain across various fields beyond cryptocurrency. Below, we present the related works to highlight the associated limitations and discuss how the proposed solution addresses these challenges.

Mishra et al. [20] proposed to use blockchain and IoT-based infrastructure with deep learning algorithms to build secure smart cities. Raw data from smart cities undergoes pre-processing with adaptive data cleaning using a de-noising autoencoder. The proposed solution combines blockchain functionalities with a modified DNN for anomaly-free transaction verification, with results showing higher accuracy compared to other algorithms. However, data processing and event detection in many scenarios are problematic due to the high volume of data. To address this, we propose to use an LSTM autoencoder to reduce the transaction size before including it in a block. Our experiments confirm the effectiveness of the LSTM autoencoder to encode data with a favorable compression ratio. This ensures efficient transaction processing and optimal storage utilization.

Nguyen et al. [21] proposed a neural network-based learning engine for a smart home communication network. The system consists of IoT, application, blockchain and cloud layers. The IoT layer facilitates device communication, the application layer provides a user interface, the blockchain layer records communication details and transactions, including smart contracts and generates Quality of Service (QoS) parameters. The cloud layer then categorizes transactions into smart, moderate, and avoided transactions, using a swarm-based data selection algorithm, demonstrating high prediction accuracy in detecting false authentications. This solution faces data transfer delays between the blockchain and cloud, which impacts overall efficiency. Our solution overcomes data transfer delays by storing encoded data directly within the blockchain, eliminating the need for external components such as the cloud.

Al-Shammari et al. [22] developed a specialized blockchain framework for smart healthcare applications. It utilized an intrusion detection dataset in the edge network to process internet of medical things (IoMT) data. The proposed DNN model performed better than state-of-the-art machine learning algorithms on an intrusion detection system (IDS) dataset. IoMT sensors, though capable of collecting sensitive data, face limitations in real-time analysis. An integrated IDS on the edge server ensures reliable data transmission, with validated data proceeding to the blockchain network. However, the proposed system lacks assurance regarding data security during transfer to the blockchain. Our approach ensures secure data transfer to the blockchain by integrating an anomaly detection model into the smart contract. Only the encoded version of non-anomalous data is stored along with a cryptographic hash in the ledger. Data can only be decoded using the system's decoding algorithm.

8   Conclusion and Future Work

This paper focuses on securing IoT communications and data storage by using a BIoT framework incorporating an anomaly detection-enabled endorsement nodes. The proposed method addresses the security challenges in heterogeneous IoT networks, aligned with the CIA security triad model. The non-anomalous transactions are encoded before being stored in the blockchain ledger, addressing privacy concerns and improving storage utilization. The proposed solution was implemented on Hyperledger Fabric, and evaluation results showed that the implemented model discarded 2970 anomalous transactions based on the calcualted error threshold of 0.781 and achieved a compression ratio of 47.81 for encoded transactions with SNR 18.12 dB.

However, a key limitation lies in accurate data pre-processing during the DNN model training phase to avoid incorrect pattern detection. In the future, we will create a prototype to assess the real-world performance of the proposed solution, and perform a hands-on evaluation of its functionality. We also plan to optimize the proposed solution with cache technology for quick query responses to ensure efficient interactions, especially in dynamic real-world environments.

References

1. Shaaban, S., El Badawy, H.M., Hashad, A.: Performance evaluation of the IEEE 802.11 wireless LAN standards. In: Proceedings of the World Congress on Engineering, vol. 1 (2008)
2. Rokonuzzaman, M., Akash, M.I., Khatun Mishu, M., et al.: IoT-based distribution and control system for smart home applications. In: 2022 IEEE 12th Symposium on Computer Applications & Industrial Electronics (ISCAIE 2022), Penang, Malaysia, pp. 95–98, 21–22 May 2022
3. Ati, M., Al Bostami, R.: Protection of data in edge and cloud computing. In: 2022 IEEE International Conference on Computing (ICOCO 2022), Kota Kinabalu, Malaysia, pp. 169–173, 14–16 November 2022
4. Bouzidi, M., Gupta, N., Cheikh, F.A., et al.: A novel architectural framework on IoT ecosystem, security aspects and mechanisms: a comprehensive survey. IEEE Access 10, 101362–101384 (2022)
5. Aldahmani, A., Ouni, B., Lestable, T., et al.: Cyber-security of embedded IoTs in smart homes: challenges, requirements, countermeasures, and trends. IEEE Open J. Veh. Technol. 4, 281–292 (2023)

6. Sharma, A., Chauidhary, A., Bhatt, D.P.: Blockchain based secure communication in IoT environment. In: 2023 International Conference on Artificial Intelligence and Smart Communication (AISC), Greater Noida, India, pp. 182–186 (2023)
7. Yadav, A.K., Vishwakarma, V.P.: Adoptation of blockchain of things (BCOT): opportunities & challenges. In: 2022 IEEE International Conference on Blockchain and Distributed Systems Security (ICBDS), Pune, India, pp. 1–5 (2022)
8. Pan, J., Wang, J., Hester, A., et al.: Edgechain: an edge-IoT framework and prototype based on blockchain and smart contracts. IEEE Internet Things J. 6(3), 4719–4732 (2019)
9. Androulaki, E., Barger, A., Bortnikov, V., et al.: Hyperledger fabric: a distributed operating system for permissioned blockchains. In: Proceedings of the Thirteenth EuroSys Conference (EuroSys '18), New York, NY, Association for Computing Machinery, USA (2018)
10. Nakaike, T., Zhang, Q., Ueda, Y., et al.: Hyperledger fabric performance characterization and optimization using GoLevelDB benchmark. In: 2020 IEEE International Conference on Blockchain and Cryptocurrency (ICBC), pp. 1–9 (2020)
11. Frikha, T., Ktari, J., Zalila, B., et al.: Integrating blockchain and deep learning for intelligent greenhouse control and traceability. Alex. Eng. J. 79, 258–273 (2023)
12. Xu, X., Sun, G., Luo, L., et al.: Latency performance modeling and analysis for Hyperledger Fabric blockchain network. Inf. Process. Manag. 58, 102436 (2021)
13. Wright, C.B.: A peer-to-peer electronic cash system. SSRN Electron. J. (2008)
14. Nandanwar, H., Katarya, R.: A systematic literature review: approach toward blockchain future research trends. In: 2023 International Conference on Device Intelligence, Computing and Communication Technologies, (DICCT), Dehradun, India, pp. 259–264 (2023)
15. Hochreiter, S., Schmidhuber, J.: Long short-term memory. Neural Comput. 9(8), 1735–1780 (1997)
16. Yang, Z., Xu, B., Luo, W., et al.: Autoencoder-based representation learning and its application in intelligent fault diagnosis: a review. Measurement 189, 110460 (2022)
17. Maleki, S., Maleki, S., Jennings, N.R.: Unsupervised anomaly detection with LSTM autoencoders using statistical data-filtering. Appl. Soft Comput. 108, 107443 (2021)
18. Phantawee, N.: Pump Sensor Data, Kaggle. https://www.kaggle.com/datasets/nphantawee/pumpsensor-data. Cited 2023
19. Sun, J., Dong, P., Qin, Y., et al.: Improving bandwidth utilization by compressing small-payload traffic for vehicular networks. Int. J. Distrib. Sens. Netw. 15, 155014771984305 (2019)
20. Mishra, S., Chaurasiya, V.K.: Blockchain and IoT based infrastructure for secure smart city using deep learning algorithm with dingo optimization. Wirel. Pers. Commun. 132, 17–37 (2023)
21. Nguyen, D.C., Aouedi, O., Menon, S., et al.: Blockchain and machine learning inspired secure smart home communication network. Sensors, Basel, Switzerland 23 (2023)
22. Al-Shammari, N.K., Syed, T.H., et al. An Edge-IoT framework and prototype based on blockchain for smart healthcare applications. Eng. Technol. Appl. Sci. Res. (2021)

Enhancing Intrusion Detection System Using Machine Learning and Deep Learning

R. Madhusudhan$^{(\boxtimes)}$, Shubham Kumar Thakur, and P. Pravisha

Department of Mathematical and Computational Sciences,
National Institute of Technology, Karnataka, Mangaluru, India
madhu@nitk.edu.in

Abstract. Intrusion Detection Systems (IDS) are critical components in ensuring the security and integrity of computer networks by identifying and thwarting malicious activities. With the ever-evolving nature of cyber threats, it is imperative to develop IDS models that are capable of accurately detecting and classifying network intrusions. This research aims to enhance the performance of IDS systems by harnessing the power of various machine learning algorithms, namely Random Forest, Neural Networks, and One-Dimensional Convolutional Neural Networks(1D-CNN). The widely used CICIDS2017 dataset is employed to train and evaluate the proposed models, following an extensive preprocessing phase to extract pertinent information and eliminate extraneous noise. The accuracy of each algorithm is thoroughly assessed, and a comparative analysis is conducted to determine the optimal approach for building an effective IDS model. The results unequivocally demonstrate that it exhibits exceptional performance and robust capabilities in accurately detecting and classifying network intrusions. These empirical findings significantly contribute to the advancement of IDS technologies, empowering organizations to bolster network security and effectively mitigate the ever-growing array of cyber threats.

Keywords: Intrusion Detection System · Machine Learning · Deep Learning · 1D-CNN · Network Security

1 Introduction

In today's rapidly evolving cybersecurity landscape, enhancing Intrusion Detection Systems (IDS) is crucial to keeping pace with the increasing sophistication of cyber threats and the emergence of new attack vectors. Traditional IDS approaches, such as signature-based detection methods, have limitations in detecting novel and sophisticated threats [15]. Signature-based IDS can easily detect known attacks, but it fails to work for new attacks where patterns are not known or not updated in the database [12]. Therefore, there is a pressing need to enhance IDS models using advanced techniques like machine learning and deep learning to bolster their effectiveness. Intrusion detection systems are designed

© The Author(s), under exclusive license to Springer Nature Switzerland AG 2024
L. Barolli (Ed.): AINA 2024, LNDECT 201, pp. 326–337, 2024.
https://doi.org/10.1007/978-3-031-57870-0_29

to monitor network traffic, identify potential security breaches, and promptly respond to malicious activities [1]. The IDS is one of the most important components of the security infrastructure that can prevent cyber threats from various types of attackers [17]. The fundamental objective of IDS is to detect anomalies, signs of unauthorized access, and suspicious behavior within a network or a host. IDS scans network traffic to identify and report a violation based on the preconfigured customized detection levels [9]. Traditional IDS approaches have predominantly relied on signature-based detection methods, which involve matching network traffic patterns against predefined signatures or rules to identify known attacks [16]. While effective against well-known threats, these rule-based approaches struggle to detect emerging and sophisticated attacks. The static nature of signature-based IDS hampers their ability to adapt and detect novel threats effectively [6].

Advanced techniques such as machine learning and deep learning have garnered a significant amount of interest as potential solutions to the limitations of conventional IDS. By harnessing the capabilities of algorithms, these techniques can identify anomalies, categorize network traffic, and analyze patterns. Machine learning algorithms can adapt and generalize from data, enabling them to detect previously unseen attack patterns. Deep learning models, particularly neural networks, excel at automatically learning complex representations and patterns from data, making them promising candidates for enhancing IDS [8]. The random forest method is an ensemble learning approach that incorporates multiple decision trees to get accurate predictions. Neural networks, particularly deep learning models, have revolutionized the field of machine learning and demonstrated exceptional performance in various domains [13]. One-Dimensional Convolutional Neural Networks (1D-CNNs) have emerged as powerful tools for analyzing sequential data, such as network traffic [2]. The continuous enhancement of IDS models will remain essential to effectively combat emerging threats and ensure the resilience of digital infrastructures [10].

The objective of this article is to explore the potential of Random Forests, Neural Networks, and 1D-CNN in enhancing the performance of IDS systems. Furthermore, the article proposes hybrid approaches that leverage the strengths of multiple techniques to improve detection accuracy and robustness. The article makes the following scientific contributions:

– Development of an innovative and effective Enhancing Intrusion Detection System (IDS) framework capable of detecting various types of attacks within real-world network traffic.
– A comparative analysis is conducted to evaluate the performance of Random Forest, Neural Networks, and 1D-CNN in enhancing IDS, considering their strengths and weaknesses.

The rest of the article is organized as follows: Sect. 2 describes the literature review. Section 3 presents the proposed framework. In Sect. 4, the simulation and results are discussed. The conclusion and future work are presented in Sect. 5.

2   Literature Review

Intrusion Detection Systems (IDS) have attracted significant attention in research circles, with numerous studies exploring the integration of machine learning (ML) techniques to bolster their capabilities. A significant amount of research has examined the efficacy of Random Forest in intrusion Detection Systems (IDS) applications. These applications use Random Forest to identify a wide range of network threats, resulting in remarkable detection rates and reducing the number of false positives. Several studies explored the application of Random Forest in anomaly-based IDS, achieving promising results in terms of accuracy and computational efficiency. Zhong et al. [3] proposed methods to improve the efficiency of the intrusion detection system used on IoT servers. The proposed system employs multiple methods to detect anomalies in the network, such as flow identification and intrusion detection. The findings of the study underscored the model's proficiency in handling imbalanced datasets and its robustness against the dimensionality of features. The authors highlight a few limitations, including the integrity of the data used to generate the input and output.

The application of Softmax models for intrusion detection has also received attention, and several researchers have proposed an ensemble approach based on Softmax for network anomaly detection. Al-Turaiki et al. [4]. proposed two novel deep learning-based models to solve the binary and multiclass classification of network attacks. The models can be integrated into network systems to identify unusual events, such as new attacks and violations, that occur within the network of an organization. The study showcased the effectiveness of Softmax in handling intricate network traffic patterns and achieving enhanced detection accuracy compared to conventional methods. The models demonstrated exceptional performance in terms of accuracy and precision. Neural networks, including sophisticated deep learning architectures, have emerged as a significant area of focus in IDS research. Several researchers employed a multi-layer perceptron (MLP) neural network for intrusion detection, highlighting its efficacy in identifying network attacks.

Azzaoui et al. [5] proposed a deep neural network-based IDS that was specifically designed for network traffic classification. The study emphasized the ability of neural networks to capture intricate patterns within network traffic data. The efficiency of the proposed methodology was evaluated through the execution of numerous sets of experiments involving various hyperparameter combinations. The evaluation was conducted using the CICIDS2017 and NSL-KDD datasets with standard performance metrics. The results of these experiments were subsequently compared with the results of other shallow and deep ANN models. The result of the experiment demonstrated good accuracy, demonstrating the efficiency of the test. For further improvement of the model, the authors emphasize the importance of future research that will expand the range of hyperparameters and examine deeper neural networks. The application of 1D-CNN in IDS has shown tremendous promise in harnessing a 1D-CNN model to detect network attacks and leveraging its capability to capture spatial and temporal dependencies in network traffic. Qazi et al. [6] proposed a deep learning architecture based on a one-dimensional convolutional neural network to identify network intrusions. The architecture that was proposed was able to identify four distinct forms of network incursions that are most frequently encountered. These include DoS Hulk, DDoS, and DoS GoldenEye, which are all examples of active attack types, and PortScan, which is an example of a passive attack type. The evaluation was conducted using the CICIDS2017 dataset, and the results demonstrated superior performance compared to traditional machine-learning algorithms.

Zhang et al. [11] proposed a flow-based intrusion detection model known as SGM-CNN. The model combines convolutional neural networks and imbalanced class processing. The authors researched how the model's performance was affected by varying the number of convolution kernels and the learning rate. The evaluation of the model was conducted using the UNSW-NB15 and CICIDS2017 datasets. The experimental findings demonstrated improved accuracy, offering a practical solution to the problem of imbalanced intrusion detection. Zhang et al. [14] proposed TIKI-TAKA, a novel framework for assessing the robustness of DL-based network intrusion detection systems against adversarial attacks. The authors utilized three well-known malicious traffic detectors that make use of neural networks and deployed five cutting-edge attack variants. The framework includes three defense methods to improve NIDS robustness against adversarial manipulations: model voting ensembling, ensembling adversarial training, and query detection. The results of the experiments show that the proposed approaches are capable of restoring intrusion detection rates to over one hundred percent against the majority of malicious traffic. Additionally, the model can prevent attacks that could have potentially catastrophic implications, such as botnets.

In summary, the existing research highlights the potential of machine learning models, such as 1D-CNN, Random Forest, Softmax, and Neural Networks, to enhance the performance of IDS. In this study, the collective impact of Random Forest, Neural Networks, and 1D-CNN in the development of IDS will be examined. Moreover, a critical yet often overlooked issue in IDS evaluation data imbalance will be addressed. This study will contribute to advancing machine learning (ML)-driven intrusion detection systems (IDS), providing valuable insights for future developments in this field.

3   Proposed Framework

An innovative and efficient Enhancing Intrusion Detection System (IDS) framework that can identify a wide range of real-world network traffic-based attacks has been proposed in this article. This framework can recognize a broad variety of network traffic-based attacks that occur in the real world. Figure 1 shows a flow diagram of the methods that are utilized in the proposed framework. The proposed framework consists of two major phases: model training and model prediction. The following subsection will provide an overview of the methods that were applied in each step, specifically Random Forest, Neural Networks (NN), and 1D-CNN.

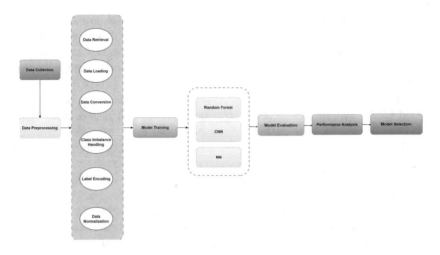

Fig. 1. Flow Diagram of methodologies used in the Proposed framework

3.1   Dataset

In the proposed framework, a 1D-CNN model is utilized on the CICIDS2017 dataset for intrusion detection. The CICIDS2017 dataset is a large-scale dataset with 2.8 million records that covers various network protocols and includes both benign and malicious traffic. This data is captured from a realistic network environment, representing real-world conditions. The dataset is labeled with different classes for network attacks and normal traffic. The dataset also provides a comprehensive set of features and attributes for each network flow.

3.2   Data Preprocessing

A series of essential preprocessing steps will be executed to prepare the network traffic data for analysis and model development in the network intrusion detection system (IDS). These steps were designed to enhance the quality, consistency, and suitability of the data for further processing and analysis. Following is a comprehensive summary of each step that was carried out:

i. *Data Retrieval* - During the data retrieval stage of preprocessing, the 'glob' module is used to collect CSV files containing network traffic data from a specified directory. The files were merged into a single Pandas DataFrame labeled 'combined csv' to guarantee a cohesive dataset for preprocessing.

ii. *Data Loading* - After combining the network traffic data into the "combined csv" DataFrame, the next step was the "load data" process. During this step, the columns that are not relevant and handle missing values are removed. Columns such as 'Flow Packets/s', 'Flow Bytes/s', and 'Label' were removed as they were deemed unnecessary for the IDS model. Missing values in the remaining columns were filled with the mean value to ensure data completeness and accuracy.

iii. *Data Conversion* - After removing unnecessary columns and handling missing values, the data conversion step transformed the dataset into a numeric format suitable for calculations and modeling. This involved using functions like "astype" and "pd.to_numeric" to ensure all features were represented as numeric values.

iv. *Class Imbalance Handling* - Addressing class imbalances in intrusion detection systems is crucial for the accurate detection of network intrusions. Oversampling and undersampling techniques help balance class representation, allowing the model to learn from both frequent and rare intrusions. This improves the performance, sensitivity, and overall robustness of the intrusion detection system.

v. *Label Encoding* - In the intrusion detection system model, the label encoding step is a crucial phase that converts string labels to numerical representations. It enables efficient handling, compatibility with machine learning algorithms, and accurate analysis of network traffic data.

vi. *Data Normalization* - Data normalization is a vital preprocessing step in our IDS model that brings network traffic data features to a consistent scale. It improves stability, convergence, and interpretability by eliminating bias caused by varying scales. Normalization ensures fair analysis, prevents feature dominance, and enhances the performance and understanding of the IDS model.

3.3 Model Training

In the model training section of IDS, three models are employed: Random Forest, Neural Networks, and 1D-CNN, each chosen for its suitability in detecting network intrusions. Each of the models is trained on a dataset that has been meticulously prepared and goes through a series of preprocessing steps. An evaluation metric is used to determine the most optimized solution [7].

i. *Random Forest* is a powerful machine learning algorithm used for various applications, including network intrusion detection. In the proposed IDS, Random Forest is employed as one of the models, and its training process and evaluation measures are described in detail. The method incorporates multiple decision trees that have been trained on various data and features, aggregating their predictions to make accurate predictions.

ii. *Neural networks* (NNs) are shown in machine learning models inspired by the human brain and known for their ability to learn complex patterns and make accurate predictions from diverse datasets. The training and optimization of NN models involves adjusting hyperparameters, such as the network architecture, learning rate, and regularization techniques, to achieve the best performance.

iii. *Convolutional Neural Networks* CNNs are well-suited for IDS systems due to their ability to effectively analyze visual data, such as network traffic. They can automatically learn and extract relevant features from raw data, eliminating the need for manual feature engineering. CNNs excel at capturing local dependencies and spatial relationships in the input data, which is crucial for detecting network intrusions. A 1D-CNN model is utilized on the CICIDS2017 dataset for intrusion detection.

4   Simulation and Results

The performance of Random Forest (RF), a 3-layer Neural Network (NN), a 5-layer Neural Network (NN), and a 1D-CNN model is analyzed using 2 convolutional layers and 1 fully connected layer. The k-fold evaluation technique was employed to obtain robust performance measures for each model by dividing the dataset into folds for training and testing. Different cases were considered in the experimental setup to evaluate the models' accuracy, covering diverse scenarios and datasets.

4.1   Result of Random Forest

In the study, the Random Forest algorithm is applied with 5-fold cross-validation to evaluate the performance of the model. Various metrics are used to assess the accuracy of the model, including validation accuracy, balanced validation accuracy, and balanced test accuracy. The primary metric of interest was balanced test accuracy. A validation accuracy of 99.21% was achieved. This metric represents the accuracy of the model when it was tested on the validation dataset. It indicates the overall correctness of the model's predictions. The balanced validation accuracy was measured at 88.15%. This metric takes into account the imbalance in the dataset and provides a fair evaluation of the model's performance. The disproportionate representation of various classes in the dataset ensures that accuracy is not skewed. The balanced test accuracy, which is the primary metric, was recorded at 84.41%. This metric indicates the accuracy of the model on an independent test dataset, considering the class imbalance. It is crucial in real-world scenarios where class distributions may be uneven.

4.2   Result of Neural Network

In the study, a 3-layer neural network (NN) and a 5-layer NN are utilized. For each network, an initial run is conducted with batch normalization and dropout techniques. Then a 5-fold evaluation is performed for both networks, employing batch normalization and dropout in each evaluation fold. This facilitated a comprehensive evaluation of their performance. Finally, the results were compared with the 3-layer and 5-layer NNs to determine whether the architecture yielded better accuracy and performance. This systematic approach provided valuable insights into the effectiveness of each network and helped us identify the most suitable architecture. In the first run of the 5-layer neural network with batch normalization and dropout, the following settings were used: learning rate: 0.001 (1e-3), regularization parameter: 0.00001 (1e-5), batch normalization was used, dropout was used with rates of 0.1 and 0.3, and number of epochs: 20.

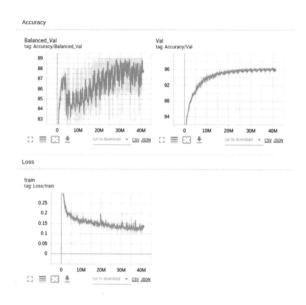

Fig. 2. NN 3 layer: 1st run with batch norm + dropout

The performance evaluation of the first run of a 3-layer neural network with batch normalization and dropout is shown in Fig. 2. After training, the neural network achieved a validation accuracy of 89.12%, showing a slight improvement compared to previous runs. The network's performance was further evaluated on a test set, resulting in an accuracy of 91.14%. The incorporation of batch normalization and dropout approaches into the architecture of the network proved to be advantageous. Batch normalization was able to normalize activations, which improved stability and efficiency during training. Dropout, on the other hand, introduced regularization, which prevented overfitting and improved generalization. The combined effect of these techniques led to higher accuracy on both validation and test sets, indicating improved performance and robustness. These findings highlight the effectiveness of batch normalization and dropout in enhancing neural network performance, making them valuable additions to the model. In the 5-fold evaluation of the neural network with batch normalization and dropout, the following settings were used: learning rate: 0.001, regularization parameter: 0.00001, batch normalization and dropout (rates of 0.1 and 0.3) were used, and number of epochs: 20

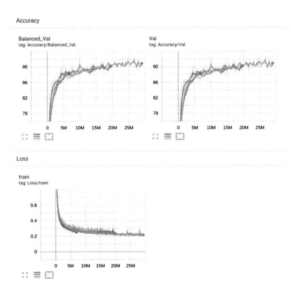

Fig. 3. NN 3 layer: 5-Fold evaluation

The performance evaluation of the 5-fold evaluation of the neural network with batch normalization and dropout is shown in Fig. 3. The balanced test accuracy achieved across the 5 folds was 84.19%. This result indicates the overall performance of the model, considering all classes in the dataset. The incorporation of batch normalization and dropout has contributed to the network's effectiveness. In the 5-fold evaluation of the 5-layer neural network (NN), the best hyperparameter settings were found to be a learning rate (lr) of 1e-4 and a regularization strength (reg) of 1e-5. The network was trained for a total of 60 epochs. The performance evaluation of the 5-fold evaluation is shown in Fig. 4. The evaluation results showed a balanced test accuracy of 85.57% across the five folds. These results indicate the performance of the 5-layer NN under the specified hyperparameter settings and provide valuable insights into its effectiveness in handling the given task.

4.3   Result of Convolutional Neural Network

The 5-fold evaluation was conducted on a 2-layer 1D-CNN model to assess its performance. The model consists of two convolutional layers designed for analyzing sequential data. The dataset was divided into five subsets, and the model was trained and evaluated five times, using a different fold as the validation set each time. Figure 5 shows the performance evaluation of the 5-fold evaluation performed on a 2-layer 1D-CNN model. The purpose of the evaluation was to determine the accuracy of the model, and the findings revealed that the model achieved an average accuracy of 87.11%. This evaluation provides valuable insights into the model's generalization ability and suitability for the task at hand.

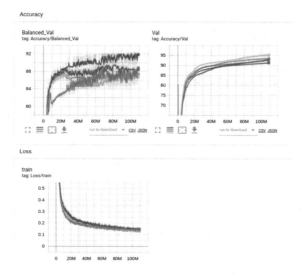

Fig. 4. 5 layer NN: 5-fold Evaluation

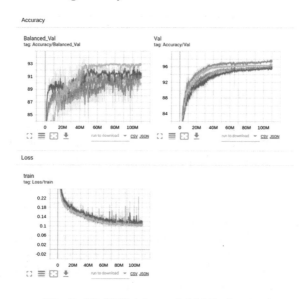

Fig. 5. 1D-CNN 2 layer: 5-fold Evaluation

4.4 Evaluation

Based on the 5-fold accuracy results, the best-performing model among the classifiers is the 1D-CNN with 2 convolutional layers and 1 fully connected layer. It achieved an overall 5-fold accuracy of 87.11%, surpassing the accuracies of the other classifiers. The Random Forest (RF) classifier achieved an accuracy of 84.41%. Although it performed reasonably well, the 1D-CNN model outperformed it. The 3-layer neural network (NN) achieved accuracies of 84.19% and 84.87% in different runs. These accuracies are comparable to RF, but they fall short of the accuracy that the 1D-CNN model achieves. The 5-layer neural network (NN) achieved an accuracy of 85.57%, which is higher than the accuracies of RF and the 3-layer NN models but still lower than the accuracy of the 1D-CNN model. Based on these results, the 1D-CNN model with 2 convolutional layers and 1 fully connected layer stands out as the best model in terms of classification accuracy. Its ability to capture relevant patterns and extract meaningful features from the data, even without spatial or sequential relationships, contributes to its superior performance. The accuracy of the 5-fold evaluation is shown in Table 1.

Table 1. Classifier 5-fold Accuracy

Classifier	5-fold Accuracy (%)
RF	84.41
NN 3 layer	84.19
NN 5 layer	85.57
1D-CNN	87.11

5 Conclusion and Future Work

This study successfully enhanced an Intrusion Detection System (IDS) model using machine learning (ML) and deep learning (DL) techniques. Different machine learning and deep learning models were tested on the CICIDS 2017 dataset. The results showed that deep learning models, especially the 1D-CNN, did better than the old RF classifier, with an average accuracy of 86.11%. By leveraging DL techniques, complex patterns and representations were extracted from network traffic data, leading to improved detection capabilities. This highlights the potential of DL models for enhancing IDS performance and ensuring network security. The future scope of intrusion detection systems (IDS) involves fine-tuning deep learning models, exploring ensemble techniques, and incorporating feature engineering to extract valuable information from network traffic data. Real-time detection capabilities can be enhanced by integrating streaming data processing techniques, while adversarial robustness becomes crucial to ensuring IDS reliability against attacks. By focusing on these areas, IDS models can be advanced, resulting in more accurate and effective systems that address the evolving challenges of network security.

References

1. Li, Y., Zhang, L., Cao, J.: A survey of intrusion detection systems based on machine learning and deep learning in the IoT environment. J. Ambient Intell. Humaniz. Comput. (2021)
2. Haddadouche, N., Benameur, A., Mokdad, L., Alajlan, N.: Deep neural networks for intrusion detection: a comprehensive review. Int. J. Netw. Manag. (2020)
3. Zhong, M., Zhou, Y., Chen, G.: Sequential model-based intrusion detection system for IoT servers using deep learning methods. Sensors **21**(4), 1113.48 (2021)
4. Al-Turaiki, I., Altwaijry, N.: A convolutional neural network for improved anomaly-based network intrusion detection. Big Data **9**, 233–252 (2021)
5. Azzaoui, H., Boukhamla, A.Z.E., Arroyo, D., Abdallah, B.: Developing new deep-learning model to enhance network intrusion classification. Evol. Syst. (2022)
6. Qazi, E.U.H., Almorjan, A., Zia, T.: A one-dimensional convolutional neural network (1D-CNN) based deep learning system for network intrusion detection. Appl. Sci. **12**(16), 7986 (2022)
7. Alzubaidi, L., Zhang, J., Humaidi, A.J., et al.: Review of deep learning: concepts, CNN architectures, challenges, applications, future directions. J. Big Data **8**, 1–74 (2021)
8. Fu, Y., Du, Y., Cao, Z., Li, Q., Xiang, W.: A deep learning model for network intrusion detection with imbalanced data. Electronics **11**(6), 898 (2022)
9. Ashiku, L., Dagli, C.: Network intrusion detection system using deep learning. Procedia Comput. Sci. **185**, 239–247 (2021). ISSN 1877-0509
10. Vaiyapuri, T., Sbai, Z., Alaskar, H., Alaseem, N.A.: Deep learning approaches for intrusion detection in IIoT networks-opportunities and future directions. Int. J. Adv. Comput. Sci. Appl. **12**(4) (2021)
11. Zhang, H., Huang, L., Wu, C.Q., Li, Z.: An effective convolutional neural network based on SMOTE and Gaussian mixture model for intrusion detection in imbalanced dataset. Comput. Netw. **177**, 107315 (2020)
12. Kocher, G., Kumar, G.: Machine learning and deep learning methods for intrusion detection systems: recent developments and challenges. Soft Comput. **25**, 9731–9763 (2021)
13. Liu, Y., Zhang, M.: Neural network methods for natural language processing. Comput. Linguist. **44**, 193–195 (2018)
14. Zhang, C., Costa-Perez, X., Patras, P.: Adversarial attacks against deep learning-based network intrusion detection systems and defense mechanisms. IEEE/ACM Trans. Netw. **30**(3), 1294–1311 (2022)
15. Dini, P., Elhanashi, A., Begni, A., Saponara, S., Zheng, Q., Gasmi, K.: Overview on intrusion detection systems design exploiting machine learning for networking cybersecurity. Appl. Sci. **13**(13), 7507 (2023). https://doi.org/10.3390/app13137507
16. Kumar, V.: Signature based intrusion detection system using SNORT. Int. J. Comput. Appl. Inf. Technol. **1**, 7 (2012)
17. Lee, S.W., Mohammadi, M., Rashidi, S., Rahmani, A.M., Masdari, M., Hosseinzadeh, M.: Towards secure intrusion detection systems using deep learning techniques: comprehensive analysis and review. J. Netw. Comput. Appl. **187**, 103111 (2021). ISSN 1084-8045

Integration of Machine Learning with Quantum Annealing

Hadi Salloum[1,2(✉)], Hamza Shafee Aldaghstany[1,2], Osama Orabi[1,2], Ahmad Haidar[1,2], Mohammad Reza Bahrami[2,3], and Manuel Mazzara[2]

[1] Qdeep, Innopolis, Russia
h.salloum@qdeep.net
[2] Innopolis University, Innopolis, Russia
[3] Samarkand International University of Technology, Samarkand, Uzbekistan

Abstract. The hasty advancement of Quantum Computing (QC) technologies has led to offer plethora of promising offering promising opportunities for solving complex optimization problems, especially Quantum Annealing (QA). This paper presents an investigation into the fusion of ML and QA. It also explores and elucidates the capabilities of ML empowered by QA techniques. Furthermore, it introduces an implementation strategy that increases the potential of QA through the integration of ML, and vice versa. This research advances our understanding of utilizing QA paradigms to enhance optimization methodologies within the realm of ML.

Keywords: Quantum Annealing · Machine Learning · D-Wave · Optimization · Quantum Computing

1 Introduction

The dynamic landscape of QC has sparked enormous interest in using quantum paradigms to redefine computing methods in various fields [13,26]. In this broad spectrum, QA (a sophisticated optimization technique in the field of quantum computing) has become an important topic due to its potential to solve complex problems [30].

QA emerged in the late 1990s and was originally conceived by respected researchers such as T. Kadowaki and H. Nishimori. It developed within the broader framework of quantum computing. This method, based on principles such as quantum tunneling and adiabatic quantum computing (AQC), aims to solve optimization problems efficiently by using quantum fluctuations to explore solution spaces more effectively than classical methods [14].

The fundamental introduction of D-Wave systems in the early 2000s marked an important milestone in the development of QA. These systems, particularly the D-Wave One and its specialized QA processors, were initially met with skepticism regarding their quantum nature and performance. However, they have now gained attention due to their potential to effectively address optimization problems [2].

In addition, D-Wave announced the launch of its next-generation system, Advantage, in February 2019 (see Fig. 1). This improved architecture featured an impressive expansion of the total number of quantum bits (qubits) to over 5000 and adopted

© The Author(s), under exclusive license to Springer Nature Switzerland AG 2024
L. Barolli (Ed.): AINA 2024, LNDECT 201, pp. 338–348, 2024.
https://doi.org/10.1007/978-3-031-57870-0_30

the Pegasus Graph topology, which increased the connections per qubit to 15. D-Wave claimed that the Advantage architecture offers a remarkable 10x speedup in resolution time compared to the 2000Q product offering. Following this, an incremental successor to, the Advantage Performance Update, is expected to deliver an additional 2x speedup over the Advantage and an astounding 20x speedup over the 2000Q system, among other significant improvements [9,33].

(a) D-Wave Advantage™ Quantum Computer

(b) D-Wave Advantage™ QPU

Fig. 1. D-Wave Advantage^{TM} Quantum Computer and QPU [19]

At the same time, the development of ML spans several decades and its origins can be traced to groundbreaking work on neural networks and pattern recognition in the 1950s and 1960s. In the decades that followed, remarkable advances were made in the refinement of ML. Learning algorithms and methods culminating in the establishment of supervised learning, unsupervised learning and reinforcement learning as primary paradigms, each tailored to specific learning scenarios [12,23,31].

The convergence of QA and ML gained significant momentum at the beginning of the 21st century. This convergence was driven by researchers' shared recognition of the transformative potential of QC techniques, including annealing. This recognition came from the perspective of revolutionizing computing efficiency and addressing the complex optimization challenges associated with various ML taskss [1,5,21].

This paper explores the synergy between ML and QA and opens up advanced possibilities for tackling complex optimization problems. Presents a strategy for optimizing the potential of QA using ML techniques and vice versa. This research pioneers the use of QC paradigms to strengthen optimization methods in ML.

2   Elucidating the Fundamentals of Quantum Annealing

QA represents a unique variant within the AQC domain that differs from gate-based QC models. It is presented as an advanced iteration based on the classic simulated annealing algorithm, specifically designed to address optimization challenges classified as NP-hard. This approach cleverly uses quantum phenomena such as superposition, entanglement and the remarkable aspect of quantum tunneling to navigate complex problem landscapes [20, 30].

Essentially, QA is based on the principles of the adiabatic theorem, which orchestrates the step-wise evolution of a quantum system from an easily fabricated initial state to the ground state of a Hamiltonian problem that encodes the core of the optimization problem. The adiabatic theorem ensures the system, commencing in its ground state $|\psi_0\rangle$ and evolving under a time-dependent Hamiltonian $H(t)$, remains within its instantaneous ground state $|\psi(t)\rangle$ throughout the evolution process, expressed as:

$$|\psi(t)\rangle \approx |\psi_0\rangle \quad \text{for all } t. \tag{1}$$

This basic principle guarantees the initialization of the system in a recognizable quantum state, specifically the ground state of the initial Hamiltonian H_0. It then ensures that the system continues to remain in this state while gradually transitioning to the ground state of the Hamiltonian problem H_p over time.

$$H_p = \sum_i w_i \sigma_i, \tag{2}$$

Here, w_i means the weights assigned to different terms within the objective function, while σ_i represents operators acting on the qubits.

Over time, the system converges steadily and adiabatically toward the ground state of H_p, providing the desired solution to the optimization problem. This approach skillfully addresses NP-hard optimization challenges and exploits quantum phenomena such as superposition and entanglement. Adiabatic evolution, guided by the adiabatic theorem, orchestrates a seamless transition of a quantum system from an easily fabricated initial state to the ground state of a Hamiltonian problem, thereby encoding the underlying optimization task [11, 16, 17, 24].

3   Quantum Annealing in D-Wave Quantum Computer

QC, particularly using the example of D-Wave quantum computers, uses the principles of QA to enable novel approaches to problem solving [28]. At the heart of these quantum processors are qubits, the quantum counterparts of classical bits. Unlike classical bits, which exist in states 0 or 1, qubits exploit their quantum nature to exist in a superposition of both states simultaneously. Through the QA process, these qubits transition from superposition states to the final classical states 0 or 1, as shown in Fig. 2, thus contributing to the formation of classical states within the quantum system.

The underlying physics behind the qubit states is represented by energy diagrams that illustrate the progression of the qubit states during the QA process. Initially, the system exists in a single-valley energy state and has a singular minimum. As QA progresses, this energy landscape transforms into a double-well potential by creating a

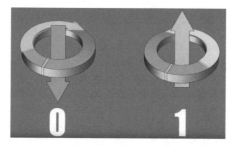

Fig. 2. A qubit's state is implemented as a circulating current [32]

barrier. Here the valleys represent states 0 and 1, and the qubit settles into one of these valleys after the annealing is complete (see Fig. 3).

Fig. 3. Energy diagram changes over time as the quantum annealing process runs and a bias is applied [32].

External magnetic fields, called biases, play a crucial role in changing the probability that a qubit will end up in a particular state. By modifying the double-well potential, the biases affect the probability of the qubit transitioning to state 0 or 1, demonstrating the inherent controllability of these quantum systems. The true power of qubits lies in their connections, made possible by couplers. These couplers allow qubits to influence each other and align themselves to similar or opposite states. The programmability of correlation strengths between coupled qubits via coupling strengths as well as biases define the problem landscape within the D-Wave quantum computer. The introduction of couplers induces entanglement, a quantum physics phenomenon in which two qubits are connected and share a common quantum state. This entanglement creates four possible states that represent different combinations of the two qubits. During the annealing process, the states of the qubits traverse this landscape of potential states before converging to a final state once the annealing process is complete [11,32].

As the number of qubits increases, the complexity of the system increases exponentially. For example, two qubits host four potential states, while three qubits host eight, illustrating the increasing computational complexity with qubit additions. This exponential increase in available states presents computational challenges in defining the energy landscape for problem solving and highlights the complexity of scaling quantum systems [16].

To implement this complex calculation process, several basic steps are crucial:

1. System description and initialization: In this first phase, a QC platform equipped with coupled rf-SQUID qubits is selected. The qubits must be configured to perform QA correctly. The construction involves establishing the initial state of the qubits and ensuring proper connectivity for interactions during the annealing process.
2. Determining the Energy Scale: It is important to understand the characteristic energy scales of the system. This step involves evaluating various interactions, such as qubit-qubit couplings and qubit-resonator interactions .It is critical to map the energy landscape of the problem onto these qubit interactions to ensure that the problem's Hamiltonian is compatible with the available energy scales.
3. Freezing point estimation: The identification of critical parameters that influence freezing behavior, such as: noise sources, decoherence times and energy relaxation time scales, is essential. Characterizing the freezing point involves determining the critical operating temperature at which quantum coherence and entanglement degrade due to ambient noise.
4. Annealing offset: Annealing offsets, which represent the energy gaps between the ground and excited states of the qubits during annealing, must be carefully calculated and optimized. In this step, numerical simulations or analytical methods are used to optimize these compromises and ensure efficient state transitions during the annealing process.
5. Variation of the global annealing schedule: The parameterization of the global annealing schedule includes the definition of parameters such as annealing duration, ramp profiles and intermediate time steps. Variations in this schedule are examined by changing ramp rates, total annealing times, and intermediate retention times in different phases, allowing a more detailed examination of the solution space [32].

4   Enhanced Machine Learning Using Quantum Annealing

In classical ML, parameter optimization in neural networks or clustering tasks often requires extensive computational resources to navigate the huge solution space. QA leveraging quantum fluctuations offers a potential way to efficiently traverse these complex spaces and potentially offer accelerated solutions to such tasks.

QA has proven its applicability in various ML domains. In classical hybrid quantum systems, ML tasks leverage QA by offloading certain aspects of optimization to quantum processing units to improve speed and accuracy. Researchers have leveraged QA features to train ML classifiers and validate their effectiveness on various datasets suitable for real-time applications. It has been particularly studied for feature selection, with the aim of detecting the most relevant features in high-dimensional datasets. In addition, QA promises the optimization of hyperparameters in ML models, thereby contributing to high model performance.

The efficiency of QA in tackling these complicated problems has sparked interest in extending classical ML algorithms with quantum-inspired techniques. This section addresses the potential of QA techniques and implementation approaches to extend and optimize critical ML tasks [15, 22, 25, 27].

4.1   Accelerating Hyperparameter Optimization

Hyperparameter optimization significantly impacts model performance, but examining large parameter spaces presents computational challenges. QA is revolutionizing this process by examining numerous hyperparameter settings simultaneously. For example, in natural language processing tasks, QA encodes hyperparameter optimization as an objective function, leveraging its ability to efficiently traverse complex landscapes. This approach reduces computational costs and accelerates the identification of superior hyperparameter settings that are critical to improving model accuracy and robustness [25].

Implementation Approach

To improve the implementation of hyperparameter optimization using quantum annealing, we proceed in different steps:

1. **Formulating Objective Function for Quantum Annealer:** Define an objective function representing model performance concerning hyperparameters.
2. **Encoding Optimization Problem:** Encode the hyperparameter optimization problem into a suitable Ising model or QUBO formulation.
3. **Leveraging QA for Exploration:** Utilize the quantum annealer to navigate through the parameter space concurrently, seeking optimal hyperparameter configurations efficiently.

4.2   Optimized Feature Selection

Feature selection is pivotal for model interpretability and performance but becomes challenging with high-dimensional datasets. QA offers a powerful approach to streamline this process by efficiently exploring vast feature spaces. In medical image classification, for instance, QA formulates feature selection as an optimization problem. By encoding feature subset selection into a QA framework, it effectively identifies subsets of features significantly contributing to predictive accuracy [10, 27].

Implementation Approach

To effectively implement feature selection leveraging QA, the following steps are essential:

1. **Problem Formulation:** Frame feature selection as an optimization problem to maximize predictive performance while minimizing computational complexity.
2. **Encoding Feature Subset Selection:** Map the feature selection problem to an appropriate Ising model or QUBO formulation.
3. **Exploration using QA:** Leverage QA to efficiently explore feature combinations and identify optimal or near-optimal feature subsets contributing significantly to model performance.

4.3 Refined Clustering Algorithms

Traditional clustering algorithms struggle with high-dimensional data and complex computations. QA optimizes clustering tasks by efficiently exploring configurations defining distinct clusters. For retail customer segmentation based on various attributes, QA frames clustering optimization as an objective function. By encoding clustering configurations into a QA framework, it efficiently identifies well-defined clusters, aiding targeted marketing strategies and offering deeper insights into customer behaviors [15].

Implementation Approach

To implement refined clustering algorithms using QA, a systematic approach includes:

1. **Objective Definition for Clustering:** Define an objective function to optimize cluster formation, emphasizing distinctiveness and intra-cluster coherence.
2. **Mapping to QA Formulation:** Encode the clustering optimization problem into a suitable Ising model or QUBO formulation.
3. **QA Exploration:** Utilize QA to navigate through potential cluster assignments, seeking configurations optimizing the defined objective function, resulting in well-defined clusters.

4.4 Streamlined Regression Model Optimization

Regression models require parameter tuning for precise predictions, posing challenges with high-dimensional parameter spaces. QA streamlines this by efficiently exploring parameter spaces. For predicting stock prices using market indicators, QA encodes parameter tuning into its framework, rapidly identifying parameter configurations minimizing prediction errors for more accurate predictions [8].

Implementation Approach

To effectively optimize regression models using QA, a structured approach includes:

1. **Objective Formulation:** Frame the regression model optimization as an objective function, minimizing prediction errors or optimizing performance metrics.
2. **Encoding Parameter Tuning:** Map the problem of parameter optimization to an Ising model or QUBO formulation.
3. **QA for Exploration:** Utilize QA to efficiently explore parameter spaces, identifying optimal or near-optimal parameter configurations that enhance prediction accuracy.

4.5 Optimized Reinforcement Learning

Reinforcement learning involves training agents to make sequential decisions in dynamic environments. QA efficiently navigates complex decision-making spaces. In autonomous robot navigation, QA optimizes policy discovery by encoding policy optimization into its framework, efficiently identifying near-optimal policies for effective decision-making [7, 18].

Implementation Approach

Implementing optimized reinforcement learning using QA involves the following steps:

1. **Policy Optimization Objective Definition:** Formulate an objective function aiming to maximize cumulative rewards in various states.
2. **QA Framework Encoding:** Map the policy optimization problem into a suitable Ising model or QUBO formulation.
3. **Efficient Policy Discovery:** Leverage QA to explore state-action spaces efficiently, identifying near-optimal policies for effective decision-making by reinforcement learning agents.

5 Optimization of Quantum Annealing Using Machine Learning

Optimizing the parameters and configurations of Quantum Annealers is crucial for achieving superior performance and solution quality in quantum computations. However, this process often involves intricate adjustments and explorations within the QC realm. Leveraging ML techniques can significantly expedite and refine the optimization of QA parameters, manifesting in the following ways:

1. **Hyperparameter Exploration:** ML algorithms, such as Bayesian optimization or genetic algorithms, efficiently traverse the parameter space of Quantum Annealers. By exploring various configurations (e.g., annealing schedules, qubit couplings), these algorithms identify settings that maximize quantum performance metrics or solution quality.
2. **Predictive Models for Performance Estimation:** Trained on historical QA performance data, ML models predict expected performance for different parameter configurations. Guiding the search process, these models narrow the exploration space towards configurations likely to yield superior quantum outcomes.
3. **Automated Feedback Loop:** Implementing a feedback loop where ML models learn from QA outcomes enables continuous adaptation and refinement of parameter settings.
4. **Transfer Learning for Generalization:** ML models trained across various quantum tasks transfer knowledge across similar problems or annealers. This knowledge transfer accelerates parameter optimization by applying insights gained from optimizing one QA to improve others.

Efficiently managing scheduling and anneal offsets in Quantum Annealers is crucial for faster convergence and more accurate results. Reinforcement Learning techniques provide dynamic adjustments to these parameters, enhancing QA. Utilizing Reinforcement Learning involves enhancing QA in more detailed ways:

1. **Dynamic Scheduling with Reinforcement Learning Policies:** Reinforcement Learning agents learn optimal scheduling policies by formulating scheduling problems as Markov Decision Processes (MDPs). Exploring action sequences efficiently schedules tasks, minimizing idle time and maximizing QA utilization.

2. **Anneal Offset Optimization via Reinforcement Learning Exploration:** Reinforcement Learning agents explore anneal offset spaces to minimize energy transitions during quantum computation. Learning from rewards based on energy transitions, Reinforcement Learning algorithms optimize the offset parameter to minimize errors and improve solution quality.

3. **Adaptive Parameter Adjustment using Reinforcement Learning:** Reinforcement Learning models dynamically adjust scheduling sequences and anneal offsets based on feedback from QA performance, ensuring optimal settings under varying quantum conditions.

4. **Policy Gradient Methods for Annealing Optimization:** Leveraging policy gradient methods, Reinforcement Learning models learn adaptable annealing policies. These policies dynamically influence the anneal offset, enhancing the stability and accuracy of quantum computations.

Integrating Reinforcement Learning for QA parameter optimization represents a major leap in QC. By leveraging ML's adaptability, Quantum Annealers can dynamically fine-tune settings, significantly boosting performance and accuracy. This approach has vast potential for exploring versatile QA strategies across different domains [3,4,6,29].

6   Discussion

The convergence of QA with ML introduces a transformative approach to address complex optimization challenges. QA enhances tasks like hyperparameter optimization, feature selection, clustering, regression, anomaly detection, and reinforcement learning by efficiently exploring vast solution spaces, leading to faster and more accurate solutions.

Moreover, the optimization of quantum annealers using ML techniques presents an innovative approach to fine-tune and optimize quantum computations. By leveraging ML algorithms, such as Bayesian optimization, genetic algorithms, and reinforcement learning, quantum annealers can be dynamically adjusted, leading to superior performance and solution quality.

The proposed integration of reinforcement learning techniques for dynamic adjustments to scheduling, anneal offsets, and parameter adjustments in quantum annealers represents a groundbreaking advancement. Reinforcement learning policies optimize scheduling, anneal offsets, and parameter adjustments, significantly enhancing the stability and accuracy of quantum computations.

Overall, the marriage between QA and ML, along with the optimization of quantum annealers using ML techniques, signifies a promising direction in advancing computational capabilities, paving the way for tackling increasingly complex real-world problems.

7   Conclusion

The fusion of quantum annealing and machine learning represents a pivotal approach in solving intricate optimization problems. Quantum annealing accelerates tasks across domains, while optimizing quantum annealers using machine learning techniques refines quantum computations. This convergence holds immense potential to reshape computational methodologies, providing unprecedented solutions for complex real-world challenges across scientific and industrial domains. The future presents an exciting era where quantum computing and machine learning intertwine, revolutionizing problem-solving capabilities.

References

1. Adachi, S.H., Henderson, M.P.: Application of quantum annealing to training of deep neural networks. arXiv preprint arXiv:1510.06356 (2015)
2. Albash, T., Rønnow, T.F., Troyer, M., Lidar, D.A.: Reexamining classical and quantum models for the D-wave one processor: the role of excited states and ground state degeneracy. Eur. Phys. J. Spec. Top. **224**(1), 111–129 (2015)
3. Ayanzadeh, R., Halem, M., Finin, T.: Reinforcement quantum annealing: a hybrid quantum learning automata. Sci. Rep. **10**(1), 7952 (2020)
4. Barbosa, A., Pelofske, E., Hahn, G., Djidjev, H.N.: Using machine learning for quantum annealing accuracy prediction. Algorithms **14**(6), 187 (2021)
5. Bian, Z., Chudak, F., Macready, W.G., Rose, G.: The Ising model: teaching an old problem new tricks. D-wave Syst. **2**, 1–32 (2010)
6. Chen, Y.-Q., Chen, Y., Lee, C.-K., Zhang, S., Hsieh, C.-Y.: Optimizing quantum annealing schedules with Monte Carlo tree search enhanced with neural networks. Nat. Mach. Intell. **4**(3), 269–278 (2022)
7. Crawford, D., Levit, A., Ghadermarzy, N., Oberoi, J.S., Ronagh, P.: Reinforcement learning using quantum Boltzmann machines. arXiv preprint arXiv:1612.05695 (2016)
8. Date, P., Potok, T.: Adiabatic quantum linear regression. Sci. Rep. **11**(1), 21905 (2021)
9. Willsch, D., et al.: Benchmarking advantage and D-wave 2000Q quantum annealers with exact cover problems. Quant. Inf. Process. **21**(4), 141 (2022)
10. Ferrari Dacrema, M.M., Moroni, F., Nembrini, R., Ferro, N., Faggioli, G., Cremonesi, P.: Towards feature selection for ranking and classification exploiting quantum annealers. In: Proceedings of the 45th International ACM SIGIR Conference on Research and Development in Information Retrieval, pp. 2814–2824 (2022)
11. Grant, E.K., Humble, T.S.: Adiabatic quantum computing and quantum annealing. In: Oxford Research Encyclopedia of Physics (2020)
12. Salloum, H., Mazzara, M., Bahrami, M.R.: Correction for the classical conditions for a collision in a three-body system using general relativity and machine learning. In: International Conference on Mathematical Modeling in Physical Sciences. Springer, Heidelberg (2024)
13. Salloum, H., Alawir, M., Asekrea, S., Alatasi, M.A., Bahrami, M.R., Mazzara, M.: Quantum advancements in securing networking infrastructures. In: Barolli, L. (ed.) AINA 2024, vol. 201, pp. xx–yy. Springer, Heidelberg (2024)
14. Kadowaki, T., Nishimori, H.: Quantum annealing in the transverse Ising model. Phys. Rev. E **58**(5), 5355 (1998)
15. Kumar, V., Bass, G., Tomlin, C., Dulny, J.: Quantum annealing for combinatorial clustering. Quant. Inf. Process. **17**, 1–14 (2018)

16. McMahon, D.: Quantum Computing Explained. Wiley, Hoboken (2007)
17. McGeoch, C.C.: Adiabatic Quantum Computation and Quantum Annealing: Theory and Practice. Springer, Heidelberg (2022)
18. Mott, A., Job, J., Vlimant, J.-R., Lidar, D., Spiropulu, M.: Solving a Higgs optimization problem with quantum annealing for machine learning. Nature 550(7676), 375–379 (2017)
19. Harden, M.: D-Wave sticks with its approach to quantum computing. TechCrunch (2019). https://techcrunch.com/2019/11/15/d-wave-sticks-with-its-approach-to-quantum-computing/
20. Neukart, F., Compostella, G., Seidel, C., Von Dollen, D., Yarkoni, S., Parney, B.: Traffic flow optimization using a quantum annealer. Front. ICT 4, 29 (2017)
21. Neven, H., Denchev, V.S., Rose, G., Macready, W.G.: QBoost: large scale classifier training with adiabatic quantum optimization. In: Asian Conference on Machine Learning, pp. 333–348. PMLR (2012)
22. Nath, R.K., Thapliyal, H., Humble, T.S.: A review of machine learning classification using quantum annealing for real-world applications. SN Comput. Sci. 2, 1–11 (2021)
23. Rogers, S.K., Kabrisky, M.: An Introduction to Biological and Artificial Neural Networks for Pattern Recognition, vol. 4. SPIE Press (1991)
24. Santoro, G.E., Tosatti, E.: Optimization using quantum mechanics: quantum annealing through adiabatic evolution. J. Phys. A: Math. Gener. 39(36), R393 (2006)
25. Sagingalieva, A., et al.: Hyperparameter optimization of hybrid quantum neural networks for car classification. arXiv preprint arXiv:2205.04878 (2022)
26. Eddin, S., Salloum, H., Shahin, M.N., Salloum, B., Bahrami, M.R., Mazzara, M.: Quantum microservices: transforming software architecture with quantum computing. In: International Conference on Advanced Information Networking and Applications. Springer, Heidelberg (2024)
27. Von Dollen, D., Neukart, F., Weimer, D., Bäck, T.: Quantum-assisted feature selection for vehicle price prediction modeling. arXiv preprint arXiv:2104.04049 (2021)
28. Wang, Y., Wu, S., Zou, J.: Quantum annealing with Markov chain Monte Carlo simulations and D-wave quantum computers. Stat. Sci. 362–398 (2016)
29. Wauters, M.M., Panizon, E., Mbeng, G.B., Santoro, G.E.: Reinforcement-learning-assisted quantum optimization. Phys. Rev. Res. 2(3), 033446 (2020)
30. Yulianti, L.P., Surendro, K.: Implementation of quantum annealing: a systematic review. IEEE Access 10, 12345–12356 (2022)
31. Ali, Z.: Fundamentals of neural networks. In: Intelligent Control Systems Using Soft Computing Methodologies, pp. 17–38. CRC Press (2001)
32. D-Wave Systems Documentation. https://docs.dwavesys.com
33. D-Wave Systems. Systems and Solutions. D-Wave Systems. https://www.dwavesys.com/solutions-and-products/systems

From Context to Forecast: Ontology-Based Data Integration and AI for Events Prediction

Jefferson Amará, Victor Ströele$^{(\boxtimes)}$, Regina Braga, and José Maria N. David

Graduate Program in Computer Science, Federal University of Juiz de Fora (UFJF),
Juiz de Fora, Brazil
jnamara@ice.ufjf.br,
{victor.stroele,regina.braga,jose.david}@ufjf.br

Abstract. This study focuses on event detection and prediction using long short-term memory (LSTM) algorithms implemented in ontology-based sensor data integration. By integrating data from various sources, we facilitate the creation of semantically enriched and contextually integrated data, thereby enhancing the capability for more holistic predictions. The study introduces a framework that leverages ontology models for data integration, fostering the development of semantic context within a sensor network. A feasibility study, conducted with actual sensor data obtained from hydrometric and hydrological stations, underscores the framework's proficiency in abstracting the data integration process, constructing context, correlating events, and predicting their occurrences. The feasibility study results demonstrate the framework's effectiveness and highlight the potential of combining ontologies and artificial intelligence to enhance data interpretation.

1 Introduction

For event detection and prediction, contextualized data is a crucial factor [1]. In the IoT and big data era, multiple and heterogeneous data sources are common, and integrating them is an important task for establishing data context [2]. Concerning sensor data sources, the integration of data for event prediction has garnered the attention of numerous researchers, who have proposed solutions and applications across various fields, including digital health, smart cities, logistics and supply chain, environmental risk, and disaster prevention [3–6].

In the context of sensor data sources, event prediction aims to foresee specific events or anomalies, allowing for proactive decision-making and timely responses. Numerous approaches to address this objective can be found in the literature [7]. Several studies have comprehensively reviewed the most common methods for handling event prediction [8]. The limitation and challenge observed in many existing studies is the absence of solutions that effectively consider data from diverse, integrated, and contextualized sources before applying AI methods. As a result, event predictions are often confined to a single data source, leading to incomplete insights and hindering the holistic view of predictions [9].

Furthermore, event prediction systems can enhance their ability to detect events by considering a network of integrated sensor data [10]. As highlighted by [11], the

L. Barolli (Ed.): AINA 2024, LNDECT 201, pp. 349–361, 2024.
https://doi.org/10.1007/978-3-031-57870-0_31

predicted events may exhibit correlations and mutual influences. In [8], authors further assert the existence of intricate dependencies among events.

Drawing upon the key insights, limitations, challenges, and suggested avenues identified in the previously cited works, the following research question was formulated:

RQ - How to semantically integrate data from multiple sources to detect and predict events of interest from these data?

This paper proposes a framework that continues the effort described in the previous work [10]. The framework aims to process, integrate, contextualize data, and detect, correlate, and predict events from multiple data sources. Ultimately, we aim to identify events of interest based on correlated events detected in integrated data, which might remain undetected or experience delayed detection when using a single data source.

To achieve this, we employed a Local Concept Schema (LCS), ontology, and AI models to integrate data from various sources, detecting (i) events based on data from a specific source and (ii) correlated events based on context from multiple data sources. Ontological terms and rules were employed to uncover semantic meaning, allowing for the proposal of a common context among detected events.

The methodological process encompassed five main steps:

- A Tertiary Study was conducted to comprehend the state-of-the-art and identify gaps and paths in semantic data integration, event detection, and prediction.
- An ontology model was developed for data integration, context building, and event detection.
- An artificial intelligence model was implemented for event prediction.
- An effort to design and build the processing of the proposed models through coding.
- An evaluation phase was conducted to assess the effectiveness of the proposed framework in integrating, detecting, and predicting events of interest.

The framework comprising the ontology and AI models was applied and assessed using real data from hydrometric and hydrological sensor stations. We define the ontology as a standardized representation for integrating data from diverse sources. The ontology also established context among these sources and detected events of interest. An AI model based on the integrated data predicted events on these sources. As a result, we initially contextualized and integrated data from these sensors, subsequently detecting and predicting flood events. The contexts were constructed based on the ontology model, while the predictions relied on the AI model. Shared contexts were identified based on location and flow direction among these stations, illustrating that events occurring at one station could influence events at another station.

In summary, the proposal makes the following contributions: (i) allows the integration of data from different sensors, (ii) provides the abstraction of sensor data structure, allowing add new data sources, (iii) relates events from different sensors, (iv) detect events of interest based on ontology definitions, (v) predict events of interest based on context and AI patterns recognition.

The paper is organized as follows: Sect. 2 presents key concepts on semantic data integration, event prediction, and the related work; Sect. 3 describes the framework for data integration and event prediction; Sect. 4 evaluates the proposal in a real-context environment (hydrological and hydrometric sensors); Sect. 5 presents final remarks.

2   Data Integration and Event Prediction

Data integration is critical in managing the diversity and abundance of available information. It offers a strategic approach to unify disparate sources, improve data quality, enrich context, and facilitate Big Data analysis. This integration not only supports informed decision-making but also enhances the effectiveness of AI systems [3]. Data integration contributes to data accuracy by providing a holistic view, empowering organizations and individuals to extract meaningful insights from the extensive data [2].

However, this process of data integration is not without its challenges. The data landscape presents obstacles such as managing source diversity (databases, applications, cloud, social networks, and IoT), handling massive data volumes, adapting to various data formats, and ensuring real-time integration for decision-making [9].

In this complex data landscape, the role of ontologies emerges as a potential solution [12]. Ontologies aid in defining semantic relationships between disparate datasets, contributing to standardization and semantic clarity. They address the challenge of heterogeneity by harmonizing data with diverse structures and meanings [4]. In sensor data, which includes information captured by IoT devices and environmental sensors, ontologies play a crucial role in overcoming the diversity in formats, structures, and contexts.

Sensor data integration goes beyond physical unification; it involves semantic harmonization to ensure a consistent understanding of the data [3]. Applying ontologies in sensor data integration allows for creating semantic structures defining and relating underlying concepts [5,12]. The World Wide Web Consortium (W3C) and Open Geospatial Consortium (OGC) propose the Semantic Sensor Network (SSN) as an ontology standardization for sensor data, formalizing the representation of terms and concepts [13].

A noteworthy aspect of sensor data integration is incorporating Artificial Intelligence (AI), particularly predictive capabilities. When combined with ontologies that semantically structure sensor data, advanced machine learning algorithms enable meaningful predictions [6]. Long Short-Term Memory (LSTM) networks, a type of recurrent neural network (RNN), find application in analyzing sensor data due to their proficiency in capturing long-range dependencies and temporal patterns [14].

This predictive capability enhances operational efficiency and allows for implementing preventive measures in response to anticipated events. The synergy between semantically integrated data sources, facilitated by ontologies, and the predictive capabilities of AI, exemplified using LSTM networks for sensor data, presents a powerful approach to navigating the challenges of the current Data Era.

2.1   Related Work

Considering this paper's research areas, a tertiary study was carried out. The objective was to identify articles addressing processes for semantic sensor data integration, event detection, and event prediction. Also, we were looking for related papers dealing with flood event prediction once our feasibility study relied on data from hydrometric and hydrological stations. The tertiary study aimed to identify specific studies to help understand the state-of-art [17].

We identified five tertiary studies relying on these topics. Two bring directions, patterns, and ongoing strategies on event prediction methods [1,8]. One focuses on edge technologies for disaster management [15]. Finally, another two of them propose solutions for sensor network AI-based disaster detection [7,16].

In [1], the study identifies the main mature and classical approaches to event prediction. It introduces a comprehensive taxonomy of event prediction approaches that transcends application domains. This cross-disciplinary perspective enables a better understanding of event prediction problems and opens roads for designing and developing advanced and context-independent techniques.

In [8], the authors systematically survey technologies, applications, and evaluations related to event prediction in the big data era. The study emphasizes that event prediction methods often need to forecast multiple facets of events, requiring the joint prediction of these heterogeneous yet correlated outputs. The authors also highlight the necessity of addressing complex dependencies among the prediction outputs. In contrast to conventional isolated tasks in machine learning and predictive analysis, event prediction involves events that can correlate with and influence each other.

In [16], the authors address the detection of flood-induced disasters by building a wireless sensor network and a decision model based on a Support Vector Machine (SVM). The network-based decision model observes changes in weather conditions compared to historical information at a specific location. The main limitations of the presented solution include the reliance on relative values to determine disaster detection thresholds, which may not be applicable in different locations and may require the collection of new data for each specific location. The authors suggest that future research could concentrate on devising methods to establish disaster detection thresholds that are more widely applicable and less reliant on specific historical data from a particular location.

In [7], the authors present a study on flood prediction modeling using long short-term memory (LSTM). The article suggests that the synchronized sequential input and output (SSIO) architecture is more adept at capturing long-term dependencies than the sequential input and single output (SISO) architecture. However, the discussion regarding the applicability of the proposed solution to alternative historical datasets and data integration is not addressed.

Building upon the insights from related works, we introduce a framework that extends the reviewed approaches, intending to advance abstraction in data integration and contextualized AI prediction. The objective is to enhance the detection and prediction of events of interest. In this regard, we propose an ontology extended from SSN to facilitate the abstraction of the data integration step. The presented ontology serves to standardize and integrate data by employing a canonical semantic representation for sensor data.

3   Data Integration and Event Detection Framework

This section presents a framework for sensor data integration and event detection and prediction. The framework input is data from multiple sensors from the same knowledge domain. The framework integrates data from the multiple sensors, correlate context and predict future events based on the integrated data. The output is the events detected. The following subsection outlines the key aspects of our approach.

The framework is structured and implemented through distinct layers, as shown in Fig. 1, providing an insight into the data flow, layer composition, and component interactions. We first describe each layer and the development process in this section. Section 4 evaluates the framework in real-world sensor data from hydrometric and hydrological stations.

In this work, the data preprocessing and enrichment were carried out with an ontology model extended from SSN, which can extract syntactic, semantic, and context knowledge from sensor data. This model considers pre-defined context to define sensor relationships and helps detect events based on the sensor network.

LSTM (Long Short-Term Memory) is applied for event prediction. LSTM is a Recurrent Neural Network (RNN) architecture widely used in Deep Learning. It captures long-term dependencies, making it ideal for sequence prediction tasks.

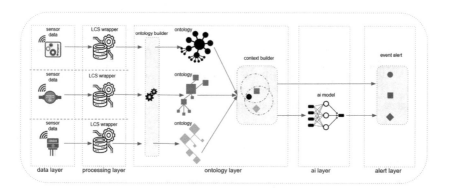

Fig. 1. Framework Overview

The **Data Layer** represents data acquisition facilitated by the sensors. At this stage, the data exhibits structure, syntax, and semantics heterogeneity, containing raw attribute values obtained from the monitored entities.

The **Preprocessing Layer** receives sensor data from the Data Layer and is tasked with establishing a canonical representation of this information. The *LCS wrapper* components construct this representation based on a *Local Concept Schema* (LCS), defining the transformation rules for converting sensor data into a canonical model.

Event classification rules and context between sensors are established within this layer with an ontology model extended from the SSN model, as shown in Fig. 2. Based on LCS definitions performed by the Preprocessing Layer, the **Ontology Layer** is responsible for process those results instantiating classes, data properties and object properties as show in Table 1. The instances are stored and subject to inference and reasoning processes.

In the ontology layer, rule-based event detection is defined by considering predefined thresholds for each data source. This allows the classification of observations as

(a)

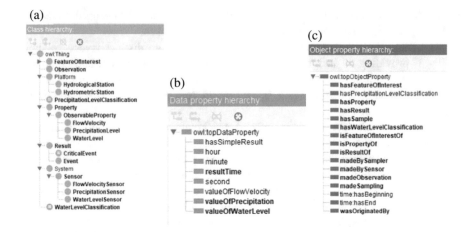

(b)

(c)

Fig. 2. (a) Class (b) Data Property (c) Object Property

Table 1. Details of Extended Ontology Classes

Class	Description	Instantiated by
Sensor	Represents a physical device capable of observing a property or phenomenon	Automatic extraction process
Observation	Represents an act of observing a property or phenomenon	Automatic extraction process
Platform	Represents a hosting entity for sensors, providing a context for their observations	Automatic extraction process
FeatureOfInterest	Represents the entity or property being observed	Predefined in ontology
ObservableProperty	Represents the property or phenomenon that is being observed	Predefined in ontology
Result	Represents the outcome or output of an observation or actuation	Predefined in ontology

events of interest when they surpass these thresholds. The results of this threshold-based event detection are then directly forwarded to the alert layer. The Semantic Web Rule Language (SWRL) is employed in this layer to express contextual rules within the ontology. Table 2 provides some of the implemented rules, such as the 'location_context_for' property, triggered when an event e occurs at station $s1$, and $s1$ is within the influence area of $s2$. Defining context is a crucial step, as the AI input layer relies on it.

The **AI Layer** stands out as a pivotal component, harnessing artificial intelligence techniques for advanced data analysis and event prediction. Leveraging a Long Short-Term Memory (LSTM) neural network, a recurrent neural network (RNN) type, this layer establishes connections between sensor data in the same context. The primary goal is to enhance the system's predictive capabilities based on the amalgamated information from these sensors.

Table 2. Context rule sample

SWRL Rules		
occurred_in_station(?e,?s1) ∧ is_influence_area_for(?s1,?s2)	⇒	is_location_context_for(?e,?s2)
occurred_at_time(?e,?t) ∧ is_in_range_time_to_influence(?t,?s)	⇒	is_temporal_context_for(?e,?s)
is_location_context_for(?e,?s) ∧ is_temporal_context_for(?e,?s)	⇒	is_influenced_by(?s,?e)

The LSTM neural network excels in handling sequential and time-series data, making it an effective choice for analyzing sensor readings over time. Trained on historical data from sensors, the network learns intricate patterns and dependencies among various environmental factors.

Our proposal works with a context-aware input selection where the LSTM model processes data from sensors based on the context. We adapt the standard LSTM model to handle this context-dependent input selection. We introduced a context definition that determines the input (x_t) on Layer 1.

This model consists of two LSTM layers with 64 hidden units in each layer. The output of the first LSTM layer ($h_t^{(1)}$) serves as the input to the second LSTM layer. The superscripts denote each LSTM layer's parameters (weights and biases). The input at each time step (x_t) is processed by the first layer, and the hidden states ($h_t^{(1)}$) become the input to the second layer. The final hidden state ($h_t^{(2)}$) output is passed through an additional output layer for binary classification.

Layer 1

$$i_t^{(1)} = \sigma(W_{ii}^{(1)}x_t + b_{ii}^{(1)} + W_{hi}^{(1)}h_{t-1}^{(1)} + b_{hi}^{(1)})$$
$$f_t^{(1)} = \sigma(W_{if}^{(1)}x_t + b_{if}^{(1)} + W_{hf}^{(1)}h_{t-1}^{(1)} + b_{hf}^{(1)})$$
$$o_t^{(1)} = \sigma(W_{io}^{(1)}x_t + b_{io}^{(1)} + W_{ho}^{(1)}h_{t-1}^{(1)} + b_{ho}^{(1)})$$
$$g_t^{(1)} = \tanh(W_{ig}^{(1)}x_t + b_{ig}^{(1)} + W_{hg}^{(1)}h_{t-1}^{(1)} + b_{hg}^{(1)})$$
$$c_t^{(1)} = f_t^{(1)} \odot c_{t-1}^{(1)} + i_t^{(1)} \odot g_t^{(1)}$$
$$h_t^{(1)} = o_t^{(1)} \odot \tanh(c_t^{(1)})$$

Layer 2

$$i_t^{(2)} = \sigma(W_{ii}^{(2)}h_t^{(1)} + b_{ii}^{(2)} + W_{hi}^{(2)}h_{t-1}^{(2)} + b_{hi}^{(2)})$$
$$f_t^{(2)} = \sigma(W_{if}^{(2)}h_t^{(1)} + b_{if}^{(2)} + W_{hf}^{(2)}h_{t-1}^{(2)} + b_{hf}^{(2)})$$
$$o_t^{(2)} = \sigma(W_{io}^{(2)}h_t^{(1)} + b_{io}^{(2)} + W_{ho}^{(2)}h_{t-1}^{(2)} + b_{ho}^{(2)})$$
$$g_t^{(2)} = \tanh(W_{ig}^{(2)}h_t^{(1)} + b_{ig}^{(2)} + W_{hg}^{(2)}h_{t-1}^{(2)} + b_{hg}^{(2)})$$
$$c_t^{(2)} = f_t^{(2)} \odot c_{t-1}^{(2)} + i_t^{(2)} \odot g_t^{(2)}$$
$$h_t^{(2)} = o_t^{(2)} \odot \tanh(c_t^{(2)})$$

In the first LSTM layer (Layer 1), $i_t^{(1)}$, $f_t^{(1)}$, and $o_t^{(1)}$ represent the input gate, forget gate, and output gate at time step t. These gates, calculated using sigmoid activation, regulate the flow of information. The candidate cell state $g_t^{(1)}$, determined by the hyperbolic tangent activation function, signifies new information that could be stored in the cell state. The updated cell state $c_t^{(1)}$ considers the forget and input gates, along with the candidate cell state and the previous cell state ($c_{t-1}^{(1)}$), denoted by $\odot$ for element-wise multiplication. The updated hidden state $h_t^{(1)}$ is a function of the output gate and the hyperbolic tangent of the updated cell state. The second LSTM layer (Layer 2) is analogous, adding that the final hidden state ($h_t^{(2)}$) it is directed through an additional layer for binary classification.

The **Alert Layer** selectively considers events classified as *events of interest* and, guided by ontology-defined detection and context conditions and LSTM predictions, triggers an "event alert" concerning another sensor station. The current context rules encompass *location* and *temporal* aspects, where *location* pertains to stations in the same area and flow direction, and *temporal* relates to stations whose influence is observed at another station with or without a time delay.

4   Feasibility Study

This section describes a feasibility study using real use case scenarios. The framework was used to integrate data and detect and predict events from data produced by sensors in hydrological and hydrometric monitoring stations. A feasibility study attempts to characterize a technology to ensure it does what it claims to do and is worth developing [19].

Our framework was evaluated considering both the semantic data integration and event prediction. In the Introduction Section, one Research Question was stated (RQ), and two secondary research questions (SRQ) were derived. The first secondary research question (SRQ1) is important to measure the capacity of semantically integrating data from multiple sources. The second secondary research question (SRQ2) investigates whether the decision-making system is achieving its goals by detecting and predicting events of interest.

(i) RQ - Can the framework semantically integrate data from multiple data sources to detect and predict events of interest from these data?
   – SRQ1 - Does using the proposed framework and implemented ontology model allow the integration of sensor data and add new data sources?
   – SRQ2 - Are the events of interest correctly detected and predicted?

This evaluation followed the GQM (Goal - Question - Metric) model [18], which determines that the study's objectives should be defined, followed by the research questions and metrics for evaluating the research questions. The scope of this evaluation and the Goal are described as follows: "**To analyze** the framework **for the purpose of** semantic data integration and event prediction **in relation to** sensor data **under the point of view of** software development decision-makers **in the context of** data from hydrological and hydrometric sensors". The metrics M1 and M2 aim to answer SRQ1, and M3 to answer SRQ2. The metrics are:

M1: The data representation of multiple and different sensors standardized as a semantic canonical representation.
M2: Number and diversity of sensor data sources integrated.
M3: AI model performance (precision, recall, true positives and false negatives).

4.1   Data Description

We obtained sensor data from hydrological and hydrometric domains. The data from hydrometric stations are available at https://wateroffice.ec.gc.ca/, and data from hydrological stations are available at https://acis.alberta.ca/acis/. Both are public data sources made available by the Government of Canada and Alberta. We also use data to validate event detection. The data is from a public data source available at https://www.gdacs. org/ and is made available by the Global Flood Detection System, which monitors floods worldwide using near-real-time satellite data.

The feasibility study considered data from eight hydrological stations (Jumpingpound Ranger, Elbow Ranger, Elbow Auto, Priddis, Calgary Springbank, Cop Upper, and Calgary International Airport), which monitor precipitation levels, and nine hydrometric stations (05BH004, 05BH010 05BH015, 05BJ001, 05BJ004, 05BJ008, 05BJ010, 05BH004 and 05BM904), which monitor the water level of the riverbed to which they are associated. The data is from 2005 to 2023.

The data provided by the hydrological sensors contains the following attributes: *"Station Name", "Date (Local Standard Time)", "Precip. (mm)", "Precip. Accumulated (mm)"*, among others. The hydrometric sensors have the attributes: *ID, PARAM, Date, and Value*, where the *ID* feature represents the station identification and *PARAM* means the parameter monitored by the station (water level or flow rate).

4.2 Experimental Results

We assume that the primary events of interest are linked to detecting high-water levels in hydrometric stations, as these events often signify potential flooding, posing a threat to people and buildings in the affected region. To identify these events, we leverage data from various sensors integrated within the framework of 'location' and 'time' aspects.

This section presents the outcomes obtained when the previously described sensor data undergoes analysis within the framework. We aim to semantically integrate data from diverse sensors and detect and predict events, addressing the two research questions introduced in the Introduction Section.

Integrating data from hydrological and hydrometric sensors considers factors such as their installation location, flow value and direction, mutual influence, and the temporal aspects of this influence. Figure 3 illustrates the diagram depicting the locations of stations within the network of integrated sensors. Each dotted arrow represents contextual relationships established based on the flow direction context.

For instance, by integrating data from hydrological sensors at Cop Upper with hydrometric sensors at 05BH004 (considering they share the same location context), precipitation data can be a feature for predicting water levels at 05BH004.

At this point, the framework cannot measure how much each hydrological station influences the hydrometric station, although there are mutual influences between hydrometric stations. To define the influence context from hydrological stations, we defined if there is influence or not based on the previous simultaneity of high precipitation and high-water level and station metadata. We also based our context rules in the study [20] on which the authors investigate causes, assessment, and damages in the area where data originated. Then, based on these data, we defined the context rules, as exhibited in Table 2. The definition of more accurate context rules depends on domain experts or context detection automation, which are points to be improved in future works.

To assess Metric M1, we compared the input data from hydrological and hydrometric station sensors with their representations post-data integration. The instances of

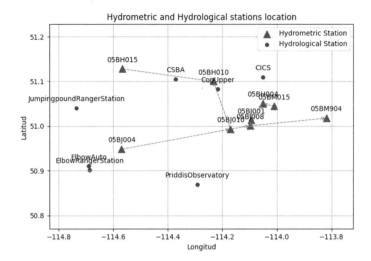

Fig. 3. Network location of sensor stations

integrated data representation preserve the original values from each inputted dataset while incorporating semantic values related to context.

Following our feasibility study, where data from eight hydrological and nine hydrometric sensors were successfully integrated, Metric M2 is deemed accomplished. This success is attributed to incorporating all contextual combinations between hydrological and hydrometric sensors and among hydrometric sensors themselves.

Tables presented in Fig. 4 and Fig. 5 represent the performance metrics of the classification model for two different hydrometric stations(e.g., 05BH004 and 05BJ001) and help us to answer the Metric M3. The classification task involves distinguishing between "Non-Critical" and "Critical".

Precision is the ratio of true positive predictions to the total predicted positives. Recall, also known as sensitivity or true positive rate, is the ratio of true positive predictions to the total actual positives. The F1-score is the harmonic mean of precision and recall, providing a balanced measure. Support is the number of actual occurrences of each class in the specified dataset.

Class	Precision	Recall	F1-Score	Support
Non-Critical	1.00	1.00	1.00	1029
Critical	0.93	0.87	0.90	15

Class	Precision	Recall	F1-Score	Support
Non-Critical	1.00	0.99	1.00	1797
Critical	0.83	1.00	0.91	45

Fig. 4. Station 05BH004   **Fig. 5.** Station 05BJ001

Examining, for example, the results for Hydrometric Station 05BH004, we observe high precision (1.00) for the "Non-Critical" class, indicating accurate predictions when this class is identified. The model demonstrates strong performance with F1-scores of 1.00 and 0.90 for "Non-Critical" and "Critical" classes, respectively. However, the recall for the "Critical" class is slightly lower (0.87), suggesting that the model may overlook some instances of this class, which is undesirable. Upon analyzing the two missed events at this station, we have identified the following contextual information:

- Context 231:
  $[CICS : precipitation\_value : 0; 05BH004 : water\_level\_value : 4.12887]$
- Context 232:
  $[CICS : precipitation\_value : 0; 05BH004 : water\_level\_value : 3.99932]$

This implies a potential bias in the model towards prioritizing the precipitation feature over the water level feature. Consequently, the model may miss critical events when heavy rain has ceased but high water levels persist in a specific location. Ongoing adjustments to the model are essential to rectify this behavior as part of our continuous research efforts.

Even though the AI model did not predict the critical event, the "double-check" implementation, which includes both the ontology rule and the AI model, ensures that the ontology-rule threshold still detects it. This guarantees that the event is triggered within the Alert layer, facilitating affordable and reliable decision-making.

In summary, from the results, we depict that if we only monitor the immediate on-site sensor data, we can ascertain that the water level is high. Still, this information proves insufficient for making informed decisions. By tracking the evolving measurements from the on-site sensor, we can discern that the water level is rising, providing us with a degree of advance notice to implement necessary measures.

However, if we monitor both the immediate data and the progression of the on-site sensor alongside additional sensors within the same context, we can forecast that the water level will rise. This comprehensive approach provides a more robust foundation for anticipating and implementing preventive measures.

4.3 Discussion

Considering the results and the metrics we adopted, we can answer the SRQs:

(i) RQ - Can the framework semantically integrate data from multiple data sources to detect and predict events of interest from these data?
   - SRQ1 - Does using the proposed framework and implemented ontology model allow the integration of sensor data and add new data sources?
   - SRQ2 - Are the events of interest correctly detected and predicted?

SRQ1) Do the use of the proposed framework and implemented ontology model allow the integration of sensor data and add new data sources? *Totally.* Despite being hydrological or hydrometric sensors, we demonstrate that we can integrate data from different sensors using ontologies as a canonical model. Furthermore, the framework can add new data sources by including a file containing LCS definitions.

SRQ2) Are the events of interest being correctly detected and predicted? *Totally*. We show that the framework can detect and predict events from different sensors with expanded context using integrated data and context definitions in the analyzed data. Although the context attribute enrichment depends on the domain expert, the framework detects and predicts the events through ontology-based rules or AI models.

Thus, by answering these two secondary research questions, we can answer the main research question once the proposal's core is *Semantic Data Integration* and *Event detection and prediction based on context*. **RQ) Can the framework semantically integrate data from multiple data sources in order to detect and predict events of interest from these data?** *Totally*. The framework promoted semantic data integration and, based on context, detected and predicted events from different data sources.

5 Conclusions and Future Works

This paper proposes using ontologies for semantic data integration based on context to event detection and prediction. As a continuous implementation effort after the previous work [10], we used data from monitoring sensors from hydrological and hydrometric stations. The experimental results indicate the proposal's feasibility once it can semantically integrate data and detect and predict events, including anticipating detection based on context.

In future work, we intend to evolve to support automating the context detection attributes, building more complex context rules, implementing provenance models for event detection, and improving the AI model for dealing with bias and better performance. With these new issues, we will conduct a new evaluation cycle with new and different domains and data sources.

References

1. Gmati, F.E., Chakhar, S., Chaari, W.L., Xu, M.: A taxonomy of event prediction methods. In: Wotawa, F., Friedrich, G., Pill, I., Koitz-Hristov, R., Ali, M. (eds.) IEA/AIE 2019. LNCS (LNAI), vol. 11606, pp. 12–26. Springer, Cham (2019). https://doi.org/10.1007/978-3-030-22999-3_2
2. Tan, W.-C.: Deep data integration. In: Proceedings of the 2021 International Conference on Management of Data, p. 2 (2021)
3. Krishnamurthi, R., et al.: An overview of IoT sensor data processing, fusion, and analysis techniques. Sensors **20**(21), 6076 (2020)
4. Liu, J., et al.: Towards semantic sensor data: an ontology approach. Sensors **19**(5), 1193 (2019)
5. Thirumahal, R.: Semantic integration of heterogeneous data sources using ontology-based domain knowledge modeling for early detection of COVID-19. SN Comput. Sci. **3**(6), 428 (2022)
6. Chukkapalli, S.S.L., et al.: Ontologies and artificial intelligence systems for the cooperative smart farming ecosystem. IEEE Access **8**, 164045–164055 (2020)
7. Li, W., Kiaghadi, A., Dawson, C.: Exploring the best sequence LSTM modeling architecture for flood prediction. Neural Comput. Appl. **33**, 5571–5580 (2021)

8. Zhao, J.: Event prediction in the big data era: a systematic survey. ACM Comput. Surv. (CSUR) **54**, 1–37 (2021)
9. Amnai M., et al.: A survey on the challenges of data integration. In: 2022 9th International Conference on Wireless Networks and Mobile Communications (WINCOM), pp. 1–6 (2022)
10. Amara, J., Ströele, V., Braga, R., Bauer, M.: Sensor data integration using ontologies for event detection. In: Barolli, L. (ed.) AINA 2023. LNNS, vol. 661, pp. 171–183. Springer, Cham (2023). https://doi.org/10.1007/978-3-031-29056-5_17
11. Matsubara, Y., et al.: Fast mining and forecasting of complex time-stamped events. In: Proceedings of the 18th ACM SIGKDD International Conference on Knowledge Discovery and Data Mining, pp. 271–279 (2012)
12. Asfand-E-Yar, M., Ali, R.: Semantic integration of heterogeneous databases of same domain using ontology. IEEE Access **8**, 77919 (2020)
13. Compton, M., et al.: The SSN ontology of the W3C semantic sensor network incubator group. J. Web Semant. **17**, 25–32 (2012)
14. Wang, L., et al.: A deep learning-based high-temperature overtime working alert system for smart cities with multi-sensor data. Nondestruct. Test. Eval. (2023)
15. Aboualola, M., et al.: Edge technologies for disaster management: a survey of social media and artificial intelligence integration. IEEE Access (2023)
16. Al Maghayreh, E., Alomari, A., Eldos, T., Saleh, A., Jararweh, Y.: Wireless sensor networks applications: a comprehensive survey. Comput. Mater. Continua **63**(3), 1261–1292 (2020)
17. Keele, S., et al.: Guidelines for performing systematic literature reviews in software engineering. Technical report, ver. 2.3, EBSE Technical report, EBSE (2007)
18. Caldiera, G.V.R.B., Rombach, H.D.: The goal question metric approach. In: Encyclopedia of Software Engineering, pp. 528–532 (1994)
19. dos Santos, R.P.: Managing and monitoring software ecosystem to support demand and solution analysis. Ph.D. thesis, Universidade Federal do Rio de Janeiro (2016)
20. Pomeroy, J.W., et al.: The 2013 flood event in the South Saskatchewan and Elk River basins: causes, assessment and damages. Can. Water Resour. J./Revue canadienne ressources hydriques **41**(1–2), 105–117 (2013)

FEVER: Intelligent Behavioral Fingerprinting for Anomaly Detection in P4-Based Programmable Networks

Matheus Saueressig[1], Muriel Figueredo Franco[1(✉)], Eder J. Scheid[1],
Alberto Huertas[2], Gerome Bovet[3], Burkhard Stiller[2], and Lisandro Z. Granville[1]

[1] Institute of Informatics (INF) – Federal University of Rio Grande do Sul (UFRGS),
Porto Alegre, Brazil
{msaueressig,mffranco,ejscheid,granville}@inf.ufrgs.br

[2] Communication Systems Group CSG, Department of Informatics IfI – University of Zurich
UZH, Zürich, Switzerland
{huertas,stiller}@ifi.uzh.ch

[3] Cyber-Defence Campus, Armasuisse Science and Technology, Thun, Switzerland
gerome.bovet@armasuisse.ch

Abstract. The evolving computer network landscape has enabled programmability in various network aspects, including Software-defined Networking (SDN) for control plane programmability and the introduction of the Programming Protocol-independent Packet Processors (P4). P4, a vendor-independent protocol, allows programmability on the data plane, offering flexibility for new services and applications. However, this flexibility introduces the need for automated solutions to monitor and manage the security of evolving networks and services. In this work, we propose FEVER, a framework utilizing P4-based telemetry and network device (switch) resource consumption to create fingerprints of network and P4 application behaviors. FEVER provides a comprehensive approach to identifying network anomalies through various metrics. The framework was evaluated in a virtualized scenario using unsupervised Machine Learning (ML) algorithms to detect diverse P4 program behaviors and traffic overload, demonstrating its potential for early detection of malicious activities in programmable networks. The results indicate high accuracy in identifying misbehavior and detecting sudden changes in P4 programs affecting the network.

1 Introduction

Society and companies' increasing demands for high-speed and reliable communication propel the evolution of computer networks. Programmable networks have emerged as a pivotal approach to these evolving requirements. This approach provides the flexibility and adaptability necessary to accommodate the diverse services and applications that enable the evolution of communications [13]. Examples of programmable networks and facilitating technologies include the concepts of Software-defined Networking (SDN) [1] and Network Functions Virtualization (NFV) [3].

L. Barolli (Ed.): AINA 2024, LNDECT 201, pp. 362–373, 2024.
https://doi.org/10.1007/978-3-031-57870-0_32

Programmable networks allow behaviors and services to be changed rapidly and fashionably. SDN improves network management by decoupling the control plane from the data plane, thus building more flexible and efficient networks by running intelligent controllers out of the switches [1]. However, SDN is still dependent on protocols like OpenFlow, which might be negatively impacted by the different implementations of data path hardware that vary according to vendors. This makes complex the management of different types of switches [7]. There are also approaches emerging as an ally to add programmability for the data plane. The Programming Protocol-independent Packet Processors (P4) is a protocol and vendor-independent solution [4] that defines new packet processing protocols without needing specific hardware support. It enables the development of protocols tailored to specific network requirements. Also, it allows the monitoring of insightful metrics using frameworks that allow for collecting and reporting network states by only using the data plane [16].

One critical aspect of computer networks is cybersecurity, which is not an exception for programmable networks and P4 implementations. Currently, there are research efforts towards security solutions to detect and mitigate cyberattacks, such as P4-based firewalls [17] and detection of Distributed Denial-of-Service (DDoS) using Machine Learning (ML) [12]. Also, there are considerable efforts for network verification based on assertions to identify faulty P4 programs [19] and bug-free programs [6]. Although such solutions are promising, the existing solutions focus on network traffic and targeting specific cyberattacks, thus not covering the detection of possible malicious behaviors from the switch and P4 program perspective. Therefore, there is still room for novel approaches correlating vast amounts of data and metrics available for a fine-grained analysis of networks and P4 program behaviors during run-time, thus allowing for detecting malicious behavior in networks considering both traffic and P4 programs running. Behavioral fingerprinting can be an ally for such analysis as it can provide a deeper understanding of device behavior beyond traditional monitoring metrics. Examples of scenarios where behavioral fingerprinting was successfully applied include the detection of ransomware in resource-constrained devices [15] and, on programmable networks, for identifying operating systems running on hosts [2].

Therefore, we advocate that intelligent behavioral fingerprinting [15] can be used to model patterns and characteristics of network devices and P4 applications during routine operation. After that, patterns could be used to identify anomalies in the traffic of P4 programmable switches or behavior changes during the execution of a Firewall developed as a P4 program, among others. We assume that each device and program may exhibit unique traffic flow characteristics, protocol usage, resource consumption, and responses to varying network conditions. By capturing and analyzing such unique behavioral patterns, behavioral fingerprinting can offer a more comprehensive and context-aware perspective for network management activities. Based on that, it is possible to identify malicious behaviors on programmable networks even before they become a more complex problem, including cyberattacks at early phases and malicious changes in P4 programs.

Thus, in this work, we propose FEVER, a framework for behavioral fingerprinting of programmable switches running P4 applications to identify anomalies by combining resource consumption and network telemetry. For that, the framework *(i)* identifies a set of metrics (*e.g.*, network-centric and resources consumption) to be used for the behavioral fingerprinting of programmable networks and *(ii)* describes a clear path to generate the fingerprints. Furthermore, the FEVER framework consists of two unsupervised ML algorithms [18] implemented and trained to identify anomalous traffic (*e.g.*, elephant flows and DDoS) coming from different hosts and modifications on P4 program code that change their behaviors. The feasibility and performance of FEVER is evaluated in a realistic virtualized scenario composed of virtual P4-based switches and P4 programs. The proposed ML algorithms are evaluated based on the F1-score and application scenarios emulated using Mininet and bmv2 switches, followed by a discussion of key findings. The results show that our frameworks can identify different P4 program behaviors and traffic overload in specific switches by looking at the behavior fingerprinting generated based on switches' resource consumption.

The rest of this document is organized as follows: Sect. 2 presents related work. The FEVER framework is described in Sect. 3, followed by evaluation and discussion in Sect. 4. Finally, in Sect. 5, the conclusion and future work are presented.

2   Related Work

We conducted a systematic literature, particularly of applications for detecting and mitigating cyberattacks and managing cybersecurity in programmable networks. Work focused on three aspects was identified: *(i)* network availability, *(ii)* network security, and *(iii)* privacy.

P4 programs have been developed to ensure greater network availability and perfect functioning from a performance and security point of view. Such applications include protecting against impersonation attacks by filtering malicious traffic [9], identifying DDoS attacks by statistically analyzing traffic flows [5], and also verifying P4 programs using static checks [6] and assertions [19] to identify possible execution faults. Furthermore, P4 has also been used to identify devices' Operational Systems (OS) and react to them accordingly, such as dropping packets or defining rate-limit for specific OS types [2]. However, most of this work focuses primarily on traffic analysis only or has too specific use cases for cyberattack detection. Such solutions also have limitations in identifying dynamic changes on the network, such as when a P4 program is maliciously replaced or changed by network operators or even when a potential cyberattack is imminent (*i.e.*, anomalous behaviors started). Thus, we argue that identifying anomalies at an earlier stage requires an intelligent monitoring of devices individually to prevent anomalies from propagating throughout the network.

Therefore, although there are several emerging applications for P4 programs and programmable networks, there is still a need for automated solutions that allow for proper monitoring and management of the security of existing networks and services. Current challenges include limitations in memory usage and accessibility for P4 program development and using collected metrics for effective performance and security applications [8]. There is also the need to make the network more robust and autonomous, which involves combining telemetry and ML elements to create infrastructures that adapt to the needs of the network and can predict possible failures. Such elements can support better detection of cyberattacks and anomalies while improving the detection performance of interested behaviors.

ML can be an ally to address such issues due to its potential to understand complex data patterns and adapt to heterogeneous scenarios based on different training datasets [18]. The opportunity has therefore arisen to implement approaches based on ML to analyze statistical metrics and resource usage behavior in order to identify anomalies before such behavior becomes a problem for the operation of services in programmable networks [11], such as a DDoS attack or a malicious change to a network program. To do this, we can use the concept of behavioral fingerprinting, which, although there are applications of the concept in other scenarios (*e.g.*, malware detection in IoT scenarios) [15], is still is underinvestigated in the context of programmable network security.

3 The FEVER Framework

FEVER is proposed to explore programmable device behavior fingerprinting for misbehavior detection in programmable switches, traffic, and network programs running as part of the network (*i.e.*, P4-based programs). Behavior fingerprinting can be defined as a collection of metrics of an object with expected values over time. The behavioral fingerprint of a network device might include information about how it handles traffic, processes packets, uses computational resources, and responds to various commands and requests. This data is collected through monitoring techniques and can be used to establish a baseline of normal behavior for the device and network applications running. When anomalies occur (*e.g.*, unusual traffic patterns, unexpected responses, or deviations from the established baseline), it may indicate the presence of malicious activity, network attacks, or hardware/software issues. Therefore, by continuously monitoring and comparing the device's current behavior to its behavioral fingerprint, network administrators can detect and respond to potential threats or network performance problems proactively.

Figure 1 shows the FEVER architecture, with the different components and steps described. The architecture is divided into three modules: *(i)* Data Analysis, which represents the tasks related to the definition of a possible range of metrics and also processing collected data, *(ii)* Behavior Fingerprinting, which determines the scenarios to be considered (*e.g.*, normal and anomalous) and train the models to identify the different behaviors of P4 programs and devices; and, finally, the *(iii)* Testbed represents the environment used to monitor the behaviors in order to create datasets for behavior fingerprinting and also detect anomalies. Each of these modules and their respective components are described below.

In the *Data Analysis*, the first step consists of defining candidate behavioral metrics to be monitored, such as memory and Central Processing Unit (CPU)-related resources, In-band network telemetry framework, and network policies. Such metrics might change according to the environment in which the programmable network is built (*e.g.*, Intel Tofino or a Mininet-based emulation). After defining metrics, in the next step, it is necessary to select those that are relevant to detect different behaviors. For

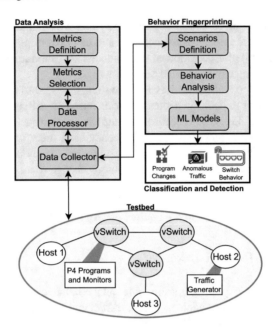

Fig. 1. FEVER Architecture

that, applying different techniques to identify the correlation and dependence between metrics is possible. Possible techniques include heatmaps to understand the relationship between metrics and linear plots to verify the behavior of two or more metrics under regular and anomalous circumstances.

The tasks involved in the collection and processing of data are also part of the *Data Analysis* module. The *Data Processor* is a component that receives data collected by the *Data Collector* and refines the data to make it more digestible by the *Metrics Selection* and other modules. Such refinements include removing noise from the raw data and preparing to plot the results for behavior analysis. The *Data Collector* also communicates with the *Behavior Fingerprinting* module to collect data from the monitors and whether additional data processing is needed.

For the FEVER framework, we consider both in-device and externally collected behavior sources. Examples of metrics mapped and being considered in our implementation are shown in Table 1. The resource consumption is divided in terms of CPU and Random-Access Memory (RAM), which also include different levels of granularity, such as CPU migrations, Instructions and cycles, page faults, and Resident Set Size (RSS). Such metrics allow us to understand behavior better and generate the fingerprints. Additional metrics can also be considered, including Ternary Content Addressable Memory (TCAM), a high-speed memory widely used in networking devices and programmable switches. To collect resource data from the switches, we have written a shell script that runs the Linux *perf* and *proc* commands simultaneously, while *iperf* was integrated into the script to create different flow behaviors according to the test needs, such as a normal flow of packets between all switches and specific elephant flows.

Table 1. Overview of Metrics Considered for Behavioral Fingerprinting

Metric	Example of Usage	Monitoring Method
CPU	The CPU usage is related to an abnormal increase or decrease of instructions in case of an anomaly. Bugs and malicious attacks can increase the usage of CPU	proc
RAM	RAM monitoring helps us to detect anomalies if they allocate memory to do malicious activity or due to misconfiguration	perf
Queue Depth	Anomalies might affect packet processing by slowing it, thus increasing the congestion. The Package Queue may increase if an anomaly changes the behavior of the switch	INT
Syscalls	Analysis of additional events running in background can improve the detection performance	perf
Processes	It is important to identify processes that represents virtual switches or that perform core functions	ps aux
Packet Header	Analysis of per-packet headers to statistically characterize switch behavior	INT and Wireshark

Figure 2 shows an example of collected metrics selection using a heatmap. It can be observed that there is only one light diagonal in the matrix with the value 1, showing that the high correlation is only between the feature itself. The features with high correlation can be maintained since they did not correlate to other values more than once and due to the high granularity of the values. After the heatmap analysis, a line plot can be provided for a more in-depth analysis of overlapping metrics.

The fingerprinting is finally built in the last module. For that, the scenarios are defined to determine which behaviors have to be considered for the generation of the training dataset and traffic generation. For example, P4 program actions and behaviors can be monitored to identify program changes and bugs, while switch resource consumption can be analyzed to understand anomalous behavior. The *ML Models* are then generated to identify program changes, abnormal traffic before it disrupts the network, and anomalous detection by monitoring the behaviors of network switches. Different ML techniques can be used for behavioral fingerprinting because they can detect complex data patterns and handle multi-variate data.

An instance of the FEVER is implemented as a proof-of-concept, running on Mininet and providing ML models based on One-Class Support Vector Machine (OCSVM) and Local Outlier Factor (LOF) as unsupervised learning for behavior fingerprinting. These ML techniques were used because of their simplicity and potential to identify outliers based on normal behaviors. Also, scripts were implemented for monitoring virtual switches and generating traffic according to specific scenarios. The source-code of the FEVER and datasets are publicly available at [10].

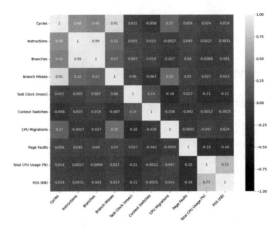

Fig. 2. Heatmap for Metrics Correlation and Selection

4   Evaluation

To validate our approach, we have developed a data pipeline focused on monitoring how anomalies (*e.g.*, unexpected traffic and changes on P4 programs) alter the CPU and memory usage in individual switch devices and compared this situation with a regular data flow. For this experiment, we have created a fingerprint of P4-enabled switch considering the CPU and memory metrics, thus, analyzing the resources usage of processes related to bmv2 while running a Mininet network. We run a topology of three switches and three hosts, configured as the testbed shown in Fig. 1. All the switches run a P4 program called Multi-Hop Route Inspection (MRI), which enables users to monitor the routes taken by packets and their associated queue lengths. A Python script was developed to set up the Mininet environment and configure the hosts and bmv2 switches. As described in the framework, *iperf* was also used to create the traffic flows to simulate behaviors.

The monitoring scripts implemented by FEVER collect metrics each second and save the collected data to a Comma-Separated Values (CSV) file for each switch for one hour. Each virtual switch's Process ID (PID) is identified for that. As Mininet treats each switch as a single process, it is possible to identify the PID corresponding to the switch activity and monitor it using *proc* and *perf* commands. Thus, after one hour of running the simulation, we gathered the monitored data from CSV files generated with the PID of the switches being monitored. Each experiment consisted of running the desired P4 program (from the p4lang tutorial repository) [14] for one hour with modifications to generate the desired traffic.

The behaviors considered consist of (*i*) normal traffic, (*ii*) anomalous traffic, and (*iii*) modified P4 behavior. Normal traffic was defined as simple requests that use less than 10% of network capacity sent from two clients to a server. The (*ii*) anomalous traffic is a high-traffic flow in which the hosts flood the network using the maximum available throughput. For the (*iii*) modified behavior, we consider an MRI with modifications, such as a conditional branch and arithmetic operations.

The experiments have to be run until the behaviors can be precisely modeled. For each round, we create a heat map of all metrics and drop the metrics with high correlation, such as all metrics with 99% of correlation identified using heatmaps (*cf.* Fig. 2). Next, a manual analysis is performed to verify potential clear behaviors and outliers, thus, allowing the human in the loop to calibrate the model and understand its feasibility. In case of simple scenarios, a basic set of rules can be defined, but when this manual verification shows complex correlations, our ML models must be used. We ran our experiments three times for each behavior to create our training dataset for normal, anomalous, and modified behaviors. Therefore, the experiments were run nine times, with a 1-hour duration each. All features were selected and extracted following the FEVER framework as introduced in Sect. 3. We used 3,240 samples of normal behavior for training and 6,840 samples of all behaviors (normal, anomalous traffic, and modified P4 program) for evaluation.

The performance of the ML models is analyzed using the well-established metrics *recall, precision, and F1-score*. F1-score is calculated as the harmonic mean of precision and recall, where precision is the number of true positives divided by the total values predicted as positive (*i.e.*, true and false positives) and the recall is calculated as the true positives divided by the sum of true positives and false negatives.

4.1 Detection of Anomalous Traffic

This experiment shows the performance of FEVER to detect abnormal data flow increases through a switch to identify possible flooding attacks earlier, such as Ping Flood, SYN/ACK Flood, and HTTP/HTTPS Flood. We have used the *iperf* tool to create anomalous traffic to send UDP packets. The default behavior of *iperf* when using UDP is to send data as fast as possible, without any specific rate limiting, *i.e.*, it floods the network with UDP packets to measure the maximum throughput. We have considered this scenario ideal since it would not disrupt the network but would send enough traffic to be detected. In a real-world scenario, detecting an increase in flow before a DDoS caused by the data flood would be beneficial.

Both Host 1 and Host 2 behave like clients, and Host 3 is the server. Clients send UDP packets to the service, creating a high-traffic flow through Switch 3, the one connected to Host 3. Since the high-traffic flow was directed to Host 3, we analyzed only Switch 3 to detect the anomaly. The RSS was identified as the best metric for such a scenario based on initial analysis (*e.g.*, heatmaps, overlapping, and line plots) of metrics collected. After the heatmap analysis, line plots were provided to identify metrics with overlapping or unidentifiable anomalous behavior, which have to be automatically identified by ML algorithms. The dropping of such overlapping increases the accuracy of detection algorithms, but it is not a realistic task in real environments since abnormal behavior is unknown in real environments. Therefore, the ML models must be used to understand metrics even when overlapping happens.

Figure 3 shows the overall performance of both ML models implemented using different features for anomalous traffic identification, including all features combined. Even though RSS alone has a maximum performance, it is dangerous to rely on it alone since this could indicate an overfitting of the model. Anomaly detection relies on robustness and identifying unknown patterns incompatible with normal behavior. Using only one feature might incapacitate the ability of different scenarios to be analyzed, and it generates trained models for specific situations rather than a generalized approach. Using all features has a high F1-Score, so it would not compromise the overall analysis, and it would guarantee the possibility of more data from unknown patterns being collected, thus identifying anomalies.

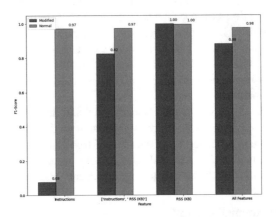

Fig. 3. F1-score for the Detection of Anomalous Traffic using Different Features

The OCSVM achieved an maximum accuracy for identifying anomalous behavior in terms of high traffic in a determined switch. Since it is an ML algorithm designed for situations where only examples of the normal class are available during the training phase, it learns a boundary around the regular instances. It classifies any value outside this boundary as an anomaly. The magnitude difference between the RSS memory and lack of overlapping helped the algorithm ideally detect the anomaly. However, relying only on features high relevance by dropping other features might overfitting the model since the nature of an anomaly is unknown. OCSVM have learned a boundary that effectively captures this separation. This can be verified if we drop the RSS feature (the most relevant metric). In this scenario, OCSVM performs poorly, with a recall of 0.43 and a precision rate of 0.56.

The LOF also scored the same as the OCSVM, thus, achieving a perfect accuracy for scenarios where the anomaly is known. LOF, in particular, is designed to identify local deviations from most data points. Therefore, if anomalies form distinct clusters or have noticeable local differences, LOF can excel in detecting them. As we have isolated regions detected by the LOF, the overlapping did not interfere in the model performance. In conclusion of such an experiment, memory analysis is shown to be a reliable source to detect flow changes and, thus, interesting to prevent flood attacks or network disruptions since, with more packets being processed, more memory needs to be allocated to operate the switch processes properly.

4.2 Detection of Changes on P4 Programs

We also created behavior fingerprinting for two scenarios: *(i)* when the same program is running but with possible malicious modifications and *(ii)* when a different program is running (*i.e.*, program identification). Figure 4 shows the overall F1-score for identifying changes in a P4 program (*i.e.*, original MRI but with a basic math operation and new variables included in the source-code). The F1-score is provided for both LOF and OCSVM using different features (*e.g.*, CPU instructions and RSS combinations) to identify if the running program is correct (normal) or modified according to scenario *(i)*. It shows that the RSS still plays an important role when identifying modified behavior, where CPU instructions can be used as an ally to identify normal behavior.

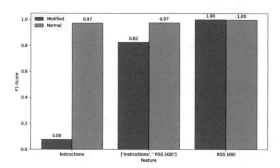

Fig. 4. F1-score for the Detection of Changes in the Original P4 Implementation of the MRI

Besides that, we have checked if the behavioral fingerprinting can also identify which program is running. For that, different metrics can be used since the program's resource consumption vary according to their functionalities and processing demands. Figure 5 shows that each metric's percentage varies according to a pre-defined baseline. For that, we have used the resource consumption of a *Basic Forward* as a baseline and analyzed how much (in terms of percentage) each of the following P4 programs surpasses the baseline resource usage: *MRI, Link Monitor, and Explicit Congestion Notification (ECN)*. All P4 programs, including the baseline, are the same as those available at the p4lang repository [14].

As can be seen, all metrics vary for the programs, especially those related to memory and CPU usage. This can be used then to create a fingerprint to identify specific programs running in the network and also highlight if malicious changes are made in the switch, such as when a malicious operator replaces P4 programs to affect the network service chain.

By using such metrics to determine the behavior fingerprint of each P4 program, we were able to classify 100% the correct problem, thus identifying what is running in the switch by only looking at the resource consumption. When dropping some metrics (*e.g.*, RSS, page faults, and CPU migrations), our model achieves a performance of 95% accuracy in identifying the correct program running.

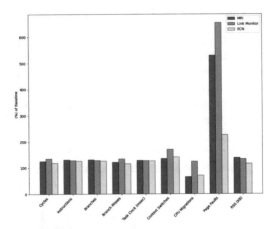

Fig. 5. Analysis of Consumption Behaviors of Different P4 Programs

5   Conclusions and Future Work

This work proposed FEVER, a framework for the behavioral fingerprinting of programmable networks, including detecting the misbehavior of P4-based switches and P4 programs. ML-based models are employed together with statistical processing to understand metrics that highlight potential anomalies based on a given traffic and a set of P4 programs running. This allows the identification of *(i)* anomalous traffic, *(ii)* malicious changes in the P4 program's code, and the *(iii)* replacement of P4 programs to disrupt the network functionality and associated services. In real-world scenarios, all of these anomalies can happen in parallel, thus making clear the need for automated ML models ready to infer from different data patterns.

In conclusion, our experiments provide essential insights into the fingerprinting of programmable networks, and our FEVER framework has proven to be a consistent methodology for analyzing the behavior of programmable switches and P4 programs, which can be adapted to real-life scenarios. As a limitation, it is essential to note that our evaluations are conducted in an emulated environment; therefore, due to different abstractions and technical aspects, there are challenges to implementing it in real-world devices, as discussed along the work.

Future work includes *(i)* investigation of the sensitivity of detection of small changes on P4 program codes and *(ii)* analysis of additional metrics for behavioral fingerprinting, including full integration of INT framework and syscalls to provide more information to represent complex behaviors better. Furthermore, implementation on real-world scenarios composed by Tofino switches is envisioned.

Acknowledgements. This work was supported by The São Paulo Research Foundation (FAPESP) under the grant number 2020/05152-7, the PROFISSA project.

References

1. Badotra, S., Panda, S.N.: Software-defined networking: a novel approach to networks. In: Gupta, B., Perez, G., Agrawal, D., Gupta, D. (eds.) Handbook of Computer Networks and Cyber Security: Principles and Paradigms, pp. 313–339. Springer, Cham (2020). https://doi.org/10.1007/978-3-030-22277-2_13
2. Bai, S., Kim, H., Rexford, J.: Passive OS fingerprinting on commodity switches. In: IEEE 8th International Conference on Network Softwarization (NetSoft), pp. 264–268 (2022)
3. Bondan, L., et al.: FENDE: marketplace-based distribution, execution, and life cycle management of VNFs. IEEE Commun. Mag. **57**(1), 13–19 (2019)
4. Bosshart, P., et al.: P4: programming protocol-independent packet processors. SIGCOMM Comput. Commun. Rev. **44**(3), 87–95 (2014)
5. Ding, D., Savi, M., Siracusa, D.: Tracking normalized network traffic entropy to detect DDoS attacks in P4. Trans. Dependable Secure Comput. **19**(6), 4019–4031 (2021)
6. Dumitrescu, D., Stoenescu, R., Negreanu, L., Raiciu, C.: BF4: towards bug-free P4 programs. In: SIGCOMM 2020, Virtually, USA, pp. 571–585 (2020)
7. Goswami, B., Kulkarni, M., Paulose, J.: A survey on P4 challenges in software defined networks: P4 programming. IEEE Access **11**, 54373–54387 (2023)
8. Hauser, F., et al.: A survey on data plane programming with P4: fundamentals, advances, and applied research. J. Netw. Comput. Appl. **212**, 103561 (2023)
9. Li, G., et al.: NETHCF: enabling line-rate and adaptive spoofed IP traffic filtering. In: IEEE 27th International Conference on Network Protocols (ICNP 2019), Chicago, USA, pp. 1–12 (2019)
10. Saueressig, M., Franco, M.F.: FEVER-P4 repository (2024). https://github.com/ComputerNetworks-UFRGS/FEVER-P4
11. Saueressig, M., Franco, M.F., Scheid, E.J., Granville, L.Z.: An approach for behavioral fingerprinting of P4 programmable switches. In: XX Escola Regional de Redes de Computadores (ERRC 2023), Porto Alegre, Brazil, pp. 22–60 (2023)
12. Musumeci, F., Ionata, V., Paolucci, F., Cugini, F., Tornatore, M.: Machine-learning-assisted DDoS attack detection with P4 language. In: IEEE International Conference on Communications (ICC 2020), Virtually, pp. 1–6 (2020)
13. Nunes, B.A.A., Mendonca, M., Nguyen, X.N., Obraczka, K., Turletti, T.: A survey of software-defined networking: past, present, and future of programmable networks. IEEE Commun. Surv. Tutor. **16**(3), 1617–1634 (2014)
14. Open Networking Foundation: P4Language (2023). https://github.com/p4lang
15. Sánchez, P.M.S., Valero, J.M.J., Celdrán, A.H., Bovet, G., Pérez, M.G., Pérez, G.M.: A survey on device behavior fingerprinting: data sources, techniques, application scenarios, and datasets. IEEE Commun. Surv. Tutor. **23**(2), 1048–1077 (2021)
16. Tan, L., et al.: In-band network telemetry: a survey. Comput. Netw. **186**, 107763 (2021)
17. Teng, L., Hung, C.H., Wen, C.H.P.: P4SF: a high-performance stateful firewall on commodity P4-programmable switch. In: IEEE/IFIP Network Operations and Management Symposium (NOMS 2022), Budapest, Hungary, pp. 1–5 (2022)
18. Usama, M., et al.: Unsupervised machine learning for networking: techniques, applications and research challenges. IEEE Access **7**, 65579–65615 (2019)
19. Wang, Q., Pan, M., Wang, S., Doenges, R., Beringer, L., Appel, A.W.: Foundational verification of stateful P4 packet processing. In: 14th International Conference on Interactive Theorem Proving (ITP 2023). Schloss-Dagstuhl-Leibniz Zentrum für Informatik, pp. 1–32 (2023)

Improved Purpose-Based Concurrency Control to Reduce the Electric Energy Consumption of a Server Cluster

Tomoya Enokido[1]([✉]), Dilawaer Duolikun[2], and Makoto Takizawa[3]

[1] Faculty of Business Administration, Rissho University, Tokyo, Japan
eno@ris.ac.jp
[2] Department of Advanced Sciences, Faculty of Science and Engineering,
Hosei University, 3-7-2, Kajino-cho, Koganei-shi, Tokyo 184-8584, Japan
[3] Research Center for Computing and Multimedia Studies, Hosei University,
3-7-2, Kajino-cho, Koganei-shi, Tokyo 184-8584, Japan
makoto.takizawa@computer.org

Abstract. In order to provide context- and energy-aware information systems, the EEPO-VM (Energy-Efficient Purpose Ordering with Virtual Machines) scheduler is newly proposed to serialize transactions based on the significancy of purposes assigned to the transactions in addition to reducing the total processing electric energy consumed by a server cluster which supports multiple virtual machines. We show a transaction assigned a more significant purpose can be preferentially performed than the other transactions and the total processing electric energy (TPE) consumption of a server cluster which supports multiple virtual machines can be reduced in the EEPO-VM scheduler through a simulation.

Keywords: Concurrency control · Significancy of roles · Significancy of purposes · Transactions · Object-based systems

1 Introduction

An application is composed of multiple objects [1,2] in object-based systems. An object is an encapsulation of data and methods to manipulate the data like a database. In order to utilize a provided application service, each subject or user initiates a transaction [3,4] on a client and the transaction issues methods to manipulate objects in an application. Consistency among objects has to be maintained to properly perform an application while conflicting transactions are concurrently performed on each object. Hence, conflicting transactions have to be serialized [3]. The *TO* (*Time-stamp Ordering*) [3] scheduler and *2PL* (*Two-Phase Locking*) protocol [3,4] are widely used to serialize conflicting transactions. In the TO scheduler and 2PL protocol, conflicting transactions are serialized based on the time-stamp order of transactions and the first-in-first-out (FIFO) manner, respectively.

L. Barolli (Ed.): AINA 2024, LNDECT 201, pp. 374–384, 2024.
https://doi.org/10.1007/978-3-031-57870-0_33

The concept of significancy among *roles* [5,6] is proposed in our previous studies [7–10]. A role is defined to be a set of access rights on objects in a *Role-Based Access Control (RBAC)* model [5,6] and shows a job function in each organization. The *RO (Role Ordering)* scheduler [9,10] is proposed based on the concept of significancy among roles. A transaction issued by a subject granted a more significant role is preferentially performed than the other transactions on each object in the RO scheduler. Hence, in the RO scheduler, conflicting transactions are serialized based on the significancy of roles granted to subjects. The *EERO-VM (Energy-Efficient Role Ordering in Virtual Machine environment)* [7] scheduler is proposed based on the RO scheduler. In the EERO-VM scheduler, the total processing electric energy (TPE) consumption of a server cluster which supports multiple virtual machines is reduced by not performing *meaningless* methods [7,8] on each object in addition to serialize conflicting transactions based on the significancy of roles granted to subjects.

In this paper, we consider a cluster of servers which support virtual machines. The *EEPO-VM (Energy-Efficient Purpose Ordering with Virtual Machines)* scheduler is newly proposed based on the significancy among *purposes* [11,12] assigned to transactions. Each transaction issued by a subject is assigned a purpose. A purpose is defined to be a subset of roles granted to the subject. A transaction assigned a more significant purpose is preferentially performed on each object by using the EEPO-VM scheduler. In addition, meaningless methods are not performed on each object. As a result, conflicting transactions are serialized based on the significancy of purposes assigned to transactions and the TPE consumption of a server cluster can be reduced. Evaluation results show transactions assigned more significant purposes are preferentially performed other transactions in the EEPO-VM scheduler. In addition, the TPE consumption of a server cluster which supports virtual machines is shown to be more reduced in the EEPO-VM scheduler compared with the RO scheduler.

In Sect. 2, the significancy among purposes assigned to transactions is discussed. In Sect. 3, the EEPO-VM scheduler is newly proposed based on the significancy among purposes. Evaluation results of the EEPO-VM scheduler are shown in Sect. 4.

2   System Model

2.1   Server Cluster

A server cluster $\mathbf{S}$ is a set $\{s_1, ..., s_n\}$ ($n \geq 1$) of physical servers. Let $\mathbf{C}$ be a set $\{cl_1, ..., cl_l\}$ ($l \geq 1$) of clients. Each server s_t holds the number tc_t (≥ 1) of homogeneous cores. Let $\mathbf{core}_t$ be a set $\{core_{1t}, ..., core_{tc_t t}\}$ of cores in a server s_t. Let tt_t (≥ 1) be a set of threads in each core $core_{gt}$. Each server s_t holds the total number tnt_t ($= tc_t \cdot tt_t$) of threads. Let $\mathbf{thread}_t$ be a set $\{thread_{1t}, ..., thread_{tnt_t t}\}$ of threads in a server s_t. A set $\mathbf{vm}_t = \{vm_{1t}, ..., vm_{tnt_t t}\}$ of virtual machines is supported by a server s_t. Each virtual machine vm_{kt} is allocated to one thread th_{kt} in a server s_t. Application data is encapsulated with methods as an object [1,2]. Let $\mathbf{O}$ be a set $\{o_1, ..., o_v\}$ ($v \geq 1$) of objects and each object is allocated to one virtual machine in a server cluster $\mathbf{S}$.

A *subject* issues a *transaction* [3] which is an atomic sequence of methods to manipulate objects. An *access right* α is a pair $\langle o, m \rangle$ of an object o and a method m. A role r is defined to be a set of *access rights* in the RBAC model [5,6]. If a role r granted to a subject sub includes an access right $\langle o, m \rangle$, a subject sub can manipulate an object o by a method m. Let $subR_i$ be a families $\{r_1, ..., r_q\}$ of roles granted to a subject sub_i which initiates a transaction T_i. Let $time(T_i)$ be time when the transaction T_i is initiated. Let $id(cl_s)$ be an identifier of the client cl_s where transaction T_i is initiated. An unique identifier $id(T_i) = \langle time(T_i), id(cl_s) \rangle$ is given to each transaction T_i. Here, $id(T_i) < id(T_j)$ iff (1) $time(T_i) < time(T_j)$ or (2) $time(T_i) \parallel time(T_j)$ and $id(cl_i) < id(cl_j)$.

2.2   Significancy of Roles

Class and *function* types of methods are supported by each object. A class method creates and drops an object. By a function method, data in an object is manipulated. Each object supports *output*, *full change*, and *partial change* types of function methods. In a full change method, a whole data is fully changed in an object while only a part of data is changed in an object by a partial change method. An output method derives data in an object.

A method m_1 *semantically* (*sem-*) *dominates* m_2 ($m_1 \succcurlyeq m_2$) iff m_1 is significant than m_2 in an application. m_1 is *sem-equivalent* with m_2 ($m_1 \cong m_2$) if $m_1 \succcurlyeq m_2$ and $m_2 \succcurlyeq m_1$. m_1 is *sem-significant* than m_2 ($m_1 \succcurlyeqq m_2$) if $m_1 \succcurlyeq m_2$ and $m_1 \ncong m_2$. m_1 and m_2 are *sem-uncomparable* ($m_1 \parallel m_2$) iff neither $m_1 \succcurlyeq m_2$ nor $m_2 \succcurlyeq m_1$.

[**Significancy among methods**] A method m_1 is *more significant* than m_2 ($m_1 \succ m_2$) iff (1) m_1 is a class method and m_2 is a function method, (2) m_1 is a change method and m_2 is an output method, (3) m_1 is a full change method and m_2 is a partial change method, or (4) m_1 and m_2 are the same type methods and $m_1 \succcurlyeqq m_2$.
A method m_1 is *significantly* (*sig-*) *equivalent* with m_2 ($m_1 \equiv m_2$) iff m_1 and m_2 are a same type and $m_1 \cong m_2$. m_1 *sig-dominates* m_2 ($m_1 \succeq m_2$) iff $m_1 \succ m_2$ or $m_1 \equiv m_2$. m_1 and m_2 are *sig-uncomparable* ($m_1 \parallel m_2$) iff neither $m_1 \succeq m_2$ nor $m_2 \succeq m_1$.

Let se_h be a security class [13,14] given to an object o_h. The security class se_1 *can flow into* se_2 ($se_1 \mapsto se_2$) iff data in the object o_1 can be brought to the object o_2. se_1 precedes se_2 ($se_1 \prec se_2$) iff $se_1 \mapsto se_2$ but $se_2 \not\mapsto se_1$. se_1 and se_2 are equivalent ($se_1 \equiv se_2$) iff $se_1 \mapsto se_2$ and $se_2 \mapsto se_1$. se_2 dominates se_1 ($se_1 \preceq se_2$) iff $se_1 \prec se_2$ or $se_1 \equiv se_2$.

[**Significancy among objects**] An object o_1 is *more significant* than an object o_2 ($o_1 \succ o_2$) iff $se_1 \succ se_2$.
o_1 and o_2 are *sig-equivalent* ($o_1 \equiv o_2$) iff $se_1 \equiv se_2$. o_1 *sig-dominates* o_2 ($o_1 \succeq o_2$) iff $o_1 \succ o_2$ or $o_1 \equiv o_2$. o_1 and o_2 are *sig-uncomparable* ($o_1 \parallel o_2$) iff neither $se_1 \succeq se_2$ nor $se_2 \succeq se_1$.

[Significancy among access rights] An access right α_1 is *more significant* than an access right α_2 $(\alpha_1 \succ \alpha_2)$ iff (1) $o_1 \succ o_2$, (2) $m_1 \succ m_2$ and $o_1 \equiv o_2$, or (3) $\alpha_1 \succ \alpha_3$ and $\alpha_3 \succ \alpha_2$ for some access right α_3.
α_1 and α_2 are *sig-equivalent* $(\alpha_1 \equiv \alpha_2)$ iff 1) $o_1 \equiv o_2$ and $o_1 \neq o_2$, or 2) $m_1 \equiv m_2$ and $o_1 = o_2$. α_1 *sig-dominates* α_2 $(\alpha_1 \succeq \alpha_2)$ iff $\alpha_1 \succ \alpha_2$ or $\alpha_1 \equiv \alpha_2$. α_1 and α_2 are *sig-uncomparable* $(\alpha_1 \parallel \alpha_2)$ iff neither $\alpha_1 \succeq \alpha_2$ nor $\alpha_2 \succeq \alpha_1$.

An access right α_1 is *maximally reachable* from another access right α_2 $(\alpha_1 \hookleftarrow \alpha_2)$ iff $\alpha_1 \succeq \alpha_2$ and there is no access right α_3 such that $\alpha_2 \succeq \alpha_3 \succeq \alpha_1$ in a set $\boldsymbol{\alpha}$ of access rights.

[Significancy among roles] A role r_1 *sig-dominates* r_2 $(r_1 \succeq r_2)$ iff (1) there is an access right $\alpha_1 \in r_1 - r_2$ for some access right α_2 in r_2 such that $\alpha_1 \hookleftarrow \alpha_2$ in $r_1 \cup r_2$ and (2) there is no access right $\alpha_2 \in r_2$ for every access right $\alpha_1 \in r_1$ such that $\alpha_2 \hookleftarrow \alpha_1$ in $r_1 \cup r_2$.
r_1 and r_2 are *sig-equivalent* $(r_1 \equiv r_2)$ iff $r_1 \succeq r_2$ and $r_2 \succeq r_1$. r_1 and r_2 are *sig-uncomparable* $(r_1 \parallel r_2)$ iff neither $r_1 \succeq r_2$ nor $r_2 \succeq r_1$.

Notations "$r_1 \sqcup r_2$" and "$r_1 \sqcap r_2$" show a *least upper bound* and *greatest lower bound*, respectively, of a pair of roles r_1 and r_2. $r_1 \sqcup r_2 = r_3$ iff $r_3 \succeq r_1$, $r_3 \succeq r_2$, and there is no role r' such that $r_3 \succeq r' \succeq r_1$ and $r_3 \succeq r' \succeq r_2$. $r_1 \sqcap r_2$ $= r_3$ iff $r_1 \succeq r_3$, $r_2 \succeq r_3$, and there is no role r' such that $r_1 \succeq r' \succeq r_3$ and r_2 $\succeq r' \succeq r_3$.

[Significancy among role families] A family FR_1 of roles *sig-dominates* another family FR_2 of roles $(FR_1 \succeq FR_2)$ iff $\sqcap_{r \in FR_1} r \succeq \sqcup_{r \in RR_2} r$.
FR_1 and FR_2 are *sig-equivalent* $(FR_1 \equiv FR_2)$ iff $FR_1 \succeq FR_2$ and $FR_2 \succeq FR_1$. FR_1 and FR_2 are *sig-uncomparable* $(FR_1 \parallel FR_2)$ iff neither $FR_1 \succeq FR_2$ nor $FR_2 \succeq FR_1$.

2.3   Purposes-Oriented Dominant Relation

Each transaction T_i issued by a subject sub_i is assigned a *purpose* [11,12] $subP_i$ $(\subseteq subR_i)$ which is a subset of roles assigned to the subject sub_i. A subject sub_i is *more significant* than a subject sub_j on a role r $(sub_i \succ_r sub_j)$ iff sub_i grants r to sub_j or $sub_i \succ_r sub_k \succ_r sub_j$ for some subject sub_k. sub_i and sub_j are *equivalent* on r $(sub_i \equiv_r sub_j)$ iff $sub_i \succ_r sub_j$ and $sub_j \succ_r sub_i$. sub_i and sub_j are *independent* with respect to r $(sub_i \parallel_r sub_j)$ iff neither $sub_i \succ_r sub_j$ nor $sub_j \succ_r sub_i$.

[Purpose-oriented dominant relation] A subject sub_i *purpose-oriented dominates* (*P-dominates*) sub_j $(sub_i \succeq_P sub_j)$ with respect to purposes of transactions T_i and T_j iff (1) $subP_i \succeq subP_j$ or (2) $sub_i \succ_r sub_j$ for some role $r \in subP_i$ $\cap subP_j$ and $sub_j \not\succ_r sub_i$ for every role $r \in subP_i \cap subP_j$ if $subP_i \parallel subP_j$.

[Significancy among transactions] A transaction T_i *P-dominates* another transaction T_j $(T_i \succeq_P T_j)$ iff $sub_i \succeq_P sub_j$.
T_i and T_j are *P-equivalent* $(T_i \equiv_P T_j)$ iff $T_i \succeq_P T_j$ and $T_j \succeq_P T_i$. T_i and T_j are *P-independent* $(T_i \parallel_P T_j)$ iff neither $T_i \succeq_P T_j$ nor $T_j \succeq_P T_i$.

Let H be a *history* of a set $\mathbf{T}$ $(= \{T_1, ..., T_k\})$ $(k \geq 1)$ of transactions.

[Purpose-oriented (p-) precedent relation] A transaction T_i *purpose-oriented (p-) precedes* T_j in H ($T_i \to_H T_j$) iff 1) $T_i \succeq_p T_j$, or 2) $id(T_i) < id(T_j)$ if $T_i \parallel_P T_j$ or $T_i \equiv_P T_j$.

A history H which is a totally ordered set $\langle \mathbf{T}, \to_H \rangle$ is serializable iff the p-precedent relation $\to_H$ is acyclic.

[Purpose-oriented (P-) partition] A history $H = \langle \mathbf{T}, \to_H \rangle$ is *P-partitioned* into the subhistories $H_f = \langle \mathbf{T}_f, \to_{H_f} \rangle$ ($f = 1, ..., d$):

1. $\mathbf{T} = \mathbf{T}_1 \cup \cdots \cup \mathbf{T}_d$ and $\mathbf{T}_f \cap \mathbf{T}_g = \phi$ for every pair of subhistories H_f and H_g.
2. For every pair of transactions T_i and T_j in H_f, $T_i \to_H T_j$ if $T_i \to_{H_f} T_j$.
3. For every pair of transactions T_j in H_f and T_j in H_g, if $T_i \to_H T_j$ for some pair of transactions T_i and T_j, there are no pair of transactions such that $T_j \to_H T_i$.

[Purpose-oriented (P-) serializable] A history $H = \langle \mathbf{T}, \to_H \rangle$ is *P-serializable* iff the history H is P-partitioned into the subhistories $H_1, ..., H_d$.

2.4   DAVM and PCDAVM Models

In this paper, class and change methods are classified into a *write (wm)* method. Output methods are classified into a *read (rm)* method. The *DAVM (Data Access in Virtual Machine environments)* model [15] is proposed in our previous studies. Let $wm_i^{kt}(o_h)$ and $rm_i^{kt}(o_h)$ be a write and read methods, respectively, issued by a transaction T_i to manipulate an object o_h supported by a virtual machine vm_{kt}. $WM_t(\tau)$ and $RM_t(\tau)$ indicate sets of write and read methods, respectively, being performed in a server s_t at time τ. Let $maxRrate_t$ [B/sec] and $maxWrate_t$ [B/sec] be the maximum read rate and write rate, respectively, supported by a server s_t. At time τ, the read rate $Rrate_i^{kt}(\tau)$ ($\leq maxRrate_t$) [B/sec] of a read method $rm_i^{kt}(o_h)$ is calculated as $degR_t(\tau) \cdot maxRrate_t$ where $0 \leq degR_t(\tau) \leq 1$. At time τ, the write rate $Wrate_i^{kt}(\tau)$ ($\leq maxWrate_t$) of a write method $wm_i^{kt}(o_h)$ is calculated as $degW_t(\tau) \cdot maxWrate_t$ where $0 \leq degW_t(\tau) \leq 1$. Here, $degR_t(\tau)$ and $degW_t(\tau)$ indicate degradation ratios of read and write rates in a server s_t at time τ. $degR_t(\tau) = 1/(|RM_t(\tau)| + rr_t \cdot |WM_t(\tau)|)$ where $0 \leq rr_t \leq 1$ and $degW_t(\tau) = 1/(ww_t \cdot |RM_t(\tau)| + |WM_t(\tau)|)$ where $0 \leq ww_t \leq 1$.

In the *PCDAVM (Power Consumption for Data Access in Virtual Machine environment)* model [15], the base electric power $be_t(\tau)$ [W] at time τ [15] is defined as shown in Eq. (1):

$$be_t(\tau) = \begin{cases} mine_t + bce_t + acore_t(\tau) \cdot ce_t + athread_t(\tau) \cdot te_t) & \text{if } acore_t(\tau) \geq 1. \\ mine_t, & \text{otherwise.} \end{cases}$$

$$(1)$$

Here, $mine_t$ [W] shows the minimum electric power of a server s_t. bce_t [W] is the base electric power to make at least one core active. $acore_t(\tau)$ and $athread_t(\tau)$ indicate the numbers of active cores and threads in a server s_t at time τ. The electric power te_t [W] and ce_t [W] are consumed by a server s_t to make one thread and one core active.

In the PCDAVM model, the electric power $e_t(\tau)$ consumed by a server s_t at time τ to perform read and write methods is given as Eq. (2):

$$e_t(\tau) = \begin{cases} be_t(\tau) & if \ |WM_t(\tau)| = |RM_t(\tau)| = 0. \\ be_t(\tau) + we_t & if \ |RM_t(\tau)| = 0 \ and \ |WM_t(\tau)| \geq 1. \\ be_t(\tau) + wre_t(\gamma(\tau)) & if \ |RM_t(\tau)| \geq 1 \ and \ |WM_t(\tau)| \geq 1. \\ be_t(\tau) + re_t & if \ |RM_t(\tau)| = 1 \ and \ |WM_t(\tau)| \geq 0. \end{cases} \quad (2)$$

we_t [W] and re_t [W] show the electric power of a server s_t where only write and read methods are performed in a server s_t, respectively. If read and write methods are concurrently performed in a server s_t, the server s_t consumes the electric power $wre_t(\gamma(\tau))$ [W]. Here, $wre_t(\gamma(\tau)) = \gamma(\tau) \cdot we_t + (1 - \gamma(\tau)) \cdot re_t$ where $\gamma(\tau) = |WM_t(\tau)| \ / \ |RM_t(\tau) \cup WM_t(\tau)|$. $re_t \leq wre_t(\gamma(\tau)) \leq we_t$. From time τ_1 to τ_2, a server s_t consumes the total processing electric energy (TPE) $TPE_t(\tau_1, \tau_2) = \Sigma_{\tau=\tau1}^{\tau2}(e_t(\tau) - minE_t)$ [J].

3   EEPO-VM Scheduler

A method m_1 *purpose-oriented (p-) precedes* m_2 in H_f ($m_1 \rightarrow_{H_f} m_2$) iff (1) m_1 is issued before m_2 by the same transaction T_i, (2) $T_i \rightarrow_{H_f} T_j$, and a pair of methods m_1 and m_2 are issued by transactions T_i and T_j, respectively, or (3) $m_1 \rightarrow_{H_f} m_3 \rightarrow_{H_f} m_2$ for some method m_3. Let $H_f^{o_h}$ be a *local subhistory* of methods on each object o_h in a subhistory H_f. A method m_1 p-precedes m_2 in $H_f^{o_h}$ ($m_1 \rightarrow_{H_f^{o_h}} m_2$) iff $m_1 \rightarrow_{H_f} m_2$.

[Absorption of methods] For a pair of methods m_1 and m_2 where $m_2 \rightarrow_{H_f^{o_h}} m_1$, m_1 *absorbs* m_2 iff one of the following condition holds:

- m_1 is a full change method and m_2 is a partial change method, and there is neither output nor class method m_3 such that $m_2 \rightarrow_{H_f^{o_h}} m_3 \rightarrow_{H_f^{o_h}} m_1$, or m_1 absorbs m_3 and m_3 absorbs m_2 for some method m_3.
- m_1 and m_2 are output methods, and there is neither change nor class method m_3 such that $m_2 \rightarrow_{H_f^{o_h}} m_3 \rightarrow_{H_f^{o_h}} m_1$, or m_1 absorbs m_3 and m_3 absorbs m_2 for some method m_3.
- m_1 is a class method for dropping an object o_h and m_2 is a change type method, and there is neither output nor class method m_3 such that $m_2 \rightarrow_{H_f^{o_h}} m_3 \rightarrow_{H_f^{o_h}} m_1$, or m_1 absorbs m_3 and m_3 absorbs m_2 for some method m_3.

[Meaningless methods on an object] A method m is *meaningless* on an object o_h iff the method m is absorbed by another method m' in $H_f^{o_h}$.

An *EEPO-VM (Energy Efficient Purpose Ordering scheduler with Virtual Machines)* is newly proposed to not only make transactions P-serializable but also reduce the TPE consumed by a server cluster equipped with virtual machines. First, a transaction T_i issues a *begin* request b_i to all objects which the transaction T_i manipulates. Then, the transaction T_i manipulates the objects by issuing methods and lastly issues an *abort* (ab_i) or *commit* (com_i) request to the objects. Each client cl_s periodically sends a partition message p_s to all objects in **O** for making P-partitions. A partition message p_s includes a partition number $p_s.n$. $p_s.n = p_s.n + 1$ each time a partition message p_s is sent by a client cl_s.

A set $Q_h = \{Q_{h1}, ..., Q_{hl}\}$ of *receipt queues* for each client cl_s, an *P-partition queue* PSQ_h, and a *method order queue* MOQ_h are manipulated by each object o_h. Suppose a transaction T_i is created on a client cl_s. A method issued by the transaction T_i is enqueued into a receipt queue Q_{hs}. If a partition message p_s which includes the same partition number $p_s.n$ exists in every receipt queue Q_{hs}, every begin requests preceding the partition message p_s are moved from every receipt queue Q_{hs} to the P-partition queue RSQ_h. After that, a partition message p where $p.n = p_s.n$ is enqueued into the MOQ_h and RSQ_h queues. A P-partition H_f is composed of a subset $\mathbf{T}_f$ of transactions whose begin requests are stored between p and p' where $p.n + 1 = p'.n$ in RSQ_h. In the P-partition H_f, begin requests are totally ordered in the p-precedent relation $\rightarrow_{H_f}$. Each method m issues by a transaction T_i whose begin request b_i is between p and p' in RSQ_h is moved from Q_{hs} to between p and p' in MOQ_h, and sorted in the p-precedence relation $\rightarrow_{H_f}$ by the **Psort**(m, MOQ_h, p, p') function. Methods in Q_{hs} are processed by **P_Partition** procedure as shown in Algorithm 1.

Algorithm 1. P_Partition procedure

Require: PSQ_h, MOQ_h, $Q_h = \{Q_{h1}, ..., Q_{hl}\}$.
Ensure: P-partitioned MOQ_h.
  procedure P_PARTITION(PSQ_h, MOQ_h, $Q_h = \{Q_{h1}, ..., Q_{hl}\}$)
    for every Q_{hs} in Q_h **do**
      for every method m_i in Q_{hs} **do**
        if b_i is between p and p' in RSQ_h **then**
          Psort(m_i, MOQ_h, p, p');
        end if
      end for
    end for
  end procedure

Let $\mathbf{Tr}_h$ and $\mathbf{M}_h$ be sets of transactions and methods, respectively, being performed on an object o_h. Meaningless methods and conflict relations among methods are checked by using the following functions:

- **Meaningless_check**(m): **True** if a method m is a meaningless method in $H_f^{o_h}$. Otherwise, **False**.
- **Compatible**$(m, \mathbf{M}_h, \mathbf{Tr}_h)$: **True** if a method m does not conflict with every method in $\mathbf{M}_h$ and a transaction T issuing a method m does not conflict with every transaction in $\mathbf{Tr}_h$. Otherwise, **False**.

In EEPO-VM scheduler, methods in MOQ_h are performed on an object o_h by the **Perform** procedure as shown in Algorithm 2:

Algorithm 2. Perform procedure

Require: PSQ_h, MOQ_h, $\mathbf{Tr}_h$, $\mathbf{M}_h$.

  procedure PERFORM(PSQ_h, MOQ_h, $\mathbf{Tr}_h$, $\mathbf{M}_h$)

    $m \leftarrow$ a top method in MOQ_h;

    if $m = p$ **then**

      if $\mathbf{Tr}_h = \phi$ and $\mathbf{M}_h = \phi$ **then**

        every b_i preceding m ($= p$) in RSQ_h is removed;

        m is removed from RSQ_h and MOQ_h;

      end if

    else if m is a commit or abort message of a transaction T **then**

      $\mathbf{Tr}_h \leftarrow \mathbf{Tr}_h - \{T_i\}$;

      m is removed from MOQ_h;

    else   ▷ m is a method issued by a transaction T

      if **Compatible**(m, $\mathbf{M}_h$, $\mathbf{Tr}_h$) **then**

        if $T \notin \mathbf{Tr}_h$ **then**

          $\mathbf{Tr}_h \leftarrow \mathbf{Tr}_h \cup \{T\}$;

        end if

        if **Meaningless_check**(m) **then**

          m is removed from MOQ_h;

        else

          $\mathbf{M}_h \leftarrow \mathbf{M}_h \cup \{m\}$;

          m is perform on an object o_h;

          m is removed from MOQ_h;

        end if

      end if

    end if

  end procedure

---

4   Evaluation of EEPO-VM Scheduler

We evaluate the EEPO-VM scheduler in terms of the TPE consumption [J] of a server cluster $\mathbf{S}$ equipped with virtual machines compared with the RO scheduler [10]. In addition, we confirm the more significant transactions with respect to purposes can be preferentially performed on each object in the EEPO-VM scheduler.

We consider a server cluster $\mathbf{S}$ composed of five homogeneous servers s_1, ..., s_5 ($n = 5$). Every server s_t holds a dual-core CPU ($tc_t = 2$) and two threads runs on each core ($tt_t = 2$). Tables 1 shows parameters in $DAVM$ model [15] and $PCDAVM$ model [15]. Every server s_t holds four threads and each virtual machine vm_{kt} runs on one thread $thread_{kt}$ in a server s_t. Hence, twenty virtual machines run in a server cluster $\mathbf{S}$. One object o_h is allocated to each virtual machine vm_{kt}. Hence, a set $\mathbf{O} = \{o_1, ..., o_{20}\}$ ($v = 20$) of objects are allocated in a server cluster $\mathbf{S}$. Each object o_h supports an output, full change, and partial change methods. Data size of each object o_h is randomly defined between 40 and 90 [MB]. Five subjects sub_1, ..., sub_5 issue transactions. Three roles r_1, ..., r_3 where $r_1 \succeq r_2 \succeq r_3$ are defined and every role is owned by sub_1. Here, $subR_1 = \{r_1, r_2, r_3\}$, $subR_2 = subR_3 = \{r_2, r_3\}$, and $subR_4 = subR_5 = \{r_3\}$.

Transactions issued by the subjects sub_1, sub_4, and sub_5 are assigned a purpose $subP_1 = subP_4 = subP_5 = \{r_3\}$. Transactions issued by the subjects sub_2 and sub_3 are assigned purposes $subP_2 = subP_3 = \{r_2, r_3\}$. Here, $subP_2 = subP_3 \succeq_P subP_1 = subP_4 = subP_5$. There are five clients cl_1, ..., cl_5. Each subject sub_i creates transactions on a client cl_i ($i = 1, ..., 5$). Each subject sub_i issues the same number ω ($0 \leq \omega \leq 300$) of transactions. Let nt ($= \omega \cdot 5$) ($0 \leq nt \leq 1,500$) be the total number of transactions issued to the server cluster $\mathbf{S}$. The starting time of each transaction T_i is randomly decided between 1 and 360 [sec] in simulation time. Five methods are randomly selected from sixty methods on the twenty objects by each transaction T_i and issued to objects in the server cluster $\mathbf{S}$.

Table 1. Parameters of PCDAVM and DAVM models.

bce_t	ce_t	te_t	re_t	we_t	$mine_t$	$maxWrate_t$	$maxRrate_t$	ww_t	rr_t
1.1 [W]	0.6 [W]	0.5 [W]	1 [W]	4 [W]	17 [W]	85.3 [MB/sec]	98.5 [MB/sec]	0.077	0.667

Figure 1 shows the TPE consumed by the server cluster $\mathbf{S}$ in the EEPO-VM and RO schedulers. For $0 \leq nt \leq 1,500$, the TPE consumed by the server cluster $\mathbf{S}$ in the EEPO-VM scheduler is lower than the RO scheduler since meaningless methods are not performed on each object in the EEPO-VM scheduler. Figure 2 shows the average execution time $avgT_i$ [sec] ($i = 1, ..., 5$) of each transaction issued by each subject sub_i in the EEPO-VM scheduler. A transaction which is assigned a more significant purpose is preferentially performed than the other transactions in the EEPO-VM scheduler.

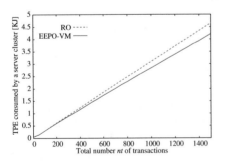
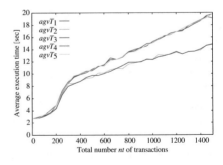

Fig. 1. TPE consumption [KJ].   **Fig. 2.** $avgT_i$ in the EEPO-VM scheduler.

The EEPO-VM scheduler can reduce the TPE consumed by a server cluster $\mathbf{S}$ which supports multiple virtual machines than the RO scheduler. Transactions assigned more significant purposes are preferentially performed than the other transactions in the EEPO-VM scheduler.

5   Concluding Remarks

The EEPO-VM scheduler is newly proposed to serialize conflicting transactions based on the significancy of purposes in addition to reduce the TPE consumed by a server cluster which supports multiple virtual machines. The evaluation results show transactions assigned more significant purposes can be more preferentially performed on each object in the EERO-VM scheduler. The TPE consumed by a server cluster which supports multiple virtual machines is shown to be more reduced in the EEPO-VM scheduler compared with the RO scheduler in the evaluation.

References

1. Enokido, T., Duolikun, D., Takizawa, M.: An energy-efficient quorum-based locking protocol by omitting meaningless methods on object replicas. J. High Speed Netw. **28**(3), 181–203 (2022)
2. Enokido, T., Duolikun, D., Takizawa, M.: Energy consumption laxity-based quorum selection for distributed object-based systems. Evol. Intell. **13**, 71–82 (2020)
3. Bernstein, P.A., Hadzilacos, V., Goodman, N.: Concurrency Control and Recovery in Database Systems. Addison-Wesley, Boston (1987)
4. Gray, J.N.: Notes on data base operating systems. In: Bayer, R., Graham, R.M., Seegmüller, G. (eds.) Operating Systems. LNCS, vol. 60, pp. 393–481. Springer, Heidelberg (1978). https://doi.org/10.1007/3-540-08755-9_9
5. Ferraiolo, D.F., Kuhn, D.R., Chandramouli, R.: Role based access control. Artech House (2005)
6. Sandhu, R.S., Coyne, E.J., Feinstein, H.L., Youman, C.E.: Role-based access control models. IEEE Comput. **29**(2), 38–47 (1996)
7. Enokido, T., Duolikun, D., and Takizawa, M.: Energy-efficient role-based concurrency control with virtual machines. In: Proceedings of the 18th International Conference on P2P, Parallel, Grid, Cloud and Internet Computing (3PGCIC-2023), pp. 81–91 (2023)
8. Enokido, T., Duolikun, D., and Takizawa, M.: Energy-efficient role ordering scheduler. In: Proceedings of the 20th International Conference on Network-Based Information Systems (NBiS-2017), pp. 78–90 (2017)
9. Enokido, T., Barolli, V., Takizawa. M.: A legal information flow (LIF) scheduler based on role-based access control model. Comput. Stand. Interfaces **31**(5), 906–912 (2009)
10. Enokido, T.: Role-based serializability using role ordering schedulers. J. Interconnect. Netw. **7**(4), 437–450 (2006)
11. Enokido, T., Takizawa, M.: A purpose-based synchronization protocol for secure information flow control. Comput. Syst. Sci. Eng. **25**(2) (2010)
12. Enokido, T., Takizawa, M.: Energy-efficient purpose ordering scheduler. In: Proceedings of the 14th International Conference on Broad-Band Wireless Computing, Communication and Applications (BWCCA-2019), pp. 137–149 (2019)
13. Nakamura, S., Enokido, T., Takizawa, M.: Time-based legality of information flow in the capability-based access control model for the Internet of Things. Concurr. Comput. Pract. Exp. **33**(23) (2021)

14. Denning, D.E., Denning, P.J.: Cryptography and Data Security. Addison-Wesley Publishing Company, Boston (1982)
15. Enokido, T. and Takizawa, M.: The power consumption model of a server to perform data access application processes in virtual machine environments. In: Proceedings of the 34th International Conference on Advanced Information Networking and Applications (AINA-2020), pp. 184–192 (2020)

Proposal for a Resource Allocation Model Aimed at Fog Computing

André D'Amato[1]([✉]) and Mario Dantas[2]

[1] Universidade Tecnológica Federal do Paraná (UTFPR), Apucarana, Brazil
andredamato@utfpr.edu.br
[2] Universidade Federal de Juiz de Fora (UFJF), Juiz de Fora, Brazil
mario.dantas@ice.ufjf.br

Abstract. The emergence of fog computing has presented challenges in effectively allocating resources within this environment. Addressing user satisfaction, many of these challenges can be mitigated through the quality of experience paradigm, which incorporates various contextual parameters. To optimize resource utilization, leveraging the quality of context paradigm can significantly enhance system performance. Consequently, this paper introduces a model aimed at dynamically enhancing individual user experiences while concurrently boosting overall system performance within the fog computing environment through quality of context considerations. Experimental results demonstrate tangible enhancements in runtime job execution and noticeable improvements in the overall system performance upon the implementation of our proposed model.

Keywords: Distributed System · Job Management · Resource Allocation · Quality of Experience · Throughput · Quality of Context · Users satisfaction

1 Introduction

In recent years, the allocation of remote resources provided by computational clouds has become a crucial subject due to the increasing demand for such services. A computational cloud receives data generated by geographically dispersed user devices at the network edge, processes this data, and delivers the service. These devices can vary widely, including personal computers, smartphones, in-vehicle multimedia devices, video games, and residential application sensor hubs. Generally, users are geographically distant from data centers, resulting in significant delays in data transmission. However, with the massive growth of data over the last years, the demand for computational resources provided by data centers continues to escalate. It's predictable that the cloud computing model might undergo significant impact, as the increase in communication latency when billions of devices are connected could adversely affect applications, severely degrading users' Quality of Service (QoS) and Quality of Experience (QoE) [5].

The fog computing paradigm can be a solution to this problem. In fog computing, part of the data processing that would be sent to a cloud can occur among nearby personal devices situated at the network edge. Consequently, latency issues can be mitigated, as part of the processing takes place near the users' devices. In the fog computing model, edge devices can form small local data centers that support multi-location and elasticity [10]. Therefore, it's possible to state that fog computing allows a reduction in data sent to the cloud, consequently reducing communication latency and the amount of data processed by it. Although fog computing is a good solution for addressing problems arising from cloud computing, this paradigm presents several challenges.

Fog computing is a solution primarily designed to handle Internet of Things (IoT)-related applications [10], which typically involve processing information collected from one or more sources in real-time. As a result, decisions need to be made to meet user needs [10], maintaining QoS and consequently QoE. However, relying exclusively on edge resources is not always possible, as some computational and data storage requirements may exceed the capacity of edge devices. Additionally, a resource configuration closer to the user might lack sufficient capacity to meet that user's request due to availability issues or even memory and processing limitations.

Moreover, the technological diversity of edge computing devices and the growing user demand pose challenges in establishing resource allocation to favor both the environment and individual application needs. Edge devices impose a high level of heterogeneity, making resource allocation policies difficult to implement and establishing technologies capable of handling different device types challenging. While resource allocation in any data center should meet user demands, it's crucial to maintain maximum load balancing so that resources can be shared among other users. Therefore, resource scaling in a fog environment should achieve the best fit between QoE and load balancing. In this context, an approach based on Quality of Context (QoC) is proposed for decision-making, focusing on user QoE considered from the total task completion time (*makespan*).

The rest of this article is organized as follows: Sect. 2 covers the theoretical foundations used in the proposed model, presented in Sect. 3 and Sect. 4; Sect. 5 discusses the conducted experiment and presents the results; Sect. 6 addresses related works, and Sect. 7 presents final considerations and future work.

2   Basic Concepts and Foundational Theory

2.1   Fog Computing

Just as cloud computing is a paradigm that extends the functionalities of grid computing by adding service models, fog computing is a paradigm aimed at extending the functionalities of cloud computing to computational devices at the network edge [10]. The concept of Fog Computing, or fog computing, was developed by Cisco in 2012 [10], and this paradigm is focused on the Internet of Things market. Environments related to applications involving the Internet of Things concept consist of heterogeneous devices, locally or geographically distributed. These devices are commonly physical embedded components that can be interconnected with other virtual (or hardware) resources for control, data processing, and analysis.

In addition to providing support for the Internet of Things, fog computing aims for greater integration between devices located at the network edge and the high-capacity processing centers formed by clouds and grid computing. With this, processing loads generated by applications running in the core of the network can be distributed among personal devices in advance, thus minimizing the workload exerted in the core and expanding processing and data storage capacity.

2.2   Quality of Context (QoC)

Context is a set of information that describes the state of a component used for interaction between an application and its respective users [3]. Context information is commonly used in ubiquitous environments. In such environments, context parameters used are information obtained by sensors that monitor events and verify the conditions of the system's environment. This work addresses context parameters considering two dimensions: available computational capacity in the environment and application characteristics. Information regarding computational capacity refers to the computational power of resources present in the distributed system, such as processor type, processing rate, number of processes, network transmission rate, and network latency. Application characteristics are information that describes the computational needs and data model of the application. The data model is information regarding the category of the application, for example, whether it's parallel or sequential in nature, CPU-oriented, or network-oriented.

A system is context-aware if it utilizes context information to provide relevant services to the user. The relevance directly depends on the task executed in the system [3]. In other words, context information is used to provide services tailored to user needs. Good context modeling also helps reduce the computational complexity of context-sensitive applications.

2.3   Relationship Between QoE and QoC

The initial models of QoE in the computational context were based on purely subjective metrics induced by technology acceptance models. However, acceptance models addressed metrics collected through opinion tests. However, the application of opinion tests is widely criticized by the scientific community, as the results generated by this process are considered less meaningful.

Qualitative metrics addressed in opinion tests have limitations compared to quantitative metrics. For example, qualitative metrics may classify the quality of a certain service as good or bad. However, quantitative metrics, besides classifying the service quality, can reveal how good or bad the service is. Quantitative units have quantifiable metrics based on numerical scales, thereby enabling a clear understanding of the quality of what is being measured. The most recent models of Quality of Experience tend to link QoE metrics with QoS metrics, as the latter are more representative and easily collected through hardware and software agents embedded in the network.

From quantitative studies of QoE through QoS metrics, mathematical models were established to estimate users' perception of services in the realm of computational networks. For this reason, nowadays, it is possible to estimate QoE by collecting QoS-related data (such as jitter, delay, packet loss ratio). In this scenario, QoE emerges as a differentiated paradigm and an important factor in solving various challenges related to service adoption and proper resource utilization, as QoE has a strong relationship with QoS parameter adjustments. QoS parameters determine the context in which services will be provided to users. The context information used by quantitative QoE models usually refers to resource management provided by network service infrastructures, such as data transmission control, jitter control, and packet loss rate control [4]. QoE-oriented QoS approaches establish contextual information, enabling the prediction of QoC, to adapt services and predict QoE according to the most relevant parameters to meet user expectations.

3   Proposed Model

This work introduces a load balancing strategy centered on leveraging users' implicit knowledge for resource allocation. By processing rules based on this knowledge, the aim is to optimize system performance and user satisfaction. The approach uses the Quality of Context (QoC) paradigm to process contextual information tied to users' requested resources, generating knowledge for more suitable resource allocation. This model employs the DIKW information hierarchy, defining distinct roles for each data level: Wisdom, Knowledge, Data, Quality of Experience (QoE), Quality of Service (QoS), and Quality of Context (QoC). Wisdom involves gathering application data to predict users' QoE, while knowledge quantifies the quality of resource allocation based on QoC parameters. The data level manages resource-specific details influencing system performance, and the QoE level oversees application execution and user satisfaction.

Ultimately, decisions regarding resource allocation are driven by the QoC layer, impacting resource availability in the environment through the QoS level.

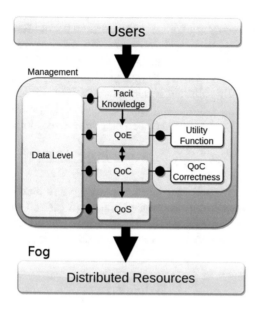

Fig. 1. Overview of the proposed model.

As mentioned earlier, users' tacit knowledge is used to guide resource allocation in the proposed model. Tacit knowledge refers to the knowledge acquired throughout users' lives. Specifically, in the proposed model, this knowledge refers to users' ability to accurately specify the resources needed to execute their application. The goal is to optimize system performance, enhance user experience, and enable interaction with the environment so that users can customize their application execution context. Figure 1 illustrates the proposed model regarding data flow management, defined by managing applied data levels. Load balancing management primarily utilizes users' tacit knowledge as a heuristic for resource allocation. Thus, the QoC paradigm is intended to process context information provided by the user to improve performance and QoE.

Contrary to inexperienced users, adept users can guide resource allocation optimization through context information concerning the required resources. Therefore, users need to specify information regarding their application's needs, such as memory quantity, processing capacity, number of processes, network speed, disk space, and operating system. These details correspond to users' tacit knowledge, which, through a utility function (as shown in Fig. 1), estimates QoE according to the user's request and what the user received. As depicted in Fig. 1, the QoC corresponding to resource allocation reflects the applied QoS. QoS is determined by the amount of each allocated resource in the environment.

The proposed model necessitates calculating context using conditional probability, as a context is determined by conditional information that relies on each other. For instance, in a specific context, CPU power x to execute Job1 is available; allocating this CPU depends on the number of available processes adequate to perform the job according to the application's needs. Each context information is determined as a Context Parameter (CP). In the mentioned example, the MIPS (millions of instructions per second) rate of the CPU depends on the number of available processes. In the proposed model, the probability of a specific resource pc_i successfully fulfilling a request on a determined node with capacity pc_j is calculated by Eq. 1. This equation calculates the probability of a quantity of a specific resource pc_i to execute correctly in a given node with capacity pc_j.

$$P(pc_i|pc_j) = 1 - \frac{P(pc_j|pc_i) * pc_i}{pc_j} \tag{1}$$

The probability for each requested resource pc_i is defined by Eqs. 2 and 3, where AVR_{grid} is the simple average of the total capacity of the environment concerning the requested resource, considering all involved devices. The AVR_{grid} metric is defined by Eq. 4, where **n** is the number of resources associated with the **PC**.

$$P(pc_i) = \frac{pc_i}{AVR_{grid}} \tag{2}$$

$$P(pc_j|pc_i) = \frac{pc_i}{pc_j} \tag{3}$$

$$AVR_{grid} = \frac{\sum_{j=0}^{n} pc_j}{n} \tag{4}$$

The availability of a pc_i is determined by the sum of the probabilities of all context parameters CP_j provided. Since the resulting probability must be considered from a specific individual element, the individual occurrence needs to be considered for all possible **n** PCs. Therefore, the availability probability (PD) of a particular PC is determined by Eq. 5.

$$PD(pc_i) = \frac{1}{n} * \sum_{j=1}^{n} P(pc_i|pc_j); P(pc_i|pc_j)\epsilon PCC(pc_i) \tag{5}$$

The QoC is used in the proposed model to measure the quality of allocation for a specific resource. Thus, the QoC can be used for two purposes: improving resource allocation and investigating system performance. In this sense, the QoC is used for both decision-making and identifying potential system malfunctions.

4   Components of the Proposed Model

Figure 2 provides an overview of the components of the proposed model. The model utilizes a hierarchical management scheme determined by: a global manager; and local managers, which can be geographically distributed. The objective in utilizing this type of management is to allow greater freedom for local resource schedulers to define their own management policies. Globally, decisions are made from the context manager, which is the component responsible for forwarding tasks to the local fog systems. Each fog domain has a Resource Management System (SGR) Resource Management System, RMS, which has its own local resource schedulers and manages each local fog environment.

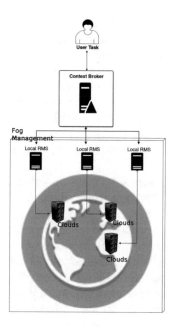

Fig. 2. Architecture of the proposed model.

At the top of Fig. 2, users who will submit applications to the computational fogs are represented. The context manager *context broker* is responsible for determining the best execution context and generating the job submission file from the context indicated by the user and the current state of the system's resources. The context information indicated by the users considered are respectively:

- Application category;
- Required communication speed (links);
- Required processing speed;
- Number of processes;
- Estimated execution time.

The context information about the state of the environment considered are respectively:

- Communication speed of the links;
- Communication latency of the links;
- Available processing capacity;
- Number of available processes.

From the context information, the Quality of Context (QoC) for each resource is calculated, and the resource servers are ranked. The resources specified by the users are described in a file composed of expressions in the form of attribute-value tuples, determining the amount of requested resources. These expressions consist of attributes and values, where each attribute represents the resource and the value of the attribute represents the requested amount. For example, a request for 20 MIPS processing capacity would be expressed by the tuple (mips 20). The resource management system compares the values of the necessary attributes with the values of the available resources to determine if the request can be satisfied.

If there are enough resources to meet the requirements of the submitted application, the local SGR(s) of the resources receive the data for the execution of the submitted task, and the task is then placed in a waiting list.

If there are insufficient resources to meet the requirements of the submitted application, the application must wait indefinitely until the necessary resources are released in the system. The proposed model uses priority to ensure that the most qualified users have access to resources. Therefore, if a lower-priority application is occupying the resources required by a higher-priority application, the lower-priority application will be placed in the waiting list, and the higher-priority application will be served. Local resource management systems must have policies to resolve possible cases of starvation of applications that are in the waiting queue for resources.

5   Experimental Environment and Results

In order to investigate the performance of the proposed model, several experiments were conducted in the SimGrid simulator. The environment used as a platform for conducting the tests was the platform.xml file available alongside the simulator. The platform.xml file defines 90 heterogeneous nodes arranged in different locations and connected by links with different transmission capacities, creating a suitable environment to apply the previously defined policies. SimGrid is an ideal testing platform for resource allocation policies involving networks, as it has simulation classes that implement point-to-point communication mechanisms. The connection between the various nodes of a computational grid can be specified in SimGrid. Therefore, to conduct this experiment, links were specified to connect all nodes with the aim of applying different routing strategies available in SimGrid: Dijkstra, Round-Robin, or both (Dijkstra + Round-Robin) [9]. The strategy Dijkstra + Round-Robin was considered in the conducted tests and compared with the context-based strategy, exactly as proposed in the model. As an input file, the experimental environment received task descriptions, consisting of the following information:

- Category: Application category, CPU or network-oriented, sequential or parallel;
- Process_amount_req: Number of processes required by the user;
- Computation_amount_req: Estimated computation in FLOPS by the user;
- Communication_amount_req: Estimated communication in bits by the user;
- Execution_time: Estimated execution time by the user.

Workloads were synthesized by generating random values for the aforementioned parameters, aiming to simulate the unpredictable behavior of a real distributed system. A workload with 3000 tasks was generated, distributed randomly among 20 fictitious users represented by numbers (IDs) from 1 to 20. The tasks (or jobs) were synthesized using the random function of the GCC library. System clock time was used to calculate and generate random numbers.

5.1 Results

The graph in Fig. 3 shows the total execution times or makespan (wait + runtime) for all workloads using the proposed model and using the Round-Robin policy along with Dijkstra-based routing. In the graphs of Figs. 3 and 4, the thinner line (light gray) represents the standard deviation generated by the Round-Robin + Dijkstra algorithm, and the thicker line (dark gray) represents the standard deviation of the proposed model.

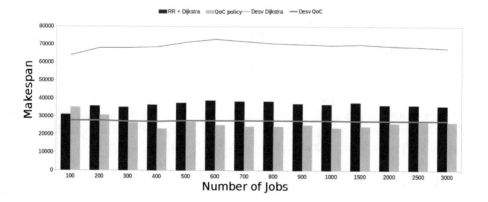

Fig. 3. Average total execution time (makespan).

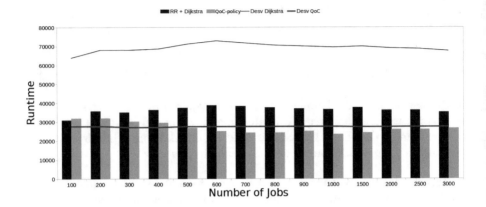

Fig. 4. Average total execution time excluding wait time.

The **X** axis shows the workloads with different numbers of tasks, while the **Y** axis shows the average execution time in seconds. According to the graph in Fig. 3, the proposed model achieved a lower average performance than the Round-Robin algorithm for executing a workload of 100 tasks. However, for the remaining workloads, the proposed model achieved a higher average performance. This occurred because for a workload of 100 jobs, the environment did not adapt to the abrupt context change, as all free resources were occupied without the environment being able to adapt before the completion of all tasks. Therefore, the proposed model manages to distribute appropriately and adapt to workloads with a large number of tasks. The same situation occurs with the execution time in Fig. 4.

In all executions, the standard deviation obtained by the proposed model was lower than the Round-Robin algorithm's standard deviation, revealing a better load distribution and environment performance. Furthermore, the lower standard deviation of the proposed model represents less dispersion in the resulting execution times. The objective of this experiment was to analyze the performance of the proposed model when subjected to thousands of jobs with different characteristics considering a well-defined interconnection grid in the SimGrid environment. Based on the results obtained from the experiment, it was concluded that the proposed model performs better when compared to the Round-Robin algorithm with routing based on the Dijkstra algorithm.

6   Related Work

Messina [8] proposes an agent-based model to provide SLA negotiation. The authors use an ontology to establish the necessary resources for each submission. Ontologies are employed to provide knowledge to the environment. Consequently, it becomes possible to establish a more tailored resource allocation according to the input parameters of the application. The ontologies generate knowledge about the semantic rules to perform the matching between the requested resource and the available resource. The work proposed in [8] uses an ontology to provide knowledge; however, this approach does not handle situations not foreseen by the ontology well. The context-based strategy of the model proposed in this article provides the best configuration for application execution based on the updated state of the system. The approach proposed in this article allocates resources considering not only the application's performance but also the system's performance, which does not occur in the work proposed by [8].

Das employs a resource scheduling policy based on artificial intelligence [2]. However, Das uses the Teaching-Learning Based Optimization (TLBO) algorithm as a basis for their proposal. The justification given by the authors for adopting the learning algorithm in the context of resource allocation in computational grids is that TLBO is considered a lightweight and effective algorithm to find global solutions for optimization problems. Another artificial intelligence-based work is presented in [7]. The authors employ an approach focused on estimating values for various data transmission parameters, such as latency and link utilization. The approach used in [7] aims solely to provide service guarantees based on QoS prediction through fuzzy logic. Parameters or formulations for controlling global performance are not specified.

In [12], a dynamic resource allocation method called DRAM is proposed for load balancing in fog environments. The method relies on a scheduler capable of dynamically migrating services to achieve load balancing for fog computing systems. Negative aspects related to migration, QoS degradation, and QoE are not considered in the work of [12]. In the works [6,11], scheduling strategies with real-time constraints are discussed. In [11], aspects of QoE and QoS are addressed but not extensively enough to propose directives to measure and improve these attributes. In [6], the authors develop work aiming to investigate how utility is affected by performance parameters in fog environments focused on health applications. To evaluate the use of a fog datacenter, the resources of the iFogSim tool were utilized. In the work [1], a new resource allocation algorithm based on stable matching (FSMRA) is proposed to benefit users and providers in the fog environment. However, the authors do not clearly demonstrate how aspects involving user satisfaction and environment performance are treated.

7 Conclusion

This work has presented a resource allocation model tailored for fog computing environments. Fog computing allows preprocessing of information before it is directed to cloud computing. To achieve this, various network edge devices form local clusters of devices with a high degree of heterogeneity in terms of computational resource capacity. Resource allocation in such an environment can be significantly improved by contextual information. Contextual information enables the selection of resources appropriate to user requirements. Thus, our model enhanced the performance of simulated systems, which correspond to the key characteristics of a fog environment. The experimental results obtained from simulations revealed that the model enhances system performance, considering the test execution time and user satisfaction. User satisfaction was assessed using a quantitative prediction model presented in Sect. 3. As future work, we aim to conduct experiments in real environments. For this purpose, the implementation of a real scheduler utilizing the policy addressed in this work is required, as well as the implementation of the proposed architecture.

References

1. Battula, S.K., Garg, S., Naha, R.K., Thulasiraman, P., Thulasiram, R.: A micro-level compensation-based cost model for resource allocation in a fog environment. Sensors (2019)
2. Das, D., Pradhan, R., Tripathy, C.R.: Optimization of resource allocation in computational grids. Int. J. Grid Comput. Appl. **6**(1), 1–18 (2015)
3. Dey, A.K.: Providing Architectural Support for Building Context-aware Applications. PhD thesis, Atlanta, GA, USA (2000). AAI9994400
4. Fiedler, M., Hossfeld, T., Tran-Gia, P.: A generic quantitative relationship between quality of experience and quality of service. Network IEEE **24**(2), 36–41 (2010)
5. Hong, C.-H., Varghese, B.: Resource management in fog/edge computing: a survey on architectures, infrastructure, and algorithms. ACM Comput. Surv. **52**, 1–37 (2019)
6. Khattak, H.A., Arshad, H., ul Islam, S., Ahmed, G., Jabbar, S., Sharif, A.M., Khalid, S.: Utilization and load balancing in fog servers for health applications. EURASIP J. Wirel. Commun. Netw. **2019**, 1–12 (2019)
7. Kolomvatsos, K., Anagnostopoulos, C., Marnerides, A.K., Ni, Q., Hadjiefthymiades, S., Pezaros, D.P.: Uncertainty-driven ensemble forecasting of QoS in software defined networks. In: 2017 IEEE Symposium on Computers and Communication (ISCC), pp. 908–913, June 2017
8. Messina, F., Pappalardo, G., Santoro, C., Rosaci, D., Sarne, G.: An agent based negotiation protocol for cloud service level agreements. In: WETICE Conference (WETICE), 2014 IEEE 23rd International, pp. 161–166, June 2014
9. Möhring, R.H., Schilling, H., Schütz, B., Wagner, D., Willhalm, T.: Partitioning graphs to speedup Dijkstra's algorithm. J. Exp. Algorithmics **11** (2007)
10. Shekhar, S., et al.: Urmila: dynamically trading-off fog and edge resources for performance and mobility-aware IoT services. J. Syst. Architect. **107**, 101710 (2020)
11. Talaat, F.M., Ali, S.H., Saleh, A.I., Ali, H.A.: Effective load balancing strategy (ELBS) for real-time fog computing environment using fuzzy and probabilistic neural networks. J. Netw. Syst. Manag. 1–47 (2019)
12. Xu, X., et al.: Dynamic resource allocation for load balancing in fog environment. Wirel. Commun. Mob. Comput. **2018** (2018)

Implementation of a Cloud Based Voice Recognition Motor Control System for Omnidirectional Wheelchair

Keita Matsuo[✉] and Leonard Barolli

Department of Information and Communication Engineering, Fukuoka Institute of Technology (FIT), 3-30-1 Wajiro-Higashi, Higashi-Ku, 811-0295 Fukuoka, Japan
{kt-matsuo,barolli}@fit.ac.jp

Abstract. In recent years, there are designed and developed various robots which can collaborate with a wide range of devices to assist humans. Also, industrial robots have many applications in factory operations. Moreover, robots such as vacuum cleaners, security bots, therapy robots, and wheelchair robots offer valuable assistance in various human tasks. Thus, the robots and related technologies can improve the quality of life. Worldwide there are more than one billion people with disabilities who rely on wheelchairs for their mobility. In this paper, we implment a cloud based voice recognition motor control system for omnidirectional wheelchair. Also, we implemented a measurement system for the voice control of omnidirectional wheelchair. The average voice recognition time for each direction is "Front": 0.981 [s], "Back": 0.942 [s], "Right": 0.960 [s], and "Left": 0.954 [s]. From these results, we conclude that the cloud based voice recognition system can be used for control of omnidirectional wheelchair.

1 Introduction

The World Health Organization (WHO) [2] reports that there are more than 1 billion individuals with disabilities worldwide, constituting roughly 15% of the global population. In recent years, numerous facilities and devices are designed and implemented to support people with disabilities and elderly people. One of these devices used in many applications is the wheelchair.

Today, there is a variety of wheelchairs designed to enhance the daily lives of individuals such as wheelchairs equipped with navigation systems [5,14,15]. These advanced wheelchairs are capable of assisting users by guiding them to their desired destinations. They can be operated using a head movement interface, which relies on inertial sensors to control the wheelchair movements [7].

A Brain-Computer Interface (BCI) is used for disabled individuals who have limited or no control over their limbs, including those experiencing cerebromedullospinal disconnection. This interface allows them to control the wheelchair using their brain signals, providing newfound mobility and independence [3,19]. The Human Machine Interface (HMI) is another iterface for wheelchair control. The

HMI systems utilize piezoelectric sensors to detect movements on the face and tongue, enabling users to navigate the wheelchair with these subtle gestures [4].

Additionally, there are eye control interfaces, such as those based on electrooculography (EOG). These interfaces enable users to operate the wheelchair by tracking their eye movements and signals [6,8].

In our previous work, we implemented omnidirectional wheelchair robot and considered their applications as mobile routers for providing network connection [16–18] and for assisting people who use wheelchairs [9,13]. Furthermore, we have used the implemented omnidirectional wheelchair for different sports such as wheelchair tennis and badminton [10,11].

In our research work, we have considered various interfaces such as voice recognition, touch screen interfaces, small finger mouse, interfaces for detecting subtle finger movements, and various joysticks [12,13]. However, we did not consider voice recognition interface because of some issues such as misrecognition and recognition time. With the recent advancement of cloud services, it has become easier to implement voice recognition for the control of the omnidirectional wheelchair.

In this paper, we use Azure voice cognitive service provided by Microsoft and implment a cloud based voice recognition motor control system for omnidirectional wheelchair. Also, we implemented a measurement system for the voice control of omnidirectional wheelchair. We evaluate the system performance by experiments. The experimental results show that the cloud based voice recognition system can be used for control of omnidirectional wheelchair.

The rest of this paper is structured as follows. In Sect. 2, we introduce the proposed omnidirectional wheelchair. In Sect. 3, we explain the implementation of the cloud based voice recognition motor control system. In Sect. 4, we describe experimental results. Finally, conclusions and future work are given in Sect. 5.

2   Proposed Omnidirectional Wheelchair

In this section, we present the omnidirectional wheelchair which is able to provide a convenient environment for users. In Fig. 1(a) is shown the implemented omnidirectional wheelchair and in Fig. 1(b) is shown the omniwheel. The omniwheel is able to rotate not only front and back but also to move in right and left directions using 28 small tires. The omnidirectional wheelchair can move in a narrow space while maintaining a consistent direction, reducing the time required to reach a destination without the need for frequent directional changes.

Figure 2 shows the operation of omnidirectional wheelchair by joystick controller in front of a refrigerator and a sink. But this operation is little bit inconvenient. If the user will able to use both hands as shown in Fig. 3 the operation environment will be better for wheelchair users. For this reason, we decided to implement a voice recognition interface for omnidirectional wheelchair. The details of the interface will be described in the next section.

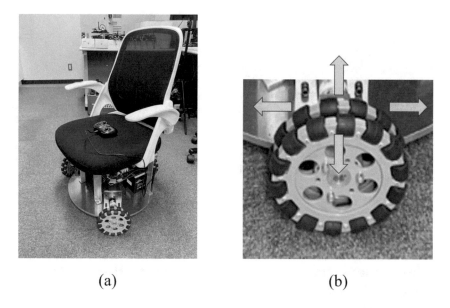

(a)   (b)

Fig. 1. Implemented omnidirectional wheelchair and omniwheel.

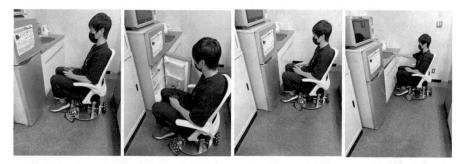

Fig. 2. A snapshot of using omnidirectional wheelchair with joystick controller in front of refrigerator and sink.

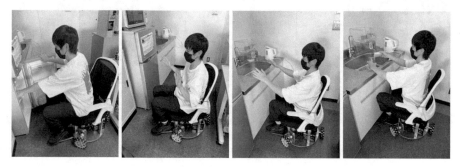

Fig. 3. A snapshot of using omnidirectional wheelchair without joystick controller in front of refrigerator and sink.

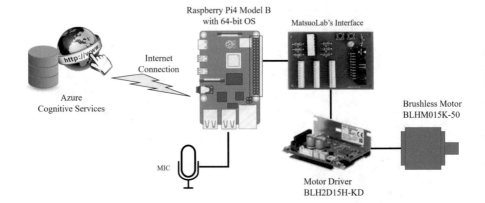

Fig. 4. Architecture of implemented cloud based voice recognition system.

3 Implementation of Cloud Based Voice Recognition Motor Control System

In this section, we explain cloud based voice recognition system for motor control which we implemented using Microsoft Azure [1]. Azure is as a cloud service provided by Microsoft and have various basic services such as computing, storage, data base, network, security, backup, monitoring and so on. In addition, there are other applicable services for AI, machine learning, IoT, big data, and recovery of disaster. Azure have now more than 200 services.

In Fig. 4 is shown the architecture of implemented cloud based voice recognition system. In order to evaluate the proposed system, we implemented a brushless motor control unit as shown in Fig. 5.

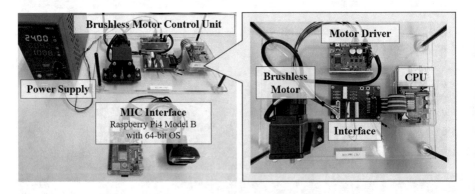

Fig. 5. A snap shot of implemented Motor Control Unit.

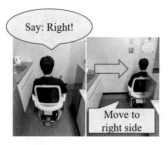

Fig. 6. Image of controlling omnidirectional wheelchair with cloud based voice recognition.

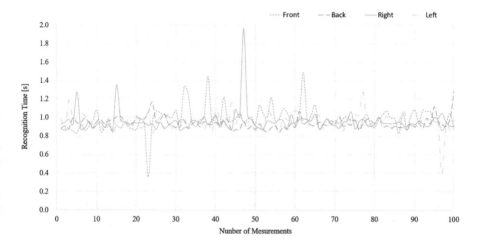

Fig. 7. Experimental results of voice recognition time for controlling omnidirectional wheelchair.

4 Experimental Results for Controlling Omnidirectional Wheelchair with Cloud Based Voice Recognition

We controlled the omnidirectional wheelchair by using the cloud based voice recognition. We measured the voice recognition time to control omnidirectional wheelchair. We present the image of omnidirectional wheelchair with cloud based voice recognition in Fig. 6. We considered four voice recognition words: 'left', 'right', 'front', 'back'.

In Fig. 7 are shown the experimental results of voice recognition time for controlling the omnidirectional wheelchair on the cloud. From the experimental results, we can see that the maximum voice recognition time is within 2.0 [s] and the average recognition time is about 1.0 [s]. These results show that we are able to use the cloud based voice recognition system to control the implemented omnidirectional wheelchair because now the maximum speed of omniderectional wheelchair is about 1.6 [km/h]. Even it takes about one second for voice recognition, there are not safety issues because the wheelchair speed is slow.

Table 1 shows the average recognition time and Word Error Rate (WER) for each word. The error rates of "Back" and "Right" are higher than other words. One reason for these results is that Japanese speakers had difficulties with the pronunciation of "R", "L" and "Back". We believe that the recognition accuracy would significantly improve when the words are pronaunced by native speakers.

Table 1. Average time of voice recognition and word error rate.

Items	Front	Back	Right	Left
Average Time of Recognition	0.981 [s]	0.942 [s]	0.960 [s]	0.954 [s]
Word Error Rate	2.0 [%]	10.0 [%]	17.0 [%]	2.0 [%]

5   Conclusions and Future Work

In this paper, we implemented a cloud based voice recognition motor control system using Azure which can send the control signal to the omnidirectional wheelchair for about one second. The average voice recognition time for four directions was "Front": 0.981 [s], "Back": 0.942 [s], "Right": 0.960 [s] and "Left": 0.954 [s]. From these results, we conclude that cloud base voice recognition system can be used for control of omnidirectional wheelchair. However, if we use the omnidirectional wheelchair with the voice recognition system for sports we need to reduce the voice recognition time in order to ensure a safe environment.

In the future work, we will decrease the total control time of the omnidirectional wheelchair robot and improve the WER. Also, we would like to improve the control of omnidirectional wheelchair.

Acknowledgements. This work is supported by JSPS KAKENHI Grant Number JP22K11598.

References

1. Microsoft azure. https://portal.azure.com/
2. World health organization. https://www.who.int/
3. Ansari, M.F., Edla, D.R., Dodia, S., Kuppili, V.: Brain-computer interface for wheelchair control operations: an approach based on fast fourier transform and on-line sequential extreme learning machine. Clin. Epidemiol. Glob. Health **7**(3), 274–278 (2019)
4. Bouyam, C., Punsawad, Y.: Human–machine interface-based wheelchair control using piezoelectric sensors based on face and tongue movements. Heliyon **8**(11) (2022)
5. Chatterjee, S., Roy, S.: Multiple control assistive wheelchair for lower limb disabilities & elderly people (2021)

6. Choudhari, A.M., Porwal, P., Jonnalagedda, V., Mériaudeau, F.: An electrooculography based human machine interface for wheelchair control. Biocybern. Biomed. Eng. **39**(3), 673–685 (2019)
7. Gomes, D., Fernandes, F., Castro, E., Pires, G.: Head-movement interface for wheelchair driving based on inertial sensors. In: 2019 IEEE 6th Portuguese Meeting on Bioengineering (ENBENG), pp. 1–4 (2019)
8. Kaur, A.: Wheelchair control for disabled patients using EMG/EOG based human machine interface: a review. J. Med. Eng. Technol. **45**(1), 61–74 (2021)
9. Matsuo, K., Barolli, L.: Design and implementation of an omnidirectional wheelchair: control system and its applications. In: 2014 Ninth International Conference on Broadband and Wireless Computing, Communication and Applications, pp. 532–535. IEEE (2014)
10. Matsuo, K., Barolli, L.: Prediction of RSSI by Scikit-learn for improving position detecting system of omnidirectional wheelchair tennis. In: Barolli, L., Hellinckx, P., Enokido, T. (eds.) BWCCA 2019. LNNS, vol. 97, pp. 721–732. Springer, Cham (2020). https://doi.org/10.1007/978-3-030-33506-9_66
11. Matsuo, K., Kulla, E., Barolli, L.: Implementation of a collision avoidance system for machine tennis game. In: Barolli, L. (eds.) Advances in Networked-based Information Systems. NBiS 2023. LNDECT, vol. 183, pp. 150–158. Springer, Cham (2023). https://doi.org/10.1007/978-3-031-40978-3_17
12. Mitsugi, K., Matsuo, K., Barolli, L.: A comparison study of control devices for an omnidirectional wheelchair. In: Barolli, L., Amato, F., Moscato, F., Enokido, T., Takizawa, M. (eds.) WAINA 2020. AISC, vol. 1150, pp. 651–661. Springer, Cham (2020). https://doi.org/10.1007/978-3-030-44038-1_60
13. Mitsugi, K., Matsuo, K., Barolli, L.: Evaluation of a user finger movement capturing device for control of self-standing omnidirectional robot. In: Barolli, L., Woungang, I., Enokido, T. (eds.) AINA 2021. LNNS, vol. 227, pp. 30–40. Springer, Cham (2021). https://doi.org/10.1007/978-3-030-75078-7_4
14. Ngo, B.V., Nguyen, T.H., Ngo, V.T., Tran, D.K., Nguyen, T.D.: Wheelchair navigation system using EEG signal and 2D map for disabled and elderly people. In: 2020 5th International Conference on Green Technology and Sustainable Development (GTSD), pp. 219–223. IEEE (2020)
15. Pradivta, I.W.N.B., Arifin, A., Arrofiqi, F. and Watanabe, T.: Design of myoelectric control command of electric wheelchair as personal mobility for disabled person. In: 2019 International Biomedical Instrumentation and Technology Conference (IBITeC), vol. 1, pp. 112–117 (2019)
16. Toyama, A., Mitsugi, K., Matsuo, K., Barolli, L.: Implementation of a moving omnidirectional access point robot and a position detecting system. In: Barolli, L., Poniszewska-Maranda, A., Park, H. (eds.) IMIS 2020. AISC, vol. 1195, pp. 203–212. Springer, Cham (2021). https://doi.org/10.1007/978-3-030-50399-4_20
17. Toyama, A., Mitsugi, K., Matsuo, K., Barolli, L.: Implementation of a moving omnidirectional access point robot and a position detecting system. In: Barolli, L., Poniszewska-Maranda, A., Park, H. (eds.) IMIS 2020. AISC, vol. 1195, pp. 203–212. Springer, Cham (2021). https://doi.org/10.1007/978-3-030-50399-4_20
18. Toyama, A., Mitsugi, K., Matsuo, K., Kulla, E., Barolli, L.: Implementation of an indoor position detecting system using mean BLE RSSI for moving omnidirectional access point robot. In: Barolli, L., Yim, K., Enokido, T. (eds.) CISIS 2021. LNNS, vol. 278, pp. 225–234. Springer, Cham (2021). https://doi.org/10.1007/978-3-030-79725-6_22
19. Zubair, Z.R.S.: A deep learning based optimization model for based computer interface of wheelchair directional control. Tikrit J. Pure Sci. **26**(1), 108–112 (2021)

Towards Priority VM Placement in Fog Networks

Asma Alkhalaf[1,2(✉)] and Farookh Khadeer Hussain[1]

[1] Faculty of Engineering and IT, University of Technology Sydney, Sydney, Australia
asma.alkhalaf@student.uts.edu.au, farookh.hussain@uts.edu.au
[2] Qassim University, Buraydah, Saudi Arabia

Abstract. Recently, fog computing has emerged as a solution to reduce congestion in the network. By situating computational nodes in close proximity to the
end-user, fog computing enhances Quality of Service (QoS). However, the number
of jobs and services is numerous, placing significant demand on fog nodes which
inherently possess limited capacity. Therefore, it is crucial to discuss the design
of a system where critical jobs are given priority over normal jobs while avoiding normal jobs starvation. In this paper, we study the applicability of existing
scheduling algorithms to address this challenge. Our findings reveal that existing
algorithms fall short in adequately addressing the placement of critical jobs without compromising their QoS. Consequently, we encourage the development of a
custom-built algorithm tailored to ensure the allocation of resources for critical
jobs while safeguarding the delay requirements of normal jobs.

Keywords: priority scheduling · VM placement · fog computing

1 Introduction

Fog computing has emerged as a promising paradigm to address the growing demands
of distributed computing applications, particularly those requiring low latency and
improved Quality of Service (QoS). Placing computational nodes closer to users at
the edge of the network reduces latency and enhances the user experience. However,
this decentralised approach presents unique challenges, especially when it comes to
prioritising critical applications.

Fog computing hosts a diverse set of applications, each with its unique QoS requirements. Some applications, such as real-time monitoring and control systems and gaming,
are latency-sensitive, while others may have more relaxed timing constraints. Balancing
the needs of these diverse workloads poses a significant challenge. Therefore, prioritising
critical applications without neglecting non-critical ones requires an efficient resource
scheduling approach.

As with the cloud, virtualisation is a primary tool in fog to facilitate running applications. Container-based and virtual machine (VM) based virtualisations are growing with
more services bringing the processing capability closer to the user. In addition, a virtual
machine (VM) can be used in fog as a micro VM representing an extension of the user's

whole VM in the cloud [1]. A vital issue for fog service providers, especially small-scale fog service providers, is efficiently placing VMs in their datacentre, particularly considering limited resources. One crucial factor to remember is the differences in priority levels between applications. Hence, higher priority VMs must be placed first to avoid violations in the QoS. Various priority-based placement algorithms are available for the cloud environment [2, 3]. However, they are not practical for the unique architecture of fog as resource availability is limited and requires an on-the-fly allocation decision.

To address the issue of placing priority VMs in a fog datacentre (FDC), in this paper, we explored the use of first-fit, best-fit and priority algorithms as possible solutions. We aim to answer whether traditional scheduling algorithms have limitations in handling the placement of priority VMs in the fog environment.

This paper is organised as follows: Sect. 2 discusses the related work. Section 3 presents the proposed solution. Section 4 explains how algorithms are evaluated. Finally, we conclude the paper in Sect. 5 and discuss future work.

2 Related Work

2.1 Literature Review

The literature discussed resource allocation and priority placement in fog, focusing on task scheduling. Some solutions implemented multilevel queues to accommodate different priorities. The authors in [4] have proposed designing a puffer with three levels of queues, one for each priority. Tasks are assigned to each queue based on deadline constraints, and the queue with the highest priority is processed first. The dynamic approach avoids neglecting older tasks by considering the wait time and remaining execution time as factors when deciding the priority.

In [5], tasks are arranged in queues based on their delay sensitivity, and then a Lyapunov-based optimisation framework is used to decide whether to process tasks in fog or cloud. The approach reduces energy consumption and overall wait time. Another solution in [6] utilises the hierarchy in fog-cloud design and processes tasks according to their execution deadline. Each task is assigned a priority and then placed in the appropriate queue. Queues are processed by their priorities, and task priority can be altered to prevent starvation.

Other solutions proposed a multilayer fog architecture where each layer has different time constraints. Using fuzzy logic in [7], the task's priority is determined and then the task is offloaded to the appropriate layer. It is worth mentioning that the authors suggest processing delay-sensitive tasks in the cloud as fog resources are not sufficient. In [8], the authors propose a middle layer between IoT devices and fog nodes in which a task is evaluated and assigned a priority based on its execution delay. Critical ones are processed in the fog, whereas less sensitive tasks are sent to the cloud.

Deciding the priority of tasks by classifying them based on their source was proposed in [9]. This task classification and prioritisation approach decreases the latency for critical tasks in an e-health system. Another resource allocation approach in [10] classified sensory tasks as high priority. It proposed placing them in the fog layer while other tasks are sent to the cloud for processing.

Other solutions aimed towards building a dynamic load balancing system to prevent delays for critical tasks. The authors in [11] are utilising blockchain technology to develop a trust-based priority algorithm with the aim of preventing malicious nodes from becoming overloaded and then causing delays to critical tasks. Tasks deadline violations are monitored, and reputation values are assigned accordingly. The priority is based on the restriction on response time so that tasks with higher restrictions are prioritised. The scalable solution in [12] actively monitors the ratio of fog nodes to the waiting time for priority tasks. It dynamically adjusts the size of the datacentre to optimise power consumption while minimising task processing delays. When the waiting time for priority tasks surpasses a predefined threshold, the system scales up by adding more nodes. An upper limit on the maximum node count is introduced to prevent network congestion.

Developing a resource scheduling algorithm was another approach proposed in the literature. In [13], the authors aim to process all tasks at the fog layer to minimise the average delay. Tasks are offloaded to fog nodes from IoT devices with consideration of the fog node's queue length. A standard task can be rejected when resources are unavailable, whereas a priority task will be scheduled based on the severity of its delay requirements. Another solution in [14] is proposing a non-preemptive priority algorithm in which a task priority is decided by its dependents. In [15], a hierarchical solution that relies on a broker to assign tasks to the cloud or fog is proposed. The assignment is based on the task's priority which is determined by its time sensitivity. The approach utilises ant colony to optimise task scheduling and conserve energy.

2.2 Critical Analysis

Upon reviewing the current work in the literature, it is evident that the placement of high priority VMs has received minimal attention. The majority of solutions discussed above (Table 1) focus merely on the problem of scheduling tasks into fog nodes, while others deal with identifying relevant factors to a task's criticalness. However, a solution has yet to consider the problem of handling the placement of high priority VMs. This paper explores the use of available scheduling algorithms for the scheduling of priority VMs and evaluates their performance. We investigate other approaches as a means to solve this problem. This paper focuses on first-fit, best-fit, and priority approaches to address this crucial issue. The working of these approaches is explained in the next section.

Table 1. Existing scheduling approaches in fog

Paper	Scheduling target (Task/VM)	Proposed approach	Critical factors	Scheduling objectives
[4]	Tasks	Multi-level queueing	Deadline and wait time	QoS for high priority
[5]	Tasks	Classification	Latency	Energy and wait time
[9]	Tasks	Classification	Not mentioned	Reduce latency
[10]	Tasks	NA	Sensory	Optimisation for hp
[11]	Tasks	Blockchain for trust	Latency	Improve QoS
[6]	Tasks	Multi-level queueing	Latency	Not mentioned
[13]	Tasks	Resource scheduling	NA	Reduce average delay
[12]	Tasks	Scalability	NA	QoS & energy
[14]	Tasks	Non-preemptive resource scheduling	Number of dependents tasks	Reduce delay
[15]	Tasks	Priority scheduling algorithm	NA	Delay and energy
[7]	Tasks	Fuzzy logic	NA	QoS
[8]	Tasks	Classification	Execution delay	QoS

3 Proposed Solution and Illustration

In this paper, we propose to examine the existing scheduling algorithms to assess their applicability in scheduling VMs in the fog environment. We established a set of evaluation criteria focusing on key performance aspects. These criteria include VM wait time, rejection rate of all VMs, rejection rate of priority VMs, and adaptability to different priority distributions. VM wait time and processing time are crucial to measure the efficiency of algorithms, while rejection rate indicates their ability to prioritise VMs. The adaptability assesses their performance under various priority distribution scenarios. These criteria were selected based on their relevance to the problem.

To address the problem identified in the previous section, we propose to use three traditional scheduling algorithms: first-fit, best-fit, and priority algorithms to test their performance efficiency in placing priority VMs in fog. In Fig. 1, we provide a reference implementation of a fog datacentre (FDC) with Fog Consortium Node (FCN) as the central management node. We assume VMs are submitted either directly by the user to the FDC or through a migration request via the cloud, which are then managed in a queue. The resource manager controls the availability of all nodes in the FDC and provides mapping information to the provisioner. The provisioner plays a crucial role in optimising the VM allocation to the physical machines within the FDC, ensuring efficient resource utilisation. We assume the scheduler will be loaded into the Provisioner module to make the decision on which waiting VM to be placed, whereas the Resource Manager takes the decision on the physical node selection.

As a general overview of the scheduling process, once a VM is submitted to the FDC, it will be added to the waiting queue. From there, the scheduler will pull it according to the implemented policy and place it at a physical node. The first-fit scheduler follows a simple approach of first come first served by placing VMs on the order they arrived in the first available machines. On the other hand, the best-fit aims to increase the utilisation of resources by choosing the best available for the VM. We define the best available resource as the one with the highest percentage of utilisation.

In contrast, the priority algorithm follows a more complicated approach depending on the FDC load and the arriving VM's priority. If the load is below the threshold, it places the VM using best-fit policy. On the other hand, when the load exceeds the threshold, it checks the VM priority and then makes the decision. If the waiting VM has normal priority, it rejects the request. While in the case that it has a high priority, it triggers the preemptive policy where it looks for low priority VMs and suspends them to allow the high priority VM to be placed.

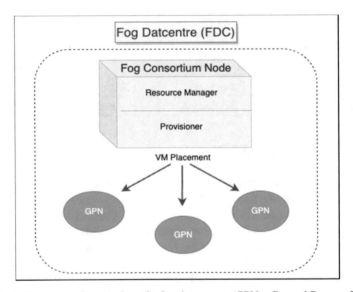

Fig. 1. Reference implementation of a fog datacentre. (GPN – General Purpose Node)

In Fig. 2, we provide an overview of the scheduling process with the following list explaining the breakdown of the components and how they interact in the system:

- *VM Submission*: The user submits a VM to the system, or the VM is migrated from another FDC. These VMs are added to the "Waiting VMs Queue" for processing.
- *Waiting VM queue*: The VM queue holds VMs waiting to be placed. VMs are organised in a first-come, first-served manner until they are scheduled for execution. The arrival of new VMs is assumed to follow a Poisson distribution.

- *VM Scheduling*: The Scheduler component decides which VM from the queue should be placed next. It determines the allocation of VMs to specific physical machines based on various factors, such as available resources, priorities, and load balancing.

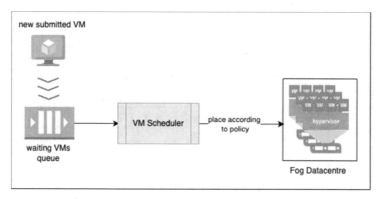

Fig. 2. VM scheduling process

3.1 Illustration

To further explain the applicability of the algorithms to our context, we illustrate the working of the scheduler using the three above algorithms. Let's assume that we have five different VMs: VM_1 , . . , VM_5 with priorities associated to each as p= {2,1,2,1,2} and processing requirements as VC= {8, 14 10, 14, 14}. We also have four hosts (Hs): H_1 , ..., H_4 with following available CPU cores C= {16, 8, 16, 14}.

We assume these VMs are arriving in a sequential order with the arrival pattern as explained below forming a queue.

VM_1, VM_2, VM_3, VM_4, VM_5

Scheduling these VMs using first-fit would be such as:

VM_1 -> H_1, VM_2 -> H_2, VM_3 -> H_4

On the other hand, using best-fit results in the following placement:

VM_1 -> H_2, VM_2 -> H_1, VM_3 -> H_3, VM_4 -> H_4

We notice that the placement requests from VM_4 and VM_5 using first-fit and VM_5 using best-fit were rejected as the system load will have crossed the threshold.

Using the priority algorithm, the placement would be as follows:

VM_1 -> H_2, VM_3 -> H_3, VM_4 -> H_4, VM_2 -> H_1

but then the placement will change to

VM_2 -> suspended, VM_5 -> H_1

Upon the arrival of VM_5, we noted that the system load exceeded capacity; therefore, the new VM could not be placed. However, since some VMs are low priority, the algorithm searches for one occupying similar or greater resources to remove and then place the new one. Hence, VM_2 was suspended to place VM_5, which is of a higher priority than the suspended one.

Table 2. Comparison of the placement using the traditional algorithms

VM		Selected host by the scheduling algorithm		
ID	Priority	first-fit	best-fit	priority
VM_1	2	H_1	H_2	H_2
VM_2	1	H_3	H_1	H_1 -then > VM suspended
VM_3	2	H_4	H_3	H_3
VM_4	1	failed	H_4	H_4
VM_5	2	failed	failed	H_1

3.2 Discussion

Using the tree traditional algorithms best-fit, first-fit, and priority for the same scenario, we can see how an individual VM is being scheduled using these algorithms Table 2. While best-fit and priority placed 4 VMs out of 5, first-fit placed only 3. Both best-fit and first-fit failed to accommodate all high-priority VMs. As we see, VM_5 was rejected by both, but it was successfully placed by the priority algorithm. This result of successfully placing all priority VMs can be counted as an outstanding achievement for the priority algorithm. However, it came with the high price of suspending another VM. Therefore, we can clearly see the issue with the priority algorithm, which is almost immediate termination for normal VMs with no resumption plan, creating the starvation problem. We define VM starvation as the fact that we have suspended the VM because there is a higher priority VM, but there is no plan to reschedule it. This definition excludes normal VMs that cannot be placed initially.

We can conclude from our illustration that there are differences in the working between these three and that they yield different results, affecting their rejection rate of all VMs and priority VMs in particular. In the next section, we present our evaluation of these algorithms on a real-world VM dataset.

4   Implementation and Validation

In this section, we discuss our evaluation approach to the three algorithms. The experiments were conducted on a standard desktop computer equipped with an Intel Core i7 processor, 16 GB of RAM, and running Windows 10 operating system. We used Cloudsim Plus simulator [16] to implement the algorithms and evaluate them.

We utilised a publicly available VMs dataset [17] to generate a diverse dataset of 90 VMs with various values for RAM consumption ranging from 24% to 64% of total physical RAM and processing capabilities ranging between 1 to 14 cores. This dataset was generated to include a variety of priority distribution scenarios, including random arrival time of priority VMs and various sizes for priority VMs. To ensure data consistency across experiments, we generated the dataset once and used it to test all algorithms.

Each scheduling algorithm was implemented according to their algorithmic description. We formed three VM datasets of sizes 10, 15, and 22 with different arriving patterns of VMs with a mix of high priority and low priority and three DC datasets with sizes 7, 10, and 14. For each algorithm, we conducted 18 independent runs, 6 per VM dataset and calculated the overall average values to reduce the impact of potential outliers. To assess each algorithm's performance, we recorded the placement accuracy and rejection rate for each run and computed the average.

The results of the experiments are presented in Fig. 3, where we track the number of accepted VMs and the rejected VMs while highlighting the ratio of rejected priority VMs to all rejected VMs. For first-fit, the average rejection rate was (38%) for all VMs, whereas (37%) of submitted priority VMs were rejected. In best-fit, the rejection rate for all VMs is (31%) and (30%) for all priority VMs. Last, for the priority algorithm, it was (33%) for all VMs and (6%) for priority ones.

Breaking down the results by FDC size in Table 3, we notice that the rejection rate of high priority VMs increases as the FDC size and the number of arrived VMs increase. Also, we note that best-fit performs better overall, but the priority algorithm has the best performance when it comes to priority VMs only.

The results of our experimental evaluation reveal several notable insights into the performance of the scheduling algorithms under examination. Two of the tested approaches have failed to consider priority, so the gap becomes very clear. These findings align with the expected theoretical behaviours of these algorithms. In addition, the simple application of the priority algorithm is insufficient, as we will face starvation if the priority VMs rate is constantly high. Our results demonstrate that the existing algorithms fall short when trying to place priority VMs. Therefore, we can confirm a demand for a specialised algorithm that satisfies the QoS of both priority and normal VMs.

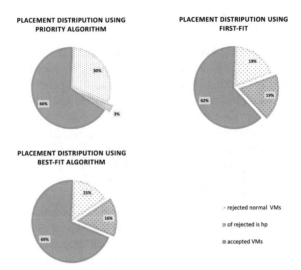

Fig. 3. Average placement and rejection by each algorithm and demonstration of the percentage of high priority VMs among all rejected VMs

Table 3. Placement results analysis

DC size	VM count	Overall rejection rate			HPVM rejection rate		
		ff	bf	pr	ff	bf	pr
7	10	37%	29%	32%	30%	22%	2%
10	15	36%	29%	32%	39%	33%	2%
14	22	39%	34%	37%	44%	36%	15%

5   Conclusion

We explored the problem of prioritising the placement of critical VMs in the fog environment. While there has been prior work on VM prioritisation in cloud computing, these methods are unsuitable for fog environments due to differences in resources. Similarly, current work in fog focuses on task prioritisation, but due to core differences in resource allocation between tasks and VMs, these approaches are not applicable.

In this paper, we have examined the effectiveness of traditional scheduling algorithms in solving the problem of placing priority VMs in the fog datacentre with its limited resources. Existing algorithms need to be revised to solve the problem as they may not be tailored for fog environments with its diverse workloads and strict QoS requirements. Therefore, the need for customised algorithms that align with the unique challenges of fog computing is evident, prompting the research and development of a suitable scheduling algorithm to prioritise VMs and applications based on their criticality. In our future work, we plan to develop a solution to serve systems that prioritise their subscribed users while accommodating partial load from other types of users.

Acknowledgments. The first author wishes to express gratitude to Qassim University for awarding a scholarship, which facilitated the completion of this research.

References

1. Bittencourt, L.F., Lopes, M.M., Petri, I., Rana, O.F.: Towards virtual machine migration in fog computing. In: 2015 10th International Conference on P2P, Parallel, Grid, Cloud and Internet Computing (3PGCIC), 4–6 November 2015, pp. 1–8 (2015). https://doi.org/10.1109/3PGCIC.2015.85

2. Kinger, K., Singh, A., Panda, S.K.: Priority-aware resource allocation algorithm for cloud computing. Presented at the Proceedings of the 2022 Fourteenth International Conference on Contemporary Computing, Noida, 2022. https://doi.org/10.1145/3549206.3549236

3. Savitha, S., Salvi, S.: Perceptive VM allocation in cloud data centers for effective resource management. In: 2021 6th International Conference for Convergence in Technology (I2CT), 2–4 April 2021, pp. 1–5 (2021). https://doi.org/10.1109/I2CT51068.2021.9417960

4. Liao, J.X., Wu, X.W.: Resource allocation and task scheduling scheme in priority-based hierarchical edge computing system. In: 2020 19th International Symposium on Distributed Computing and Applications for Business Engineering and Science (DCABES), 16–19 October 2020, pp. 46–49 (2020). https://doi.org/10.1109/DCABES50732.2020.00021

5. Hazra, A., Adhikari, M., Amgoth, T., Srirama, S.N.: Joint computation offloading and scheduling optimization of IoT applications in fog networks. IEEE Trans. Netw. Sci. Eng. **7**(4), 3266–3278 (2020). https://doi.org/10.1109/TNSE.2020.3021792

6. Adhikari, M., Mukherjee, M., Srirama, S.N.: DPTO: a deadline and priority-aware task offloading in fog computing framework leveraging multilevel feedback queueing. IEEE Internet Things J. **7**(7), 5773–5782 (2020). https://doi.org/10.1109/JIOT.2019.2946426

7. Chakraborty, C., Mishra, K., Majhi, S.K., Bhuyan, H.K.: Intelligent latency-aware tasks prioritization and offloading strategy in distributed fog-cloud of things. IEEE Trans. Indust. Inf. **19**(2), 2099–2106 (2023). https://doi.org/10.1109/TII.2022.3173899

8. Vambe, W.T., Sibanda, K.: A fog computing framework for quality of service optimisation in the Internet of Things (IoT) ecosystem. In: 2020 2nd International Multidisciplinary Information Technology and Engineering Conference (IMITEC), 25–27 November 2020, pp. 1–8 (2020). https://doi.org/10.1109/IMITEC50163.2020.9334083

9. AlZailaa, A., Chi, H.R., Radwan, A., Aguiar, R.: Low-latency task classification and scheduling in fog/cloud based critical e-health applications. In:0 ICC 2021 - IEEE International Conference on Communications, 14–23 June 2021, pp. 1–6 (2021). https://doi.org/10.1109/ICC42927.2021.9500985

10. Sangulagi, P., Sutagundar, A.: Agent based dynamic resource allocation in sensor cloud using fog computing. Int. J. Emerg. Technol. **10**(2), 122–128 (2019)

11. Cao, S., et al.: Delay-aware and energy-efficient IoT task scheduling algorithm with double blockchain enabled in cloud-fog collaborative networks. IEEE Internet of Things J. **11**(2), 3003–3016 (2023). https://doi.org/10.1109/JIOT.2023.3296478

12. Bhushan, S., Mat, M.: Priority-queue based dynamic scaling for efficient resource allocation in fog computing. In: 2021 IEEE International Conference on Service Operations and Logistics, and Informatics (SOLI), 11–12 December 2021, pp. 1–6 (2021). https://doi.org/10.1109/SOLI54607.2021.9672442

13. Tran-Dang, H., Kim, D.S.: Task priority-based resource allocation algorithm for task offloading in fog-enabled IoT systems. In: 2021 International Conference on Information Networking (ICOIN), 13–16 January 2021, pp. 674–679 (2021). https://doi.org/10.1109/ICOIN50884.2021.9333992

14. Fellir, F., Attar, A.E., Nafil, K., Chung, L.: A multi-agent based model for task scheduling in cloud-fog computing platform. In: 2020 IEEE International Conference on Informatics, IoT, and Enabling Technologies (ICIoT), 2–5 February 2020, pp. 377–382 (2020). https://doi.org/10.1109/ICIoT48696.2020.9089625

15. Xu, J., Hao, Z., Zhang, R., Sun, X.: A method based on the combination of laxity and ant colony system for cloud-fog task scheduling. IEEE Access 7, 116218–116226 (2019). https://doi.org/10.1109/ACCESS.2019.2936116

16. Filho, M.C.S., Oliveira, R.L., Monteiro, C.C., Inácio, P.R.M., Freire, M.M.: CloudSim plus: a cloud computing simulation framework pursuing software engineering principles for improved modularity, extensibility and correctness. In: 2017 IFIP/IEEE Symposium on Integrated Network and Service Management (IM), 8–12 May 2017, pp. 400–406 (2017). https://doi.org/10.23919/INM.2017.7987304

17. Anoep, S., et al.: The Grid Workloads Archive. http://gwa.ewi.tudelft.nl/

A Security Evaluation of Chaos Attribute-Based Access Control (ABAC) for Cloud Computing

Omessead BenMarak[1]([⊠]), Anis Naanaa[2], and Sadok Elasmi[1]

[1] Higher School of Communications of Tunis, University of Carthage, Ariana, Tunisia
{omessead.benmbarak,elasmi}@supcom.tn
[2] National Engineering School of Tunis, University of Tunis, Tunis, Tunisia
anisn2000@yahoo.fr

Abstract. This article aims to introduce a cryptographic solution to enhance the security and reliability of the Attribute-Based Access Control (ABAC) model. Building upon the existing structure of attribute-based encryption, our proposal presents a new encryption model named 'Chaos-ABAC'. This model relies on the implementation of chaotic algorithms for both transmitted data encryption and decryption. We demonstrate the resilience of this approach against various falsification attempts, ensuring heightened semantic security against attacks aimed at divulging information from plaintext. Moreover, we demonstrate the scalability of this model, allowing us to enhance its cryptographic features beyond those of existing models.

1 Introduction

Promotions in cloud computing technology have facilitated the transfer of vast amounts of sensitive information across networks. In this context, ensuring the security of user access and identities has become increasingly crucial. Access control models serve as effective techniques for safeguarding private information in such scenarios, and various methods have been proposed to address communication security concerns.

Several traditional access control models such as Discretionary Access Control (DAC) [1], Mandatory Access Control (MAC) [2], and Role-Based Access Control (RBAC) [3] fall short in meeting these genuine needs. Attribute-Based Access Control (ABAC) [4] is deemed highly suitable for systems operating in open environments like Cloud Computing. This model enables users to access data based on a set of attributes. However, the ABAC model carries specific security limitations [5,6]. Because subject attributes integrate attribute-sharing functionalities, this raises the potential for privacy breaches, such as unintentional disclosure to third parties or the aggregation of sensitive information within less secure environments. To enhance security, Attribute-Based Access Control (ABAC) models employing encryption algorithms have emerged. Examples include Distributed Access Control in Cloud (DAAC) [7] and Access Control for Multi-Authority Cloud Storage Systems Based on Temporal Attribute (TAAC) [8,9].

Despite the numerous research and development efforts dedicated to addressing this issue, as well as the exploration of various algorithms to tackle it, our proposal revolves around integrating chaotic systems into ABAC access control models. The various approaches to utilizing chaotic signals in cryptography are currently centered [6] on two main research directions. The first involves using chaos to encrypt messages for transmission [10,11], while the second entails employing chaos to establish a shared secret as a communication key among authorized parties. These two methods operate independently and are mutually compatible, enabling their seamless integration within a unified final system. Chaotic systems are dynamic systems that, under specific initial conditions, exhibit chaotic behavior. This chaotic behavior is characterized by the following attributes:

- Sensitivity to initial conditions: Chaotic systems display exponential sensitivity to minor alterations in their initial conditions, potentially leading to significant differences in the system's output.
- Bounded: States demonstrating chaotic behavior within the system have defined limits.
- Deterministic: Chaotic systems lack inherent randomness; given the initial conditions and parameters, it is possible to simulate the system.
- Aperiodic: Chaotic behavior in these systems is non-periodic; there is no regular repetition in their dynamics.

The rest of the paper is as follows: Section 2 describes the preliminaries of chaotic-based encryption, where we define the logistic map and the usefulness of nominal density in data distribution. Section 3 delineates the standard structure of ABAC control models, while the subsequent part provides a detailed overview of the workflow of the chaos-ABAC model. Following this, the results and discussion segment will involve testing the safety of the proposed technology in comparison with other models, and we will propose solutions to enhance security. Finally, we will conclude with a Conclusion.

2   Chaotic Based Encryption

In various contemporary applications, chaotic signals find utility either in transmitting information or encrypting data. Our specific interest lies in the encryption of data intended for transmission, particularly in the realm of secure communication. As highlighted earlier, deterministic chaos can produce dynamic behaviors that mimic randomness. Consequently, leveraging these behaviors as carriers of messages in the field of telecommunications would be advantageous. The fundamental concept involves concealing data using chaotic signals and transmitting them to the receiver through a public channel. The receiver then retrieves the encrypted text. The principle of the transmission system hinges on the synchronization of the parameter sets for both chaotic generators at the transmission and reception ends.

2.1   Logistic Map

This method [12] offers the possibility of encrypting a message using the simple one-dimensional logistic map defined in the interval E by the Eq. 1:

$$X_{n+1} = b \cdot (1 - X_n) \tag{1}$$

where $X_n \in [0, 1]$ and the control parameter b are chosen so that Eq. 1 exhibits chaotic behavior.

2.2   Normal Density

The concept of normal density in chaotic schemes refers to the distribution of values that the system takes within a specific range of repetition. In the context of chaotic systems, a probabilistic approach can be used to study the probability density of the values that the system assumes at each iteration. The normal density ρ is often calculated as the ratio of the number of occurrences of a particular value i to the total number of processes N. Mathematically, this can be represented as:

$$\rho(i) = \frac{y(i)}{N}; \tag{2}$$

where y(i) is the number of iterations that result in a value within interval i, N is the total character of loops, and $\rho(i)$ is the normal density associated with that period. Using more technical language, if you have a sequence generated by a chaotic system, you can divide the total gaps of possible values into sub-intervals and then count the integer of occurrences of each sub-interval over the iterations. The normal density is then calculated as the relative probability of occurrence in each sub-interval.

From this plot, we have three curves for three initial condition values: 0.001, 0.9, and 1. Assigning to Fig. 1, for initial states greater than or equal to 1, there is no chaotic behavior, so we specify the first conditions from the interval between 0 and strictly less than 1.

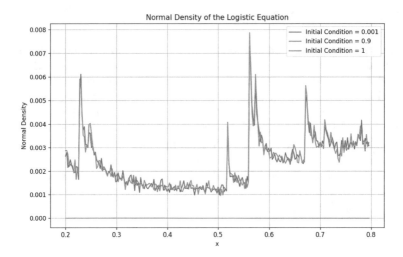

Fig. 1. Normal density with chaotic behavior for b = 3.8.

So, Fig. 2 illustrates how to calculate and represent the normal density for the chaotic logistic equation where:

- Maximum number of iterations: $X_{max} = 65000$
- Number of bins: S = 256
- Initial condition: $X_0 = 0.43$
- Selected interval: $[0.2, 0.8]$

3   Standard ABAC Model

The foundational architecture outlined by the National Institute of Standards and Technology (NIST) [13] for the distributed Attribute-Based Access Control (ABAC) model serves as the cornerstone of our subsequent investigations. Figure 3, illustrates the standard ABAC model's reference architecture. The ABAC model comprises four primary constituents:

- **Policy Enforcement Point (PEP)**: This component receives, interprets, and forwards authenticated subject requests to the Policy Decision Point (PDP), responding following the PDP's determinations.
- **Policy Decision Point (PDP)**: Responsible for evaluating access policies extracted from the Policy Administration Point (PAP) based on attribute values acquired through inquiries to the Policy Information Point (PIP), ultimately deciding access authorization.
- **Policy Information Point (PIP)**: Provides attribute values upon receipt of a request from the PDP context, sourcing these values from services such as LDAP, environmental sensors, or other repositories.
- **Policy Administration Point (PAP)**: Functions to manage (e.g., create, deploy and maintain) and retrieve policies that the PDP subsequently evaluates, ensuring policy verification before integration into the system.

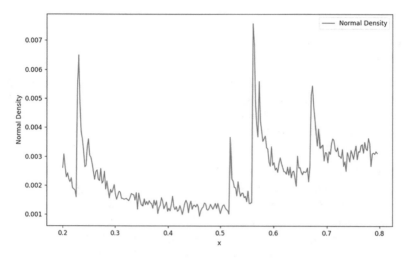

Fig. 2. Normal density with Chaotic behavior b = 3.8. From this plot, we can identify the extremities of the normal density for the logistic map equation within the range [0.2, 0.8], The minimum value $\rho = 0.001$, indicating that each of the 256 intervals with width ϵ will be visited at least 65 times. The maximum value $\rho = 0.0075$, indicating that at most one interval will be visited 488 times during the 65000 iterations.

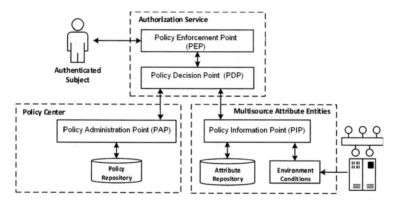

Fig. 3. Design diagram of the ABAC system.

4   Workflow of Chaos-ABAC Model

For our proposed model, we define:

- **Subject Attributes:** Represented by $S_1, S_2, S_3, ...S_n$, denoting user-specific attributes.
- **Environment Attributes:** Represented by $E_1, E_2, E_3, ...E_m$, denoting user-specific attributes.
- **Object Attributes:** Represented by $O_1, O_2, O_3, ...O_q$, defining object-related attributes.

- **Action:** Represented by A indicating the intended action on the object.
- **Token Generation:** Each attribute S_i, O_j, O_k A encrypted to generate corresponding tokens J_i, J_j, J_k, J_A.

Let F represent the set of all subjects, objects, environments, and actions, denoted as $\in F = \{P, S, O, E, A\}$. This set represents the entities in the Chaos-ABAC model, where P signifies the policy employed by the data owner.. Using these entities, the encryption scheme related to the Chaos-ABAC model is formally defined as follows:

1. **Token Verification:** The cloud owner verifies the received tokens J_i, J_j, J_k, J_A to ascertain their validity based on predefined access policies.
2. **Access Authorization Message:** Upon successful token validation, an access authorized message M_{access} = "Access Authorized" is created.
3. **Message Encryption:** The owner encrypts the message M_{access} using a combination of tokens and the logistic map, with the initial condition, parameter b as well as the number of iterations shared securely between the owner and user to obtain $C = Encrypt(M_{access}, b, x_0, C_i)$. An access authorized message M_{access} = "Access Authorized" is created.
4. **Synchronization of Chaotic Sequence:** By initializing the chaotic system with the same parameters, the owner generates a sequence of pseudo-random numbers that synchronizes with the sequence generated during encryption. This synchronization is vital to perform the decryption process correctly.
5. **Message Transmission and Decryption:** The encrypted message C is transmitted to the user, who decrypts it using the same me parameter b, x_0, C_i used during encryption. The user applies the synchronized chaotic sequence to reverse the modifications or operations applied during encryption. Using the received tokens and the logistic map iterations synchronized with the encryp-

Table 1. Comparaison between existing ABE-ABAC schemes and Chaos-ABAC

	CP-ABE [14]	KP-ABE [15]	MA-ABE [16]	Chaos-ABAC
ABAC Model	Centralized	Centralized	Distributed	Distributed
Attribute Type	Subject, Object	Subject, Object	Subject, Object	Subject, Object, Action, Environment
Presentation of attribute	Private Key	Ciphertext	Private Key	Initial Condition, Parameter, Nomber of iterations
Policy Provider	Data Owner	User Manager	Data Owner	Data Owner
Authorization of attribute value	KDC	Data Owner	Multiple Authorities	Multiple attribute Authorities

tion process, the user performs a reverse operation to decrypt the message $M_{decrypted} = decrypted(M_{access}, b, x_0, C_i)$.

5   Results and Discussion

5.1   Comparisons with Related Work

A comprehensive comparison was conducted between KP-ABE, CP-ABE, and MA-ABE within the context of ABAC alongside our Chaos-ABAC scheme across five distinct facets in Table 1. Firstly, CP-ABE and KP-ABE are limited to endorsing a centralized ABAC model, whereas both MA-ABE and our ABAC model proficiently support the distributed ABAC model. Secondly, about the variety of supported attribute types, our scheme encompasses four distinct attribute categories: subject, object, action and environmental attributes in contrast to the remaining three schemes which solely accommodate subject and object attributes. Furthermore, while CP-ABE and MA-ABE associate attribute credentials with the user's private key, in KP-ABE, these credentials are affixed to the ciphertext. However, our scheme intricately embeds these credentials within attribute tokens, dynamically altering them in response to temporal changes, initial conditions, parameters, and the number of iterations of our chaos algorithm. Regarding access policy, the protocol varies across different attribute-based access control mechanisms. In KP-ABE, the user manager defines the policy before private key generation. Conversely, for CP-ABE, MA-ABE, and Chaos-ABAC, the policy originates from the data owner before encryption and becomes inseparably linked to the ciphertext. Moreover, the entity responsible for authorizing attribute credentials differs among these systems: the Key Distribution Center (KDC) assumes this role in CP-ABE, the data owner in KP-ABE, multiple authorities in MA-ABE, and multiple attribute authorities within our devised scheme.

5.2   Security Analysis of the Proposed Model

The utility and validity of a cryptography algorithm or technique hinge upon its robustness against attacks. While no analysis can encompass all possible forms of attacks, an encryption system must be designed to withstand at least some of the most commonly known attacks. Attack techniques aim to compromise the algorithm and retrieve plain texts without accessing the encryption key.

Initially, we carried out a time calculation to determine the duration an attack might require to retrieve a message between the data owner and the user while varying the number of attributes each time. The time required for an attack to retrieve a message in an ABAC system without using encryption algorithms is depicted by a curve shown in Fig. 4. From this figure, it is evident that the message retrieval time for an attack is close to zero for Five attributes and increases to 5.2×10^{-6} s for 50 attributes. Even with the increase in attributes, the attack does not take much time for message retrieval.

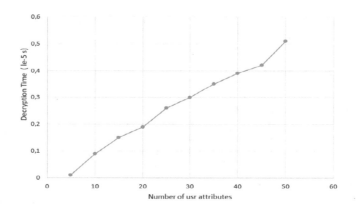

Fig. 4. Decryption time for a message in an ABAC system under attack.

Afterward, we tested our Chaos-ABAC system by calculating the decryption time of a message for various attribute values under attack. In comparison to the previous figure, Fig. 5 illustrates that the message recovery time by the attacker significantly increases compared to no attack, even when using ABE or CP-ABE (previous contributions we made [5]). For Five attributes, the decryption time is 0.21×10^{-3} s, which escalates to 1.75 ms for 50 attributes while maintaining the same hardware characteristics.

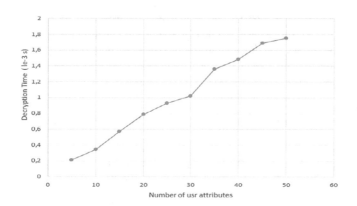

Fig. 5. Decryption time for Chaos-ABAC With keystream Attack.

These discoveries suggest that the model is still susceptible to attacks. To tackle this vulnerability, we will explore a mathematical breakdown of how the attack retrieves the transmitted message.

If we assume that K is the key that is generated by x_0 and b, and a plaintext $P = p_1p_2...$, then the keystream $K = k_1k_2 \ldots$ is created using the Logistic Map equation.

$$C = e_{k1}(p_1)e_{k2}(p_2)... = c1c2... \tag{3}$$

Various attack techniques can be employed to retrieve the keystream utilized by the algorithm, such as the **Chosen-Plaintext Attack** method [17]. To demonstrate the keystream generation, we utilize a two-symbol source, $S_2 = \{s_1, s_2\}$, and seek the encrypted text of messages crafted exclusively from either s_1, ors_2. This assumes that for ($i \in \{1, 2\}$), i represents the symbol linked to each permissible sub-interval, while x denotes the iteration exploring the prohibited sub-intervals.

In these conditions, we conduct a chosen plaintext attack by requesting the encrypted text for the following plaintext message:
$P = (s_1 s_1 s_1 s_1 s_1 s_1 s_1 s_1 s_1 s_1 s_1 s_1 s_1 s_1 \ldots \ldots)$. For instance, if we receive the corresponding encrypted text $C = (2133313133133\ldots)$, we can confidently ascertain that the 6th symbol in the keystream is s_1, as well as the 9th symbol, and also the 11th symbol. Upon considering the entire message, a partial understanding of the keystream is attained:
$K = xxxxxs_1 xxs_1 xs_1 xs_1 xs_1 xxs_1 xs_1 xxs_1 xs_1 xs_1 xxs_1 xs_1 x \ldots$. Presently, it is confirmed that the letter x represents an iteration beyond the range of $[0.2, 0.8]$, excluding the initial x which corresponds to the initial condition.

Any ciphertext with identical values x_0 and b will rely on the same key stream, which is only determined by the parameter b and the initial condition x_0. This susceptibility to cryptanalysis suggests the vulnerability of such encryption. This methodology is promising for broader applications in higher-order sources. The frequent appearance of the symbol x in the key flow is explainable through the Fig. 2, illustrating the immutable natural density represented by the logistic map equation. Regions near the extremes experience more frequent visits. Consequently, opting for the interval $[0.2, 0.8]$ leads to discarding half of the iterations.

5.3   Proposed Techniques

This result shows that Chaos' encryption method is not sufficient. We subsequently propose a modification to increase security and resist attacks. To address this issue, we are attempting to enhance the security of the initial condition x_0, the parameter b, as well as the number of iterations.

The authors [11] proposed a solution to secure the number of iterations. As the number of iterations, C_i, serves as a critical parameter for attacking the encryption algorithm, our approach involves hiding C_i. The concept is to covertly mask C_i using a pseudo-random number, which will then function as a ciphertext. This strategy renders it infeasible for an attacker to retrieve the chaotic iterations from the ciphertext, effectively thwarting any successful decryption attempt on the transmitted information. Employing the confidential function f(.) that produces the pseudo-random number following the equation specified below

$$C_i' = C_i + f(X_0^{(i)}) \tag{4}$$

For parameter b and the initial condition, we use a key shared between the owner and the user. The logistic map's iterations based on the attributes or a specific key influence the encryption process to increase security and randomness. Encryption involves non-linear operations using the logistic map outputs, making it computationally intensive and challenging to reverse-engineer without the proper parameters and keys.

6 Conclusion

Aiming at the security threats of existing ABAC systems that we will use in the Cloud Computing environment, we designed a new secure ABAC model, called Chaos-ABAC, to support provable decision-making on access policies. This combination of attribute-based access control with the logistic map-based encryption scheme fortifies data protection and permission management within cloud environments. We test the security of this model, this representation must be reinforced by other encryption techniques to be valid.

References

1. Lipton, R.J., Snyder, L.: A linear time algorithm for deciding subject security. J. ACM **24**, 455–464 (1977). https://doi.org/10.1145/322017.322025
2. McCune, J.M., Jaeger, T., Berger, S., Cáceres, R., Sailer, R.: Shamon: a system for distributed mandatory access control. In: Proceedings of the 2006 22nd Annual Computer Security Applications Conference (ACSAC 2006), Miami Beach, FL, USA, 11–15 December 2006, pp. 23–32 (2006)
3. Komlenovic, M., Tripunitara, M., Zitouni, T.: An empirical assessment of approaches to distributed enforcement in role-based access control (RBAC). In: Proceedings of the First ACM Conference on Data and Application Security and Privacy, CODASPY 2011, San Antonio, TX, USA, 21–23 February 2011 (2011)
4. Sharma, N.K., Joshi, A.: Representing attribute based access control policies in OWL. In: Proceedings of the 10th IEEE International Conference on Semantic Computing, pp. 333–336 (2016)
5. Zhu, Y., Yu, R., Ma, D., Chu, W.C.-C.: Cryptographic attribute-based access control (ABAC) for secure decision making of dynamic policy with multiauthority attribute tokens. IEEE Trans. Reliab. **68**(4), 1330–1346 (2019)
6. BenMbarak, O., Naanaa, A., ElAsmi, S.: New secure access control model for cloud computing based on Chaotic systems. In: International Wireless Communications and Mobile Computing (IWCMC) (2023)
7. Ruj, S., Nayak, A., Stojmenovic, I.: DACC: distributed access control in clouds. In: Conference on Trust, Security and Privacy in Computing and Communications, pp. 91–98 (2011)
8. Shen, J., Zhou, T., Chen, X., Li, J., Susilo, W.: Anonymous and traceable group data-sharing in cloud computing. IEEE Trans. Inf. Forensics Secur. **13**(4), 912–925 (2018)

9. Kapse, G.V., Thakare, V.M., Sherekar, S., Kapse, A.V.: Multi-authority data access control for cloud storage system with attribute-based encryption. IOSR J. Comput. Eng. (IOSR-JCE) 53–56 (2019)
10. Mulwa, O.: Chaos Theory and its Potential for Cryptography, Section Securtity (2022). https://www.section.io/engineering-education/chaos-cryptography/
11. Rebhi, N., Ben Farah, M.A., Kachouri, A., Samet, M.: Analyse De Sécurité d'une Nouvelle Méthode De Cryptage Chaotique. In: 4th International Conference: Sciences of Electronic, Technologies of Information and Telecommunications, Tunisia, 25–29 March 2007 (2007)
12. Wang, F., Cui, G.: A new image encryption algorithm based on the logistic chaotic system. In: 3rd International Conference on Computer (2010)
13. Mell, P., Grance, T.: The NIST definition of cloud computing, Gaithersburg (2011). https://nvlpubs.nist.gov/nistpubs/Legacy/SP/nistspecialpublication800-145.p
14. Goyal, V., Pandey, O., Sahai, A., Waters, B.: Attribute-based encryption for fine-grained access control of encrypted data. In: Proceedings of the 13th ACM Conference on Computer and Communications Security, pp. 89–98 (2006)
15. Bethencourt, J., Sahai, A., Waters, B.: Ciphertext-policy attribute-based encryption. In: Proceedings of the IEEE Symposium on Security and Privacy, pp. 321–334 (2007)
16. Chase, M.: Multi-authority attribute based encryption. In: Vadhan, S.P. (ed.) TCC 2007. LNCS, vol. 4392, pp. 515–534. Springer, Heidelberg (2007). https://doi.org/10.1007/978-3-540-70936-7_28
17. Biryukov, A.: Chosen plaintext attack. In: van Tilborg, H.C.A., Jajodia, S. (eds.) Encyclopedia of Cryptography and Security. Springer, Boston (2011). https://doi.org/10.1007/978-1-4419-5906-5_557

LoRa and Cloud-Based Multi-robot Pesticide Spraying for Precision Agriculture

Nirali Sanghvi and Rajdeep Niyogi[✉]

IIT Roorkee, Roorkee 247667, India
{nirali_s,rajdeep.niyogi}@cs.iitr.ac.in

Abstract. With the increase in global population and demand for food, there is a need to transit from our traditional agricultural practices towards modern agricultural techniques using sensors, robots, cloud computing etc, which on a broader view is known as Agriculture 4.0. Using chemical pesticides is a very important process in agricultural activity but it is not only dangerous but it also has hazardous effect on humans who come in its regular contact. In this paper, we propose a methodology using multiple robots for spraying pesticides over an agricultural field using cloud computing. Long Range (LoRa) Network is used because of its long range, low power and penetration capabilities. Webots, an open-source simulator that supports multi-robot programming, is used for the simulations. Using four robots instead of one caused a time reduction of nearly 72%. The coverage obtained ranges from 87.6% to 96.85%.

1 Introduction

Agriculture is one of the most significant aspect for the survival of human beings. It not only produces food but also provides employment to large number of people. With the increase in population, there will be a rise in demand for agricultural products. It is estimated that by 2050, the world population will reach 9.8 billion. This surge in demand can potentially strain traditional farming methods [1]. To effectively address this challenge and meet the increasing demand, it is necessary to have a transition from traditional farming practices to more advanced and automated approaches [2]. Some of the essential tasks in the agricultural process, such as land preparation, sowing, irrigation, fertilization, and pesticide application, require a large workforce. All these processes require a lot of manpower and labor but still they may not meet the increasing demand. So, integrating technology into agriculture becomes crucial to enhance productivity, reduce labor-intensive processes, and optimize time management.

The fourth industrial revolution, also known as Industry 4.0, has initiated a transformative era across various fields such as engineering, industry, and management. Agriculture has also undergone a significant evolution, having a transition from Agriculture 3.0 to the current era of Agriculture 4.0. Agriculture 3.0 started in the 20 th century from the development of computer programs and robots. Now, agriculture along with Industry 4.0 is the new revolution Agriculture 4.0. With the emerging techniques like Cloud Computing, Artificial Intelligence, Remote Sensing, etc, the agricultural activities are

L. Barolli (Ed.): AINA 2024, LNDECT 201, pp. 426–436, 2024.
https://doi.org/10.1007/978-3-031-57870-0_38

made more automated and less human dependent [3]. Precision agriculture, smart agriculture all are the results of technology used in agriculture. Agricultural robots are equipped with sensors and intelligent systems that are versatile and also, learn quickly.

Agricultural robots or agribots are defined as robots that are used for agricultural purposes. Various agricultural robots and systems have been developed till now and are being developed for various agricultural processes. Agribots are made to increase efficiency across different farming operations. For instance, seed-spreading robots are good at precise seed placement, weeding robots effectively remove unwanted vegetation, pesticide-spraying robots streamline the application of protective substances, and crop-harvesting robots automate the harvesting process [4]. While much research has been dedicated to autonomous robots performing specific tasks, the future of agribot research should emphasize multi-robot systems in agriculture. This shift aims to create systems that are less dependent on human intervention, promising increased efficiency and productivity in the agricultural sector. The main aim of agricultural robots is to substitute humans and perform the agricultural tasks with more efficiency and precision. The most demanding applications of agricultural robotics are weed control and pesticide spraying [5]. Almost 40 percent of the agricultural produce is lost due to pests and diseases [6]. Figure 1 shows the diseases in plants which may destroy the entire yield and thus the need of pesticides to prevent it.

So, application of chemical pesticides is one of the most important tasks in agriculture. If chemical pesticides are sprayed more than required, it may not only have adverse affect on the crop but also degrade the quality of soil and air. Hence, it needs to be ensured that right amount of pesticide is sprayed to the required position and minimum amount of it is wasted. by using manual methods of pesticide spraying, only 0.1 percent of the pesticide reaches the targeted position [7]. Also, manual spraying techniques have adverse effects on the health of the person doing it. Techniques like blanket spraying where the entire farm is covered with pesticides is not preferable as the use of pesticide for it is in large amounts leading to its wastage and also having adverse affects in crops and environment. So, having an efficient mechanism for spraying which has minimal human interference and more precision is needed. LoRa (Long Range) [8] is a chirp spread spectrum (CSS) based modulation technique. It provides a long range data communication along with low power consumption which is the important for most of the IOT applications. It has less development cost and very strong penetration range which means it can transmit through walls and vegetation in an effective manner which is ideal for our application [11].

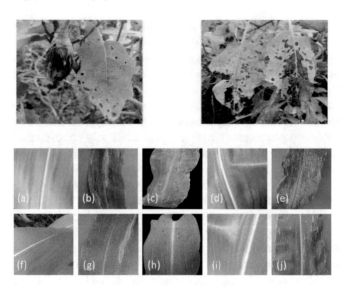

Fig. 1. Diseases in plants [9, 10].

In our work, we propose a cloud based approach that uses Long Range (LoRa) network for communication and aims to achieve optimal coverage and minimal wastage of pesticide using multi-robot. The approach proposed will have minimal human interference thus preventing humans from side-effects of chemical pesticides and can also be used for large fields.

The rest of this article is structured as follows. Related work is discussed in Sect. 2. The proposed approach and algorithm is given in Sect. 3. Section 4 discusses the results obtained. Conclusion and future work is given in Sect. 5.

2   Related Work

All agricultural robots have some common components like a vision system, a control system, a mobile system and mechanical actuators [2]. The vision system is composed of cameras like RGBD cameras, thermal cameras which help in detecting and identifying the targeted position. A control system consists of a controller or computer which can control the actions taken by the system. The mobile system is responsible for providing motion to the robot so that it can move through the field and complete the task that is assigned to it. The mechanical actuators physically executing the tasks to accomplish the operation. Like for spraying, we need sprayers, nozzles, etc., for plucking fruits we need a robotic arm and so on.

The utilization of chemical fertilizers and pesticides is a crucial yet potentially hazardous aspect of agriculture. For a good yield, it is necessary that the plant is healthy and free from diseases and also it has to be protected from the attack of pests. In [12] it is said that about 30%-35% of the damage can be prevented by spraying pesticides. However, using pesticides in excessive amounts may have various side affects not only on the plant or the quality of soil and air but it can also increase the resistance power of the pests. Different methods are being used for the purpose of spraying pesticides. Some of the most common used methods are backpack sprayers, aerial method, etc. In backpack sprayers, the pesticide is sprayed manually. This method is less efficient of larger areas and also has adverse affect on the health of the farmer. Aerial sprayers are developed these days wherein the pesticides are sprayed using air-crafts, helicopters or drones. In [13], an aerial automated sprayer has been developed. [14] also demonstrates an un-mmaned aerial vehicle that can be used. However, in aerial spraying, there is an issue of spray drift which is very severe. Spray drift is defined as the movement of droplets beyond the targeted area [15]. Ground based sprayers are also used wherein the sprayer is attached to tractors or other ground based and then the pesticides are sprayed. Now, with the modernization of agriculture, autonomous sprayers are also introduced wherein human interference is negligible.

Autonomous sprayers make use of sensors so as to perceive the target and complete its task. Also, variable rate spraying is used for it. Variable rate spraying makes sure that the right amount of pesticide is sprayed at the targeted location. Variable rate spraying is also given importance so as to reduce the deposition of pesticides and thus minimize its adverse effects on the environment. Some of the features of variable rate spraying are they allow to achieve either full spray or semi-spray or no-spray as per requirement leading to optimal use of pesticides [16]. It also saves money and can be used with different plants depending on that plants requirement. [15] reviews various types of real time sensor based variable rate sprayers.

Different sensor systems have been developed using different sensors like infrared sensor system, ultrasonic sensor system, LIDAR sensor system, machine vision system, etc. Each of these systems use the sensors to gather the data, then analyze it and take the desired action. An infrared based sensor system has been implemented in [17] where the sprayer has been mounted on the tractor and the target is detected automatically. In [18], a proposed sprayer system incorporates an ultrasonic sensor detection system. This system works in conjunction with a variable rate spraying mechanism achieved through a controlled unit for regulating flow rates, a micro-controller, and a spray delivery system. An automatic adjustable site specific spraying device has been proposed in [20] that is based on variable rate spraying and it can also deal with variable size targets. It is equipped with a direct nozzle for spraying the pesticide, a camera to capture the image and then spray the pesticide at the targeted position. [19] also proposes an variable rate sprayer in which the sprayer uses profile based variable rate spraying. It means the amount of pesticide to be sprayed is dependent on the density of leaves and the canopy of the tree.

Using multi-robots in the field of agriculture is still being explored. The major reason for it is achieving coordination and communication among them and overall maintaining all of them. But, using multi-robots may not only decrease the time required for overall completion of the task but it will also help in increasing efficiency. [22] proposes an communication protocol to achieve robot to robot communication in an agricultural field and also an collision avoidance algorithm for the same. [23] reviews the use of multi-robot for precision agriculture. [21] proposes using both LoRa and MQTT protocol for smart farms.

3   Proposed Approach

3.1   Methodology

In this section, we discuss the proposed approach for our system. We here consider each sprayer to be homogeneous. The sprayer that we will be using for our proposed system will be variable rate sprayer as proposed in [20] that is, it will spray variable quantity of pesticide as required. Figure 2 gives an diagrammatic overview of the system proposed.

Components in the Proposed Methodology:

1. *Autonomous Vehicles:* These autonomous vehicles help the sprayer to move from one place another thus providing mobility to it. It assists the sprayer to move which helps in spraying pesticide over the entire field thus obtaining optimal coverage.
2. *LoRa based sensors:* These sensors help to collect data of the climatic conditions on the field are called as LoRa nodes. They collect this data and pass on the data packets in the network for further processing.
3. *LoRa gateway:* The data from the LoRa sensors is obtained at the LoRa gateway. LoRa gateway acts as a connecting link to pass on the data from LoRa end nodes to further processing.
4. *Cloud:* The cloud provides us the computing power required for our proposed methodology.
5. *Controller:* Controller is the responsible for passing forward the decision obtained from the cloud so that it can be executed.
6. *End devices:* End devices send notification to the end user.

There is a network of LoRa based sensors that is LoRa nodes like temperature sensors, humidity sensors in the field. Now, in the field, we have our multiple sprayers mounted on autonomous vehicles. These sprayers are not connected to each other nor they have any information about the location of other sprayers. But, this sprayers are mounted with different sensors like a GPS sensor which can be used to send its location into the system, a camera which will be used to detect its target, a controller so as to control the flow of pesticide at the targeted location depending on requirement which is obtained by analyzing all the factors. Now, all this data gathered is sent to the cloud for processing using LoRa gateway. Now, these packets of data are transmitted to the cloud for further processing. The cloud uses the data obtained from the LoRa gateway to make relevant decisions for the next action to be performed. This decision is stored in the cloud and then passed on further for execution. This data is either passed on to controller or end device depending on the decision. If there is any decision that needs users' intervention then it is passed on to the used through end devices like mobiles or desktops. Then, as soon as the user takes the decision, the decision made is passed on to the cloud for further processing as per the decision made by the user. If further processing does not need human intervention, then the processed data is passed on to the controller which sends data to the controllers over the field about the action to be executed. Then these controllers execute the desired action.

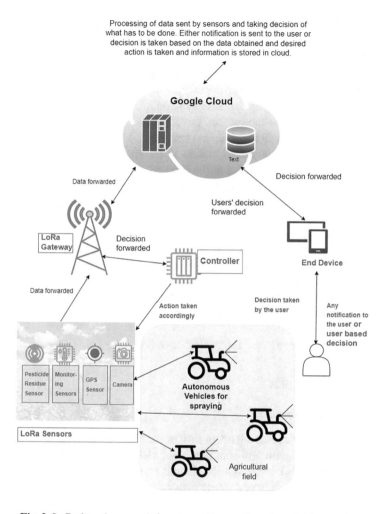

Fig. 2. LoRa based approach for our multi-agent based pesticide spraying.

3.2 Algorithm

The algorithm used for decision making in cloud with respect to our environment is given below. The input to the system is the sensor data obtained and the output of the system will be the decision taken. The algorithm works are follows:

Step 1: All the robotic sprayers start their motion from their initial position. Their positions are sent to the cloud using LoRa gateway.

Step 2: As the robotic sprayers start the motion, they start to detect their target which is the plant.

Step 3:

(a) As the plant is detected, the sprayer turns on and it starts to spray the pesticide towards the targeted position for a programmed duration and time.

(b) Now, the location of the robotic sprayers that is sent to cloud is continuously analyzed in the cloud. It two robotic sprayers are in the same row or are in close vicinity of each other then:

(i) The quantity of pesticide in each of the sprayers is checked and the one with more amount of pesticide continues the motion.

(ii) If the robotic sprayers have same amount of pesticide then the sprayer with higher priority continues the motion and the other one's direction is changed so as to cover the area where the pesticide is not yet sprayed.

Step 4: If the pesticide is still in present in the sprayers then they continue their work but if it is about to finish in any of the sprayers then:

(a) If there is another sprayer in the close vicinity of that location and only a small area is yet left to be covered, then that sprayer along with other sprayers eventually finishes the task.

(b) If there is still a large area to be covered, then the user is notified about this situation and the sprayer reaches to the nearest location where it can be refilled. After refilling it goes to the nearest area that is till yet not covered and starts its work again.

Step 5: Steps 1,2,3, and 4 are repeated for all the sprayers till the entire field is sprayed with pesticides. Now, when the entire field is covered then the user is notified that the task has been completed and all the robotic sprayers stop spraying and move to their destinations.

By executing the above steps, the goal of covering entire field is achieved.

4   Simulation Results

4.1   Simulation Environment

Webots [24] is an open source platform that is used to perform real life simulations. It is used to model, program and simulate robot and thus the results obtained can be analyzed. It also supports multi-robot simulations and also the simulations can be transferred to a real robot which is one of the major advantages of Webots. [25] compares various existing simulators like V-Rep, Webots, etc. over different factors like the type of visualization, ease of programming, ability to run algorithms etc. Webots provides 3D simulations, has a smooth learning curve and also supports multiple languages which makes Webots suitable for our experiment. The simulations have been performed on a system with 11th Gen Intel(R) Core(TM) i5-1155G7 processor with 16 GB RAM, 2.5 GHz CPU, and 64-bit Windows operating system.

To conduct these experiments, we utilized the R2023b version of the Webots simulator. The environment size is taken as $100 \times 100, 200 \times 200, 500 \times 500$ and 1000×1000 m which is equivalent to 1 hectare. The number of sprayers deployed n is taken as: $N = 1, 2, 3, 4$. Figure 3 shows $100 \text{ m} \times 100 \text{ m}$ environment created in Webots and that was used for the simulations. It is a field wherein the crops are arranged in a row pattern. For simulation purposes, robot uses is Pioneer 3-AT robot, which is supported by Webots. As shown in Fig. 4, Pioneer 3-AT is an outdoor robot used for navigation, mapping etc. which will be suitable for our field environment. Figure 5 show the placement of the sonar sensors that are used for detecting the distance around the circumference of the robot.

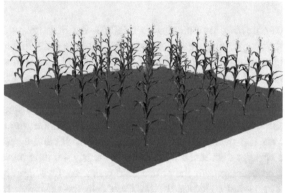

Fig. 3. Environment setup done in Webots.

Fig. 4. Adept's Pioneer 3-AT used for simulations. [26].

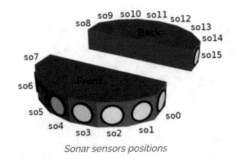

Fig. 5. Placement of sensors in Adept's Pioneer 3-AT [26].

4.2 Results

The time taken to achieve communication is nearly 1–2 s. So after running simulations, the time required for spraying pesticide over the entire field using a single robot is 7.66 h. During the entire duration, the user was informed on the instances when the pesticide was over and when the entire field was covered. Also, the nearly 87.60% entire field was sprayed with pesticide giving optimal coverage.

Now we consider two robotic sprayers over the field. Both the sprayers are starting from different initial locations. So, after running the simulation, the time required by two robotic sprayers to traverse the entire field is 3.58 h. It can be seen that by adding one more robotic sprayer decreased the timing by nearly 4 h thus increasing efficiency of the job done. The overlap over the entire field was zero. Overlapping area is defined as the area that is covered twice or the area wherein the pesticide has been sprayed more than one time. The coverage obtained in this case is nearly 92.50% which is quite acceptable. By adding one more robotic sprayer that is using three robotic sprayers over the field, the time required by all of them to cover the entire field is nearly 2.18 h and the coverage obtained is nearly 95.86 % with no overlap. Increasing more number of robots for this environmental dimension is not optimal. So, it can be observed that by our proposed approach for multiple robotic sprayers, the efficiency of the work is significantly improved with minimal human interference. For 200 × 200 m of land, the time required is 16.29 h, 9.57 h, 6.71 h and 4.67 h and coverage of 89.06%, 91.13%, 95.25% and 95.05% respectively using $N = 1,2,3,4$ robots.

Similarly for 500 × 500 m of land, the time required is 40.52 h, 22.95 h, 16.28 h and 11.42 h sand coverage of 90.05%, 92.25%, 96.05% and 96.35% respectively using $N = 1,2,3,4$ robots. And for 1000 × 1000 m of land, the time required is 77.92 h, 41.85 h, 26.66 h and 21.49 h and coverage of 92.13%, 93.45%, 96.55% and 96.85% respectively using $N = 1,2,3,4$ robots. Figure 6 shows the total time taken by the robotic sprayers to cover the entire area.

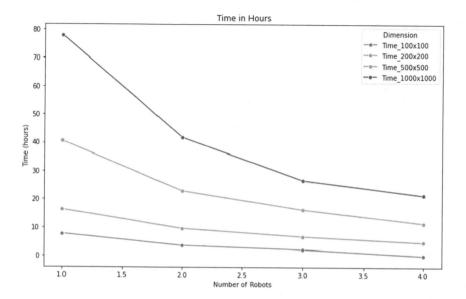

Fig. 6. Total time taken by the robot to spray pesticide over entire field.

5   Conclusions

In this paper, we proposed a methodology for multi-robot pesticide spraying. With this methodology, there is an increase in efficiency with minimal wastage of pesticides and also there will be negligible or no spray drift. This makes sure that pesticide reaches the targeted location making it more effective. Also, with this approach the human contact with the pesticides is minimal preventing them from hazardous effects of chemical pesticides. Using our proposed approach e obtained nearly 72% reduction in time while using single robot and four robots for the same area. The coverage obtained is in the range of 87.6% to 96.8%.

We simulated the proposed methodology in Webots which is a open-source simulator in which a realistic environment can be created, supports multi-robot programming. For our future work, we also aim at using the same mechanism for weed removal, etc. agricultural purposes. We aim to implement this methodology practically in the agricultural fields at Roorkee, India.

Acknowledgement. The second author was in part supported by a research grant from Google.

References

1. Ayaz, M., Ammad-Uddin, M., et al.: Internet-of-Things (IoT)-based smart agriculture: toward making the fields talk. IEEE Access **7**, 129551–129583 (2019)
2. Cheng, C., Fu, J., et al.: Recent advancements in agriculture robots: benefits and challenges. Machines **11**(1), 48 (2023)

3. Zhai, Z., Martínez, J.F., et al.: Decision support systems for agriculture 4.0: survey and challenges. Comput. Electron. Agric. **170**, 105256 (2020)
4. Reddy, N.V., Reddy, A., et al.: A critical review on agricultural robots. Int. J. Mech. Eng. Technol. **7**(4), 183–188 (2016)
5. Shamshiri, R.R., Weltzien, C., et al.: Research and development in agricultural robotics: a perspective of digital farming. Int. J. Agric. Biol. Eng. **11**(4), 1–14 (2018)
6. Varandas, L., Faria, J., et al.: Low-cost IoT remote sensor mesh for large-scale orchard monitorization. J. Sens. Actuator Netw. **9**(3), 44 (2020)
7. Bhardwaj, T., Sharma, J.P.: Impact of pesticides application in agricultural industry: an Indian scenario. Int. J. Agric. Food Sci. Technol. **4**(8), 817–822 (2013)
8. Alliance, LoRa. LoRa specification. LoRa Alliance (2015)
9. Bi, C., Xu, S., Hu, N., et al.: Identification method of corn leaf disease based on improved Mobilenetv3 model. Agronomy **13**(2), 300 (2023)
10. Anand, R., Veni, S., Aravinth, J.: An application of image processing techniques for detection of diseases on brinjal leaves using k-means clustering method. In: International Conference on Recent Trends in Information Technology (ICRTIT) (2016)
11. Zourmand, A., Hing, A.L.K., et al.: Internet of things (IoT) using LoRa technology. In: IEEE International Conference on Automatic Control and Intelligent Systems (I2CACIS) (2019)
12. Cho, S.I., Ki, N.H.: Autonomous speed sprayer guidance using machine vision and fuzzy logic. Trans. ASAE **42**(4), 1137–1143 (1999)
13. Vardhan, P.H., Dheepak, S., et al.: Development of automated aerial pesticide sprayer. Int. J. Eng. Sci. Res. Technol. **3**(4), 458–462 (2014)
14. Garre, P., Harish, A.: Autonomous agricultural pesticide spraying UAV. In: IOP Conference Series: Materials Science and Engineering, vol. 455. IOP Publishing (2018)
15. Abbas, I., Liu, J., et al.: Different sensor based intelligent spraying systems in agriculture. Sens. Actuators A: Phys. **316**, 112265 (2020)
16. Wei, D., Weimin, D., Xiongku, H.: Technologies and evaluation methodology of variable spray. J. China Agric. Univ. **14**(3), 94–102 (2009)
17. Xiongkui, H., Aijun, Z., Yajia, L., et al.: Precision orchard sprayer based on automatically infrared target detecting and electrostatic spraying techniques. Int. J. Agric. Biol. Eng. **4**(1), 35–40 (2011)
18. Jeon, H.Y., Zhu, H.: Development of a variable-rate sprayer for nursery liner applications. Trans. ASABE **55**(1), 303–312 (2012)
19. Nan, Y., Zhang, H., et al.: Low-volume precision spray for plant pest control using profile variable rate spraying and ultrasonic detection. Front. Plant Sci. **13**, 1042769 (2023)
20. Berenstein, R., Edan, Y.: Automatic adjustable spraying device for site-specific agricultural application. IEEE Trans. Autom. Sci. Eng. **15**(2), 641–650 (2017)
21. Yoon, H.W., Kim, D.J., et al.: L & M farm: a smart farm based on LoRa & MQTT. In: International Conference on Omni-layer Intelligent Systems (COINS) (2020)
22. Ünal, İ, Kabaş, Ö., et al.: Adaptive multi-robot communication system and collision avoidance algorithm for precision agriculture. Appl. Sci. **13**(15), 8602 (2023)
23. Ribeiro, A., Conesa-Muñoz, J.: Multi-robot systems for precision agriculture. In: Bechar, A. (ed.) Innovation in Agricultural Robotics for Precision Agriculture. PPA, pp. 151–175. Springer, Cham (2021). https://doi.org/10.1007/978-3-030-77036-5_7
24. Michel, O.: Cyberbotics ltd. webots^TM: professional mobile robot simulation. Int. J. Adv. Robot. Syst. **1**(1), 5 (2004)
25. Ramli, N.R., Razali, S., Osman, M.: An overview of simulation software for non-experts to perform multi-robot experiments. In: International Symposium on Agents, Multi-Agent Systems and Robotics (ISAMSR) (2015)
26. https://www.cyberbotics.com/doc/guide/pioneer-3at?version=cyberbotics:R2019a-rev1

DeFog: Adaptive Microservice Scheduling on Kubernetes Clusters in Cloud-Edge-Fog Infrastructures

Athanasios Prountzos and Euripides G. M. Petrakis[✉] [ID]

School of Electrical and Computer Engineering, Technical University of Crete (TUC),
73100 Chania, Crete, Greece
aprountzos@tuc.gr, petrakis@intelligence.tuc.gr

Abstract. DeFog is an innovative microservice placement and load-balancing approach for Kubernetes multi-cluster Cloud-Fog-Edge architectures to minimize application response times. Applications are modeled as Service Oriented Architectures (SOA) comprising multiple interconnected (micro) services. As the resources of the Edge and the Fog are limited, choosing among services to run on the Edge or the Fog is the problem this work is dealing with. DeFog focuses on dynamic (i.e., adaptive) and decentralized service placement with zero downtime, eliminating the need for coordination among the clusters. Several placement policies are tested on two realistic SOA applications to select the one that reduces application latency.

Keywords: Fog Computing · Edge Computing · Kubernetes · Microservices · Service Placement

1 Introduction

The Cloud-Fog-Edge architecture paradigm provides a new model for addressing the challenges of Cloud computing and of the Internet of Things (e.g., network latency, bandwidth limitations, etc.). All three layers arranged in a hierarchy complement each other and offer a range of benefits for diverse applications. Edge computing introduces computation and storage at Edge devices, minimizing the data transfer to the Fog or the Cloud. The Fog layer acts as a crucial bridge between the Cloud and the Edge, augmenting computational resources and services nearer to the Edge. At one end, running the application at the Edge, or the Fog would be optimal in terms of latency but, might not be feasible in terms of adequacy of resources at these layers. On the other end, running the application on the Cloud might not be optimal in terms of response time (i.e., communication cost increases moving from the Edge to the Fog and the Cloud). The placement of (micro)services is just one aspect of this broader problem, which also encompasses issues related to network latency, security, and transfer costs [3, 12].

© The Author(s), under exclusive license to Springer Nature Switzerland AG 2024
L. Barolli (Ed.): AINA 2024, LNDECT 201, pp. 437–448, 2024.
https://doi.org/10.1007/978-3-031-57870-0_39

DeFog deals with the great challenge of optimizing service placement and load balancing decisions in a three-tier Cloud-Fog-Edge environment leveraging a distribution of Lightweight Kubernetes (K3s) clusters. Service Mesh [7] manages service discovery and load balancing for service instances (i.e., copies of the same service) at different infrastructure layers. This approach prevents any service from being overwhelmed and has zero downtime (i.e., applications continue to run while the service is being placed). By offloading a portion of requests to connected clusters, the system maintains better stability for each service and prevents resource usage from exceeding a predefined percentage (e.g., 80%) of its service limits. Unlike previous studies that mainly used synthetic environments and reported results based on simulations (rather than real applications and real deployments), DeFog reports results using realistic Service Oriented Architectures (SOA) applications deployed on a distribution of real Kubernetes deployments. Overall, DeFog provided more meaningful and practical information about the actual performance of all methods tested.

Least Frequently Used (LFU) [8] is the reference service placement strategy. LFU dictates that a service receiving many requests might replace the least frequently used service in a cluster closer to the end-user (i.e., in the Fog or the Edge) provided that, after removing that service, adequate compute resources exist to accommodate the new service in the cluster. We propose alternative strategies that consider various metrics such as Requests per Second (RPS), response latency, and resource utilization. The results of this study reveal that a replacement policy that uses individual microservice latency as the crucial factor influencing service placement, outperformed LFU by at least 10% in application response time.

Section 2 presents work related to service placement in Cloud-Fog-Edge environments. In Sect. 2 we bestow an overview of the technological background that forms the foundation of DeFog. In Sect. 3, we delve into the design and implementation phase of Defog. Section 5 presents a detailed analysis of the performance of the various placement policies followed by conclusion and issues for future research in Sect. 6.

2   Related Work

Placing services in distributed systems has attracted significant research interest in recent years [3,12]. Most of these methods are either static (i.e., service placement decisions do not adapt to application resource requirements at runtime) or centralized (i.e., they introduce a central node to orchestrate the placement and operation of the infrastructure). Most importantly, the performance of all these methods has been evaluated using simulators and not real infrastructures running real or realistic use cases. Table 1 presents a comparison of related works.

Apat, Sahoo, and Maiti [1] address energy consumption challenges in Fog by formulating a service placement scheme that maximizes resource utilization. Shaik and Baskiyar [13] develop a distributed approach considering computation

Table 1. Service placement in Cloud - Fog -Edge: comparison.

Method	Infrastructure	Placement Orchestra-tion	Platform	Application	Placement Type
[1]	Cloud-Fog	Centralized	Simulator	Simulation	Static
[13]	Fog	Distributed	PFogSim Simulator	Simulation	Dynamic
[2]	Cloud-Fog	Centralized	FogTorchPI Simulator	Simulation	Static
[6]	Fog	Distributed	iFogSim Simulator	Simulation	Dynamic
[9]	Fog	Distributed	iFogSim Simulator	Simulation	Static
[4]	Cloud-Fog	Centralized	Simulator	Simulation	Static
[8]	Cloud-Fog	Distributed	Icarus Simulator	Simulation	Dynamic
DeFog	Cloud-Fog-Edge	Distributed	Kubernetes (K3s)	iXen, eShop	Dynamic

and communication costs. Brogi, Forti, Guerrero, and Lera [2] address multi-service application management in heterogeneous Fog infrastructures. They propose a centralized approach that combines genetic algorithms with Monte Carlo simulations to optimize the placement process and improve system performance.

Mahmud, Ramamohanarao and Buyya [6] focus on ensuring the QoS of applications by meeting service delivery deadlines, and optimizing resource utilization. The placement strategy dynamically relocates modules to optimize the number of Fog nodes that are computationally active. Pallewata, Kostakos and Buyya [9] introduce a decentralized microservices-based IoT application placement policy. Both [6] and [9] model and evaluate their proposed policies in an iFogSim-simulated environment.

Farhadi et al. [4] propose a two-time-scale framework that optimizes both service placement and request scheduling by formulating the problem as a mixed integer linear program (MILP). They aim to achieve near-optimal performance in terms of service placement and request scheduling. Ascigil et al. [8] focuses on uncoordinated strategies for service placement in Edge-Clouds. They propose a set of techniques that enable uncoordinated resource allocation in the Edge-Cloud environment. Both studies, [4] and [8], use a combination of synthetic and trace-driven simulations.

3   Tools and Architecture

Kubernetes (k8s)[1] automates the deployment and management of containerized applications (such as SOA) but may not be the best option for resource constraint environments. K3s[2] operates with lower resource requirements while still enjoying the benefits of high availability and security offered by K8s. A K3s

---

[1] https://kubernetes.io/.
[2] https://docs.k3s.io/.

cluster consists of a set of worker machines called Agent nodes that run containerized applications, and one or more Server nodes that run the lightweight control plane that manages the nodes and Pods in the cluster.

A Server node in K3s is the control plane that manages the worker nodes. It consists of the API server, SQLite, Scheduler, controller manager, and Tunnel proxy components. The API server provides a management interface for the entire cluster. SQLite is the default storage (instead of etcd) for cluster state and configuration data. The scheduler is responsible for assigning workloads to the worker nodes. The controller manager ensures that the desired state of the cluster is maintained. The Tunnel proxy is responsible for managing and securing network traffic between the server and the worker nodes in the cluster. The Agent node is the worker node that executes the containerized applications. This node comprises containerd, which functions as the container runtime, and kubelet, which is responsible for managing the containers and facilitating communication with the API server. Additionally, kube-proxy manages network proxy and load balancing functions for services running within the node. Flannel is a lightweight network fabric that allows networking between containers across different nodes.

Linkerd Service Mesh[3] introduces an abstraction layer between services for applications deployed across multiple nodes or clusters. Linkerd's data layer handles network traffic between services so that each service can discover where other services are placed. It consists of a set of micro-proxies ("sidecar proxies") that are deployed alongside each service instance in each Pod. When a request is made to a Pod, the micro-proxy intercepts the traffic and determines if it needs to be routed to another Pod within the cluster. The Linkerd-init container is also added to each Pod; it executes before any other containers to perform one-time initialization tasks such as fetching certificates and setting up service discovery.

The control plane is made up of a set of components that work together to manage and configure the proxies that form the data plane, including the identity, destination, and proxy-injector services. The identity service acts as a certificate authority that accepts certificate signing requests (CSRs). These certificates are used for mutual TLS (mTLS) between the proxies so that communication between proxies is secure and authenticated. Linkerd provides also a Viz visualization tool and Prometheus. The Service Mirror and the Gateway components implement multi-cluster functionality. They work by mirroring service information between clusters, which ensures that services in one cluster can easily discover and communicate with services in another cluster. The Service Mirror component watches for updates to services in a target cluster and mirrors that information locally on a source cluster. The Gateway allows for services in one cluster to communicate with services in another cluster by establishing encrypted connections between them. Finally, TrafficSplit enables traffic distribution across multiple service instances based on predefined proportions (i.e., weights).

Figure 1 illustrates two connected K3s clusters with Linkred's data and control plane. Cluster A serves as the source cluster and Cluster B as the target.

---

[3] https://linkerd.io/.

A microservice is mirrored at Cluster A, and a virtual IP address is created by Cluster A to provide a consistent response to Pods resolving the mirrored service. The Gateway in Cluster B functions as an ingress controller, routing incoming requests to the appropriate service. TrafficSplits enables routing and percentage distribution of traffic to mirror services without requiring additional configuration of the traffic source. DeFog can seamlessly connect microservices running on multiple clusters on each of the Cloud, Fog, or Edge interconnected layers.

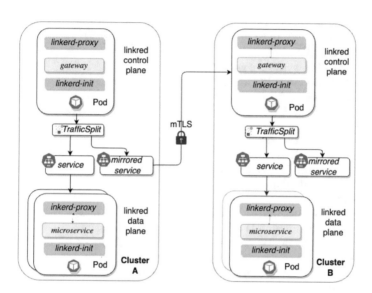

Fig. 1. Linkerd's multicluster functionality with mirroring and Trafficsplit.

DeFog allows multiple service instances (i.e., copies) to be instantiated in multiple clusters, facilitating load balancing and traffic redistribution. If a stressed service is detected, DeFog adjusts the weights of TrafficSplit, diverting a proportion of traffic to connected clusters. This dynamic load balancing mechanism ensures that the system efficiently handles varying workloads while maintaining performance. This flexibility optimizes resource utilization, enhances system responsiveness, and ensures a reliable service environment.

4   Service Placement

DeFog employs a continuous evaluation process that takes into account various system requirements. By analyzing metrics such as RPS, response latency, and RAM usage, DeFog dynamically determines the optimal placement of services based on their importance and resource demands. This evaluation process considers the resource requests and limits for CPU and RAM defined in the configuration files of the services. The main process of DeFog is described by Algorithm 1. DeFog initiates a loop that lists the deployed microservices (D), and the mirrored ones (R) whose requests are being forwarded to connected clusters along with resource requirements. Then, the service placement algorithm is executed at regular intervals (e.g., every 30 s). The rest of the time, the process collects resource usage for each deployed service, which combined with the resource limits set in the Kubernetes configuration file, calculates a weight that is used to update TrafficSplit.

Algorithm 1. DeFog main process (runs on each cluster).

1: **procedure** DEFOG( )
2:     Establish Prometheus connection
3:     Create kubectl configuration
4:     Collect cluster data
5:     Period = 30 seconds   ▷ repeat every 30 seconds
6:     **while** true **do**
7:         R, D ← getServices( )   ▷ D: deployed services, R: requested services
8:         **if** period **then**
9:             Run service placement algorithm(R, D)   ▷ one of LRU, RLSD, LFU-RAM, RLSD-RAM
10:        **else**
11:            currentResources, resourcesLimits ← gatherData( )
12:            **for** Services in D **do**
13:                weight = calculateTrafficSplitWeight(service, currentResources, resourcesLimits)
14:                updateServiceTrafficSplit(weight)
15:            **end for**
16:        **end if**
17:    **end while**
18: **end procedure**

After DeFog requests the deployment of a service, if its image does not already exist within the node, the container runtime of the node downloads the specified image from the container registry (i.e., Docker Hub). Once the image is available on the node, Kubernetes comes into action and creates a Pod with the necessary configuration for the service. The scheduler component, which is responsible for determining Pod placement, automatically assigns the Pod to the (single) node of the cluster. Finally, the container runtime starts the service within the Pod, making it fully operational and accessible within the cluster.

4.1   Service Placement Policies

The implementation of effective service placement policies becomes paramount in ensuring optimal system functionality. DeFog tests and implements a modified version of the Least Frequently Used (LFU) policy and proposes three new policies that dynamically places microservices on a hybrid infrastructure, taking into account comprehensive evaluations based on performance metrics such as RPS (Requests Per Second), Response Latency and resource utilization.

Least Frequently Used (LFU) is a cache eviction algorithm used to determine which items in a cache should be removed when the cache reaches its maximum capacity. LFU focuses on the dynamic placement of services, considering their Requests Per Second (RPS) metric as a key factor. The core concept behind LFU is that items that are least frequently accessed should be evicted first. When implementing the LFU strategy, DeFog aims to deploy services that have higher RPS values, indicating higher usage frequency. If there are available resources in the cluster, DeFog proceeds with deploying these services. However, if the cluster's resources are limited and cannot accommodate a requested service, DeFog compares the RPS values of the requested service with the already deployed services (ordered by RPS) and evicts a deployed service if possible. Algorithm 2 is an overview of the LFU policy. D represents the already Deployed services and R represents the Requested services to a given cluster. To perform the ranking, the RPS metrics are collected for each of the services in D and R. Services with bigger RPS are ranked higher.

Algorithm 2. Least Frequently Used (LFU) replacement algorithm.

```
 1: procedure DEFOG(R,D)
 2:     sort R, D by RPS
 3:     while R not empty do
 4:         serviceToDeploy ← R[0]
 5:         if serviceToDeploy in D then
 6:             R.pop(0)
 7:             continue
 8:         else if checkAvailableResources(serviceToDeploy) == True then
 9:             Deploy serviceToDeploy
10:             R.pop(0)
11:         else if D not empty then
12:             serviceToDelete ← D[Last]
13:             if serviceToDelete.requests < serviceToDeploy.requests then
14:                 Delete serviceToDelete
15:                 D.remove(Last)
16:             end if
17:         end if
18:     end while
19: end procedure
```

The *Response Latency-based Service Deployment (RLSD)* policy follows a similar approach to the LFU algorithm but with a focus on the Response Latency metric of each service. By considering response latency as a crucial factor, it prioritizes the deployment of services with lower Response Latency values. If these services are deployed closer to the user end, it will enable faster response times and improved performance. If there are available resources in the cluster, DeFog proceeds with deploying these services. If the cluster's resources are limited and cannot accommodate new services, DeFog employs a decision-making process. It compares the Response Latency values of the new service with the already deployed services in the cluster. if necessary, RLSD evicts a deployed service with higher latency to make room for the deployment of a service with lower latency. Like the LFU algorithm, the algorithm defines D as the set of already deployed services and R as the set of requested services within a specific cluster. To determine the ranking, response latency metrics are collected for each service in both D and R. Services with lower response latency are assigned higher priority in the ranking.

In low-resource systems like Fog or Edge computing, the implications of high RAM usage are amplified. Services consuming substantial RAM tend to exacerbate the competition for limited resources, resulting in resource contention. Considering the significant impact of RAM usage on latency, DeFog introduces two variation to the LFU and RLSD policies, referred to as *LFU-RAM* and *RLSD-RAM* respectively. These algorithm variations maintain the strategic approach of LFU and RLSD in terms of service deployment based on RPS or Response Latency rankings. A key distinction lies in the eviction process when the cluster's resources reach their limits. If the cluster lacks sufficient resources to accommodate the new service, the algorithm compares the requested RAM of the new service with the service that currently utilizes the most RAM resources in the cluster. If the new service requires less RAM, the algorithm evicts the deployed service with the highest RAM usage. This eviction process frees up resources and creates space for the new service to be deployed. This eviction process creates space and resources for other services that can utilize them more efficiently, alleviating resource contention, and ultimately improving overall latency performance.

4.2   Traffic Management Algorithm

Algorithm 3 dynamically handles traffic and intelligently splits it between deployed instances within the local cluster and mirrored services across other clusters. The current CPU and RAM utilization of the service is compared to the desired CPU and RAM limits multiplied by a threshold value. This threshold defines the maximum resources of a running Pod (i.e., microservice). Experiments reveal that a good value of this threshold is 0.8 [11].

If Pod exceeds its maximum resources, additional requests are diverted to mirror services in clusters higher in the hierarchy. The portion of requests to be diverted is determined by their weights. If the weight exceeds 1, it indicates that the service's resource utilization has exceeded the desired limits by a certain margin. This excess weight signifies the additional load that needs to be distributed among the mirrored services (i.e., if more than one, the excess traffic is distributed equally among mirror services). DeFog dynamically adjusts the traffic splitting among mirrored services based on the weight calculation. This

Algorithm 3. TrafficSplit Weight Calculation.

1: **procedure** CALCULATETRAFFICSPLITWEIGHT(service, currentResources, resourceLimits)
2:     currentCPU ←currentResources.cpu[service]
3:     currentRAM ← currentResources.ram[service]
4:     desiredCPU ←resourceLimits.cpu[service] × THRESHOLD
5:     desiredRAM ← resourceLimits.ram[service] × THRESHOLD
6:     weight ← currentCPU / desiredCPU + currentRAM / desiredRAM
7:     **if** weight > 1 **then**
8:         weight ← weight - 1
9:     **else**
10:         weight ← 0
11:     **end if**
12:     return weight
13: **end procedure**

---

ensures that services experiencing high CPU and RAM utilization are appropriately calmed by redistributing the excess load to the available computational resources distributed across multiple clusters.

5   Evaluation

Figure 2 illustrates the experimental set-up. Linkerd's Multi-cluster extension establishes links between the clusters in a hierarchical placement order, resulting in a three-layered pyramid of clusters. At the topmost layer, Cluster 1 represents the Cloud. The second layer consists of Clusters 2 and 3, representing the Fog layer. Lastly, Clusters 4 and 5, constitute the Edge layer. A delay of 50 ms was applied between the Cloud and Fog layers, while a delay of 20 ms was implemented between the Fog and Edge layers, in addition to the time for answering the request.

The architecture is realized by means of five Virtual Machines (VMs) deployed in the Google Cloud Platform (GCP) and distributed across multiple regions. Each VM represents a cluster. In each VM a K3s is installed representing a single-node cluster. The K3s control plane components are encapsulated within a single process, simplifying the operation and management of the cluster. To simulate the Cloud-Fog-Edge infrastructure, we designed each cluster with specific characteristics, such as available CPU, RAM, machine type, and placement in different cloud zones. Table 2 displays a detailed overview of the unique characteristics of each cluster. The cluster is responsible for orchestrating containerized applications' deployment, scaling, and management.

5.1   Experimental Results

The present study employs two distinct applications to evaluate and validate the proposed service placement strategies, the Google's Online Boutique (eShop) [5]

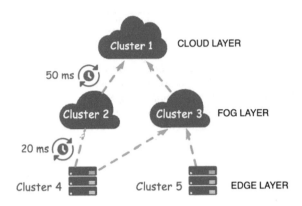

Fig. 2. Hierarchical arrangement and interconnectedness of the clusters.

Table 2. Cluster Characteristics.

Cluster attributes	Cluster 1	Cluster 2 and 3	Cluster 4 and 5
Zone	europe-west2-a	europe-west3-a/b	europe-west6-a/b
Machine	e2-standard	e2-standard	e2-custom
vCPU	8	4	2
RAM (GB)	32	16	3

and iXen [10] consisting of 12 and 15 distinct microservices respectively. All microservices are initially deployed in the cluster belonging to the Cloud layer (Cluster 1) except for the frontend of each application, which is deployed in both clusters of the Edge Layer (Cluster 4 and Cluster 5). The database (e.g., MongoDB, MySQL) and user identification services of each application should not be deployed in multiple clusters.

The load testing strategy involved three different request distributions applied in a sequence using Locust[4], each lasting 5 min, resulting in a total load testing duration of 15 min [11]. The requests are evenly distributed across the Edge Layer Clusters where the frontends of the applications reside. The total time taken for the placement of microservices is primarily influenced by the time required for each service image to be downloaded from its repository (i.e., Docker Hub) and subsequently deployed. The placement process on Edge clusters usually took around 2 min. The remaining time was dedicated to conducting the measurements.

Figure 3 summarize the average response times for each placement strategy, showcasing the differences in performance with the response latency corresponding to the initial placement (before a placement policy is applied). The findings highlight the superior performance and effectiveness of the RLSD strategy. The RLSD strategy consistently demonstrates lower overall latency compared to the

---

[4] https://locust.io.

LFU, LFU-RAM, and RLSD-RAM strategies, indicating its effectiveness in optimizing response times. LFU is the worst method. The incorporation of resource contention considerations in the LFU-RAM and RLSD-RAM strategies enable them to perform competitively.

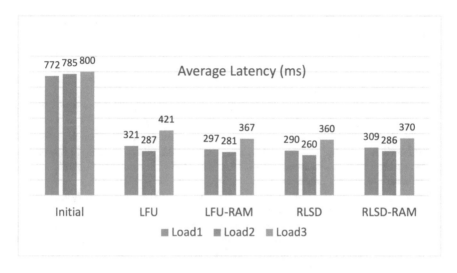

Fig. 3. Average response times for each placement strategy.

6   Conclusion and Future Work

The primary objective of this study was to reduce the application latency in a hybrid Cloud-Fog-Edge multicluster environment by strategically placing services in the most appropriate clusters. We evaluated various service placement strategies namely, LFU, RLSD, LFU-RAM, and RLSD-RAM. RLSD outperformed all other policies by emphasizing the significance of considering response latency in service placement decisions. Exploring machine learning techniques to predict workload patterns and make proactive service placement decisions is an interesting idea for future research. The threshold in the traffic splitting feature of DeFog is statically provided. If the resource limits on microservices provided by their developers aren't optimized, a static threshold may result in under-utilization or stress of the microservices. Investigating mechanisms that adjust the threshold dynamically also could improve the performance.

Acknowledgment. We are grateful to Google for the Google Cloud Platform Education Grants program.

References

1. Apat, H.K., Sahoo, B., Maiti, P.: Service placement in fog computing environment. In: International Conference on Information Technology (ICIT 2018), pp. 272–277. Bhubaneswar, India, December 2018. https://ieeexplore.ieee.org/document/8724192
2. Brogi, A., Forti, S., Guerrero, C., Lera, I.: Meet genetic algorithms in Monte Carlo: optimised placement of multi-service applications in the Fog. In: IEEE International Conference on Edge Computing (EDGE 2019), pp. 13–17. Milan, Italy, August 2019. https://ieeexplore.ieee.org/document/8812204
3. Cardellini, V., Presti, F.L., Nardelli, M., Rossi, F.: Self-adaptive container deployment in the Fog: a survey. In: International Symposium on Algorithmic Aspects of Cloud Computing (ALGOCLOUD 2019), pp. 77–102. Munich, Germany, September 2019. https://link.springer.com/chapter/10.1007/978-3-030-58628-7_6
4. Farhadi, V., et al.: Service placement and request scheduling for data-intensive applications in edge clouds. In: IEEE INFOCOM 2019 - IEEE Conference on Computer Communications, pp. 1279–1287. Paris, France, April 2019. https://ieeexplore.ieee.org/document/8737368
5. Google: e-Shop: Online Boutique (2022). https://github.com/GoogleCloudPlatform/microservices-demo
6. Mahmud, R., Ramamohanarao, K., Buyya, R.: Latency-aware application module management for fog computing environments. ACM Trans. Internet Technol. **19**(1), 1–21 (2018). https://dl.acm.org/doi/10.1145/3186592
7. Manpathak, S.: Kubernetes Service Mesh: A Comparison of Istio, Linkerd, and Consul, June 2021. https://platform9.com/blog/kubernetes-service-mesh-a-comparison-of-istio-linkerd-and-consul/
8. Ascigil, O., Phan, T.K., Tasiopoulos, A.G., Sourlas, V., Psaras, I., Pavlou, G.: On uncoordinated service placement in edge-clouds. In: 2017 IEEE International Conference on Cloud Computing Technology and Science (CloudCom 2017), pp. 41–48 (2017). https://ieeexplore.ieee.org/document/8241089
9. Pallewatta, S., Kostakos, V., Buyya, R.: Microservices-based IoT application placement within heterogeneous and resource constrained fog computing environments. In: IEEE/ACM 12th International Conference on Utility and CloudComputing (UCC 2019), pp. 71–81. Auckland, New Zealand, December 2019. https://dl.acm.org/doi/10.1145/3344341.3368800
10. Petrakis, E.G., Koundourakis, X.: iXen: secure service oriented architecture and context information management in the cloud. J. Ubiquitous Syst. Pervasive Netw. **14**(2), 1–10 (2021). https://iasks.org/articles/juspn-v14-i2-pp-01-10.pdf
11. Prountzos, T.: Dynamic Micro-Service Placement in Hybrid Cloud - Fog Infrastructures. Technical report, Diploma Thesis, School of Electrical and Computer Engineering, Technical University of Crete (TUC), Chania, Crete, July 2023. https://dias.library.tuc.gr/view/96617
12. Salaht, F.A., Desprez, F., Lebre, A.: An overview of service placement problem in fog and edge computing. ACM Comput. Surv. **53**(3), 1–35 (2020). https://dl.acm.org/doi/10.1145/3391196
13. Shaik, S., Baskiyar, S.: A scalable approach to service placement in fog/cloud environments. In: IEEE International Performance, Computing, and Communications Conference (IPCCC 2021), pp. 1–8. Austin, TX, USA, October 2021. https://ieeexplore.ieee.org/document/9679396

Challenges, Novel Approaches and Next Generation Computing Architecture for Hyper-Distributed Platforms Towards Real Computing Continuum

Francesco Lubrano[✉], Giuseppe Caragnano, Alberto Scionti,
and Olivier Terzo

LINKS Foundation, via Pier Carlo Boggio 61, Turin, Italy
{fracesco.lubrano,giuseppe.caragnano,alberto.scionti,
olivier.terzo}@linksfoundation.com

Abstract. Computing platforms are evolving from centralized cloud datacenters, hosting huge computing and storage capacity to process data coming from external sources, to a more distributed and data-oriented architectures. As the result of this paradigm shift, we assisted to the raise of Edge computing as a more efficient way of (pre-)processing data on-site. Further, modern distributed computing architectures envisage the integration of a plethora of heterogeneous systems. This complexity comes with new challenges: the effective management of the vast heterogeneous resources and the dynamic nature of these platforms, with nodes joining and leaving frequently, necessitates adaptive and resilient management strategies, resource discovery, scheduling, and dynamic decentralized orchestration, to ensure and maintain system stability and performance. This paper provides an in-depth analysis of these challenges and draws a reference architecture exposing the necessary features to overcome the complexity of managing such complex distributed heterogeneous systems. By leveraging AI scheduling algorithms and federated learning techniques, proposed architecture aims to provide a robust and efficient solution matching with the demands for processing always growing masses of data and complex workloads.

1 Introduction

Computing domain is experiencing a drastic change addressed by the wide spread of the Internet of Things (IoT), the proliferation of data analytics systems, big data techniques, and the subsequent integration of Artificial Intelligence (AI). Over the past few years, these technological advancements have not only reshaped the landscape of information processing but have also ushered in an era where data plays a pivotal role across diverse sectors. The significance of data has soared to unprecedented levels, with voluminous datasets becoming a cornerstone of decision-making processes.

© The Author(s), under exclusive license to Springer Nature Switzerland AG 2024
L. Barolli (Ed.): AINA 2024, LNDECT 201, pp. 449–459, 2024.
https://doi.org/10.1007/978-3-031-57870-0_40

Indeed, companies and industries are going through a digital transformation: internal processes and systems are ever more connected to digital platforms and this is leading to a massive increase in the data flow, not only between industries and cloud providers, but also between organizations themselves. Experts forecasts indicate an exponential growth trajectory for data volumes in the foreseeable future, underscoring the crucial role that data will continue to play in shaping our technological landscape [1,12].

This surge in data volume, while transformative, has brought forth a paradigm shift in computing platforms. Traditional centralized cloud datacenters, designed to handle substantial computing and storage capacities for processing data from external sources, are evolving towards more distributed and data-centric architectures. The emergence of the 'computing-at-the-edge' (of the network) paradigm epitomizes this shift, as it enables the on-site (pre-)processing of data, thereby enhancing the overall efficiency [2] (i.e., overall energy consumption can be reduced by limiting the movement of large bunches of data in and out cloud datacenters).

Finally, another catalyst of this evolution is the growing diffusion of edge AI. The continuous advancements of AI in its sub-fields (e.g., machine learning – ML, deep learning – DL, natural language processing – NLP, genetics algorithms, etc.), along with the development of federated learning techniques, allowed to move (AI) inference from datacenter resources to constrained hardware at the edge or even directly on IoT devices [3]. Implementing AI at the edge presents numerous advantages compared to the solely cloud-based approach. It delivers swift processing and ultra-low latency, reduces data transmission loads substantially, and enhances security and privacy. Power consumption is markedly diminished, attributed to inference algorithms operating in the order of milliwatts on far-edge devices and from few watts to tens of watts on edge devices, as opposed to the cloud datacenters where it typically runs on devices consuming tens of watts, or even hundreds of watts [4].

The evolution of the Cloud model into Cloud Edge one brings multiple advantages. It enhances resilience, enables real-time operations, improves security and privacy, allowing to mainly process data without moving to centralized datacenters. This feature is crucial for Europe as it is a chance to ensure data sovereignty and digital autonomy. Thus, European commission is fostering the research in this field, driving leadership in the field of edge computing and the IoT and, by doing so, to regain a significant role in the computing market by 2025 [13].

Deploying integrated cloud services at the Edge and managing IoT devices is the key to support the *Computing Continuum*, i.e., allowing the data seamlessly being processed on computing devices belonging to the different tiers (IoT, far-edge, edge, and datacenters); however, it dramatically increases the complexity of such computing platforms in terms of architecture, functionalities and internal management. The contemporary computing landscape envisions the integration of a myriad of distributed and heterogeneous computing nodes, diverse networks, and structured/unstructured data sources, such as IoT devices and external APIs. This integration, while promising, contributes to increase even more the complexity that poses unprecedented challenges. Effectively managing the vast and heterogeneous resources, encompassing various hardware architectures, operating systems, programming languages, and network protocols, requires adaptive and resilient management strategies. Moreover, the dynamic nature of these platforms, with resources frequently joining and leaving, demands sophisticated resource discovery, scheduling, and dynamic decentralized orchestration mechanisms to ensure and maintain system stability and performance.

This paper aims to conduct an in-depth analysis of these challenges and related works (Sect. 2) and proposes a reference architecture to address the evolving demands of modern computing platforms (Sect. 3). The approach outlined herein advocates for a hyper-distributed computing continuum platform, fortified by a distributed cognitive orchestration system. Leveraging advanced AI scheduling algorithms and federated learning, this system constantly optimizes and enhances the overall efficiency of deployments. The integration of AI-driven scheduling and federated learning serves as a dynamic adaptive mechanism, allowing the computing continuum to respond dynamically to changes in the environment, resource availability, and workload demands.

Furthermore, this cognitive orchestration system not only enhances adaptability but also contributes to self-optimization over time. This is especially critical in addressing the challenges posed by the diverse and dynamic nature of hyper-distributed computing environments. By leveraging these cutting-edge technologies, the proposed platform seeks to offer a robust and efficient solution to the intricate demands of modern computing architectures within the era of hyper-distributed systems.

2 Related Works

Over the past decade, the EU has expressed a commitment to achieve digital sovereignty and strategic autonomy in critical technology areas, including Cloud computing. Efforts are being made to reduce dependence on non-European technology providers and create a more self-reliant digital ecosystem [16]. The EU has recognized the strategic importance of cloud computing and digital infrastructures, starting several funding programs and increasing investments in cloud infrastructures, datacenters, and connectivity. This includes public and private initiatives to support the expansion of Cloud capabilities within the EU. Although several European Cloud providers are gaining a share, the market is leaded by companies such as Amazon Web Services (AWS), Microsoft Azure, and Google Cloud Platform (GCP), headquartered in the United States. These companies have played a pivotal role in shaping the Cloud industry and continue to dominate the market. Thus, EU is addressing funds and research programs towards the development of new technologies, frameworks and infrastructures that enclose edge computing capabilities, in order to become a leader in this sector [15]. From the technical point of view, technological advancements and peculiarities of Edge computing meet Europe's priorities in terms of data privacy, low-latency applications, industrial transformation, 5G integration, smart city development, innovation, and the creation of resilient digital infrastructures. Therefore, the natural interest to this research field sustained by European and private funding, allowed the proliferation of research work and projects, resulting in a huge number of interesting publications on computing continuum and on Edge. Here are reported the main publications that influenced our work.

Many research works describe various computing models, each of them focusing on Cloud, Edge or IoT. Even when a connection among these domains exists, for instance, in the case of task offloading, these computing models are presented in isolation. According to [5] and to [6], there is a concrete need to start considering computing continuum, not just the sum of Cloud, Edge, and IoT domains, but as a new fully-integrated, unified and continuous computing and deploying model. To this end, there are efforts in defining a reference architecture for this continuum. In [14] researchers released a set of guidelines, frameworks and reference architectures to deploy digital platforms, in the form of a generic architectural template including specific building blocks for each of the considered sectors (i.e., manufacturing, agriculture, energy, and healthcare). Although the resulting templates are well-defined, there are several aspects that are not covered such as, for instance, the complexity of the resource management and orchestration of these continuum platforms. Ferrer et al. [7], by recognizing this problem, proposed the adoption of the cognitive Cloud concept, thus defining adhering principles and characteristics, and proposing to enlarge the resource management and task orchestration to distributed *cognitive* resources. We identified distributed orchestration as a trend that is gaining momentum with many research works are addressing it. In [8], authors propose the concept of *seamless computing*, a model that provides a homogeneous computing environment for multi-domain applications, supporting the mobility of workloads between Cloud and Edge. Proposing distributed orchestration and management methods for computing continuum platforms is also the final result of an in-depth analysis of the challenges and research works in the computing continuum domain carried out in [9]. This result is also supported by the extremely fast diffusion of AI in IoT [10]. AI-based IoT devices (AIoT) are not simple sensors, but rather something that can have an active role in the management of the continuum platform. To this end, the NEMO project extends its MetaOS to smart devices, in order to implement a transparent integration among intelligent devices, edge nodes and cloud resources [11]. This exploration of the related works sets the stage for our research, emphasizing the need for a holistic approach to the continuum which encompasses the architectural and orchestration levels, and the seamless integration of smart IoT devices, far-edge and edge nodes, cloud resources and services, and resulting in a reference architecture for cognitive IoT-Edge-Cloud platforms.

3   Cognitive Computing Continuum – Reference Architecture

Figure 1 depicts the envisioned reference template of a cognitive computing continuum architecture, which highlights the composition of three main layers: (i) the *hyper distributed application* layer on top; (ii) the *platform* layer which integrates a vitual environment service (VES) and the cognitive framework along with the other services required to develop and deploy the applications, and (iii) the *infrastructural* layer which contains the compute, storage and networking resources distributed all over the continuum (i.e., both at the cloud and at the edge, with the latter including also IoT devices).

The infrastructural layer offers the (physical) set of compute, storage, and networking resources that are part of an hyper (i.e., massively) distributed infrastructure, and that comprehends both those resources located in the cloud datacenters and those located at the boundaries of the network. While the former are generally concentrated into single locations providing large scalable pools of virtualized resources, the latter are more (geographically) distributed and resource constrained. By nature, datacenters expose to the users almost infinite compute and storage capabilities, but the latency experienced by moving large amount of data back and forth from the end user locations makes them less appropriate to process the data deluge produced by connected smart sensors and IoT devices. Furthermore, in many cases application constraints may limit the possibility to move (large quantity of) data over the network and across different service providers, as in the case of sensitive data. To address all these challenges, cloud resources are extended with compute and storage capabilities at the *Edge*. Powerful edge compute nodes can be federated towards forming a virtual (geo-)distributed cluster, while such resources can be deployed even within harsh environments. Resource management tools traditionally used to manage the allocation of datacenter resources are extended to provide the same capabilities for the edge nodes (which include also attached IoT devices). In this context, aspects like energy efficiency, resiliency (fault tolerance), trustworthiness, and security become of paramount importance. The tight integration of all the components depicted in Fig. 1 addresses all these challenges, by being able to connect an heterogeneous set of edge devices, to prove the adaptability of the designed software platform to the large variety of computer architectures (i.e., x86, Arm, RISC-V, etc.) and configurations. Compute nodes, whose functionalities could be extended through application specific accelerators running on top of FPGA based modules, host the operating system (OS), the application control logic, the cognitive framework agent(s) and all the pertained services (discovery, communication, data monitoring, etc.). Continuous monitoring of the underlying system (i.e., collecting data from different internal sources, such as CPU load, memory pressure, monitoring of performance counters, etc.) allows to setting up a 'continuous' learning mechanism which is leveraged to adapt the computing operations over the time, as well as by distributing such knowledge to neighbourhoods, enabling them to adapt.

On top of the FPGA devices directly linked to the host processor, application specific accelerators are deployed. The high flexibility offered by such re-configurable devices (also in terms of time required for completing the reconfiguration, which is in the order of 10^{-3} s), combined with the availability of open architectures (i.e., RISC-V) provides the mean for the quick deployment

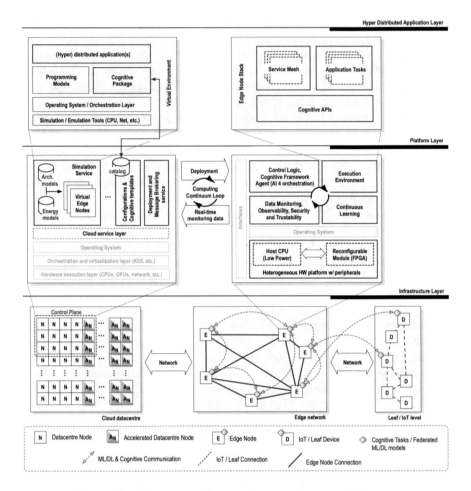

Fig. 1. Cognitive computing continuum reference architecture

and subsequent switch from one configuration to another (e.g., deploying CPU cores with a different amount of cache memory, or enabling/disabling specific ISA extensions), which in general translates into a performance and energy efficiency improvement. Interestingly, federated ML/DL techniques are adopted to reduce the pressure on the network by limiting the amount of data moved on. Indeed, parameters of learned model can be exchanged instead of the full bunch of data used locally by the nodes to be trained. These models are the key element that allow deployments to seamlessly and dynamically adapt the allocated resources to the changes imposed by the application(s) and environmental conditions, both at the resource management and application level.

Datacenter resources still remain an important asset, even in this massively distributed scenario; as such, while integrating the edge side of the continuum, the proposed reference architecture leverages on the availability of powerful nodes (also including accelerators like GPUs and FPGAs), on the one hand to manage the service layer of the cognitive-enabled software platform; on the other hand, to spin up dedicated virtualized compute resources to support the application (for instance, to perform ML/DL model aggregation in case of federated techniques). To this end, the service layer runs on top of orchestration and virtualization frameworks and operating systems. Modern frameworks will be strongly leveraged to get the best trade-off between performance and security (e.g., lightweight virtual machines as those supported by Kata Containers vs. traditional containers, which have limits in terms of isolation), putting emphasis on seamlessly being able to deploy and manage applications both on cloud datacenter resources and/or at the edge (e.g., using K3s or KubeEdge framework).

The depicted architecture comprehends a cognitive orchestration framework, that leverages ML-based and (multi-objective) optimization techniques, and dynamically adaptable allocation strategies. This allows the seamless management of compute, storage, and networking resources at the edge and in the cloud datacenters.

Finally, the presented architecture includes a service that allows to simulate with different degree of accuracy the behaviour of the target hyper distributed applications on top of distributed continuum of compute, storage and networking resources. The idea behind this service (we refer to it as the Virtual Environment Service – VES) is two fold. On the one hand, it provides the end-users (and application developers) with a tool that enables the exploitation of architectural and micro-architectural information coming from single compute nodes simulations and to tune their applications in order to adapt to environmental condition changes. On the other hand, the VES enables the creation of digital twins of the physical resources, which can be exploited for monitoring, predictive analysis and performance tuning purposes. In this regard, the development of the cognitive package (i.e., the software package that enhances edge nodes) relies on the capability of such simulated environment to reproduce the same set up as expected by the application when running in real environment.

The advantages brought by this solution are many: (i) to ease the application development phase by avoiding the needs of a physical set up to validate the application; (ii) to facilitate the tuning of the cognitive framework (i.e., ML/DL algorithms used to manage the allocation of resources and the workload prediction) on different conditions and configurations of the edge devices; (iii) to test the performance and other relevant indicators (i.e., energy consumption, security, etc.) for the application execution on new devices; and (iv) to evaluate the behaviour and tune the configuration of new devices before their deployment in field, as well as to tune the application behavior when it is running on massively distributed and heterogeneous resources.

The implementation of the VES integrates modern high-performance micro-architectural simulation (e.g., gem5 [22], Sniper [21], Marssx86 [19], Champsim [20], etc.) and emulation –no precise timing model is used (e.g., COREMU [18], Qemu [17], etc.) frameworks which provide different levels of simulation speed vs. simulation accuracy. Where possible, hardware acceleration is used to perform the simulation with a speed closer to that of the physical nodes' counterpart. VES provides the end users with the capability of configuring and launching a very large number of virtual edge nodes, connecting them and executing on top both the cognitive orchestration algorithms and the application tasks. Further, their simulation can be enhanced over the time, by integrating data coming from physical node counterparts, thus effectively creating a digital twin of the entire distributed system. To this purpose a catalogue of architectural models, as well as a catalogue of energy-efficiency models will be evaluated in order to assess the performance and energy efficiency of the whole cloud-edge configuration with different degrees of accuracy. VES enables the simulation of the entire cloud-edge network, including the operating system running on the nodes, the orchestration and abstraction tools and the ML/DL based resource management. Configurations of the simulated nodes can be exported into a dedicated *configuration and templates* catalogue, which in turns can be used to ease the deployment in field the applications along with specific configurations of the nodes. A semantic-based representation of both the application requirements and available resources in the virtual environment (as well as in the real environment) feeds the best matchmaking between the specific needs of the application and the compute resources, also enabling both end users and developers to evaluate different deployment scenarios through a "what-if" like analysis. Starting from the best matchmaking schema, an automated mechanism generates the deployment plan in a machine-readable format, which can be used to deploy the application execution units (i.e., containers, serverless functions, etc.) onto the selected resources by the management service (K3S, simulation service, etc.). To complement such solution, a programmatic mechanism (e.g., Istio [23]) to define and set up a specific communication topology among the different application units is integrated. Eventually, in order to let end-user further optimize the matchmaking between application's units and execution resources, monitoring data concerning the network usage and performance can be accessed.

4   Evaluation Results and Conclusions

The depicted architecture addresses several critical points that are affecting today's cloud landscape and demonstrates by design how heterogeneity, mobility and efficiency are managed. The definition and deployment of the AI-enabled distributed orchestration is the turning point of modern systems. Such orchestration mechanism demonstrated to be more flexible compared with legacy orchestrators and if trained correctly and continuously, can ensure optimal deployment of applications and services as well as efficient use of resources. To this end, we propose to integrate the Virtual Environment Service (VES) in the architecture itself. VES, deployed as a service of the proposed system, allows to test orchestration features and node heterogeneity but also contributes to the validation process. Through VES is possible to simulate various computing nodes and network connections, running on them orchestration agents to simulate a part of the edge or cloud nodes that participate to the orchestration process as computing nodes. The deployment of envisioned system is ongoing as well as the deployment of VES; this will allow to further validate the proposed architecture. However, preliminary performance tests have been carried out (see Table 1). As such, a small VES-testbed has been created with a couple of virtual machines whose compute resources were pinned to the physical ones of the host (cloud) servers. On top of these, emulated devices (QEMU) were run and communication performance is measured (also, between cloud and an external instance – i.e., edge).

Table 1. Communication performance (bidirectional bandwidth and latency) measured between QEMU instances running on the testing environment.

QEMU Instance-A	QEMU Instance-B	Bandwidth [Mbps]	Latency [s]
VM1 - Instance-1	VM1 - Instance-2	$\sim 7.0 \cdot 10^2$	$1.7 \cdot 10^{-3}$
VM1 - Instance-1	VM2 - Instance-1	$\sim 7.0 \cdot 10^2$	$1.7 \cdot 10^{-3}$
VM1 - Instance-1	Instance-Ext	$9.0 \cdot 10^1$	10^{-2}

In conclusion, in this paper we highlighted the great challenges behind the creation of robust, secure, energy-efficient hyper-distributed cloud-to-edge platforms, by considering the ever growing availability of heterogeneous processing architectures. To address these challenges, we depicted a reference architecture, enabling the processing through a fully cognitive computing (continuum) platform. As such, the proposed reference architecture provides an holistic solution for the great challenges posed by hyper-distributed computing environments, leveraging AI-enabled cognitive frameworks, federated learning techniques, and advanced emulation/simulation capabilities to serve also as a basis for the creation of digital twins. This comprehensive approach aligns with the evolving demands of modern computing architectures and distributed applications, providing a robust and efficient solution for the computing continuum.

References

1. Rydning, D.R.J.G.J., Reinsel, J., Gantz, J.: The digitization of the world from edge to core. Framingham: Int. Data Corp. **16**, 1–28 (2018)
2. Chang, H., Hari, A., Mukherjee, S., Lakshman, T.V.: Bringing the cloud to the edge. In: 2014 IEEE Conference on Computer Communications Workshops (INFOCOM WKSHPS), Toronto, ON, Canada, pp. 346–351 (2014). https://doi.org/10.1109/INFOCOMW.2014.6849256.
3. Singh, R., Gill, S.S.: Edge AI: a survey. Internet Things Cyber-Phys. Syst. **3**, 71–92 (2023)
4. Giordano, M., et al.: Survey and comparison of milliwatts micro controllers for tiny machine learning at the edge. In: 2022 IEEE 4th International Conference on Artificial Intelligence Circuits and Systems (AICAS), Incheon, Korea, Republic of, pp. 94–97 (2022)

5. Task Force 3: Architecture. Developing a reference architecture for the continuum - concept, taxonomy and building blocks. Zenodo (2023). https://doi.org/10.5281/zenodo.8403593

6. Jansen, M., Al-Dulaimy, A., Papadopoulos, A.V., Trivedi, A., Iosup, A.: The SPEC-RG reference architecture for the edge continuum (2022). arXiv preprint arXiv:2207.04159

7. Ferrer, A.J., et al.: Towards a cognitive compute continuum: an architecture for ad-hoc self-managed swarms. In: 2021 IEEE/ACM 21st International Symposium on Cluster, Cloud and Internet Computing (CCGrid), pp. 634–641. IEEE (2021)

8. Gogouvitis, S.V., et al.: Seamless computing in industrial systems using container orchestration. Future Gener. Comput. Syst. **109**, 678–688 (2020)

9. Kokkonen, H., et al.: Autonomy and intelligence in the computing continuum: challenges, enablers, and future directions for orchestration. arXiv preprint arXiv:2205.01423 (2022)

10. González García, C., Núñez Valdéz, E.R., García Díaz, V., Pelayo García-Bustelo, B.C., Cueva Lovelle, J.M.: A review of artificial intelligence in the internet of things. Int. J. Interact. Multimed. Artif. Intell. **5** (2019)

11. Chochliouros, I.P., et al.: NEMO: building the next generation meta operating system. In: Proceedings of the 3rd Eclipse Security, AI, Architecture and Modelling Conference on Cloud to Edge Continuum, pp. 1–9 (2023)

12. Hojlo, J.: Future of industry ecosystems: shared data and insights. IDC (2021). https://blogs.idc.com/2021/01/06/future-of-industry-ecosystems-shared-data-and-insights/

13. Lemke, M.: From HPC/cloud to edge/IoT: a major paradigm shift for Europe. HiPEAC (2022). https://www.hipeac.net/news/6995/from-hpccloud-to-edgeiot-a-major-paradigm-shift-for-europe/

14. Kung, A., et al.: Reference architectures and interoperability in digital platforms. OPENDEI (2022). https://www.opendei.eu/wp-content/uploads/2022/10/REFERENCE-ARCHITECTURES-AND-INTEROPERABILITY-IN-DIGITAL-PLATFORMS.pdf

15. EU commission. https://digital-strategy.ec.europa.eu/en/policies/iot-investing. Accessed December 2023

16. EU commission. https://ec.europa.eu/commission/presscorner/detail/en/ip_23_6246. Accessed December 2023

17. Bellard, F.: QEMU, a fast and portable dynamic translator. USENIX annual technical conference, FREENIX Track, vol. 41 (2005)

18. Wang, Z., et al.: COREMU: a scalable and portable parallel full-system emulator. In: Proceedings of the 16th ACM Symposium on Principles and Practice of Parallel Programming (2011)

19. Patel, A., Afram, F., Ghose, K.: MARSS-x86: a QEMU-based micro-architectural and systems simulator for x86 multicore processors. In: 1st International Qemu Users' Forum. Citeseer (2011)

20. Gober, N., et al.: The championship simulator: architectural simulation for education and competition. arXiv preprint arXiv:2210.14324 (2022)

21. Heirman, W., Carlson, T., Eeckhout, L.: Sniper: scalable and accurate parallel multi-core simulation. In: 8th International Summer School on Advanced Computer Architecture and Compilation for High-Performance and Embedded Systems (ACACES-2012). High-Performance and Embedded Architecture and Compilation Network of Excellence (HiPEAC) (2012)

22. Qureshi, Y.M., et al.: Gem5-X: a gem5-based system level simulation framework to optimize many-core platforms. In: 2019 Spring Simulation Conference (SpringSim). IEEE (2019)
23. Ali, S.A., Zafar, M.W.: Istio service mesh deployment pattern for on-premises. Int. J. Comput. Sci. Technol. **5**(1), 472–505 (2021)

Author Index

L. Barolli (Ed.): AINA 2024, LNDECT 201, pp. 461–462, 2024.
https://doi.org/10.1007/978-3-031-57870-0

Printed in the United States
by Baker & Taylor Publisher Services